Industrial Biotechnology Series

Probiotics, the Natural Microbiota in Living Organisms

Fundamentals and Applications

Editors

Hesham El Enshasy

Director, Institute of Bioproduct Development
Universiti Teknologi Malaysia, Johor Bahru, Malaysia

Shang Tian Yang

Department of Chemical and Biomolecular Engineering
The Ohio State University, Columbus, Ohio, USA

CRC Press
Taylor & Francis Group
Boca Raton London New York

CRC Press is an imprint of the
Taylor & Francis Group, an **informa** business

A SCIENCE PUBLISHERS BOOK

First edition published 2021
by CRC Press
6000 Broken Sound Parkway NW, Suite 300, Boca Raton, FL 33487-2742

and by CRC Press
2 Park Square, Milton Park, Abingdon, Oxon, OX14 4RN

© 2021 Taylor & Francis Group, LLC

CRC Press is an imprint of Taylor & Francis Group, LLC

Library of Congress Cataloging-in-Publication Data

Names: El Enshasy, Hesham, editor. | Yang, Shang-Tian, editor.
Title: Probiotics, the natural microbiota in living organisms :
  fundamentals and applications / editors, Hesham El Enshasy, Shang Tian
  Yang.
Other titles: Industrial biotechnology series.
Description: Boca Raton : CRC Press, [2021] | Series: Industrial
  biotechnology series | Includes bibliographical references and index. |
  Summary: "Beneficial microbes exist naturally in the body and play a
  vital role in our health. They are called probiotics. Due to growing
  concerns in using chemically synthesized medicines and antibiotics,
  research on probiotics which are natural and safe alternatives, has
  increased. For centuries, probiotics have been known for producing
  important microbiota of antimicrobial compounds that increases our
  immunity to counter the harmful effects of pathogenic organisms. In
  addition, these microbes have new potential applications in the
  treatment and negating the side effects of other diseases. The study of
  probiotic organisms and their wide potential applications in industrial
  products for human and animal uses has thus gained momentum. With the
  enormous growth of the probiotic market, many countries have developed
  new regulations and policies for this class of bioproduct"-- Provided by
  publisher.
Identifiers: LCCN 2020046672 | ISBN 9781138493605 (hardback ; alk. paper)
Subjects: MESH: Probiotics--therapeutic use | Biotechnology--methods |
  Gastrointestinal Microbiome--physiology | Gastrointestinal
  Diseases--therapy
Classification: LCC QR46 | NLM QU 145.5 | DDC 612.001/579--dc23
LC record available at https://lccn.loc.gov/2020046672

ISBN: 978-1-138-49360-5 (hbk)
ISBN: 978-0-367-68605-5 (pbk)
ISBN: 978-1-351-02754-0 (ebk)

Typeset in Palatino Roman
by Innovative Processors

Series Preface

Industrial biotechnology has a deep impact on our lives, and is the focus of attention of academia, industry and governmental agencies and become one of the main pillars of knowledge based economy. The enormous growth of biotechnology industries has been driven by our increased knowledge and developments in physics, chemistry, biology, and engineering. Therefore, the growth of this industry in any part of the world can be directly related to the overall development in that region.

The interdisciplinary *Industrial Biotechnology* book series will comprise a number of edited volumes that review the recent trends in research and emerging technologies in the field. Each volume will covers specific class of bioproduct or particular biofactory in modern industrial biotechnology and will be written by internationally recognized experts of high reputation.

The main objective of this work is to provide up to date knowledge of the recent developments in this field based on the published works or technology developed in recent years. This book series is designed to serve as comprehensive reference and to be one of the main sources of information about cutting-edge technologies in the field of industrial biotechnology. Therefore, this series can serve as one of the major professional references for students, researchers, lecturers, and policy makers. I am grateful to all readers and we hope they will benefit from reading this new book series.

Series editor
Prof. Dr. rer. Nat. Hesham A. El Enshasy
Johor Bahru, Malaysia

Preface

All living organisms we can see with naked eyes need microorganisms for their survival and healthy life. There are mounting evidences showing the important roles played by microbiota present in human and animal bodies. The living system microbiome is a matter of high interest for better understanding of macro/microorganisms interaction at all levels. Over the last decades, there is a growing interest in the application of probiotic and prebiotics in agriculture, animal feed, and human food industries. Probiotics are not limited to lactic acid bacteria (LAC) but extend to various classes of microbes. The main aim of this book is to provide a comprehensive updated overview about recent progresses in research and applications of probiotics and prebiotics in various fields. Each chapter was written by experts in the field from around the world. The book consists of thirteen chapters. It starts with an overview about probiotics and their roles in healthy life (Chapter 1). It then provides in-depth discussion on various types of microbiota and their importance and functions in affecting stomach and gastrointestinal tract (GIT) health (Chapters 2 and 3), followed by the applications of *Clostridium butyricum* in human health and industry (Chapter 4) and probiotic-based dairy products in human health (Chapters 5 and 6). The book then covers recent trends in the construction of new generation of probiotic organisms using genetic engineering tools (Chapter 7) and enzymatic production of oligosaccharides as prebiotics (Chapter 8), followed by new techniques for microencapsulation of probiotics to enhance cell viability (Chapter 9). The next three chapters highlight recent developments and applications of probiotics in agriculture (Chapter 10), aquaculture (Chapter 11), and animal feed (Chapter 12), respectively. Finally, insect gut bacteria and their roles in iron metabolism in insect is discussed in Chapter 13. The information provided in this book should be valuable for both academic and industrial researchers working in the fields of probiotic and prebiotic.

Hesham El Enshasy
Shang Tian Yang

Contents

Probiotics and Prebiotics and Their Effect on Food and Human Health: New Perspectives

Maria Antoniadou[1] and Theodoros Varzakas[2]*

[1] Dental School, National and Kapodistrian University of Athens,
15127 Athens, Greece
[2] Department Food Science and Technology, University of the Peloponnese,
24100 Kalamata, Greece

1. Introduction

Probiotics are widely used in the dairy industry to produce fermented milk products. According to FAO/WHO (2002) probiotics are live microorganisms which, when administered in adequate amounts, confer a health benefit on the host. However, their use in other food matrices, such as fruit juices or fruits, is increasing (Luckow and Delahunty, 2004, Pereira et al., 2011, Prado et al., 2008, Rivera-Espinoza and Gallardo-Navarro, 2010, Russo et al., 2015, 2016) as an alternative to probiotic dairy products for vegetarians, vegans and people who are lactose intolerant or allergic to milk proteins (Gupta and Abu-Ghannam, 2012). Most of the probiotics that are in food products are LAB species, mainly *Lactobacillus* and *Bifidobacterium* (Arques et al., 2015).

2. Probiotics and Bread

Bread is an interesting non-dairy-based vehicle for probiotics delivery given its daily consumption worldwide. The incorporation of probiotics in bread is challenging due to the high baking temperatures. Zhang et al. (2018) studied the influence of various baking conditions and subsequent storage on survival of a model strain *Lactobacillus plantarum* P8 which was systematically investigated. Bread samples with varying dough weight (5, 30, and 60 g) were baked at different temperatures (175, 205, and 235 °C) for 8 min, and the residual viability of bacteria was determined every 2 min. Under all baking

---
*Corresponding author: theovarzakas@yahoo.gr

conditions, the viability of probiotics decreased from 10^9 CFU/g to 10^{4-5} CFU/g after baking. For specific conditions a difference in bacterial viability between bread crust and crumb was observed, which was explained by the different temperature-moisture histories and microstructures developed during baking. Remarkably, during storage bacterial viability increased by 2-3 times to 10^8 CFU/g in crust and 10^6 CFU/g in crumb, respectively. The regrowth of probiotics was accompanied by a decrease in pH of the bread and an increase of the total titratable acidity. The results of this work provide valuable experimental data for further modelling and optimization studies, which then could contribute to the development of probiotic bakery products.

However, incorporation of probiotics in bread is challenging because of the high temperatures during baking that negatively affect survival of the bacteria and additional loss of bacterial viability during subsequent storage at ambient temperatures (Soukoulis et al., 2014). Sodium alginate solutions were applied with *Lactobacillus rhamnosus* GG (LGG) as a coating on the surface of pre-baked pan bread, and the crust was dried at mild conditions. The viability of LGG was found to be approximately 7.6-9.0 log CFU in 30-40 g of bread slice.

Only few studies investigated strategies to improve bacterial survival in bread. Altamirano-Fortoul et al. (2012) micro-encapsulated *Lactobacillus acidophilus* in starch, and applied it to the dough and obtained reasonably high viable counts after baking (about 10^6 CFU/bread). However, the drawback of this approach was that the probiotic coating negatively affected the physicochemical properties of the bread crust, resulting in different colors and crispness.

Only a few studies have reported on the application of probiotics in bread or other bakery products (Espitia et al., 2016, Malmo et al., 2013, Reid et al., 2007, Zhang et al., 2014).

Soukoulis et al. (2014) developed a novel approach for the development of probiotic baked cereal products. Probiotic pan breads were prepared by the application of film forming solutions based either on individual hydrogels e.g. 1% w/w sodium alginate (ALG) or binary blends of 0.5% w/w sodium alginate and 2% whey protein concentrate (ALG/WPC) containing *Lactobacillus rhamnosus* GG, followed by an air drying step at 60 °C for 10 min or at 180 °C for 1 min. No visual differences between the bread crust surface of the control sample and probiotic bread were observed. Microstructural analysis of the bread crust revealed the formation of thicker films in the case of bread prepared by the application of hydrogels ALG/WPC. The presence of WPC significantly improved the viability of *L. rhamnosus* GG throughout the air drying process and room temperature storage. During storage there was a significant reduction in *L. rhamnosus* GG viability during the first 24 h, followd by lower viable count losses during the subsequent 2-3 days of storage and growth was observed during the last days of storage (day 4-7). The use of film forming solutions on sodium alginate improved the viability of *L. rhamnosus* GG under simulated gastro-intestinal conditions, and there was no impact of the bread crust matrix on inactivation rates. The presence of the probiotic edible films did not modify or cause major shifts in the mechanistic pathway of bread staling as shown by physicochemical, thermal, texture and headspace analysis. Based on their calculations, an individual 30-40 g bread slice can deliver approx. 7.57-8.98 and 6.55-6.91 log cfu/portion before and after in-vitro digestion, meeting the WHO recommended required viable cell counts for probiotic bacteria to be delivered to the human host.

3. Enrichment of Food with Probiotics

Betoret et al. (2017) aimed to determine the effect of homogenization pressures and addition of trehalose on the functional properties of mandarin juice enriched with *Lactobacillus salivarius* spp. *salivarius*. Physicochemical and structural properties of mandarin juice were evaluated and related with quantity and stability of probiotic microorganisms as well as with its hydrophobicity. Both food matrix and processing, affected functional properties of *L. salivarius* spp. *salivarius*. Homogenization pressures and trehalose addition affected quantity and stability of probiotic microorganisms during storage. 20 MPa with 100 g/kg of trehalose allowed obtaining 10^6 colony forming units (CFU)/ml mandarin juice after ten storage days. In MRS growth, cell hydrophobicity increased with homogenization pressures, with values in the range of 67-98%. Highest cell hydrophobicity was obtained in samples homogenized at 100 MPa. Under stress growth conditions, cell hydrophobicity values were in a range 30-84%. In samples not homogenized, addition of trehalose resulted in an increased values of hydrophobicity, with highest levels in those samples with 100 g/kg of trehalose addition.

3.1 Lactic Acid Bacteria and Probiotics

Lactic acid bacteria (LAB), generally recognized as safe (GRAS) in food production are able to produce bacteriocins that are examples of natural approaches with considerable potential as food preservatives.

With regard to non-fermented meat products, like sliced cooked ham stored anaerobically (through vacuum or modified atmosphere packaging) under refrigeration, LABs represent the dominant microflora (Arvanitoyannis and Stratakos, 2012), determining a preservative effect on food.

The fermentation of meat by lactic acid bacteria (LAB) is a traditional practice in many countries of Europe. LAB are already used for probiotic use and/or their bacteriocins can be used as a food preservative. 21 LABs were isolated from ham samples and identified by PCR; in particular, *L. plantarum*-GS16 and *L. paraplantarum*-GS54 were identified as the best bacteriocin-like substance (BLS) producers, active against spoilage and pathogenic bacteria, including Gram negative. In these strains, the presence of virulence factors, such as antibiotic and gastrointestinal resistance were investigated. The kinetics of growth and the BLS biosynthesis were also evaluated at different pH values. Finally, both bacteriocin substances were partially characterized and included in the IIa class of Klaenhammer; as a potential application, both BLSs can be successfully used as food preservatives (Anacarso et al., 2017).

In order to reduce pathogenic microorganisms, different techniques have been studied, one of them being biopreservation using lactic acid bacteria (LAB). LAB are able to convert lactose and other sugars in lactic acid and could generate other final metabolites such as ethanol if a heterolactic fermentation is carried out (Li, 2004).

Some probiotic bacteria have demonstrated a good ability to reduce the level of foodborne pathogens on fresh-cut fruit. Russo et al. (2014, 2015) demonstrated that some probiotic strains have an antagonistic effect against *L. monocytogenes* on fresh-cut pineapple and melon and Siroli et al. (2015a, b) demonstrated the same effect on fresh-cut apples. *Lactobacillus rhamnosus* GG (*L. rhamnosus* GG) was demonstrated to have a bacteriostatic effect against *L. monocytogenes* and *Salmonella* on fresh-cut apple (Alegre et al., 2011) and pear (Iglesias et al., 2017).

In the study by Iglesias et al. (2018), the antagonistic capacity of the probiotic strain *Lactobacillus rhamnosus* GG was assessed in conditions simulating commercial application against a cocktail of 5 serovars of *Salmonella* and 5 serovars of *Listeria monocytogenes* on fresh-cut pear. Moreover, its effect on fruit quality, particularly on the volatile profile, was determined, during 9 days of storage at 5 °C. The, *L. monocytogenes* population was reduced by approximately 1.8 log-units when co-inoculated with *L. rhamnosus* GG. However, no effect was observed in Salmonella. Fruit quality (soluble solids content and titratable acidity) did not change when the probiotic was present. A total of 48 volatile compounds were identified using gas chromatography. Twelve of the compounds allowed the differentiation of *L. rhamnosus* GG-treated and untreated pears. Considering their odour descriptors, their increase could be positive in the flavour perception of pears treated with *L. rhamnosus* GG. The probiotic was able to control the *L. monocytogenes* population on fresh-cut pear, which serves the purpose of a vehicle of probiotic microorganisms as quality of fruit was not affected when the probiotic was present.

Members of the genus *Lactobacillus*, although best understood for essential roles in food fermentations and applications as probiotics, have also come to the fore in a number of untargeted gut microbiome studies in humans and animals. Even though *Lactobacillus* is only a minor member of the human colonic microbiota, the proportions of those bacteria are frequently either positively or negatively correlated with human disease and chronic conditions (Heeney et al., 2018). Recent findings on Lactobacillus species in human and animal microbiome research, together with the increased knowledge of probiotic and other ingested lactobacilli, have resulted in new perspectives on the importance of this genus to human health.

This possibility could be further investigated via studies designed to determine the proximity of *Lactobacillus* to the intestinal epithelium or which focus attention on other sites in the body wherein members of this genus can constitute the majority of bacteria present (e.g. vagina). Moreover, strain and/or species-specific differences (e.g. tryptophan and bile metabolism) might be useful to explain variations in the involvement of this genus, either in the prevention or mitigation of disease.

For example, findings might indicate species or strain-specific differences between autochthonous and allochthonous Lactobacillus on rheumatoid arthritis disease activity.

The production of antibacterial substances, functional characteristics like enzyme activity, resistance to low pH, tolerance to bile salt toxicity and adhesion to intestinal epithelium cells are the leading but less studied criteria for the probiotic characteristics of *L. brevis*. Auto-aggregation property of *Lactobacillus* sp. is believed to affect the adhesive property of bacteria to the intestinal epithelial cells. Bacterial adhesion depends on nonspecific physio-chemical bonding of two surfaces. The surface hydrophobicity determines the chemical and physical properties of the cell surface of bacteria (Kotzamanidis et al., 2010).

Probiotics are routinely employed in the manufacture of fermented food products by enhancing the quality, safety and sensory perception of food items. Hence, Aarti et al. (2017) described the probiotic and antioxidant potentialities of *Lactobacillus brevis* strain LAP2, isolated from Hentak (a fermented fish item of Manipur, India). The antibacterial substances production from strain LAP2 against human pathogens was observed in the range of 192.4 ± 7.78 to 300.3 ± 6.14 AU/ml. The cell free neutralized supernatant (CFNS) illustrated susceptibility to pepsin as well as variable temperature treatment.

Similarly, CFNS was effective at pH 3-7 but lost its antibacterial potential at a higher pH. The isolate was found susceptible to antibiotics and survived at pH 3.0. Similarly, the isolate showed resistivity to gastric juices (pH 3 and 4) and bile salt (0.3% w/v Sodium thioglycollate) up to 3 h and 48 h respectively. Strain LAP2 showed not only a good percentage (35-56%) of auto-aggregation and hydrophobicity (35%) but also a significant proteolytic activity. Additionally, strain LAP2 exhibited significant DPPH scavenging and hydrogen peroxide resistant property. The intrinsic characteristics of this context provide a unique knowledge on Hentak and contribute significantly to our understanding of *L. brevis* strain LAP2 in the pharmaceutical and food industries.

Certain LAB are able to produce extracellular polysaccharides (EPS) that can be either tightly associated with the cell surface forming a capsule, or loosely attached to the outer cell structures or secreted into the environment (Ruas-Madiedo et al., 2008). The major role of EPS is to protect the cells in the environment against toxic metals, host innate immune factors, phage attack and desiccation (Ryan et al., 2015; Zannini et al., 2016). Furthermore, the EPS layer is thought to be involved in the protection against adverse environmental conditions of gastrointestinal tract (GIT) including low pH, bile salts, gastric and pancreatic enzymes (Ryan et al., 2015). Additionally, it has been suggested that EPS might play a role in bacterial aggregation, biofilm formation and interaction with intestinal epithelial cells (IEC) (Dertli et al., 2015, Zivkovi et al., 2016).

EPS-producing LAB are widely used in the dairy industry since these polymers improve the viscosity and texture of the products. Besides, EPS might be responsible for several health benefits attributed to probiotic strains. However, growth conditions (culture media, temperature, pH) could modify EPS production affecting both technological and probiotic properties. In this work by Bengoa et al. (2018), the influence of growth temperature on EPS production was evaluated, as well as the consequences of these changes in the probiotic properties of the strains. All *Lactobacillus paracasei* strains used in the study showed changes in EPS production caused by growth temperature, evidenced by the appearance of a high molecular weight fraction and an increment in the total amount of produced EPS at lower temperature. Nevertheless, these changes do not affect the probiotic properties of the strains; *L. paracasei* strains grown at 20 °C, 30 °C and 37 °C were able to survive in simulated gastrointestinal conditions, to adhere to Caco-2 cells after treatment and to modulate the epithelial innate immune response. The results suggest that selected *L. paracasei* strains are new probiotic candidates that can be used in a wide range of functional foods in which temperature could be used as a tool to improve the technological properties of the product.

Cereal-based fermented products worldwide are diffused staple food resources and cereal-based beverages representing a promising innovative field in the food market. Contamination and development of spoilage filamentous fungi can result in loss of cereal-based food products and it is a critical safety concern due to their potential ability to produce mycotoxins. Lactic Acid Bacteria (LAB) have been proposed as a green strategy for the control of moulds in the food industry due to their ability to produce antifungal metabolites. Russo et al. (2017) screened eighty-eight *Lactobacillus plantarum* strains for their antifungal activity against *Aspergillus niger, Aspergillus flavus, Fusarium culmorum, Penicillium roqueforti, Penicillium expansum, Penicillium chrysogenum*, and *Cladosporium* spp. The overlaid method was used for a preliminary discrimination of the strains as no, mild and strong inhibitors.

L. plantarum isolates that displayed a broad antifungal spectrum activity were further screened based on the antifungal properties of their cell-free supernatants (CFS). CFSs from *L. plantarum* UFG 108 and *L. plantarum* UFG 121, were characterized and analyzed by HPLC because of their antifungal potential. Results indicated that lactic acid was produced at high concentration during the growth phase, suggesting that this metabolic capability, associated with the low pH, contributed to explain the highlighted antifungal phenotype. Production of phenyl lactic acid was also observed. Finally, a new oat-based beverage was obtained by fermentation with the strongest antifungal strain *L. plantarum* UFG 121. This product was either thermally stabilized and artificially contaminated with *F. culmorum*. Samples containing *L. plantarum* UFG 121 showed the best bio preservative effects, since no differences were observed in terms of some qualitative features between samples contaminated with *F. culmorum* and those which were not.

Here we demonstrate, for the first time, the suitability of LAB strains for the fermentation and antifungal biopreservation of oat-based products.

Radulovic et al. (2017) investigated the survival of free and spray-dried cells of potential probiotic strain *Lactobacillus plantarum* 564 after production and during 8 weeks of storage of soft acid coagulated goat cheese, and compositional and sensory quality of cheese. Total bacterial count of spray-dried *L. plantarum* 564 cells was maintained at a high level of 8.82 log (cfu/g) in cheese after 8 weeks of storage, while free-cell number decreased to 6.9 log (cfu/g). However, the chemical composition, PH values and sensory evaluation between control cheese sample C1 (made with commercial starter culture) and treated cheese samples C2 and C3 (made with the same starter, with the addition of free and spray-dried *L. plantarum* 564 cells, respectively) did not significantly differ. High viability of potential probiotic bacteria and acceptable sensory properties indicate that spray-dried *L. plantarum* 564 strain could be successfully used in the production of soft acid coagulated goat cheeses.

Son et al. (2017) conducted a study to evaluate the probiotic properties, including b-galactosidase and antioxidant activities, of lactic acid bacteria isolated from kimchi. Two isolates with a probiotic potential were isolated and identified by 16S RNA sequence analysis. For comparison, a commercial probiotic strain, *Lactobacillus rhamnosus* KCTC 12202BP, was used. The isolates, *Lactobacillus plantarum* Ln4 and G72, and *L. rhamnosus* KCTC 12202BP, were able to survive under artificial gastric conditions (pH 2.5 in the presence of 0.3% pepsin and 0.3% ox gall). The safety of the LAB strains was tested in terms of antibiotic resistance and production of harmful enzymes. Antibiotic resistance was assessed according to Clinical and Laboratory Standards Institute guidelines. Assessment with the API ZYM kit showed that none of the strains produced harmful enzymes, such as β-glucuronidase. Among the tested strains, *L. plantarum* Ln4 showed the strongest adhesion to HT-29 cells and the highest β-galactosidase activity (3320.99 Miller Units). *L. plantarum* Ln4 was found to have higher 1-diphenyl-2-picrylhydrazyl radical-scavenging (40.97%) and β-carotene oxidation-inhibitory activities (38.42%) than did *L. rhamnosus* KCTC 12202BP.

These results suggest that *L. plantarum* Ln4 isolated from kimchi may have a probiotic potential and could be used in functional foods.

The probiotic potential of 5 food grade Lactobacillus strains, of human origin namely *L. fermentum* (S23), *L. fermentum* NCDC77 (S8), *L. fermentum* (S25) and *L. plantarum* (S1), capable of selenium bioaccumulation (in previous studies), was investigated by Saini and Tomar (2017). The cultures, S8 and S23 maintained elevated cell numbers under conditions simulating passage through the human gastrointestinal tract, showing ability to survive in the presence of simulated gastric juice at pH 2. These cultures also survived in the presence of simulated intestinal juice with 1% bile salts. The cultures showed variable and strain dependent cell-surface hydrophobicity and auto-aggregation ability. The cultures were assayed for susceptibility to 20 antibiotics and found susceptible to 14 out of 20 antibiotics while showing variable resistance to remaining 6 antibiotics. S8 and S23 cultures showed superior antibacterial activity in comparison to the remaining 3 cultures. These findings suggest that *L. fermentum* NCDC77 (S8) and *L. fermentum* (S23) cultures, endowed with the ability to accumulate high amount of selenium, exhibited high resistance against low pH and bile. Moreover, these cultures possessed adherence and auto aggregation properties with minimum resistance to antibiotics and hence could be exploited for the development of selenium enriched functional food or nutraceuticals as a source of dietary selenium.

Several foods are carriers of probiotic bacteria. Most species of the genus *Lactobacillus* are probiotics and common inhabitants of the human gastrointestinal tract. It is known that they contribute to the health of gastrointestinal tract, but at the same time certain strains can be antagonistic against foodborne pathogens.

Iglesias et al. (2017) evaluated the effect of *Lactobacillus rhamnosus* GG on the ability of a strain of *Listeria monocytogenes* serovar 4b to survive passage through the gastrointestinal tract, and its adhesion and invasion into Caco-2 cells when it was previously habituated on fresh-cut pear stored at 10°C for 7 days. At the end of the storage, *L. rhamnosus* GG caused a reduction in the survival of *L. monocytogenes* in the gastrointestinal tract, and adhesion and invasion into Caco-2 cells. Moreover, it showed an antagonistic activity against *L. monocytogenes*.

Recent advances in the study of membrane vesicle (MV) production in Gram-positive bacteria are particularly interesting for the biotechnological food field. The Gram-positive lactic acid bacteria (LAB) are key players in various food and feed fermentation processes as starter cultures, probiotics and producers of vitamins (Hugenholtz and Smid, 2002; Shah and Patel, 2014). Among these, evidence has so far been collected for MV production from *Lactococcus lactis* (S. Alexeeva et al., unpublished) and *Lactobacillus plantarum* (Li et al., 2017). This last study demonstrated that extracellular MVs produced from *Lactobacillus plantarum* regulate up the expression of host defense genes and provide protection against the vancomycin-resistant enterococci infection, highlighting the probiotic effects delivered by MVs and the potential to develop therapeutic treatment of antimicrobial resistant infections by using MVs derived from probiotic bacteria.

Membrane vesicle (MV) production is observed in all domains of life. Evidence of MV production accumulated in recent years among bacterial species involved in fermentation processes has been observed. These studies revealed MV composition, biological functions and properties, recognizing the potential of MVs in food applications as delivery vehicles of various compounds to other bacteria or the human host. Moreover, MV producing strains can deliver benefits as probiotics or starters in fermentation processes. Next to the natural production of MVs, Liu et al. (2018) also highlighted possible methods for artificial generation of bacterial MVs and cargo loading to enhance their applicability. They believe that a more in-depth understanding of bacterial MVs opens new avenues for their exploitation in biotechnological applications (Liu et al., 2018).

Cellulose-based microgels, prepared by the sol-gel transition method, were evaluated by Li et al. (2017) to extend the loading capacity and make a more controlled release of probiotics. Granules of calcium carbonate as porogenic agents made cellulose microgels with different porosity, and the SBET of modified cellulose microgels was increased to 197 m^2/g. The microgels were comprised of a house-like interconnected network of the cellulose nanofibrils. The house-like microgels had the ability to accommodate *Lactobacillus plantarum* (*L. plantarum*) to as high as 1010 cfu/g, and released viable cells up to 3.10×10^{10} cfu/g. Furthermore, the Ca-alginate shell protection of the cellulose houses kept the loaded L. plantarum cells from liberating into the stomach but released the cells at intestinal tracts due to the effect of pH- response ability.

From the cross-sectional fluorescence image, *L. plantarum* cells mainly focused on the cellulose microgel houses instead of Ca-alginate shells. It indicated that the surface modification of the cellulose microgel houses by using alginate based hydrogel exhibited a more controlled release of *L. plantarum* cells, which could be increased to 5.80×10^8 cfu/mL in 360 minutes.

A prolonged duration of simulated intestine fluid contributed to making sure that cells reached the desired regions of the intestinal tracts. The designed composite system advances in affording better shelter for probiotic residence or other bioactive nutrients.

3.2 Probiotics and Camel Milk

Camel milk contains a greater amount of natural antimicrobial compounds than bovine milk (Elagamy et al., 1996), hence, it is a potential source from which LAB can be isolated with high probiotic potential (Abushelaibi et al., 2017a).

Several attempts have been made to isolate and characterize Streptococcus and Enterococcus strains from camel milk with major demerits (Abdelgadir et al., 2008, Akhmetsadykova et al., 2014, Davati et al., 2015, Ghali et al., 2011, Hamed and Elattar, 2013, Hassaïne et al., 2007; Jans et al., 2012, Kadri et al., 2014, Kadri et al., 2015, Kadri et al., 2015, Mahmoud et al., 2014, Soleymanzadeh et al., 2016).

The objectives of this study by Ayyash et al. (2018) were to isolate lactic acid bacteria (LAB), namely Streptococcus and Enterococcus, on M17 agar from raw camel milk and investigate their probiotic characteristics. Physiological properties, cell surface properties (hydrophobicity, autoaggregation, co-aggregation), acid and bile tolerance, bile salt hydrolysis, cholesterol removing ability, exopolysaccharide (EPS) production, hemolytic and antimicrobial activities, resistance toward lysozyme and six antibiotics, and fermentation profile (growth, pH, and proteolysis) were examined. rDNA sequencing was carried out to identify the isolates and to acquire Genbank accession numbers. LAB isolates showed cholesterol lowering and pathogens inhibition properties. Hydrophobicity and autoaggregation results revealed strong attachment capabilities of the isolated LAB. Resistance of LAB isolates to lysozyme activity and temperatures upto 60 °C were also high.

Identified LAB exhibited a promising fermentation profile. This study reveals that the isolated LAB isolates *E. faecium* KX881783 and *S. equinus* KX881778 especially and may be excellent candidates to produce functional foods to promote health benefits (Ayyash et al., 2018).

3.3 Probiotics and Chocolates

Following the contemporary trend impelled with better efficiency, a compilation of three probiotic strains microencapsulated together was incorporated in milk, semisweet and dark chocolates by Lalicić-Petronijević et al. (2017).

The determination of viability revealed that bifidobacterial (BB) were less durable than lactobacilli (LA) and streptococci (ST), but the total cell number was high (9–8 log cfu/g) during 360 days, at both tested temperatures. The main objective was to examine the effects of probiotics on the sensory and rheological properties of various types of chocolates produced, and to test the suitability of used packaging materials, as well. The obtained data showed that all evaluated samples exhibited excellent sensory quality and, among rheological parameters, yield value was more affected than the viscosity. Seal defects (leakage) of the primary packaging material assumedly contributed to the decline of viable probiotic cells.

Also, certain studies (Maillard and Landuyt, 2008) suggest that chocolate can offer a sound safeguard to probiotics during the manufacturing process and transit through the stomach, due to its composition, i.e. the amount of fat, which buffers damaging external influences.

The incorporation of probiotics in chocolate was the subject of several investigations (Chetana et al., 2013, Mandal et al., 2013; Succi et al., 2017).

4. Probiotics and Diseases

4.1 Probiotics and Reduction of Cadiovascular Disease

Probiotics can reduce cardiovascular diseases associated with high cholesterol levels, but this occurs only when they survive gastrointestinal conditions. This cholesterol-lowering effect occurs at the intestinal level, making it important for probiotics to survive the high concentrations of acid and salt in the intestinal tract and reach the colon alive (Kumar et al., 2012).

The aim of this study by Castorena-Alba (2017) was the evaluation of the ability of reference strains (RS) and food isolates strains (FIS) to survive acid and bile and the determination of in-vitro cholesterol assimilation. FIS were more tolerant to acid than RS, showing a significantly different growth response. FIS were also more resistant to the presence of bile, and to oxgall, taurocholic acid, and cholic acid, in descending order. The survival percentages ranged from 0% to 100%, presenting strain dependence. The most tolerant strains were tested for cholesterol, showing lower percentages for RS than FIS. In FIS, the percentage was negatively affected when the concentration of bile salt was increased. Therefore, it is crucial to study the viability of different strains in gastrointestinal conditions because of the strain-dependence nature.

This study by Anandharaj et al. (2015) sought to evaluate the probiotic potential of lactic acid bacteria (LAB) isolated from traditionally fermented south Indian koozh and gherkin (cucumber). A total of 51 LAB strains were isolated, among which four were identified as *Lactobacillus* spp. and three as *Weissella* spp. The strains were screened for their probiotic potential. All isolated *Lactobacillus* and *Weissella* strains were capable of surviving under low pH and bile salt conditions. GI9 and FKI21 were able to survive at pH 2.0 and 0.50% bile salt for 3 h without losing their viability. All LAB strains exhibited inhibitory activity against tested pathogens and were able to deconjugate bile salt. Higher deconjugation was observed in the presence of sodium glycocholate (P < 0.05). Strain FKI21 showed maximum auto-aggregation (79%) and co-aggregation with *Escherichia coli* MTCC 1089 (68%). Exopolysaccharide production of LAB strains ranged from 68.39 to 127.12 mg/L (P < 0.05). Moreover, GI9 (58.08 mg/ml) and FKI21 (56.25 mg/ml) exhibited maximum cholesterol reduction with bile salts. 16S rRNA sequencing confirmed GI9 and FKI21 as *Lactobacillus crispatus* and *Weissella koreensis*, respectively. This is the first study to report isolation of *W. koreensis* FKI21 from fermented koozh and demonstrates its cholesterol-reducing potential.

4.2 Probiotics and Colon Cancer

Colon cancer is a multifactorial disease associated with a variety of lifestyle factors. Alterations in the gut microbiota and the intestinal metabolome are noted during colon carcinogenesis, implicating them as critical contributors or results of the disease process. Diet is a known determinant of health, and as a modifier of the gut microbiota and its metabolism, a critical element in maintenance of intestinal health. This review by Seidel et al. (2017) summarizes recent evidence demonstrating the role and responses of the intestinal microbiota during colon tumorigenesis and the ability of dietary bioactive compounds and probiotics to impact colon health from the intestinal lumen to the epithelium and systemically. They first described changes to the intestinal microbiome, metabolome, and epithelium associated with colon carcinogenesis. This was followed by a discussion of recent evidence indicating how specific classes of dietary bioactives, prebiotics, or probiotics affect colon carcinogenesis. Finally, they briefly addressed the prospects of using multiple 'omics' techniques to integrate the effects of diet, host, and microbiota on colon tumorigenesis with the goal of more fully appreciating the interconnectedness of these systems and thus, how these approaches can be used to advance personalized nutrition strategies and nutrition research.

On the role of probiotics in cancer prevention, studies showed that selected strains of the Lactic Acid Bacteria (LAB) group were capable of binding to mutagenic amines generated by cooking a protein-rich food (Orrhage et al., 1994), degrade nitrosamines (Rowland and Grasso, 1975), reduce specific activities of bacterial β-glucuronidases, nitroreductases and azoreductases (enzymes potentially involved in transforming pro-carcinogens into active carcinogens (Goldin and Gorbach, 1977).

4.3 Probiotics and Constipation

Constipation is a frequent complaint, and probiotics could have a potentially synergistic effect on intestinal transit. Constipation is a frequent complaint involving gastrointestinal problems and corresponds to various symptoms that reduce patients' quality of life mentally and physically, such as irregular voiding, sensation of incomplete, painful or forceful voiding, hard stools and abdominal discomfort.

Probiotics may have a beneficial role in alleviating constipation symptoms, at least with certain probiotic strains (Bekkali et al., 2007, Koebnick et al., 2003, Turan et al., 2014).

Liu et al. (2017) reported that yoghurt intake improved the symptoms of slow-transit constipation, which was confirmed in 144 mice (nine groups, n = 16) with loperamide-induced constipation.

Yoghurt fermented with different probiotics was administered orally. Loperamide treatment caused a marked increase in first defecation time and a decrease in the charcoal transit ratio (P < 0.05), while loperamide treatment after intake of a new formulation of yoghurt could significantly improve defecation time and intestinal health. Significant (P < 0.001) decreases in butanoic acid content were observed in groups given three different strains of yoghurt. Yoghurt intake could also change the intestinal bacterial community composition, which was supported by operational taxonomic unit-related analysis as well as principal coordinates analysis (PCA). Their results showed conclusive evidence indicating that yogurt is an excellent functional food that improves the symptoms of slow-transit constipation.

4.4 Probiotics and Metabolic Syndrome

Metabolic Syndrome (MetS), affecting at least 30% of adults in the Western World, is characterized by three out of five variables, from high triglycerides, to elevated waist circumference and blood pressure. MetS is not characterized by elevated cholesterolemia, but is rather the consequence of a complex interaction of factors generally leading to increased insulin resistance. Drug treatments are difficult to handle, whereas well-characterized nutraceuticals may offer an effective alternative. Among these, functional foods, e.g. plant proteins, have been shown to reduce insulin resistance and triglyceride secretion. Pro- and pre-biotics, that are able to modify intestinal microbiome, reduce absorption of specific nutrients and improve the metabolic handling of energy-rich foods. Finally, specific nutraceuticals have proven to be of benefit, in particular, red-yeast rice, berberine, curcumin as well as vitamin D. All these can improve lipid handling by the liver as well as ameliorate insulin resistance. While lifestyle approaches, such as the Mediterranean diet, may prove to be too complex for a single patient, and better knowledge of selected nutraceuticals and more appropriate formulations leading to improved bioavailability will certainly widen the use of these agents, which are already in high usage for the management of such frequent patient groups (Sirtori et al., 2017).

A meta regression analysis reports that the effects of probiotics on BMI depends on the duration of intervention (≥ 8 weeks), number of species of probiotics and baseline BMI ≥ 25 kg/m^2, thus highlighting the effectiveness of probiotics in reducing BMI, especially in overweight or obese subjects (Zhang et al., 2015).

A possible mechanism explaining the effects on weight loss are the improvement of the intestinal barrier function that may reduce metabolic endotoxemia (Boutagy et al., 2016), thus ameliorating levels of lipopolysaccharides (LPS). In the adipose tissue, endotoxins from LPS trigger systemic and local inflammation with an increase in reactive oxygen species (ROS) (Boutagy et al., 2016).

4.5 Probiotics and Gastrointestinal Conditions

Alterations in the composition of the gut microbiota are associated with a number of gastrointestinal (GI) conditions, including diarrhea, inflammatory bowel diseases (IBD), and liver diseases. Probiotics are live microorganisms that may confer a health benefit to the host when consumed, and are commonly used as a therapy for treating these GI conditions by means of modifying the composition or activity of the microbiota. The purpose of this review by Parker et al. (2018) was to summarize the evidence of probiotics and GI conditions available from Cochrane, a nonprofit organization that produces rigorous and high-quality systematic reviews of health interventions. Findings from this review help provide more precise guidance for clinical use of probiotics and to identify gaps in probiotic research related to GI conditions.

4.6 Probiotics and Obesity

Obesity is recognized globally as a major mortality risk factor due to its involvement in other metabolic complications like insulin resistance, type 2 diabetes mellitus, and several cancers. Literature suggests that probiotics play a major role in the management of obesity and associated diseases. However, contradictory results have been reported in the context of probiotics role in obesity that observed no effect on body weight and feed intake (Ji et al., 2012, Kumar et al., 2012). Administration of probiotics in the form of fermented milk, lyophilized cells, Dahi (Indian traditional fermented milk) and yogurt in clinical and animal models have been studied for their anti-obesity effects.

4.7 Probiotics and Type 2 Diabetes

Lactobacillus G15 and Q14, two species of Lactobacillus used in the preparation of Chinese traditional fermented dairy foods, have the potential to prevent diabetes. Herein, Li et al. (2017) clarified their mechanisms by which gut microbiota relates to the control of diabetes. Results showed that G15 and Q14 improved the disorders in blood glucose and insulin. Moreover, G15 and Q14 promoted the enrichment of SCF-producing bacteria, and upregulated the production of acetate and butyrate. Their antidiabetic effects were closely associated with SCFAs-downstream receptors and hormone secretion. Additionally, G15 and Q14 significantly shortened the number of G-negative bacteria, which subsequently decreased the levels of LPS and inflammatory factors. G15 and Q14 improved the disorganization of the intestinal mucous and the integrity of intestinal barrier. The results indicated that Lactobacillus G15 and Q14 alleviate type 2 diabetes in a gut microbiota-dependent way via downregulating G⁻ bacteria-related LPS secretion, and upregulating SCFAs-producing bacteria-related G protein-coupled receptor 43 pathway, as well.

5. Delivery System for Probiotics

There are different formulations to deliver probiotics, varying in viability, advantages and effectiveness of bacterial cells to the human intestine (Govender et al., 2014; Martin et al., 2015). The encapsulation methods may involve immobilization in a matrix using spray drying, emulsifying?, extrusion, gelation techniques or coating of the microorganism in multilayer systems (Haffner et al., 2016, Krasaekoopt et al., 2003, Soukoulis et al., 2016). In this respect, bio-based hydrogels (coacervation process) are being widely used as efficient vehicles to deliver therapeutic agents due to their biocompatibility and biodegradability (Haffner et al., 2016; Kanmani et al., 2011; Silva et al., 2013). These systems are typically hydrophilic three-dimensional matrices capable of entrapping different molecules.

Chitosan hydrogel encapsulation systems are regarded as among the most promising matrices for protein or peptide entrapment and delivery, due its permeation-enhancing effect (Dodane et al., 1999).

Yuan et al. (2018) employed chitosan-polyphosphoric acid-(PPA) beads to entrap proteins and peptides by extruding chitosan solutions containing BSA, WPI, insulin or casein hydrolysate into a PPA solution. These proteins and peptides were selected due to their range of molecular weights and different structures. The study sought to further understand the mechanism of chitosan-PPA bead formation and to establish the usefulness of these beads to entrap and subsequently release the entrapped proteins in simulated gastrointestinal fluids.

Inhibited release of BSA, in both SGF and SIF, was achieved with low PPA concentration. Insulin and WPI were effectively retained in SGF and gradually released in SIF. Peptides from casein hydrolysate were partially (about 35%) but quickly released in SGF with no further release in SIF. Overall, these results indicate that chitosan-PPA beads show potential for lower gastrointestinal delivery of bioactive protein material.

Novel carboxymethyl cellulose-chitosan (CMC-Cht) hybrid micro- and macroparticles were successfully prepared in aqueous media either by drop-wise addition or via nozzle-spray methods by Singh et al. (2017). The systems were either physically or chemically cross linked using genipin as the reticulation agent. The macroparticles (ca. 2 mm) formed were found to be essentially of the core-shell type, while the microparticles (ca. 5 µm) were apparently homogeneous. The crosslinked particles were robust, thermally resistant and less sensitive to pH changes. On the other hand, the physical systems were pH sensitive presenting a remarkable swelling at pH 7.4, while little swelling was observed at pH 2.4. Furthermore, model probiotic bacteria (*Lactobacillus rhamnosus* GG) was for the first time successfully encapsulated in the CMC-Cht based particles with acceptable viability count. Overall, the systems developed were highly promising for probiotic encapsulation and potential delivery in the intestinal tract with the purpose of modulating gut microbiota and improving human health.

Probiotics can be protected from environmental stresses by encapsulating them within carefully-designed microcapsules (Chen et al., 2012, Cook et al., 2012, Del Piano et al., 2011, Huq et al., 2013, Sarao and Arora, 2017). Existing microencapsulation strategies are freezing, drying, spray drying, prilling and extrusion (Al-Muzafar and Amin, 2017). The materials used to fabricate these microgels are usually biopolymers that are chosen to be biocompatible and non-toxic to cells. The most commonly used biopolymers for encapsulating microorganisms are polysaccharides (such as gum acacia, pectin, and alginate) and proteins (such as gelatin and whey protein) (Li et al., 2016, Saravanan and Panduranga, 2010, Tello et al., 2015, Yeung et al., 2016).

Probiotics are used in food products, dietary supplements and pharmaceutical products because they may provide health-promoting effects in humans. To be efficacious, probiotics need to be viable in sufficient abundance within the large intestine. However, many commercial products containing probiotics suffer from a substantial loss of bacterial viability during shelf storage and during gastrointestinal transit. In this study by Yao et al. (2017), a probiotic (*Lactobacillus salivarious* Li01) was incorporated into either alginate or alginate-gelatin microgels. The morphology of the microgels was characterized by scanning electron microscopy, which indicated that they had a spherical shape and that almost all of the bacterial cells were encapsulated inside them. Probiotic viability was determined under aerobic conditions, heat treatment, and simulated gastrointestinal conditions. Encapsulation significantly enhanced the viability of the probiotic during aerobic storage. The microgels maintained their structures under simulated gastric conditions, but either eroded or swelled under simulated small intestine conditions. The alginate-gelatin microgels were the most effective at protecting the bacteria under simulated gastrointestinal conditions when compared to the alginate microgels and non-encapsulated bacteria. In conclusion, alginate-gelatin microgels have great potential for the protection and delivery of probiotics in food, supplements, and pharmaceutical products (Yao et al., 2017).

The delivery of probiotics has been a challenge because microorganisms must overcome harmful environments both at a technological and at a physiological level. Encapsulation is one of the strategies to protect microorganisms against both problems. This work by Ghibaudo et al. (2017) aimed at encapsulating probiotic *Lactobacillus plantarum* CIDCA 83114 in pectin-iron beads obtained by ionotropic gelation, enabling safe bacterial delivery and providing a source of iron and fibre.

Microorganisms in the stationary phase were suspended in a 4% w/v pectin solution at pH 5, and the suspension, dripped into a 150 mM FeSO4 solution. The beads were freeze-dried and stored for 60 days at 4 °C. The morphology of the beads was observed by scanning electron microscopy. Bacterial culturability was determined after freeze-drying, and exposure to simulated saliva, gastric and intestinal conditions. The iron and pectin releases were also investigated in the same digestive conditions.

Microorganisms were fully entrapped in smooth and spherical pectin-iron beads of ca. 1–2 mm diameter. Bacterial culturability did not decrease during storage. Encapsulation protected microorganisms against simulated digestive conditions, also enabling the complete release of iron and pectins in the gut.

The results obtained support the safe delivery of both probiotic bacteria and iron to the gut. As iron deficiency still continues to be a worldwide problem, using iron-pectin beads could be an adequate strategy to functionalize food products, contributing to attain the recommended iron intake.

6. Probiotics and Biofilm Formation

The ability of probiotic biofilm formation on carrier surfaces was demonstrated by Grossova et al. (2017). Probiotic biofilms exhibit the same properties as pathogen microbial biofilms but with higher resistance to low pH values and bile salts.

The ability of different probiotic strains (*Lactobacillus acidophilus, Bifidobacterium breve, Bifidobacterium longum*) to interact with pre-selected carriers divided into 3 categories (polymers, complex food matrices, and inorganic compounds) was tested. *Lactobacillus acidophilus* and *Bifidobacterium longum* combined with inorganic silica carrier exhibited an interaction leading to biofilm formation only. The prepared biofilm (*Lactobacillus acidophilus*) was then subjected to comparative study with planktonic bacterial culture. The ability to survive in the presence of low pH value (pH 1–3) and bile salts (0.3% solution) was evaluated. Low pH value (pH 1) had a harsh effect on free cell culture causing decreased cell viability (71.9±3.2% of viable cells). Biofilm culture exhibited higher resistance to low pH value, the viability exceeded 90%. The exposure of free cell probiotic culture to porcine bile resulted in an almost constant decrease in viability during the study period (68.2±1.1% of viable cells, after 240 min incubation).

Viability of biofilm after the exposure to bile was almost constant with a slight decrease of no more than 5% during the study.

7. Probiotics and Edible Films

In the study of Ebrahimi et al. (2018), survival of four probiotic strains (*Lactobacillus acidophilus, L. casei, L. rhamnosus and Bifidobacterium bifidum*) immobilized in edible films based on carboxymethyl cellulose (CMC) and physicochemical properties of films were investigated during 42 days of storage at 4 °C and 25 °C. Results showed a significant decrease in viability of bacterial cells during 42 days of storage at 25°C. However, viability of *L. acidophilus* and *L. rhamnosus* were in the range of recommended levels during the storage at 4 °C (10^7 CFU/g). Probiotic films caused more water vapor permeability (WVP) and opacity, and less tensile strength (TS) and elongation at break point (EB) compared to the control film. However, no significant physicochemical changes were observed among probiotic films containing different strains. Therefore, incorporation of some probiotic strains in edible coats and films could be suitable carriers at refrigerated temperatures.

8. Prebiotics

Prebiotics are defined as: "The selective stimulation of growth and/or activity/activities of one or a limited number of microbial genus (era)/species in the gut microbiota that confer health benefits to the host" (Roberfroid, 2007, Roberfroid et al., 2010). Several prebiotic substrates, in particular fructooligosaccharides (FOS), galacto-oligosaccharides (GOS), inulin and lactulose, have already obtained scientific credibility due to their inclusion in human trials where they have demonstrated prebiotic effects (Davis et al., 2010). The FOS are classified as prebiotics and are used as food ingredients because of their beneficial effects in proliferating bifidobacteria in the human colon (Roberfroid, 2007).

Inulin and other fructans that provide health benefits are accumulated in Agave salmiana. Currently, Agave species are only exploited for the production of alcoholic beverages. Few reports of beneficial effects of fructans obtained from *A. salmiana* have been presented and therefore the goal of this study became examining the specific stimulation of probiotic bacteria by *A. salmiana* powder in comparison to commercial prebiotic products with different bacteria.

The composition of *A. salmiana* that includes fructooligosaccharides (FOS) supports the growth of probiotic bacteria. Inulin Orafti GR showed almost the same stimulation to probiotic bacteria in comparison to *A. salmiana. Lactobacillus acidophilus* showed the best growth with *A. salmiana*, decreased the final pH-value to the lowest level and produced the highest concentration of lactate.

The structural heterogeneity of fructans from *A. salmiana* is useful as a prebiotic and to maintain persistence of probiotic strains in vivo. The variation of gut microbiota composition that might be caused by microbiota-targeted therapies might also be influenced. Future research should work towards optimizing the FOS content and profile in the plants by selection of plant varieties or changing agronomic and postharvest practice to develop their innovative applications for the food industry and the health promotion (Martinez-Gutierrez et al., 2017).

Inulin has been used widely to protect probiotic bacteria through microencapsulation during spray drying (Corcoran et al., 2004, Fritzen-Freire et al., 2012, Pinto et al., 2012). In addition, there is growing interest in the use of symbiotics (Tripathi and Giri, 2014). Microparticles containing inulin, hi-maize, and trehalose were produced through spray drying to encapsulate *Lactobacillus acidophilus* La-5. Afterwards, the encapsulation efficiency, thermal resistance, gastrointestinal simulation, storage stability, and the microparticles' sizes and morphology were analyzed to evaluate the protective effect against the thermal conditions of the spray dryer of the different encapsulating matrices. Inulin and himaize encapsulating matrices showed the greatest encapsulation efficiency of 93.12% and 94.26%, respectively.

Concerning thermal resistance, the trehalose encapsulating matrix provided the greatest protection for this microorganism. The microparticles produced with hi-maize showed the greatest viability in simulated gastrointestinal conditions thus providing higher protection for *Lactobacillus acidophilus* La-5. Regarding storage stability, microparticles containing trehalose showed the lowest viability losses during 120 days of storage. However, notably, at the end of 120 days of storage at room temperature (25 °C), microparticles containing inulin, hi-maize, and trehalose all kept their counts above the recommended level (> 6 logCFU/gm).

Concerning the physical characteristics of the microparticles, particle sizes were as expected for products obtained by spray drying. Scanning electron microscopy showed no ruptures or cracks on the surfaces of the microparticles, a desirable characteristic for high protection (Nunes et al., 2018).

An emerging prebiotic candidate, pectin oligosaccharide (POS), is receiving attention. POS have many functional properties, a special one being the ability to regulate gut micro flora. Several studies have concluded that POS is a better prebiotic candidate than pectin (Chen et al., 2013, Gomez et al., 2016, Olano-Martin et al., 2002). The ability of POS to cause shifts in healthy bacteria populations was found to be similar or better than FOS (Gomez et al., 2016).

Pectin oligosaccharide (POS) fractions were obtained by controlled chemical degradation of citrus peel pectin. By adjusting trifluoroacetic acid (TFA) concentration, three oligosaccharides of molecular weight (Mw) range 3000–4000 Da, 2000–3000 Da and lower than 2000 Da were obtained, as reported by Zhang et al. (2018). Varying hydrogen peroxide (H_2O_2) concentration and reaction time produced oligosaccharides of 2000–3000 Da and 3000–4000 Da. The relative proportions of acidic monosaccharide increased from 68.58% to 89.93% (TFA) and from 63.74% to 83.26% (H_2O_2) as the reaction conditions intensified. Prebiotic activity scores were used to quantify POS ability to promote selective growth of specific probiotics. Sample POSH1 from H_2O_2 degradation showed the highest prebiotic potential with prebiotic activity score 0.41 for *Lactobacillus paracasei* LPC-37 and 0.92 for *Bifidobacterium bifidum* ATCC 29521 (Zhang et al., 2018).

This study by Balthazar et al. (2017) aimed to evaluate the impact on nutritional and rheological parameters in sheep milk ice cream by fat replacement for different prebiotic fibers (inulin, fructo-oligosaccharide, galactooligosaccharide, short-chain fructo-oligosaccharide, resistant starch, soluble corn fiber, and polydextrose), and consumer perception (Pivot Profile) as well. Low caloric content, around 98 kcal/100 g, were observed due to prebiotic fibers supplementation; without loss of moisture, carbohydrates or protein values.

Sheep milk ice creams presented a pseudoplastic behavior when submitted to flow curve analysis. Moreover, inulin and fructo-oligosaccharide have proven to be a promising alternative as fat substitutes in sheep milk ice cream formulation, due to similar rheological properties, such as hardness, viscoelasticity and consistency. Sheep milk ice creams containing inulin or fructo-oligosaccharide were also perceived creamier and brighter than control full-fat sheep milk ice cream. In addition, most prebiotic ice creams were observed to be sweeter than control samples, which means those fibers can act as sweeteners. In conclusion, the replacement of sheep milk fat by prebiotics to manufacture sheep milk ice cream can be an effective alternative to improve the nutritional and physiological aspects due to low caloric value and functionality provided by prebiotics.

9. Case Study: The Effect of Probiotic and Prebiotic Addition in Carp Aquaculture

Although probiotics offer a promising alternative to the use of chemicals and antibiotics in aquatic animals (Bandyopadhyay et al., 2015), and assist the protection of cultured species from diseases, their benefits to the hosts by using them in aquaculture need to be considered. Various factors like source, dose and duration of supplementation of probiotics can affect the immunomodulatory activity of probiotics (Dawood et al., 2015a, Hai, 2015). Therefore, appropriate administration methods help to provide favorable conditions for probiotics to perform well.

Probiotics can be applied singly or in combination (Allameh et al., 2015, Chi et al., 2014; Faramarzi et al., 2011). Likewise, single supplementation of either *L. acidophilus* and/or *S. cerevisiae* to koi carp feed for protection resulted in enhanced growth performances and immune responses (Dhanaraj et al., 2010). Administration of both individual or combined *L. rhamnosus* (Lactobacillus) and *Lactobacillus sporogenes* (Sporolac) improved the health condition and disease resistance of common carp (Harikrishnan et al., 2010). Effects of different prebiotics and their levels on fish growth rate, feed efficiency, digestibility, immune responses and survival have been investigated in several studies (Collins and Gibson, 1999, Ganguly et al., 2013, Hoseinifar et al., 2014, Song et al., 2014). Synbiotics are products that contain both probiotics and prebiotics (Akrami et al., 2015, Gibson and Roberfroid, 1995) stated that the use of the synbiotic concept may give the benefit of both prebiotics and probiotics on growth of fish mainly due to the synergistic effect.

The rationale behind their combined use is that it may improve the survival of the probiotic organism, because fermentation can be implemented more effectively as its required specific substrate is readily available.

Recent attempts in fish include the use of MOS, FOS and inulin as a prebiotic, in combination with probiotics to achieve long-term health benefits via the gastro-intestinal immune system (Abdulrahman and Ahmed, 2015, Dehaghani et al., 2015).

The demand for cultured carp species has grown tremendously during the last decade due to their high market value. Recently, intensive aquaculture systems have been expanding and are emerging as one of the most practical and promising tools to meet the requirements of carp. However, in intensive fish farming, animals are subjected to stress conditions that weaken fish immune systems, leading to increased susceptibility to diseases. These diseases have resulted in production losses and remain as one of the major causes of concern for carp farmers. Recently, one of the major limiting factors in intensive fish culture is the use of dietary supplements probiotics and prebiotics. These natural ingredients enhance the immune response of fish, confer tolerance against different stressors and minimize the risk associated with the use of chemical products such as: vaccines, antibiotics and chemotherapeutics. The review by Dawood and Koshio (2016) summarizes and discusses the results of probiotic and prebiotic administration on growth performance, gut physiology, intestinal microbiota, immune response and health status of different carp species. Furthermore, this study tries to cover the gaps in existing knowledge and suggest issues that merit further investigations (Dawood and Koshio, 2016).

10. Conclusions

The use of probiotics, prebiotics in foods has been extensively reviewed. Probiotics are widely used in the dairy industry to produce fermented milk products. Their health effects have been clearly demonstrated in this chapter. Other examples in different food industries are presented along with a case study in aquaculture showing the synergistic effects and the multidisciplinary nature of these beneficial microorganisms.

References

Aarti, C., A. Khusro, R. Varghese, M. Valan Arasu, P. Agastian, N.A. Al-Dhabi et al. 2017. In vitro studies on probiotic and antioxidant properties of Lactobacillus brevis strain LAP2 isolated from Hentak, a fermented fish product of North-East India. LWT – Food Sci. Technol. 86: 438-446.

Abdelgadir, W., D.S. Nielsen, S. Hamad and M. Jakobsen. 2008. A traditional Sudanese fermented camel's milk product, gariss, as a habitat of *Streptococcus infantarius* subsp. *Infantarius*. Int. J. Food Microbiol. 127(3): 215-219.

Abdulrahman, N.M. and V.M. Ahmed. 2015. Comparative effect of probiotic (*Saccharomyces cerevisiae*), prebiotic (fructooligosaccharides FOS) and their combination on some differential white blood cells in young common carp (*Cyprinus caprio* L.). Asian J. Sci. Tech. 6: 1136-1140.

Abushelaibi, A., S. Al-Mahadin, K. El-Tarabily, N.P. Shah and M. Ayyash. 2017a. Characterization of potential probiotic lactic acid bacteria isolated from camel milk. LWT – Food Sci. Technol. 79: 316-325.

Akhmetsadykova, S., A. Baubekova, G. Konuspayeva, N. Akhmetsadykov and G. Loiseau. 2014. Microflora identification of fresh and fermented camel milk from Kazakhstan. Emirates J. Food and Agric. 26(4): 327-332.

Akrami, R., M. Nasri-Tajan, A. Jahedi, M. Jahedi, M. Razeghi Mansour and S.A. Jafarpour. 2015. Effects of dietary synbiotic on growth, survival, lactobacillus bacterial count, blood indices and immunity of beluga (*Huso huso* Linnaeus, 1754) juvenile. Aquac. Nutr. 21: 952-959.

Alegre, I., I. Vinas, J. Usall, M. Anguera and M. Abadias. 2011. Microbiological and physicochemical quality of fresh-cut apple enriched with the probiotic strain *Lactobacillus rhamnosus* GG. Food Microbiol. 28(1): 59-66.

Allameh, S.K., F.M. Yusoff, E. Ringø, H.M. Daud, C.R. Saad and A. Ideris. 2015. Effects of dietary mono- and multiprobiotic strains on growth performance, gut bacteria and body composition of Javanese carp (*Puntius gonionotus*, Bleeker 1850). Aquacul. Nutr. 22: 367-373.

Al-Muzafar, H.M. and K.A. Amin. 2017. Probiotic mixture improves fatty liver disease by virtue of its action on lipid profiles, leptin, and inflammatory biomarkers. BMC Compl. Altr. Med. 17(1): 43.

Altamirano-Fortoul, R., R. Moreno-Terrazas, A. Quezada-Gallo and C.M. Rosell. 2012. Viability of some probiotic coatings in bread and its effect on the crust mechanical properties. Food Hydrocol. 29(1): 166-174.

Anacarso, I., L. Gigli, M. Bondi, S. de Niederhausern, S. Stefani, C. Condo et al. 2017. Isolation of two lactobacilli, producers of two new bacteriocin-like substances (BLS) for potential food-preservative use. Eur. Food Res. Technol. 243: 2127-2134.

Anandharaj, M., B. Sivasankari, R. Santhanakaruppu, M. Manimaran, R. Parveen Rani and S. Sivakumar. 2015. Determining the probiotic potential of cholesterol-reducing *Lactobacillus* and *Weissella* strains isolated from gherkins (fermented cucumber) and south Indian fermented koozh. Res. Microbiol. 166: 428-439.

Arques, J.L., E. Rodriguez, S. Langa, J.M. Landete and M. Medina. 2015. Antimicrobial activity of lactic acid bacteria in dairy products and gut: Effect on pathogens. Biomed Res. Int. 2015, Article ID 584183.

Arvanitoyannis, I.S. and A.C. Stratakos. 2012. Application of modified atmosphere packaging and active/smart technologies to red meat and poultry: A review. Food Bioproc. Technol. 5: 1423-1446.

Ayyash, M., A. Abushelaibi, S. Al-Mahadin, M. Enan, K. El-Tarabily and N. Shah. 2018. *In-vitro* investigation into probiotic characterisation of *Streptococcus* and *Enterococcus* isolated from camel milk. LWT – Food Sci. Technol. 87: 478-487.

Balthazar, C.F., H.L.A. Silva, R.N. Cavalcanti, E.A. Esmerino, L.P. Cappato, Y.K.D. Abud et al. 2017. Prebiotics addition in sheep milk ice cream: A rheological, microstructural and sensory study. J. Funct. Foods 35: 564–573.

Bandyopadhyay, P., S. Mishra, B. Sarkar, S.K. Swain, A. Pal, P.P. Tripathy et al. 2015. Dietary *Saccharomyces cerevisiae* boosts growth and immunity of IMC *Labeo rohita* (Ham.) juveniles. Indian J. Microbiol. 55: 81-87.

Bekkali, N.-L.H., M.E.J. Bongers, M.M. Van den Berg, O. Liem and M.A. Benninga. 2007. The role of a probiotics mixture in the treatment of childhood constipation: A pilot study. Nutr. J. 6: 17.

Bengoa, A.A., M. Goretti Llamas, C. Iraporda, M.T. Duenas, A.G. Abraham and G.L. Garrote. 2018. Impact of growth temperature on exopolysaccharide production and probiotic properties of *Lactobacillus paracasei* strains isolated from kefir grains. Food Microbiol. 69: 212-218.

Betoret, E., L.F. Calabuig-Jimenez, R. Patrignani, Lanciotti and M. Dalla Rosa. 2017. Effect of high pressure processing and trehalose addition on functional properties of mandarin juice enriched with probiotic microorganisms. LWT – Food Sci. Technol. 85: 418-422.

Boutagy, N.E., R.P. McMillan, M.I. Frisard and M.W. Hulver. 2016. Metabolic endotoxemia with obesity: Is it real and is it relevant? Biochimie 124: 11-20.

Castorena-Alba, M.M., J.A. Vázquez-Rodríguez, M. López-Cabanillas Lomelí and B.E. González-Martínez. 2018. Cholesterol assimilation, acid and bile survival of probiotic bacteria isolated from food and reference strains. CyTA – J. Food. 16(1): 36-41.

Chen, J., R.H. Liang, W. Liu, T. Li, C.M. Liu, S.S. Wu et al. 2013. Pecticoligosaccharides prepared by dynamic high-pressure microfluidization and their *in vitro* fermentation properties. Carbohydr. Polym. 91(1): 175-182.

Chen, S., Q. Zhao, L.R. Ferguson, Q. Shu, I. Weir and S. Garg. 2012. Development of a novel probiotic delivery system based on microencapsulation with protectants. Appl. Microbiol. Biotechnol. 93(4): 1447-1457.

Chetana, R., S.R.Y. Reddy and P.S. Negi. 2013. Preparation and properties of probiotic chocolate using yoghurt powder. Food Nutr. Sci. 4: 276-281.

Chi, C., B. Jiang, X.B. Yu, T.Q. Liu, L. Xia and G.X. Wang. 2014. Effects of three strains of intestinal autochthonous bacteria and their extracellular products on the immune response and disease resistance of common carp, *Cyprinus carpio*. Fish Shellfish Immunol. 36: 9-18.

Collins, M.D. and G.R. Gibson. 1999. Probiotics, prebiotics, and synbiotics: Approaches for modulating the microbial ecology of the gut. Am. J. Clin. Nutr. 69: 1052-1057.

Cook, M.T., G. Tzortzis, D. Charalampopoulos and V.V. Khutoryanskiy. 2012. Microencapsulation of probiotics for gastrointestinal delivery. J. Control Release 162(1): 56-67.

Corcoran, B.M., F.P. Roos, G.F. Fitzgerald and C. Stanton. 2004. Comparative survival of probiotic lactobacilli spray dried in the presence of prebiotic substances. J. Appl. Microbiol. 96: 1024-1039.

Davati, N., F.T. Yazdi, S. Zibaee, F. Shahidi and M.R. Edalatian. 2015. Study of lactic acid bacteria community from raw milk of Iranian one humped camel and evaluation of their probiotic properties. Jundishapur J. Microbiol. 8(5): 1-6.

Davis, L.M., I. Martínez, J. Walter and R. Hutkins. 2010. A dose dependent impact of prebiotic galactooligosaccharides on the intestinal microbiota of healthy adults. Int. J. Food Microbiol. 144(2): 285-292.

Dawood, M.A.O. and S. Koshio. 2016. Recent advances in the role of probiotics and prebiotics in carp aquaculture: A review. Aquacult. 454: 243-251.

Dawood, M.A.O., S. Koshio, M. Ishikawa and S. Yokoyama. 2015a. Interaction effects of dietary supplementation of heat-killed *Lactobacillus plantarum* and □-glucan on growth performance, digestibility and immune response of juvenile red sea bream, *Pagrus major*. Fish & Shellfish Immunol. 45: 33-42.

Dehaghani, P.G., M.J. Baboli, A.T. Moghadam, S. Ziaei-Nejad and M. Pourfarhadi. 2015. Effect of synbiotic dietary supplementation on survival, growth performance, and digestive enzyme activities of common carp (*Cyprinus carpio*) fingerlings. Czech J. Anim. Sci. 60: 224-232.

Del Piano, M., S. Carmagnola, M. Ballarè, M. Sartori, M. Orsello, M. Balzarini et al. 2011. Is microencapsulation the future of probiotic preparations? The increased efficacy of gastro-protected probiotics. Gut Microbes 2(2): 120-123.

Dertli, E., M.J. Mayer and A. Narbad. 2015. Impact of the exopolysaccharide layer on biofilms, adhesion and resistance to stress in *Lactobacillus johnsonii* FI9785. BMC Microbiol. 15: 8.

Dhanaraj, M., M.A. Haniffa, S.A. Singh, A.J. Arockiaraj, C.M. Ramakrishanan, S. Seetharaman et al. 2010. Effect of probiotics on growth performance of koi carp (*Cyprinus carpio*). J. Appl. Aquacult. 22: 202-209.

Dodane, V., M. Amin Khan and J.R. Merwin. 1999. Effect of chitosan on epithelial permeability and structure. Int. J. Pharm. 182(1): 21-32.

Ebrahimi, B., R. Mohammadi, M. Rouhi, A.M. Mortazavian, S. Shojaee-Aliabadi and M.R. Koushki. 2018. Survival of probiotic bacteria in carboxymethyl cellulose-based edible film and assessment of quality parameters. LWT – Food Sci. Technol. 87: 54-60.

Elagamy, E., R. Ruppanner, A. Ismail, C. Champagne and R. Assaf. 1996. Purification and characterization of lactoferrin, lactoperoxidase, lysozyme and immunoglobulins from camel's milk. Int. Dairy J. 6(2): 129-145.

Espitia, P.J.P., R.A. Batista, H.M.C. Azeredo and C.G. Otoni. 2016. Probiotics and their potential applications in active edible films and coatings. Food Res. Int. 90: 42-52.

FAO/WHO. 2002. Guidelines for the evaluation of probiotics in food. Food and Agriculture Organization of the United Nations and World Health Organization. Working group report. https://www.who.int/foodsafety/fs_management/en/probiotic_guidelines.pdf

Faramarzi, M., S. Kiaalvandi and F. Iranshahi, 2011. The effect of probiotics on growth performance and body composition of common carp (*Cyprinus carpio*). J. Anim. Vet. Adv. 10: 2408-2413.

Fritzen-Freire, C.B., E.S. Prudêncio, R.D.M.C. Amboni, S.S. Pinto, A. Negrão-Murakami and F.S. Murakami. 2012. Microencapsulation of bifidobacteria by spray drying in the presence of prebiotics. Food Res. Int. 45: 306-312.

Ganguly, S., K.C. Dora, S. Sarkar and S. Chowdhury. 2013. Supplementation of prebiotics in fish feed: A review. Rev. Fish Biol. Fish. 23: 195-199.

Ghali, M.B., P.T. Scott, G.A. Alhadrami and R.A.M. Al Jassim. 2011. Identification and characterisation of the predominant lactic acid-producing and lactic acid utilising bacteria in the foregut of the feral camel (*Camelus dromedarius*) in Australia. Animal Prod. Sci. 51(7): 597-604.

Ghibaudo, F., E. Gerbino, V. Campo Dall'Orto and A. Gómez-Zavaglia. 2017. Pectin-iron capsules: Novel system to stabilise and deliver lactic acid bacteria. J. Funct. Foods 39: 299-305.

Gibson, G.R. and M.B. Roberfroid. 1995. Dietary modulation of the human colonic microbiota: Introducing the concept of prebiotics. J. Nutr. 125: 1401-1412.

Goldin, B. and S.L. Gorbach. 1977. Alterations in fecal microflora enzymes related to diet, age, lactobacillus supplements, and dimethylhydrazine. Cancer 40: 2421-2426.

Gomez, B., B. Gullon, R. Yanez, H. Schols and J.L. Alonso. 2016. Prebiotic potential of pectins and pectic oligosaccharides derived from lemon peel wastes and sugar beet pulp: A comparative evaluation. J. Funct. Foods 20: 108-121.

Govender, M., Y.E. Choonara, P. Kumar, L.C. du Toit, S. van Vuuren and V. Pillay. (2014). A review of the advancements in probiotic delivery: Conventional vs. non-conventional formulations for intestinal flora supplementation. AAPS Pharm. Sci Tech. 15(1): 29-43.

Grossova, M., P. Rysavka and I. Marova. 2017. Probiotic biofilm on carrier surface: A novel promising application for food industry. Acta Aliment 46(4): 439-448.

Gupta, S. and N. Abu-Ghannam. 2012. Probiotic fermentation of plant based products: Possibilities and opportunities. Crit. Rev. Food Sci. Nutr. 52: 183-199.

Haffner, F.B., R. Diab and A. Pasc. 2016. Encapsulation of probiotics: Insights into academic and industrial approaches. AIMS Mat. Sci. 3(1): 114-136.

Hai, N.V. 2015. Research findings from the use of probiotics in tilapia aquaculture: A review. Fish & Shellfish Immunol. 45: 592-597.

Hamed, E. and A. Elattar. 2013. Identification and some probiotic potential of lactic acid bacteria isolated from Egyptian camels milk. Life Sci. J. 10(1): 1952-1961.

Harikrishnan, R., C. Balasundaram and M.S. Heo. 2010. Potential use of probiotic- and triherbal extract-enriched diets to control *Aeromonas hydrophila* infection in carp. Dis. Aquat. Org. 92: 41-49.

Hassaïne, O., H. Zadi-Karam and N.-E. Karam 2007. Technologically important properties of lactic acid bacteria isolated from raw milk of three breeds of Algerian dromedary (*Camelus dromedarius*). Afr. J. Biotechnol. 6(14): 1720-1727.

Heeney, D., M.G. Gareau and M.L. Marco. 2018. Intestinal lactobacillus in health and disease, a driver or just along for the ride? Cur. Opin. Biotechnol. 49: 140-147.

Hoseinifar, S.H., E. Ringø, A.S. Masouleh and M.A. Esteban. 2014a. Probiotic, prebiotic and synbiotic supplements in sturgeon aquaculture: A review. Rev. Aquacult. 6: 1-14.

Hugenholtz, J. and E.J. Smid. 2002. Nutraceutical production with food-grade microorganisms. Curr. Opin. Biotechnol. 13: 497-507.

Huq, T., A. Khan, R.A. Khan, B. Riedl and M. Lacroix. 2013. Encapsulation of probiotic bacteria in biopolymeric system. Crit. Rev. Food Sci. Nutr. 53(9): 909-916.

Iglesias, M.B., M. Abadias, M. Anguera, J. Sabata and I. Vinas. 2017. Antagonistic effect of probiotic bacteria against foodborne pathogens on fresh-cut pear. LWT – Food Sci. Technol. 81: 243-249.

Iglesias, M.B., I. Viñas, P. Colás-Medà, C. Collazo, J.C.E. Serrano and M. Abadias. 2017. Adhesion and invasion of *Listeria monocytogenes* and interaction with *Lactobacillus rhamnosus* GG after habituation on fresh-cut pear. J. Funct. Foods 34: 453-460.

Iglesias, M.B., G. Echeverría, I. Vinas, M.L. Lopez and M. Abadias. 2018. Biopreservation of fresh-cut pear using *Lactobacillus rhamnosus* GG and effect on quality and volatile compounds. LWT – Food Sci. Technol. 87: 581-588.

Jans, C., J. Bugnard, P.M.K. Njage, C. Lacroix and L. Meile. 2012. Lactic acid bacteria diversity of African raw and fermented camel milk products reveals a highly competitive, potentially health-threatening predominant microflora. LWT –Food Sci. Technol. 47(2): 371-379.

Ji, Y.S., H.N. Kim, H.J. Park, J.E. Lee, S.Y. Yeo, J.S. Yang et al. 2012. Modulation of the murine microbiome with a concomitant anti-obesity effect by *Lactobacillus rhamnosus* GG and *Lactobacillus sakei* NR28. Benefic. Microbes 3: 13-22.

Kadri, Z., M. Amar, M. Ouadghiri, M. Cnockaert, M. Aerts, O. El Farricha et al. 2014. *Streptococcus moroccensis* sp. Nov. and *streptococcus rifensis* sp. Nov., isolated from raw camel milk. Int. J. System. Evol. Microbiol. 64(7): 2480-2485.

Kadri, Z., P. Vandamme, M. Ouadghiri, M. Cnockaert, M. Aerts, E.M. Elfahime et al. 2015a. *Streptococcus tangierensis* sp. Nov. and *streptococcus cameli* sp. Nov., two novel streptococcus species isolated from raw camel milk in Morocco. Antonie Van Leeuwenhoek 107(2): 503-510.

Kadri, Z., F. Spitaels, M. Cnockaert, J. Praet, O. El Farricha, J. Swings et al. 2015b. *Enterococcus bulliens* sp. Nov., a novel lactic acid bacterium isolated from camel milk. Antonie Van Leeuwenhoek 108(5): 1257-1265.

Kanmani, P., R.S. Kumar, N. Yuvaraj, K.A. Paari, V. Pattukumar and V. Arul. 2011. Cryopreservation and microencapsulation of a probiotic in alginate-chitosan capsules improves survival in simulated gastrointestinal conditions. Biotechnol. Bioproc. Eng. 16(6): 1106-1114.

Koebnick, C., I. Wagner, P. Leitzmann, U. Stern and H.J.F. Zunft. 2003. Probiotic beverage containing *Lactobacillus casei* Shirota improves gastrointestinal symptoms in patients with chronic constipation. Can. J. Gastroent. 17: 655-659.

Kotzamanidis, C., A. Kourelis, E. Litopoulou-Tzanetaki, N. Tzanetakis and M. Yiangou 2010. Evaluation of adhesion capacity, cell surface traits and immunomodulatory activity of presumptive probiotic Lactobacillus strains. Int. J. Food Microbiol. 140: 154-163.

Krasaekoopt, W., B. Bhandari and H. Deeth. 2003. Evaluation of encapsulation techniques of probiotics for yoghurt. Int. Dairy J. 13(1): 3-13.

Kumar, M., R. Nagpal, R. Kumar, R. Hemalatha, V. Verma, A. Kumar et al. 2012. Cholesterol-lowering probiotics as potential biotherapeutics for metabolic diseases. Exp. Diabet. Res. 2012: 902917.

Kumar, M., S. Rakesh, R. Nagpal, R. Hemalatha, A. Ramakrishna, V. Sudarshan et al. 2012. Probiotic *Lactobacillus rhamnosus* GG and *Aloe vera* gel improve lipid profiles in hypercholesterolemic rats. Nutrition 29: 574-579.

Lalicić-Petronijević, J., J. Popov-Raljic, V. Lazić, L. Pezo and V. Nedovic. 2017. Synergistic effect of three encapsulated strains of probiotic bacteria on quality parameters of chocolates with different composition. J. Funct. Foods 38: 329-337.

Li, M., K. Lee, M. Hsu, G. Nau, E. Mylonakis and B. Ramratnam. 2017. Lactobacillus-derived extracellular vesicles enhance host immune responses against vancomycin-resistant enterococci. BMC Microbiol. 17: 66.

Li, K.Y. 2004. Fermentation: Principles and microorganisms. pp. 612-626. *In*: Y.H. Hui, S. Ghazala, D.M. Graham, K.D. Murrell and W.K. Nip (Eds.). Handbook of vegetable preservation and processing. New York, NY: Marcel Dekker, Inc.

Li, K.K., P.J. Tian, S.D. Wang, P. Lei, L. Qu, J.P. Huang et al. 2017. Targeting gut microbiota: Lactobacillus alleviated type 2 diabetes via inhibiting LPS secretion and activating GPR43 pathway. J. Funct. Foods 38: 561-570.

Li, R., Y. Zhang, D.B. Polk, P.M. Tomasula, F. Yan, L. Liu et al. 2016. Preserving viability of *Lactobacillus rhamnosus* GG *in vitro* and *in vivo* by a new encapsulation system. J. Control. Rel. 230: 79-87.

Li, W., Y. Zhu, F. Ye, B. Li, X. Luo and S. Liu. 2017. Probiotics in cellulose houses: Enhanced viability and targeted delivery of *Lactobacillus plantarum*. Food Hydrocol. 62: 66-72.

Liu, C.-J., X.-D. Tang, J. Yu, H.-Y. Zhang and X.-R. Li. 2017. Gut microbiota alterations from different Lactobacillus probiotic-fermented yoghurt treatments in slow-transit constipation. J. Funct. Foods 38: 110-118.

Liu, Y., S. Alexeeva, K.A. Defourny, E.J. Smid and T. Abee. 2018. Tiny but mighty: Bacterial membrane vesicles in food biotechnological applications. Cur. Opin. Biotechnol. 49: 179-184.

Luckow, T. and C. Delahunty. 2004. Which juice is 'healthier'? A consumer study of probiotic non-dairy juice drinks. Food Qual. Prefer. 15: 751-759.

Mahmoud, S.F., M.M. Montaser, O. Al Zhrani and S.A.M. Amer. 2014. Isolation and molecular identification of the lactic acid bacteria from raw camel milk. J. Camel Pract. Res. 21(2): 267-273.

Maillard, M. and A. Landuyt. 2008. Chocolate: An ideal carrier for probiotics. Teknoscienze 19(3): 13-15.

Malmo, C., A. La Storia and G. Mauriello. 2013. Microencapsulation of *Lactobacillus reuteri* DSM 17938 cells coated in alginate beads with chitosan by spray drying to use as a probiotic cell in a chocolate souffle. Food Bioproc. Technol. 6(3): 795-805.

Mandal, S., S. Hati, A.K. Puniya, R. Singh and K. Singh. 2013. Development of synbiotic milk chocolate using encapsulated *Lactobacullus casei* NCDC 298. J. Food Proc. Pres. 37: 1031-1037.

Martin, M.J., F. Lara-Villoslada, M.A. Ruiz and M.E. Morales. 2015. Microencapsulation of bacteria: A review of different technologies and their impact on the probiotic effects. Innov. Food Sci. & Emerg. Technol. 27: 15-25.

Martinez-Gutierrez, F., S. Ratering, B. Juarez-Flores, C. Godinez-Hernandez, R. Geissler-Plaum, F. Prell et al. 2017. Potential use of *Agave salmiana* as a prebiotic that stimulates the growth of probiotic bacteria. LWT – Food Sci. Technol. 84: 151-159.

Nunes, G.L., M. de Araújo Etchepare, A.J. Cichoski, L. Queiroz Zepka, E.J. Lopes, J. Smanioto Barin et al. 2018. Inulin, hi-maize, and trehalose as thermal protectants for increasing viability of *Lactobacillus acidophilus* encapsulated by spray drying. LWT – Food Sci. Technol. 89: 128-133.

Olano-Martin, E., G.R. Gibson and R.A. Rastall. 2002. Comparison of the *in vitro* bifidogenic properties of pectins and pectic-oligosaccharides. J. Appl. Microbiol. 93(3): 505-511.

Orrhage, K., E. Sillerstrom, J.A. Gustafsson, C.E. Nord and J. Rafter. 1994. Binding of mutagenic heterocyclic amines by intestinal and lactic acid bacteria, Mutat. Res. 311: 239-248.

Parker, E.A., T. Roy, C.R. D'Adamo and L.S. Wieland. 2018. Probiotics and gastrointestinal conditions: An overview of evidence from the cochrane collaboration. Nutrition 45: 125-134.

Pereira, A.L.F., T.C. Maciel and S. Rodrigues. 2011. Probiotic beverage from cashew apple juice fermented with *Lactobacillus casei*. Food Res. Int. 44: 1276-1283.

Pinto, S.S., C.B. Fritzen-Freire, I.B. Munoz, P.L.M. Barreto, E.S. Prudêncio and R.D.M.C. Amboni. 2012. Effects of the addition of microencapsulated *Bifidobacterium* BB-12 on the properties of frozen yogurt. J. Food Eng. 111: 563-569.

Pothuraju, R. and S.A. Hussain. 2017. Editorial probiotics: An important player in the obesity management alone. Obesity Med. 8: 13-14.

Prado, F.C., J.L. Parada, A. Pandey and C.R. Soccol. 2008. Trends in non-dairy probiotic beverages. Food Res. Int. 41: 111-123.

Radulovic, Z., J. Miocinovic, N. Mirkovic, M. Mirkovic, D. Paunovic, M. Ivanovic et al. 2017. Survival of spray-dried and free-cells of potential probiotic *Lactobacillus plantarum* 564 in soft goat cheese. Anim. Sci. J. 88: 1849-1854.

Reid, A.A., C.P. Champagne, N. Gardner, P. Fustier and J.C. Vuillemard. 2007. Survival in food systems of *Lactobacillus rhamnosus* R011 microentrapped in whey protein gel particles. J. Food Sci. 72(1): 31-37.

Rivera-Espinoza, Y. and Y. Gallardo-Navarro. 2010. Non-dairy probiotic products. Food Microbiol. 27: 1-11.

Roberfroid, M. 2007. Prebiotics: The concept revisited. J. Nutr. 137(3 Suppl 2): 830S-837S.

Roberfroid, M.B. 2007. Inulin-type fructans: Functional food ingredients. J. Nutr. 137(11 Suppl): 2493s-2502s.

Roberfroid, M., G.R. Gibson, L. Hoyles, A.L. McCartney, R. Rastall, I. Rowland et al. 2010. Prebiotic effects: Metabolic and health benefits. Brit. J. Nutr. 104(Suppl 2): S1-S63.

Rowland, I.R. and P. Grasso. 1975. Degradation of N-nitrosamines by intestinal bacteria, Appl. Microbiol. 29: 7-12.

Ruas-Madiedo, P., A.G. Abraham, F. Mozzi and C.G. De Los Reyes-Gavilán. 2008. Functionality of exopolysaccharides produced by lactic acid bacteria. pp. 137-166. *In*: Mayo, B., López, P. and Pérez-Martínez, G. (Eds.). Molecular aspects of lactic acid bacteria for traditional and new applications. Kerala: Research Signpost.

Russo, P., M.L.V. de Chiara, A. Vernile, M.L. Amodio, M.P. Arena, V. Capozzi et al. 2014. Fresh-cut pineapple as a new carrier of probiotic lactic acid bacteria. Biomed. Res. 2014: Article ID 309183.

Russo, P., N. Pena, M.L.V. de Chiara, M.L. Amodio, G. Colelli and G. Spano. 2015. Probiotic lactic acid bacteria for the production of multifunctional fresh-cut cantaloupe. Food Res. Int. 77: 762-772.

Russo, P., M.L.V. de Chiara, V. Capozzi, M.P. Arena, M.L. Amodio, A. Rascón et al. 2016. *Lactobacillus plantarum* strains for multifunctional oat-based foods. LWT – Food Sci. Technol. 68: 288-294.

Russo, P., M. Pia Arena, D. Fiocco, V. Capozzi, D. Drider and G. Spano. 2017. *Lactobacillus plantarum* with broad antifungal activity: A promising approach to increase safety and shelf-life of cereal-based products. Int. J. Food Microbiol. 247: 48-54.

Ryan, P., R. Ross, G. Fitzgerald, N. Caplice and C. Stanton. 2015. Sugar-coated: Exopolysaccharide producing lactic acid bacteria for food and human health applications. Food Funct. 6: 679-693.

Sarao, L.K. and M. Arora. 2017. Probiotics, prebiotics, and microencapsulation: A review. Crit. Rev. Food Sci. Nutr. 57(2): 344-371.

Saravanan, M.R. and K. Panduranga. 2010. Pectinegelatin and alginate-gelatin complex coacervation for controlled drug delivery: Influence of anionic polysaccharides and drugs being encapsulated on physicochemical properties of microcapsules. Carbohydr. Polym. 80: 808-816.

Seidel, D.V., M. Andrea Azcárate-Peril, R.S. Chapkin and N.D. Turner. 2017. Shaping functional gut microbiota using dietary bioactives to reduce colon cancer risk. Seminars Cancer Biol. 46: 191-204.

Shah, N. and A. Patel. 2014. Recent advances in biosynthesis of vitamin and enzyme from food grade bacteria. Int. J Food Ferment. Technol. 4: 79-85.

Silva, J.P.S.E., S.C. Sousa, P. Costa, E. Cerdeira, M.H. Amaral, J.S. Lobo et al. 2013. Development of probiotic tablets using microparticles: Viability studies and stability studies. AAPS Pharm. Sci. Tech. 14(1): 121-127.

Singh, P., B. Medronho, L. Alves, G.J. da Silva, M.G. Miguel and B. Lindman. 2017. Development of carboxymethyl cellulose-chitosan hybrid micro- and macroparticles for encapsulation of probiotic bacteria. Carbohydr. Polym. 175: 87-95.

Siroli, L., F. Patrignani, D.I. Serrazanetti, G. Tabanelli, C. Montanari, F. Gardini et al. 2015a. Lactic acid bacteria and natural antimicrobials to improve the safety and shelf-life of minimally processed sliced apples and lamb's lettuce. Food Microbiol. 47: 74-84.

Siroli, L., F. Patrignani, D.I. Serrazanetti, S. Tappi, P. Rocculi, F. Gardini et al. 2015b. Natural antimicrobials to prolong the shelf-life of minimally processed lamb's lettuce. Postharv. Biol. Technol. 103: 35-44.

Sirtori, C.R., C. Pavanello, L. Calabresi and M. Ruscica. 2017. Nutraceutical approaches to metabolic syndrome. Ann. of Med. 49(8): 678-697.

Soleymanzadeh, N., S. Mirdamadi and M. Kianirad. 2016. Antioxidant activity of camel and bovine milk fermented by lactic acid bacteria isolated from traditional fermented camel milk (chal). Dairy Sci. Technol. 96(4): 443-457.

Son, S.-H., H.-L. Jeon, E.B. Jeon, N.-K. Lee, Y.-S. Park, D.-K. Kang et al. 2017. Potential probiotic *Lactobacillus plantarum* Ln4 from kimchi: Evaluation of b-galactosidase and antioxidant activities. LWT – Food Sci. Technol. 85: 181-186.

Song, S.K., B.R. Beck, D. Kim, J. Park, J. Kim, H.D. Kim et al. 2014. Prebiotics as immunostimulants in aquaculture: A review. Fish Shellfish Immunol. 40: 40-48.

Soukoulis, C., L. Yonekura, H.-Hui Gan, S. Behboudi-Jobbehdar, C. Parmenter and I. Fisk. 2014. Probiotic edible films as a new strategy for developing functional bakery products: The case of pan bread. Food Hydrocol. 39: 231-242.

Soukoulis, C., P. Singh, W. Macnaughtan, C. Parmenter and I.D. Fisk. 2016. Compositional and physicochemical factors governing the viability of *Lactobacillus rhamnosus* GG embedded in starch-protein based edible films. Food Hydrocol. 52: 876-887.

Succi, M., P. Tremonte, G. Pannella, L. Tipaldi, A. Cozzolino, R. Coppola et al. 2017. Survival of commercial probiotic strains in dark chocolate with high cocoa and phenols content during the storage and in a static in vitro digestion model. J. Funct. Foods 35: 60-67.

Tello, F., R.N. Falfan-Cortes, F. Martinez-Bustos, V.M. da Silva, M.D. Hubinger and C. Grosso. 2015. Alginate and pectin-based particles coated with globular proteins: Production, characterization and anti-oxidative properties. Food Hydrocol. 43: 670-678.

Tripathi, M.K. and S.K. Giri. 2014. Probiotic functional foods: Survival of probiotics during processing and storage. J. Funct. Foods 9: 225-241.

Turan, I., O. Dedeli, S. Bor and T. Ilter. 2014. Effects of a kefir supplement on symptoms, colonic transit, and bowel satisfaction score in patients with chronic constipation: A pilot study. Turkish J. Gastroent. 25: 650-656.

Yao, M., J. Wu, B. Li, H. Xiao, D.J. McClements and L. Li. 2017. Microencapsulation of *Lactobacillus salivarious* Li01 for enhanced storage viability and targeted delivery to gut microbiota. Food Hydrocol. 72: 228-236.

Yeung, T.W., I.J. Arroyo-Maya, D.J. McClements and D.A. Sela. 2016. Microencapsulation of probiotics in hydrogel particles: Enhancing *Lactococcus lactis* subsp. *cremoris* LM0230 viability using calcium alginate beads. Funct. Foods 7(4): 1797-1804.

Yuan, D., J.C. Jacquier and E.D. O'Riordan. 2018. Entrapment of proteins and peptides in chitosan-polyphosphoric acid hydrogel beads: A new approach to achieve both high entrapment efficiency and controlled *in vitro* release. Food Chem. 239: 1200-1209.

Zannini, E., D.M. Waters, A. Coffey and E.K. Arendt. 2016. Production, properties, and industrial food application of lactic acid bacteria-derived exopolysaccharides. Appl. Microbiol. Biotechnol. 100: 1121-1135.

Zhang, Q., Y. Wu and X. Fei. 2015. Effect of probiotics on body weight and body-mass index: A systematic review and meta-analysis of randomized, controlled trials. Int. J. Food Sci. Nutr. 67: 571-580.

Zhang, L., S. Huang, V.K. Ananingsih, W. Zhou and X.D. Chen. 2014. A study on *Bifidobacterium lactis* Bb12 viability in bread during baking. J. Food Eng. 122(1): 33-37.

Zhang, L., M.A. Taal, R.M. Boom, X.D. Chen and M.A.I. Schutyser. 2018. Effect of baking conditions and storage on the viability of *Lactobacillus plantarum* supplemented to bread. LWT – Food Sci. Technol. 87: 318-325.

Zhang, S., H. Hu, L. Wang, F. Liu and S. Pan. 2018. Preparation and prebiotic potential of pectin oligosaccharides obtained from citrus peel pectin. Food Chem. 244: 232-237.

Zivkovi, M., M.S. Miljkovic, P. Ruas-Madiedo, M.B. Markelic, K. Veljovic, M. Tolinacki et al. 2016. EPS-sj exopolysaccharide produced by the strain *Lactobacillus paracasei* subsp. *paracasei* BGSJ2-8 is involved in adhesion to epithelial intestinal cells and decrease on *E. coli* association to Caco-2 cells. Front. Microbiol. 7: 286.

Role of the Gut Microbiota in the Digestive Tract Diseases

Shanmugaprakasham Selvamani[1,2], Devendra Mehta[3], Chirajyoti Deb[3], Hesham El Enshasy[1,2,4] and Bassam Abomoelak[3*]

[1] School of Chemical and Energy Engineering, Faculty of Engineering, Universiti Teknologi Malaysia (UTM), Skudai, Johor Bahru, Malaysia
[2] Institute of Bioproduct Development (IBD), Universiti Teknologi Malaysia (UTM), Skudai, Johor Bahru, Malaysia
[3] Pediatric Gastroenterology Clinic Arnold Palmer Hospital Orlando Health, Orlando, FL, USA
[4] Bioprocess Development Department, Genetic Engineering and Biotechnology Research Institute, City of Scientific Research and Technology Applications (SRTA City), Universities and Research Institutes Zone, New Borg El-Arab, Alexandria 21934, Egypt

1. Introduction

Microbiota in the gut plays essential role in the gut homeostasis, especially in terms of physical structure, functionality, and metabolism (Cho and Blaser 2012). Numerous works recently highlight the role played by several bacterial species in the gut homeostasis (Hiippala et al., 2018). The intestinal microbiota plays an essential role in the intestinal barrier which is considered a crucial element for gut functionality and host defence mechanisms (Geuking et al., 2014). Generally, GI diseases are idiopathic in nature especially IBS and IBD. IBS is generally characterized by abdominal pain associated with other altered bowel functions (e.g., diarrhoea, constipation, or both). IBD is a chronic relapsing inflammatory disorder of the GI tract (Srinivasan and Akobeng, 2009). Although IBD is an idiopathic disease, some data suggests a possible role of the microbiota in genetically susceptible individuals (Zhou and Zhi, 2016). In healthy individuals, the gut microflora composition is balanced, but the balance is disturbed in case of a disease. Several studies have linked bacterial genera abundance to GI diseases. For example, *Faecalibacterium* and *Dorea* showed altered abundances in the gut of IBS patients (Jeffery

*Corresponding author: bassam.abomoelak@orlandohealth.com

et al., 2012). Similarly, in IBD patients, *Bacteroides* may be involved in the development of the disease (Zhou and Zhi, 2016).

The GI tract, especially the small intestine, offers an opportunity for the microbiota to interact with important host functions, such as food digestion and absorption of nutrients. In addition, several lymphoid tissues are localized in the intestine and are responsible for providing tight host defence against pathogens. The intestinal epithelial cells secrete lysozymes, defensins, antimicrobial peptides (AMPs), and secretory immunoglobulin A (sIgA) (Muniz et al., 2012, Yu et al., 2012). The mechanistic pathways governing microbiota-host interactions remain largely unknown, but data shows that microbiota interact with the host cells through the production of short chain fatty acids (SCFAs). SCFAs are by-products of large molecules metabolism by the commensal gut bacteria. Propionate, butyrate, and acetate are among the most abundant SCFAs and are considered an energy source for the host cells (Macfarlane and Macfarlane, 2011, Cushing et al., 2015).

Deciphering ·the microbiota-host interactions requires a precise identification of all the bacterial genera in the gut and their relative abundances. Although, bacterial culturing methods and routine molecular biology such as polymerase chain reaction (PCR) have been used successfully to achieve the identification process, a more robust and accurate identification system is needed. In the last decade, cultural-independent technologies have emerged as powerful tools in identifying the microbial involvement in idiopathic diseases including IBD and IBS (Maharshak et al., 2018). The next-generation of sequencing (NGS) and the associated bioinformatics pipelines took the lead in this field and started replacing the culturing and molecular routine techniques in clinical and diagnostic settings (Caporaso et al., 2010). The great advantages of such technologies rely on their adaptability to the different sample types (fluid, solid, or clinical biopsies). Furthermore, the availability of computational bioinformatics pipelines for the prediction of the functional profiling of the bacterial community constitutes an asset in linking the bacterial taxonomy to a predictive function (Langille et al., 2013). Taken together, the NGS and functional profiling represent promising tools in delineating the role of the microbiota in the GI tract diseases.

2. The Gastrointestinal Microbiome

The human microbiome project has revealed that the human body hosts more than one trillion microorganisms (Turnbaugh et al., 2007). The microbial composition is diverse and their abundance remains unexplained (Huttenhower et al., 2012). Microbial compositions are often addressed as "microbiota" which refers to the totality of organisms or "microbiome", which refers to the habitats of microorganisms and their collective genes (Turnbaugh et al., 2007, Barko et al., 2018). The human gastrointestinal environment is predicted to contain more than 3 million unique genes and account for the largest portion in the total human microbiome (Engen et al., 2015). In addition, the human gastrointestinal microbiome is classified as the total collection of microbiota found in the human gastrointestinal tract. It is composed of tens of trillions of microbes, including more than 1000 different species of known bacteria (van der Ark, 2017, Barko et al., 2018).

2.1. Composition of the Gastrointestinal Microbiome

The adult human gut may host more than 50 microbial phyla. The gut microbiota is dominated by strict anaerobes belonging to *Bacteroidetes, Firmicutes,* and *Proteobacteria* whereas other anaerobes including *Actinobacteria, Verrumicrobia, Acidobacteria,* and *Fusobacteria* are found in lower quantities (Sommer et al., 2013, Lee and Hase, 2014). The 'Distal' small intestine and colon are dominated by mucosa-associated bacteria from the phyla *Bacteroidetes* and *Firmicutes*. The human proximal gut is enriched with mucosa-associated bacteria including *Lactobacillus (Firmicutes), Veillonella (Firmicutes),* and *Helicobacter (Proteobacteria)*. Meanwhile, the genera of *Bacilli (Firmicutes), Streptococcaceae (Firmicutes), Actinomycinaeae,* and *Corynebacteriaceae* (both *Actinobacteria*) are abundant in the duodenum, jejunum, and ileum. Increased proportions of *Lachnospiraceae (Firmicutes)* and *Bacteroidetes* have been recorded in the colon (Bull and Plummer, 2014, Basson et al., 2016, Vandeputte et al., 2017).

The relative abundance of the genera varies by location in the gastrointestinal tract (Figure 1). The human gut regions vary in terms of physiology, pH and oxygen concentration, substrate availability, host secretions such as enzymes and bile, and by the flow rates of digested food (Bull and Plummer, 2014). The flow rate is rapid from the mouth to the caecum and much slower through the remainder of the colon. The concentration of microbiota increases steadily along the gastrointestinal tract, with small numbers in the stomach, but very high concentrations in the colon. The stomach harbours only 10^1 bacteria per gram, while the bacterial densities and diversities are higher in the duodenum (10^3/gram), jejunum (10^4/gram), ileum (10^7/gram), and colon (10^{12} bacteria/gram) (Dieterich et al., 2018).

The stomach and proximal duodenum are exceptionally inhospitable, as few bacteria can thrive in gastric acidity, proximal duodenal chyme acidity, and intermittent secretions of bile or pancreatic enzymes. The small intestine provides a more challenging environment for microbial colonizers mainly due to the short transit time (3–5 hrs) and the high bile concentrations. The large intestine, however, characterized by slow flow rates and neutral to mildly acidic pH is more hospitable. Thus the colon harbours the largest microbial community and is dominated by obligate anaerobic microbial communities. The microbiota numbers gradually change with increasing quantitative gradient and decreasing microbiota qualitative gradient where the population of aerobic bacteria progressively decreases to the benefit of strictly anaerobic bacteria (Basson et al., 2016, Vandeputte et al., 2017, Rinninella et al., 2019).

Gut microbiota also vary between epithelial and lumen environments (Basson et al., 2016). The gastrointestinal epithelium contains numerous goblet cells that secrete mucin. These glycosylated proteins form a dense protective mucosal layer which prevents microbial penetration. The environment influences the microbial variety as only specialized bacteria that are able to adhere to the mucosal layer can survive in the gastrointestinal tract (Koleva et al., 2015, Perez-Muñoz et al., 2017). The genera of *Clostridium*, *Lactobacillus*, and *Enterococcus* are well-known with adhering properties. However, other studies have reported the presence of multiple species including *Bacteroides*, *Bifidobacterium*, *Streptococcus*, *Enterobacteriaceae*, *Enterococcus*, *Clostridium*, *Lactobacillus*, and *Ruminococcus* in faecal samples. Most works have focused on isolation of microorganisms from the faecal samples. Information on the microbial composition associated with gastrointestinal tract, especially from gut mucosal layers is limited. Isolation and characterization of these mucosa-associated microbes are much more tedious and require intestinal biopsies during endoscopy or colonoscopy, compared to faecal samples which were easily obtained. It is expected that gut microbial composition could undergo minor changes due to various factors when it reaches the rectum or anus or during defaecation (Sokol and Seksik, 2010, Becker et al., 2015).

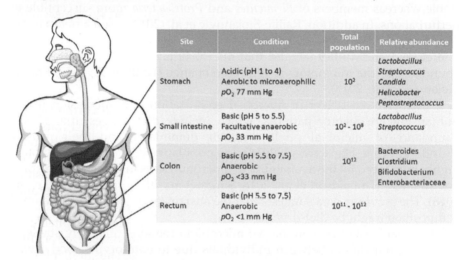

Site	Condition	Total population	Relative abundance
Stomach	Acidic (pH 1 to 4) Aerobic to microaerophilic pO_2 77 mm Hg	10^2	*Lactobacillus* *Streptococcus* *Candida* *Helicobacter* *Peptostreptococcus*
Small intestine	Basic (pH 5 to 5.5) Facultative anaerobic pO_2 33 mm Hg	$10^2 - 10^8$	*Lactobacillus* *Streptococcus*
Colon	Basic (pH 5.5 to 7.5) Anaerobic pO_2 <33 mm Hg	10^{12}	*Bacteroides* *Clostridium* *Bifidobacterium* *Enterobacteriaceae*
Rectum	Basic (pH 5.5 to 7.5) Anaerobic pO_2 <1 mm Hg	$10^{11} - 10^{12}$	

Fig. 1: The human gastrointestinal microbiota and their abundance. The microbial compositions are greatly influenced by physical conditions of the alimentary canal, such as change in the oxygen availability and the pH values of the gut content

Several studies have attempted to identify the core microbiota of the human gastrointestinal environment as it could describe the transition from a normal and healthy composition. Generally, most of the reported works indicate the gastrointestinal microbial composition to be similar up to 95% (at bacterial taxa level) among screened individuals (Amitay et al., 2018, Cheung et al., 2019). The microbial communities identified in samples collected from an individual over time are more similar to each other than those identified in two different individuals, although related persons share more bacterial strains than unrelated individuals. Studies also looked upon variation between screened individuals for further understanding of the core microbiome. However, identifying the degree of variation between individuals remains a hurdle, in the gastrointestinal microbiome. (Faith et al., 2013, Cheung et al., 2019). Arumugam et al. (2011) established the first core microbiome report with one of three distinct clusters or "enterotypes" based on metagenomic sequences. The study demonstrated that each microbial cluster was dominated by *Bacteroides, Prevotella*, and *Ruminococcus* genera. They also reported that each cluster was enriched for specific gene functions that reflected different microbial trophic chains. In a recent study, a genus-level core microbiome was reported by a Chinese cohort. The core gut microbiota had nine predominant genera where all belonged to *Firmicutes* phylum (Zhang et al., 2015).

Another study reported microbial changes in 37 adults for five years. Interestingly, the study revealed that approximately 60% to 70% of the microbial compositions were sustained over five years (Faith et al., 2013). The study reported that members of *Bacteroidetes* and *Actinobacteria* were more stable whereas members of *Firmicutes* and *Protebacteria* more susceptible to perturbations. In addition, Rajilic-Stojanovic et al. (2013) used a microarray-based approach to describe the molecular taxonomy of gut microbiota. Therefore, these studies suggest that microbial taxa are consistent and the abundance of the microbiota is subject to change (Faith et al., 2013, Rajilic-Stojanovic et al., 2013).

Another main driver of microbial colonization and succession in the human gut are dietary patterns, which include diet shapes, the relative abundance of microbial phyla on their dominancy and their relative abundance (Telle-Hansen et al., 2018). The population of specific microbiota is highly influenced by the availability of macronutrients in the gastrointestinal environment (Donaldson et al., 2016, Singh et al., 2017, Telle-Hansen et al., 2018). The previous data suggests that the sustainability of the gut microbial composition might be diet-driven.

We also know that core native microbiota remains relatively stable in adulthood but differs between individuals due to enterotypes, body mass index (BMI), exercise, lifestyle, and cultural and dietary habits (Singh et al., 2017). Accordingly, there is no unique optimal gut microbiota composition since it is different for each individual. However, a healthy host–microorganism balance is essential for optimally performing metabolic and immune functions and preventing disease development. Indeed, disturbances to the host–microbe relationship may disrupt the immune system and may manifest as a disease (Cheung et al., 2019, Rinninella et al., 2019).

2.2 The Development of Human Gastrointestinal Microbiome

The human body is a host for one trillion microorganisms and current theories suggest that humans are born in sterile condition (free of microorganisms) (Stewart et al., 2018). Previously it was believed that microbial colonization in the human gut takes place only after birth. However, recent findings have challenged this concept. Numerous works have reported findings that there exists a vertical transfer of microorganisms from the maternal body to the foetus before birth (Jiménez et al., 2008, Dominguez-Bello et al., 2010, Perez-Muñoz et al., 2017). In addition, studies on animal models as well as human samples have revealed that microbial colonization occurs during early life stages, within the first day of birth. As this period is a critical window for immunological and physiological development, various types of microorganisms are able to colonize in the gastrointestinal tract of a new born (Koleva et al., 2015, Perez-Muñoz et al., 2017). These findings are still a point of contention in the scientific community and replicability of the results remains essential.

Microbial transmission has been well described in many host microbial symbiosis', especially in invertebrate models. In animal models, the microbial transmissions are described either as "vertical transfer", which means from maternal source, or "horizontal transfer", which means, transmission occurs between members of the same species or from the environment (Bright and Bulgheresi 2010, Funkhouser et al., 2013, Perez-Muñoz et al., 2017).

Numerous studies have suggested that the early colonization of human neonate's intestines is through contact with maternal microorganisms (Barko et al., 2018). These studies support the theory that a natural pathway exists for delivering the maternal microbiota to neonates. Dominguez-Bello et al. (2010) reported that the gut of infants delivered vaginally resembles their own maternal vaginal microbiota, which is dominated by *Lactobacillus*, *Prevotella* and *Sneathia* spp. In contrast, the caesarean born infants harbour microbial communities similar to maternal skin surfaces which are dominated by *Staphylococcus*, *Corynebacterium* and *Propionibacterium* spp. (Dominguez-Bello et al., 2010, Azad et al., 2013). Similarly, the infant gut microbiome clusters were characterized between vaginally and caesarean delivered infants (Lundgren et al., 2018). The relative abundance of *Bifidobacterium*, *Streptococcus*, *Clostridium*, and *Bacteroides* were higher in vaginally delivered infants. Meanwhile, caesarean section delivered new borns differed slightly from vaginally delivered infants. They had a high abundance of *Bifidobacterium*, high *Clostridium* and low *Streptococcus* and *Ruminococcus* genera, and high abundance of the family *Enterobacteriaceae* (Lundgren et al., 2018).

Wampach et al. (2017) had revealed the process of microbial colonization and its subsequent succession from the faecal samples of vaginal and caesarean delivery. The study reported the depletion of *Bacteroidetes* phyla in caesarean delivered infants five days after birth. This had resulted in a significant increase in the genera of *Firmicutes*, as compared to vaginal delivery (Wampach et al., 2017). In addition, Liu et al. (2019) had tracked a decreased relative abundance of *Bifidobacterium* in the caesarean delivered infants. Thus, mode of delivery could cause microbial dysbiosis within the normal gut microbial establishment in the new borns (Lundgren et al., 2018, Liu et al., 2019).

Another major factor contributing to the development of an infant's gastrointestinal microbiome is the maternal breast milk (Toscano et al., 2017, Bardanzellu et al., 2018). For many infants, breast milk is the primary nutrition source during the first months of birth. Breast milk contains several components, including oligosaccharides and glycoproteins, which modulate the neonatal gut microbiota composition by favouring the growth of specific microbiota such as *Bifidobacterium* and *Lactobacillus*. There is a close relationship between the infant's gut microbiota and the mother's breast milk microbiota and human milk oligosaccharides (HMO) composition (Bode et al., 2012, Newburg and Morelli 2015).

Human breast milk also harbours a unique microbial composition which might serve as a continuous source of colonizing bacteria to the infant. It is estimated that more than a quarter of the microbes colonized in the infants' gut within the first month of birth originate from breast milk. This is also supported by the altered gut microbial composition that is observed in the non-breastfed infants (Funkhouser et al., 2013, Bardanzellu et al., 2018). Human breast milk microbiota directly seeds and shapes the microbial composition in the infant gut (Pannaraj et al., 2017). The infant gut has an abundance of antibiotic resistance genes compared to adults even when infants have not been exposed to antibiotics. This phenomenon is only possible by an abundance of *Gammaproteobacteria* in the infant gut as the *Gammaproteobacteria* often carry several resistance genes (Hu et al., 2013, Ho et al., 2018). The work of Hermansson and colleagues suggests that mode of delivery also modulates microbial composition in the human breast milk (Hermansson et al., 2019).

Bäckhed et al. (2015) has revealed that 70% of exclusively breastfed infants for a duration of 4 months, showed increased levels of *Lactobacillus* and *Bifidobacterium*, compared with formula-fed infants. In the same study, the genus of *Clostridium, Granulicatella, Citrobacter, Enterobacter, Bilophila wadsworthia*, and *Bifidobacterium adolescentis* were more dominant in the stool samples of formula-fed infants (Bäckhed et al., 2015). Growing evidence clearly proves that breastfeeding favours the development of infants' gastrointestinal microbiota and formula-feeding stimulates growth of disease-related microbial composition (Bode et al., 2012). Breast milk contains human milk oligosaccharides (HMOs) as the third most abundant component after lactose and lipids. Greater structural diversity restricts it to be digested in the upper gastrointestinal environment (Bode et al., 2012, Wang et al., 2015). Thus, HMOs serve as primary substrates for a specific cluster of microbiota found in the breast milk. This also favours colonization of these microbial clusters in the gastrointestinal environment (Marcobal et al., 2010, Bode et al., 2012, Wang et al., 2015).

Recently, maternal diet has been shown to influence the developing infant gut microbiome of caesarean borns (Lundgren et al., 2018). Higher fruit intake among pregnant women was associated with an increased abundance of *Streptococcus* and *Clostridium* group among vaginally born infants, whereas dairy intake increased the abundance of *Clostridium* cluster in them (Lundgren et al., 2018). The same study also demonstrated that consumption of fish and seafood by pregnant and breast feeding women were positively related to high abundance of *Streptococcus* in the infant gut. In the adult human, diet is also a major factor that shapes the development of gastrointestinal microbiome. Sustainability of colonic microbiota is dependent on the availability of carbohydrates that gastrointestinal microbes can utilize (Donaldson et al., 2016, Singh et al., 2017).

The human genome has a limited number of carbohydrates hydrolysing enzymes, especially polysaccharide hydrolases. Human glycoside hydrolases in upper gastrointestinal environment are unable to digest complex carbohydrates such as cellulose, resistant starch, inulin, lignin, pectin, raffinose, and oligosaccharides such as fructo-oligosaccharides (Sing et al., 2017, Telle-Hansen et al., 2018). These glycans are the primary substrates for microbial population in the large intestine and the gastrointestinal environment encodes more carbohydrates hydrolysing enzyme genes. The genera of *Bacteroides*, *Bifidobacterium* and *Ruminococcus* are primary utilizers of these glycans (Huttenhower et al., 2012, Eilam et al., 2014, Barko et al., 2018). Recent studies showed that a strict diet of 'animal-based' or 'plant-based' food results in altered gut microbial populations in humans (Zmora et al., 2018). Wu et al. (2016) compared the impact of vegan and vegetarian diets with an unrestricted control diet on the microbiome composition. The study concluded that both vegans and vegetarian diets had significantly lowered the counts of *Bifidobacterium* and *Bacteroides* species. Of the multiple host-endogenous and host-exogenous factors involved, diet emerges as a pivotal determinant of gut microbiota community structure and function (Zmora et al., 2018).

A recent study proposed three phases of microbial colonization in an average human life. A complete microbial colonization is said to occur within the first 4 years of life. It undergoes a developmental process from 3 to 14 months; a transition phase from 15 to 30 months, and reaches a stabilization period after 31 months (Stewart et al., 2018). However, maturation and stability of the gut microbial composition varies among individuals. In fact, gestational age is a major determinant of gut microbiota colonization. The microbiota composition of preterm infants (<37 weeks of gestation) is different from term counterparts. After birth, the microbial population gradually changes and increases its diversity. The guts of breastfed infants are enriched with higher levels of *Bifidobacterium* (Stewart et al., 2018). Then during the weaning period, the introduction of solid foods and the termination of milk-feeding/weaning coincide with major gut microbiota changes (Vandeputte et al., 2017). Besides the genera *Bifidobacterium* still relatively abundant, the genera of *Clostridium* and *Bacteroides* start establishing and being predominant after weaning. At the same time, there is a higher amount of bacteria within the *Firmicutes* phyla, typically found in an adult microbiota (Stewart et al., 2018).

Bergström et al. (2014) reported similar transitions of microbiota after the introduction of solid foods. Between 9 and 18 months, the count of the *Bacteroidetes* species increases with a decline in those of *Enterobacteriaceae*,

Bifidobacterium and *Lactobacillus* species. This microbial shift is presumed to be related to termination of breast and/or formula-feeding, depleting the primary fuel source for these bacteria. The rise of butyrate producing bacteria such as *Clostridium leptum* group, *E. halli*, and *Roseburia* species also contribute to the gut microbiota development. Interestingly, the study found that the longer infants were breast and/or formula-fed, the lower their levels of butyrate producing bacteria. Additionally, more and different species began to appear with the introduction of solid foods (Bergström et al., 2014). Other studies have demonstrated that the gut microbiota gradually changes within months after introduction of solid foods where the distribution of *Bifidobacterium* and *Bacteroides* greatly increased (Karlsson et al., 2011, Azad et al., 2013).

Therefore, the development and maturation of human gut microbiota constitutes a dynamic and non-random process, in which positive and negative interactions between key microbial taxa take place. Colonization of a complex microbial community in the infant gut is largely influenced by host and environmental factors. Recent studies have challenged the dogma of sterile in utero environments and 'germ-free' births. Scientific evidences have provided indications of microbial presence in the placenta, umbilical cord, amniotic fluid and breast milk. These factors are early pioneers allowing microbial colonization in the foetal stages. Then, after birth microbial compositions begin to establish and sustain in the human gut. This process is influenced by various perinatal conditions, such as mode of delivery, type of feeding, and antibiotic usage. In addition, other factors including diet, organ maturations, metabolic status, family genetics and lifestyle have been reported to impact the infant microbiota. Some of these are more difficult to determine and quantify in humans. Yet, the gut microbial composition could be maintained during adult stages to maintain the health state of the host.

2.3 Role of Gastrointestinal Microbiota

Gastrointestinal environment and its microbial composition have a symbiotic relationship. Generally, the naturally existing microorganisms in the large intestines play a vital role in further energy harvesting from undigested food particles (O'Callaghan and van Sinderen, 2016). The gastrointestinal microbiota has various carbohydrate hydrolysing active enzymes and produces various metabolites such as short-chain fatty acids (SCFAs) including propionate, butyrate and acetate (Donaldson et al., 2016, Barko et al., 2018). Most of the anaerobic microorganisms including lactic acid bacteria produce acetate, whereas propionate and butyrate are synthesized by other clusters of gut microbiota (Louis et al., 2016). These SCFAs are rapidly absorbed by epithelial cells in the GI tract where they are involved in the regulation of cellular processes such as gene expression, chemotaxis, differentiation, proliferation and apoptosis (O'Callaghan and van Sinderen 2016, Barko et al., 2018).

The gastrointestinal microbiota are also considered to be an important source of essential vitamins which are not produced by the human genome. Lactic acid bacteria being the key organisms in the de novo synthesis of numerous vitamins such as vitamin B12, vitamin K, riboflavin, biotin, nicotinic acid, panthotenic acid, pyridoxine and thiamine are not produced in other animals, plants and fungi (Morowitz et al., 2011, Rowland et al., 2018). Besides, lactic acid bacteria and *Bifidobacterium* are also the main folate producers in the human gut. Folate is a crucial vitamin needed in various metabolic processes and regulatory mechanisms (Rowland et al., 2018). Recently, Magnúsdóttir et al. (2015) had predicted genes involved in the B-vitamin biosynthesis among 256 known human gut microorganisms based on their available genome annotations. The study found that human gut microbiota has co-evolved relationships that are specific to the gut environment. This analysis suggests that human gut bacteria actively exchange B-vitamins among each other, thereby enabling the survival of organisms that do not synthesize any of these essential cofactors (Magnúsdóttir et al., 2015).

Colonization of different types of commensal microbes from breast milk and other natural resources into infants' guts shows that the gastrointestinal microbiome is actively involved in the development of the immune system (Nagai et al., 2016, Barko et al., 2018), also supported by animal studies (Gaboriau-Routhiau, Rakotobe et al., 2009, Hooper et al., 2012, Belkaid and Hand 2014). Gut colonization by a specific anaerobic and clostridia type sporulating commensal microorganism has been proven to be actively interacting with the immune system. These types of microorganisms associate with the mammalian epithelial lining in the gastrointestinal tract. This stimulates the epithelial cells to release specific immunomodulatory compounds (Nagai et al., 2016). Effect on NF-kB has also been shown in human cell lines (Quévrain et al., 2016).

3. Microbial Dysbiosis

The number of bacterial species present in the human gut has widely varied among studies. However, it is generally accepted that individuals harbour more than 1000 microbial at species-level phylotypes. In normal conditions, a healthy human gastrointestinal environment is composed of stable microbial population over a period of time (Zhong et al., 2019). However, the microbiota is sensitive to changes in lifestyle. The overall diversity of the human gastrointestinal microbiota increases steadily from birth until around 12 years of age, remaining relatively stable throughout adulthood, and then declining in later years. In adults, 60–70% of the gut microbiome is stable, with the degree of stability varying between phyla (Lynch and Pedersen 2016). Microbial numbers rapidly undergo a dynamic change due to various factors, including diet, toxins, inflammation by pathogens and exposure to drugs including antibiotics (Figure 2).

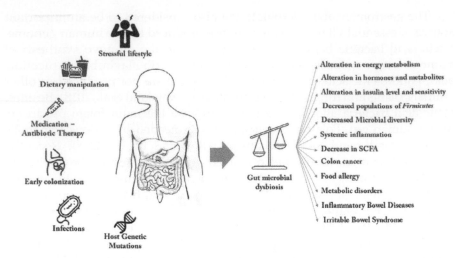

Stressful lifestyle

Dietary manipulation

Medication – Antibiotic Therapy

Early colonization

Infections

Host Genetic Mutations

Gut microbial dysbiosis

Alteration in energy metabolism
Alteration in hormones and metabolites
Alteration in insulin level and sensitivity
Decreased populations of *Firmicutes*
Decreased Microbial diversity
Systemic inflammation
Decrease in SCFA
Colon cancer
Food allergy
Metabolic disorders
Inflammatory Bowel Diseases
Irritable Bowel Syndrome

Fig. 2: Factors causing microbial dysbiosis in the human gastrointestinal environment. These factors affect the balance of gut microbiota and this leads to alteration in metabolism and various health complications

Alteration in the microbiota is described as microbial dysbiosis (Hourigan et al., 2018). Invasion of enteric pathogens is one of the most important factors leading to microbial dysbiosis because a diverse gastrointestinal microbiota enhanced resistance towards pathogenic infection and loss of diversity may promote adverse health complications. For instance, Norovirus-infected patients have shown a disrupted microbiota following infection, and could be at elevated risk for long-term health complications (Nelson et al., 2012). Several foodborne viruses have also been shown to induce inflammatory processes in the gut which can alter the microbial composition and their function (Carding et al., 2015).

Long-term diet is another major factor that influences the structure and activity of gastrointestinal microorganisms. After weaning, dietary patterns have a profound impact on shaping the childhood gut microbiota. The gut microbiota influences the development and regulation of the immune system, and energy and metabolic homeostasis of the host via the production of a vast array of metabolites such as short-chain fatty acids (SCFAs) and secondary bile acids. For instance, SCFAs produced by gut bacterial fermentation of complex dietary carbohydrates interact with G protein-coupled receptors (GPCRs) and affect adiposity and insulin resistance (David et al., 2014, Carding et al., 2015).

Antibiotic therapy is considered a factor causing dysbiosis in the gut microbiome (Langdon et al., 2016, Rosario et al., 2018). Prescription drug usages have been increasing continuously. The Center for Disease Control and Prevention has estimated that about 46.9% of the population in the United States had taken at least one prescription medication in 30 days in 2011 to 2014 compared to 39.1% during 1988 to 1994. Similarly, the European

Health Interview Survey reported that 40% of the European population also used at least one prescription drug within a two week period (Le Bastard et al., 2018). Antibiotics are well-known for causing a major impact on the microbial composition in the gut; especially chemotherapy induces specific gut microbiome dysbiosis (Zhernakova et al., 2016).

Certain therapeutic prescription drugs such as metformin which are antidiabetic have proven to cause a disturbance in the intestinal microbial composition. Prolonged use of this drug has led to increased abundance of *Escherichia* spp., *Akkermansia muciniphila*, *Subdoligranulum variabile* and decreased presence of *Intestinibacter bartlettii*. This alteration may potentially lead to adverse effects on the host metabolism, with the depletion of butyrate producer genera (Rosario et al., 2018). Similarly, Falony et al. (2016) showed that medications are associated with variation in the gut microbial composition and explain the largest total variance of beta diversity. Work done by Zhernakova et al. (2016) had demonstrated that more than 19 types of drugs including antibiotics and pain killers in daily consumption lead to perturbation in the microbial compositions. Recently, a systematic review had been performed on data from observational or intervention studies that reported medication induced gut microbiome alterations. The study reported that non antibiotic prescription drugs also have a notable impact on the intestinal microbiome (Le Bastard et al., 2018).

Recently, various factors had been reported to be associated with human gut dysbiosis. One of the studies by Grant et al. (2019) had shown that shared habitat between animals and humans could be another important factor causing a microbial shift in the gastrointestinal environment. The study compared the microbial population in the faecal samples of macaques and workers in the park. Human samples were also compared with those of healthy humans not having contact with animal hosts . Exposed workers showed significantly lower alpha diversity compared to control human groups. The study performed SourceTracker analysis which revealed that a higher percentage of microbes in the workers were potentially sourced by macaque microbiota, compared to the controls. Although the study reported no significant degree of microbial sharing between humans and macaques, hygienic issues came up due to potential exposure to occupational hazards. The study also reported that macaques in the park also had a high level of gut microbiota dispersion relative to the macaques with minimal human contact (Grant et al., 2019). Table 1 highlights recent works on other factors altering the microbial compositions in the human gastrointestinal environment.

4. Microbiota in Gastrointestinal Diseases

Recently, the gastrointestinal dysbiosis has been confirmed as associated with a wide range of diseases in humans, including metabolic, atopic, and gastrointestinal illnesses (Hourigan et al., 2018). Increasing evidence has been proving that a permanent alteration in the gut microbiome composition could alter visceral sensitivity, intestinal motility, and permeability, as well

Table 1: Recent publications reporting factors associating with gut microbiota dysbiosis

Reference	Factor(s) causing gut dysbiosis	Method(s) employed	Findings
Scher et al., 2014	Patients with Psoriatic Arthritis (PsA) & patients with skin psoriasis	16S ribosomal RNA pyrosequencing of faecal samples	• Gut microbiota less diverse than healthy persons • Both patient groups showed relative decrease in abundance of Coprococcus sp. • PsA patients showed significant reduction in Akkermansia, Ruminococcus, and Pseudobutyrivibrio spp.
Lee et al., 2019	Kidney transplant recipients associated with posttransplant diarrhoea	16S ribosomal RNA (rRNA) gene V4-V5 deep sequencing	• Post-transplant diarrhoea is not associated with common infectious diarrheal pathogens but with a gut dysbiosis. • Relative abundance of 13 commensal genera were reduced in diarrheal faecal samples. • The samples showed lower abundance of metabolic genes.
Ni et al., 2019	Primary hepatocellular carcinoma (HCC) – a type of chronic liver diseases	MiSeq sequencing of 16S rRNA gene	• Early stages of HCC did not show differences with healthy microbiota. • Stage III and IV primary HCC showed significant reduction in alpha-diversities of the microbiota. • Pro-inflammatory type Proteobacteria increased compared to healthy personnel.
Huang et al., 2019	Psoriasis (autoimmune disease)	MiSeq sequencing of 16S rRNA gene	• Relative abundances of Firmicutes and Bacteroidetes were inverted at the phylum level. • Around 16 kinds of genus level phylotype were significantly difference. • No microbial diversity and composition alteration were observed among the four types of psoriasis.
Jones et al., 2019	Obese teenagers (12–19 years) with high dietary fructose intake	16S rRNA gene sequencing of faecal samples	• High dietary fructose intake associated with significant reduction of genera Eubacterium and Streptococcus, compared to other dietary macronutrients

Table 1: *(Contd.)*

Reference	Factor(s) causing gut dysbiosis	Method(s) employed	Findings
Zhuang et al., 2019	Lung cancer	16S rDNA Illumina sequencing	• No significant decrease in alpha diversity but beta diversity greatly affected in patients compared to controls. • Patients showed elevated levels of *Enterococcus* spp.
Chen et al., 2018	Major depressive disorder (MDD)	Comparative metaproteomics analysis	• Relative abundances of 16 bacterial families from four phyla: Bacteroidetes, Proteobacteria, Firmicutes, Actinobacteria were significantly different between the MDD and healthy controls
Grant et al., 2019	Shared environment between human and long-tailed macaques (Macaca fascicularis)	V4 gene Illumina MiSeq sequencing from fecal sample both human and animal	• High variance in gut microbiota composition of macaques in contact with humans. • Human group exposed and unexposed to animal hosts, showed homogenous variance in beta diversity. • Contact between animal and human in a shared habitat showed significant alteration in gut microbiota and this could be a potential occupational hazard.
Dash et al., 2019	Helicobacter pylori infection	16S rRNA and ITS2-based microbial profiling analysis	• H. pylori infected individuals were shown to be increasingly belonging to Succinivibrio, Coriobacteriaceae, Enterococcaceae, and Rikenellaceae. • Infected patients also had increased abundance of Candida glabrata and other unclassified Fungi
Knoop et al., 2016	Oral antibiotics	16S rRNA microbial profiling analysis	• Bacterial translocation – native commensal bacteria across the colonic epithelium, promoting inflammatory responses, and predisposed to increased disease in response to coincident injury
Fodor et al., 2018	Effect of Rifaximin, a non-systemic antibiotic	V4 hypervariable region 16S ribosomal RNA gene sequencing	• After 2 weeks of rifaximin treatment, about 7 microbial taxa including Peptostreptococcaceae, Verrucomicrobiaceae, Enterobacteriaceae had significantly lower relative abundance.

as the immune response, thus promoting a proinflammatory state. Such alterations, especially in the host's immune and metabolic functions, can originate or favour the onset of several diseases such as diabetes, obesity, as well as neurological and autoimmune diseases. Recent studies have also demonstrated the participation of the microbiota in the etiopathogenesis of many gastroenterological diseases, such as irritable bowel syndrome, inflammatory bowel disease, celiac disease, non-alcoholic steatohepatitis and digestive neoplasms (Burman et al., 2016, Chiba et al., 2019)

4.1 Irritable Bowel Syndrome (IBS)

IBS is a chronic functional disorder of the gastrointestinal system that causes debilitating suffering in vast majority of people (Kruse, 1933, Chaudhary and Truelove, 1962). Alteration in brain-gut interaction has been viewed as the main cause of IBS in the absence of an organic or well-defined cause for IBS. The lack of known pathophysiological mechanisms makes it difficult to develop proper IBS diagnostics. Although, IBS affects people of all ages, children particularly suffer from constant abdominal pain and constipation or diarrhoea. Prevalence of different types of IBS ranges from 10% to 25%. About 6.7% new cases of IBS are reported per year, which leads to 25% to 45% primary care visits of adolescents (Ford and Talley, 2012, Lovell and Ford, 2012).

At the cellular and molecular levels, data have shown that IBS is a multifactorial disorder where genetic and epigenetic factors play a role in IBS symptoms (Kapeller, Houghton et al., 2008, Zhou et al., 2010, Vaiopoulou et al., 2014). In addition, data has shown that elevated levels of amino acids and phenolic compounds were found in IBS faecal samples compared to healthy controls. Preliminary data clearly showed the alterations of certain bacterial species in the IBS faecal and mucosal samples compared to healthy controls (Durban et al., 2012, Tap et al., 2017, Maharshak et al., 2018). Differential relative abundances of *Faecali bacterium* and *Dorea* genera were reported in IBS-D (diarrhoea) versus healthy controls (Maharshak et al., 2018). In a prospective study, bacterial abundance differences in paediatric subjects with Functional Abdominal Pain Disorders (FAPDs) were observed, compared to healthy controls. Data showed that 38 bacterial genera showed seasonal variations among the different seasons in FAPDs patients and corresponding controls (Figure 3).

Recently, our group showed a correlation between clinical indexes and microbiome alterations in Functional Abdominal Pain Disorders (FAPDs) patients (control group N=86, and FAPDs N=45). IBS is one of the four categories of FAPDs. An increase in major indexes such as pain, nausea, stress, and vomiting were found associated with an alteration in the gut microbiome composition. The relative abundances of *Faecalibacterium* species and other bacterial species were affected with the degree of pain index intensity in IBS pediatric patients (Figure 4). *Faecalibacterium Prausnitzii* is among the major gut bacteria (approximately 10% of total gut bacteria), and

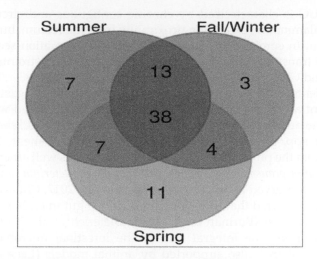

Fig. 3: Venn diagram of the most affected bacterial species in the different seasons in the control and FAPDs groups

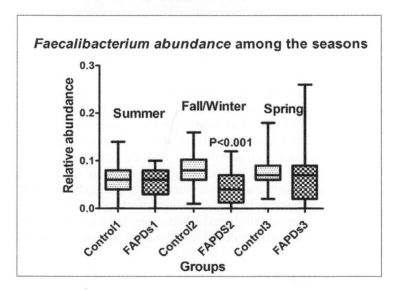

Fig. 4: The relative abundance of *Faecalibacterium* species in the two groups during the different seasons. Summer=1, Fall and Winter=2, and Spring=3

is involved in butyrate production which is considered the main food and energy source of colon cells (Duncan et al., 2017).

4.2 Inflammatory Bowel Disease (IBD)

Inflammatory bowel disease (IBD) refers to chronic inflammation of the digestive tract due to microbial dysbiosis. There are two types of IBD, Ulcerative colitis (UC) and Crohn's disease (CD) respectively (Lane et al., 2017). UC is characterized by chronic contiguous and circumferential mucosal inflammation extending proximally from the rectum, but is isolated to the colon. In comparison, the stereotypical inflammation seen in CD is patchy and transmural, and may affect any part of the gastrointestinal tract (Miyoshi and Chang, 2017).

In patients with IBD, microbial compositions of the gastrointestinal environment are known to fluctuate up and down compared with healthy controls. The number of *Fusobacterium*, adherent-invasive *Escherichia coli* and *Enterobacter* spp. are commonly observed with an increment. Meanwhile, members from the phyla *Firmicutes* and *Bacteroidetes*, as well others including *Faecalibacterium prausnitzii, Roseburia hominis, Bifidobacterium*, and *Prevotella* were usually observed to decrease (Kaplan and Ng, 2017, Chiba et al., 2019). Studies have reported that these alterations in the gut microbiota also could cause inflammation (Burman et al., 2016, Dieterich et al., 2018). Growing evidence supports the integral role of the intestinal microbiome in the pathogenesis of IBD, also supported by animal models (Lane et al., 2017, Miyoshi and Chang, 2017).

Major factors leading to IBD is similar in both UC and CD diseases. However, several factors have been found to give divergent results. Smoking and appendectomy have a divergent role in IBD; each is a protective factor for UC but a risk factor for CD. Some environmental factors are also divergent in the Western and Asian context. Usage of antibiotics in childhood would be a risk factor for IBD in Western countries, however it is found to be a protective factor in recent Asian studies. Another potential risk factor could be hygiene where an improvement in sanitation with limited exposure to microbes results in impaired immune response, causing immune-mediated diseases including IBD. Hygiene factors studied include sanitary facilities (toilet paper, water supply), family size and birth order, pets, and farm animals (Abegunde et al., 2016). Although the etiology of IBD is far from being understood, recent studies support the hypothesis that IBD results from a complex interplay between various factors (Figure 5) including genetics, immune dysregulation and environmental triggers that may exert their effect through alterations of the intestinal microbiota (Lane et al., 2017, Dieterich et al., 2018).

4.3 Crohn's Disease

Crohn's disease was first described in 1932 by Dr. Burrill B. Crohn and his colleagues Dr. Leon Ginzburg and Dr. Gordon D. Oppenheimer. Crohn's disease refers to a chronic inflammatory condition of the gastrointestinal tract. This disease is not similar to ulcerative colitis and other types of inflammatory bowel diseases. Symptoms of both diseases are quite similar, yet the areas affected in the gastrointestinal tract are different for Crohn's disease and ulcerative colitis (Hendrickson et al., 2002, Segal, 2016). Crohn's disease involves the mucosa and also deeper layers of the intestinal wall, frequently resulting in transmural inflammation. The lower end of the

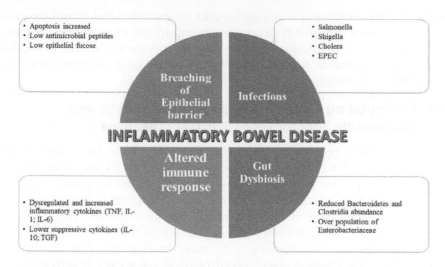

Fig. 5: Correlation between factors causing inflammatory bowel diseases

ileum and the beginning part of the large intestine are commonly affected in Crohn's disease. This disease also could affect the entire thickness of the bowel wall (Lopetuso et al., 2018). The mortality rate of Crohn's disease is high, estimated to be 1.5 times higher than the general population. A patient with clinically active Crohn's disease has a 70 to 80% chance of having active disease such as colorectal cancer, shock, volume depletion, protein-calorie malnutrition, or anaemia, in the year after diagnosis (Lo et al., 2018).

Numerous human genetic loci encoding proteins involved in host immune responses have been linked to Crohn's disease (Jostins et al., 2012). However, there is limited study on the impact of these immune response proteins on the microbial dysbiosis associated with Crohn's disease (Knights et al., 2014). Microbial dysbiosis is hypothesized as a bacterial response towards environmental stressors such as the host inflammatory response (Huttenhower et al., 2014). Gastrointestinal dysbiosis is commonly characterized by increased population of *Proteobacteria* and a decrease in *Firmicutes*, along with a decrease in community richness (Nagalingam and Lynch, 2012). Patients with Crohn's disease are exposed to antibiotics and dietary changes which are likely to affect the microbiota, but the influence of these factors and their interactions with each other are incompletely understood.

The independent effects of antibiotics, diet and inflammation are likely to reflect different mechanisms. Antibiotics are direct toxins to bacteria and may facilitate outgrowth of fungi. Changes in diet provide novel substrates supporting bacterial growth. Inflammation may cause bacterial taxa to live in an oxidative stress setting. The normal oxygen gradient in the colon influences the composition of the gut microbiota with a higher abundance of *Proteobacteria* in the microaerobic environment of the mucosal surface (Albenberg et al., 2014). Disruption of the epithelium and bleeding due to active Crohn's disease is expected to lead to greater intraluminal oxygen, which will favor outgrowth of taxa belonging to the *Proteobacteria* phylum as a consequence. In addition, some *Enterobacteriacaea* are known to exploit compounds produced during inflammation as terminal electron acceptors, promoting their outgrowth (Winter and Baumler, 2014).

4.4 Manipulating Gut Microbiome as a Diagnostic and Therapeutic Tool for Digestive Tract Diseases

For many years, prebiotic and probiotic therapy have been one of the promising tools for restoring gut homeostasis. Either in lone usage or in combination, both have proven to manipulate the defective gastrointestinal microbiome into a healthy state. Prebiotics are compounds that change the structure or metabolome of the intestinal microbiota. Inulin and oligofructose are two prebiotics that have been shown to promote the growth of beneficial *Bifidobacterium* and *Lactobacillus* spp. both in humans and in animal models. In a randomized placebo-controlled study with CD active adults had shown remissions in the disease activity index after administration of fructo-oligosaccharides (FOS). Despite some changes to immuno regulation of dendritic cells observed in those receiving FOS, no significant clinical improvements were seen, nor were there significant changes to the faecal microbiomes between the groups at baseline or after the 4-week intervention (Benjamin et al., 2011). Fermentable fibres, another form of prebiotic, are metabolized by colonic bacteria to short chain fatty acids (SCFA). SCFA is referred to as the microbial synthesis of lactate, acetate, propionate and butyrate, which are known to modulate cell proliferation, histone acetylation, gene expression and immune response (Gerasimidis et al., 2014, Dieterich et al., 2018).

Recent development of omic studies including metagenomics, proteomics and metabolomics has revealed the functional role of gastrointestinal microbiome in gut related diseases. Functional studies have highlighted microbial metabolism of nutrients, energy harvesting, pathogenesis and microbial adherence in the host gut. Omic studies defined the functional role of gastrointestinal microbiomes in healthy as well as diseased conditions . This could drive a clear outcome to restore health through various gut microbiome related therapeutics including altering diet, prebiotics, antibiotics, or by faecal transplantation. Table 2 summarizes recent therapeutic methods to alter gut microbiome to restore homeostasis.

5. Future Prospects and Conclusion

Recent studies have made great strides in understanding the presence, total composition and the functional role of commensal microorganisms that naturally exist in the human gastrointestinal environment. Their impact on human health and diseases are being well recognized. The recent application of DNA sequencing technology such as Next Generation of Sequencing

Table 2: Recent diagnostic methods of altering diets and FMT to manipulate gut microbiota as the treatment for gut related diseases

Diagnostic method	Description	Recent works
Exclusive Enteral Nutrition (EEN)	EEN is a complete exclusion diet, where patients receive 100% of their daily calorie intake from formula rather than table foods EEN has been shown to induce alterations in the microbiome as early as 1 week after initiation	Gerasimidis et al., 2014 demonstrated that 15 children with CD treated with EEN - reduced faecal diversity and increased commensal microbiota.
Specific Carbohydrate Diet (SCD)	SCD has special formulations without all grains including gluten, sweeteners except for honey, processed foods, and all milk products, except for hard cheeses and yogurt.	Suskind et al., 2016 demonstrated that SCD given to 7 children with CD – showed improvement on inflammatory markers, including faecal calprotectin. Cohen et al., 2014 reported mucosal and clinical improvement in 7 children with CD
Crohn's Disease Exclusion Diet (CDED)	CDED proposed as an improved version of EEN to increase EEN's efficiency. CDED has modified EEN where any sort of processed foods are removed.	Sigal-Boneh et al., 2014 reported that CDED reduced inflammatory markers and clinical remission.
Faecal Microbiota Transplantation (FMT)	Unfractionated faecal microbiota from a healthy donor is instilled into the intestinal tract of a sick individual with the intent of curing the disease	Nood et al., 2013 had reported FMT more efficient than vanomycin treatment to reduce *Colostridium difficile* infection. Moayeddi et al., 2015 reported 9 patients receiving FMT and 2 placebo patients were in remission within 7 weeks of treatment.

(NGS) have resulted in great advances in understanding the interaction between the gut microbiota and the human host. The gut microbiota plays an important role in the maintenance of intestinal homeostasis and the development and activation of the host immune system. With this valuable knowledge, rational strategies could be developed to manipulate the commensal microbiota in the gut for therapeutic purposes. Huge potential exists to optimize microbial-based therapies of IBDs as efforts to identify specific resident microbes and the functional pathogenic and protective roles they play in IBDs which are being studied continuously. This could then be designed into novel and new effective methods to modulate concentrations and functions of these microbes to suppress inflammation in individuals based on their unique microbial profiles. For instance, FMT has attracted attention as a new therapeutic strategy for IBD. However, although it has significant effects in the treatment of CD, its total efficacy in IBD is being perceived as controversial. There are many factors that should be taken into consideration in order to increase the success rate of FMT in IBD, such as disease state, donor selection, and the standardization of the faecal microbiota processing protocol. Moreover, gastrointestinal diseases might require precisely defined FMT consortiums to increase efficiency. Similarly, existing prebiotic and probiotic therapies were generally defined tools whose success rate could be less than their actual potential to be successful. Hence, more studies are needed to focus on the defined functional analysis of gut microbiota to overcome gastrointestinal diseases.

Acknowledgement

The FAPDs project supported by a seed grant (UL1TR001427) from the Florida State University (Medical school). The authors also would like to thank the MOHE and Universiti Teknologi Malaysia (UTM) for HICOE grant no. RJ13000.7846.4J262 that financially supported this research in Malaysia. Student author, Shanmugaprakasham Selvamani would like to extend this thanks to School of Postgraduate Studies (SPS), UTM for the financial support of Zamalah Ph.D. Scholarship session 2017/2018. Shanmugaprakasham also would like express his gratitude to Professor Dr. Hesham Ali El-Enshasy and Dr. Bassam Abomoelak for the great opportunity as well Dr. Thevarajoo, S. for being great inspirations.

References

Abegunde, A.T., B.H. Muhammad, O. Bhatti and T. Ali. 2016. Environmental risk factors for inflammatory bowel diseases: Evidence based literature review. World J. Gastroenterol. 22(27): 6296.

Albenberg, L., T.V. Esipova, C.P. Judge, K. Bittinger, J. Chen, A. Laughlin et al. 2014. Correlation between intraluminal oxygen gradient and radial partitioning of intestinal microbiota. Gastroenterology 147(5): 1055-1063, e1058.

Amitay, E.L., A. Krilaviciute and H. Brenner. 2018. Systematic review: Gut microbiota in fecal samples and detection of colorectal neoplasms. Gut Microbes 9(4): 293-307.

Arumugam, M., J. Raes, E. Pelletier, D. Le Paslier, T. Yamada, D.R. Mende et al. 2011. Enterotypes of the human gut microbiome. Nature. 473(7346): 174.

Azad, M.B., T. Konya, H. Maughan, D.S. Guttman, C.J. Field, R.S. Chari et al. 2013. Gut microbiota of healthy Canadian infants: Profiles by mode of delivery and infant diet at 4 months. Canadian Med. Assoc. J. 185(5): 385-394.

Bäckhed, F., J. Roswall, Y. Peng, Q. Feng, H. Jia, P. Kovatcheva-Datchary et al. 2015. Dynamics and stabilization of the human gut microbiome during the first year of life. Cell Host Microbe. 17(5): 690-703.

Bardanzellu, F., V. Fanos, F.A. Strigini, P.G. Artini and D.G. Peroni. 2018. Human breast milk: Exploring the linking ring among emerging components. Front. Pediatr. 6: 215.

Barko, P., M. McMichael, K.S. Swanson and D.A. Williams. 2018. The gastrointestinal microbiome: A review. J. Vet. Intern. Med. 32(1): 9-25.

Basson, A., A. Trotter, A. Rodriguez-Palacios and F. Cominelli. 2016. Mucosal interactions between genetics, diet, and microbiome in inflammatory bowel disease. Front. Immunol. 7: 290.

Becker, C., M.F. Neurath and S. Wirtz. 2015. The intestinal microbiota in inflammatory bowel disease. Inst. Lab. Anim. Res. J. (ILAR). 56(2): 192-204.

Belkaid, Y. and T.W. Hand. 2014. Role of the microbiota in immunity and inflammation. Cell. 157(1): 121-141.

Benjamin, J.L., C.R. Hedin, A. Koutsoumpas, S.C. Ng, N.E. McCarthy, A.L. Hart et al. 2011. Randomised, double-blind, placebo-controlled trial of fructo-oligosaccharides in active Crohn's disease. Gut 60(7): 923-929.

Bergström, A., T.H. Skov, M.I. Bahl, H.M. Roager, L.B. Christensen, K.T. Ejlerskov et al. 2014. Establishment of intestinal microbiota during early life: A longitudinal, explorative study of a large cohort of Danish infants. Appl. Environ. Microbiol. 80(9): 2889-2900.

Bode, L. 2012. Human milk oligosaccharides: Every baby needs a sugar mama. Glycobiology 22(9): 1147-1162.

Bright, M. and S. Bulgheresi. 2010. A complex journey: Transmission of microbial symbionts. Nat. Rev. Microbiol. 8(3): 218.

Bull, M.J. and N.T. Plummer. 2014. Part 1: The human gut microbiome in health and disease. I.M.C.J. 13(6): 17-22.

Burman, S., E. Hoedt, S. Pottenger, N.-S. Mohd-Najman, P.Ó. Cuív and M. Morrison. 2016. An (anti)-inflammatory microbiota: Defining the role in inflammatory bowel disease. Digest. Dis. 34(1-2): 64-71.

Caporaso, J.G., J. Kuczynski, J. Stombaugh, K. Bittinger, F.D. Bushman, E.K. Costello et al. 2010. QIIME allows analysis of high-throughput community sequencing data. Nat. Methods 7(5): 335-336.

Carding, S., K. Verbeke, D.T. Vipond, B.M. Corfe and L.J. Owen. 2015. Dysbiosis of the gut microbiota in disease. Microb. Ecol. Health Dis. 26(1): 26191.

Chaudhary, N.A. and S.C. Truelove. 1962. The irritable colon syndrome: A study of the clinical features, predisposing causes, and prognosis in 130 cases. QJM-Int. J. Med. 31(3): 307-322.

Chen, Z., J. Li, S. Gui, C. Zhou, J. Chen, C. Yang et al. 2018. Comparative metaproteomics analysis shows altered fecal microbiota signatures in patients with major depressive disorder. Neureport 29(5): 417-425.

Cheung, S., A.R. Goldenthal, A.-C. Uhlemann, J.J. Mann, J.M. Miller, M.E. Sublette et al. 2019. Systematic review of gut microbiota and major depression. Front. Phychiatry. 10: 34.

Chiba, M., K. Nakane and M. Komatsu. 2019. Westernized diet is the most ubiquitous environmental factor in inflammatory bowel disease. Perm. J. 23: 18-107.

Cho, I. and M.J. Blaser. 2012. The human microbiome: At the interface of health and disease. Nat. Rev. Genet. 13(4): 260.

Cushing, K., D.M. Alvarado and M.A. Ciorba. 2015. Butyrate and mucosal inflammation: New scientific evidence supports clinical observation. Clin. Transl. Gastroenterol. 6(8): e108.

Dash, N.R., G. Khoder, A.M. Nada and M.T. Al Bataineh. 2019. Exploring the impact of Helicobacter pylori on gut microbiome composition. PloS One. 14(6): e0218274.

David, L.A., C.F. Maurice, R.N. Carmody, D.B. Gootenberg, J.E. Button, B.E. Wolfe et al. 2014. Diet rapidly and reproducibly alters the human gut microbiome. Nature 505(7484): 559-563.

Dieterich, W., M. Schink and Y. Zopf. 2018. Microbiota in the gastrointestinal tract. Med. Sci. 6(4): 116.

Donaldson, G.P., S.M. Lee and S.K. Mazmanian. 2016. Gut biogeography of the bacterial microbiota. Nat. Rev. Microbiol. 14(1): 20-32.

Durbán, A., J.J. Abellán, N. Jiménez-Hernández, P. Salgado, M. Ponce, J. Ponce et al. 2012. Structural alterations of faecal and mucosa-associated bacterial communities in irritable bowel syndrome. Environ. Microbiol. Rep. 4(2): 242-247.

Eilam, O., R. Zarecki, M. Oberhardt, L.K. Ursell, M. Kupiec, R. Knight et al. 2014. Glycan degradation (GlyDeR) analysis predicts mammalian gut microbiota abundance and host diet-specific adaptations. mBio. 5(4): e01526-01514.

Engen, P.A., S.J. Green, R.M. Voigt, C.B. Forsyth and A. Keshavarzian. 2015. The gastrointestinal microbiome: Alcohol effects on the composition of intestinal microbiota. Alcohol Res. - Curr. Rev. 37(2): 223-236.

Faith, J.J., J.L. Guruge, M. Charbonneau, S. Subramanian, H. Seedorf, A.L. Goodman et al. 2013. The long-term stability of the human gut microbiota. Science 341(6141): 1237439.

Falony, G., M. Joossens, S. Vieira-Silva, J. Wang, Y. Darzi, K. Faust et al. 2016. Population-level analysis of gut microbiome variation. Science 352(6285): 560-564.

Fodor, A.A., M. Pimentel, W.D. Chey, A. Lembo, P.L. Golden, R.J. Israel et al. 2019. Rifaximin is associated with modest, transient decreases in multiple taxa in the gut microbiota of patients with diarrhoea-predominant irritable bowel syndrome. Gut Microbes 10(1): 22-33.

Ford, A.C. and N.J. Talley. 2012. Irritable bowel syndrome. BMJ: Br. Med. J. 345: e5836.

Funkhouser, L.J. and S.R. Bordenstein. 2013. Mom knows best: The universality of maternal microbial transmission. PloS Biol. 11(8): e1001631.

Gaboriau-Routhiau, V., S. Rakotobe, E. Lécuyer, I. Mulder, A. Lan, C. Bridonneau et al. 2009. The key role of segmented filamentous bacteria in the coordinated maturation of gut helper T cell responses. Immunity 31(4): 677-689.

Gerasimidis, K., M. Bertz, L. Hanske, J. Junick, O. Biskou, M. Aguilera et al. 2014. Decline in presumptively protective gut bacterial species and metabolites are paradoxically associated with disease improvement in pediatric Crohn's disease during enteral nutrition. Inflamm. Bowel Dis. 20(5): 861-871.

Geuking, M.B., Y. Köller, S. Rupp and K.D. McCoy. 2014. The interplay between the gut microbiota and the immune system. Gut Microbes 5(3): 411-418.

González-Ferrero, C., J. Irache and C. González-Navarro. 2018. Soybean protein-based microparticles for oral delivery of probiotics with improved stability during storage and gut resistance. Food Chem. 239: 879-888.

Grant, E.T., R.C. Kyes, P. Kyes, P. Trinh, V. Ramirez, T. Tanee et al. 2019. Faecal microbiota dysbiosis in macaques and humans within a shared environment. PloS One. 14(5): e0210679.

Hendrickson, B.A., R. Gokhale and J.H. Cho. 2002. Clinical aspects and pathophysiology of inflammatory bowel disease. Clin. Microbiol. Rev. 15(1): 79-94.

Hermansson, H., H. Kumar, M.C. Collado, S. Salminen, E. Isolauri, S. Rautava et al. 2019. Breast milk microbiota is shaped by mode of delivery and intrapartum antibiotic exposure. Front. Nutr. 6: 4.

Hiippala, K., H. Jouhten, A. Ronkainen, A. Hartikainen, V. Kainulainen, J. Jalanka et al. 2018. The potential of gut commensals in reinforcing intestinal barrier function and alleviating inflammation. Nutrients 10(8): 988.

Ho, N.T., F. Li, K.A. Lee-Sarwar, H.M. Tun, B.P. Brown, P.S. Pannaraj et al. 2018. Meta-analysis of effects of exclusive breastfeeding on infant gut microbiota across populations. Nat. Commun. 9(1): 4169.

Hooper, L.V., D.R. Littman and A.J. Macpherson. 2012. Interactions between the microbiota and the immune system. Science 336(6086): 1268-1273.

Hourigan, S.K., P. Subramanian, N.A. Hasan, A. Ta, E. Klein, N. Chettout et al. 2018. Comparison of infant gut and skin microbiota, resistome and virulome between neonatal intensive care unit(NICU) environments. Front. Microbiol. 9: 1361.

Hu, Y., X. Yang, J. Qin, N. Lu, G. Cheng, N. Wu et al. 2013. Metagenome-wide analysis of antibiotic resistance genes in a large cohort of human gut microbiota. Nat. Commun. 4: 2151.

Huang, L., R. Gao, N. Yu, Y. Zhu, Y. Ding, H. Qin et al. 2019. Dysbiosis of gut microbiota was closely associated with psoriasis. Sci. China Life Sci. 62(6): 807-815.

Huttenhower, C., D. Gevers, R. Knight, S. Abubucker, J.H. Badger, A.T. Chinwalla et al. 2012. Structure, function and diversity of the healthy human microbiome. Nature 486(7402): 207-214.

Huttenhower, C., A.D. Kostic and R.J. Xavier. 2014. Inflammatory bowel disease as a model for translating the microbiome. Immunity 40(6): 843-854.

Jeffery, I.B., E.M. Quigley, L. Öhman, M. Simrén and P.W. O'Toole. 2012. The microbiota link to irritable bowel syndrome: An emerging story. Gut Microbes 3(6): 572-576.

Jiménez, E., M.L. Marín, R. Martín, J.M. Odriozola, M. Olivares, J. Xaus et al. 2008. Is meconium from healthy newborns actually sterile? Microbiol. Res. 159(3): 187-193.

Jones, R.B., T.L. Alderete, J.S. Kim, J. Millstein, F.D. Gilliland, M.I. Goran et al. 2019. High intake of dietary fructose in overweight/obese teenagers associated with depletion of *Eubacterium* and *Streptococcus* in gut microbiome. Gut Microbes 10: 712-719.

Jostins, L., S. Ripke, R.K. Weersma, R.H. Duerr, D.P. McGovern, K.Y. Hui et al. 2012. Host-microbe interactions have shaped the genetic architecture of inflammatory bowel disease. Nature 491(7422): 119-124.

Kapeller, J., L.A. Houghton, H. Mönnikes, J. Walstab, D. Möller, H. Bönisch et al. 2008. First evidence for an association of a functional variant in the microRNA-510 target site of the serotonin receptor-type 3E gene with diarrhea predominant irritable bowel syndrome. Hum. Mol. Genet. 17(19): 2967-2977.

Kaplan, G.G. and S.C. Ng. 2017. Understanding and preventing the global increase of inflammatory bowel disease. Gastroenterology. 152(2): 313-321, e312.

Karlsson, C.L., G. Molin, C.M. Cilio and S. Ahrné. 2011. The pioneer gut microbiota in human neonates vaginally born at term—A pilot study. Pediatr. Res. 70(3): 282-286.

Knights, D., M.S. Silverberg, R.K. Weersma, D. Gevers, G. Dijkstra, H. Huang et al. 2014. Complex host genetics influence the microbiome in inflammatory bowel disease. Genome Med. 6(12): 107.

Knoop, K.A., K.G. McDonald, D.H. Kulkarni and R.D. Newberry. 2016. Antibiotics promote inflammation through the translocation of native commensal colonic bacteria. Gut 65(7): 1100-1109.

Koleva, P.T., J.S. Kim, J.A. Scott and A.L. Kozyrskyj. 2015. Microbial programming of health and disease starts during fetal life. Birth Defects Res. C 105(4): 265-277.

Kruse, F.H. 1933. Functional disorders of the colon: The spastic colon, the irritable colon, and mucous colitis. Cal. West. Med. 39(2): 97-103.

Lane, E.R., T.L. Zisman and D.L. Suskind. 2017. The microbiota in inflammatory bowel disease: Current and therapeutic insights. J. Inflamm. Res. 10: 63-73.

Langdon, A., N. Crook and G. Dantas. 2016. The effects of antibiotics on the microbiome throughout development and alternative approaches for therapeutic modulation. Genome Med. 8(1): 39.

Langille, M.G.I., J. Zaneveld, J.G. Caporaso, D. McDonald, D. Knights, J.A. Reyes et al. 2013. Predictive functional profiling of microbial communities using 16S rRNA marker gene sequences. Nat. Biotechnol. 31(9): 814-821.

Le Bastard, Q., G. AlⅢGhalith, M. Grégoire, G. Chapelet, F. Javaudin, E. Dailly et al. 2018. Systematic review: Human gut dysbiosis induced by non-antibiotic prescription medications. Aliment. Pharmacol. Ther. 47(3): 332-345.

Lee, J.R., M. Magruder, L. Zhang, L.F. Westblade, M.J. Satlin, A. Robertson et al. 2019. Gut microbiota dysbiosis and diarrhea in kidney transplant recipients. Am. J. Transplant. 19(2): 488-500.

Lee, W.-J. and K. Hase. 2014. Gut microbiota-generated metabolites in animal health and disease. Nat. Chem. Biol. 10(6): 416-424.

Liu, Y., S. Qin, Y. Song, Y. Feng, N. Lv, Y. Xue et al. 2019. The perturbation of infant gut microbiota caused by cesarean delivery is partially restored by exclusive breastfeeding. Front. Microbiol. 10: 598.

Lo, B., M. Prosberg, L. Gluud, W. Chan, R.W. Leong, E. Van Der List et al. 2018. Systematic review and meta-analysis: Assessment of factors affecting disability in inflammatory bowel disease and the reliability of the inflammatory bowel disease disability index. Aliment. Pharmacol. Ther. 47(1): 6-15.

Lopetuso, L.R., V. Petito, C. Graziani, E. Schiavoni, F.P. Sterbini, A. Poscia et al. 2018. Gut microbiota in health, diverticular disease, irritable bowel syndrome, and inflammatory bowel diseases: Time for microbial marker of gastrointestinal disorders. Digest. Dis. 36(1): 56-65.

Lopez-Siles, M., S.H. Duncan, L.J. Garcia-Gil and M. Martinez-Medina. 2017. *Faecalibacterium prausnitzii*: From microbiology to diagnostics and prognostics. ISME J. 11(4): 841-852.

Louis, P. and H.J. Flint. 2017. Formation of propionate and butyrate by the human colonic microbiota. Environ. Microbiol. 19(1): 29-41.

Lovell, R.M. and A.C. Ford. 2012. Global prevalence of and risk factors for irritable bowel syndrome: A meta-analysis. Clin. Gastroenterol. Hepatol. 10(7): 712-721, e714.

Lundgren, S.N., J.C. Madan, J.A. Emond, H.G. Morrison, B.C. Christensen, M.R. Karagas et al. 2018. Maternal diet during pregnancy is related with the infant stool microbiome in a delivery mode-dependent manner. Microbiome 6(1): 109.

Lynch, S.V. and O. Pedersen. 2016. The human intestinal microbiome in health and disease. N. Engl. J. Med. 375(24): 2369-2379.

Macfarlane, G.T. and S. Macfarlane. 2011. Fermentation in the human large intestine: Its physiologic consequences and the potential contribution of prebiotics. J. Clin. Gastroenterol. 45: S120-S127.

Magnúsdóttir, S., D. Ravcheev, V. de Crécy-Lagard and I. Thiele. 2015. Systematic genome assessment of B-vitamin biosynthesis suggests co-operation among gut microbes. Front. Genet. 6: 148.

Maharshak, N., Y. Ringel, D. Katibian, A. Lundqvist, R.B. Sartor, I.M. Carroll et al. 2018. Faecal and mucosa-associated intestinal microbiota in patients with diarrhoea-predominant irritable bowel syndrome. Dig. Dis. Sci. 63(7): 1890-1899.

Marcobal, A., M. Barboza, J.W. Froehlich, D.E. Block, J.B. German, C.B. Lebrilla et al. 2010. Consumption of human milk oligosaccharides by gut-related microbes. J. Agric. Food Chem. 58(9): 5334-5340.

Micah, H., F.-L. Claire and K. Rob. 2007. The human microbiome project: Exploring the microbial part of ourselves in a changing world. Nature 449(7164): 804-810.

Moayyedi, P., M.G. Surette, P.T. Kim, J. Libertucci, M. Wolfe, C. Onischi et al. 2015. Fecal microbiota transplantation induces remission in patients with active ulcerative colitis in a randomized controlled trial. Gastroenterology. 149(1): 102-109, e106.

Morowitz, M.J., E.M. Carlisle and J.C. Alverdy. 2011. Contributions of intestinal bacteria to nutrition and metabolism in the critically ill. Surg. Clin. 91(4): 771-785.

Muniz, L.R., C. Knosp and G. Yeretssian. 2012. Intestinal antimicrobial peptides during homeostasis, infection, and disease. Front. Immunol. 3: 310.

Nagai, M., Y. Obata, D. Takahashi and K. Hase. 2016 . Fine-tuning of the mucosal barrier and metabolic systems using the diet-microbial metabolite axis. Int. Immunopharmacol. 37: 79-86.

Nagalingam, N.A. and S.V. Lynch. 2012. Role of the microbiota in inflammatory bowel diseases. Inflamm. Bowel Dis. 18(5): 968-984.

Nelson, A.M., S.T. Walk, S. Taube, M. Taniuchi, E.R. Houpt, C.E. Wobus et al. 2012. Disruption of the human gut microbiota following Norovirus infection. PloS One 7(10): e48224.

Newburg, D.S. and L. Morelli. 2015. Human milk and infant intestinal mucosal glycans guide succession of the neonatal intestinal microbiota. Pediatr. Res. 77(1-2): 115-120.

Ni, J., R. Huang, H. Zhou, X. Xu, Y. Li, P. Cao et al. 2019. Analysis of the relationship between the degree of dysbiosis in gut microbiota and prognosis at different stages of primary hepatocellular carcinoma. Front. Microbiol. 10: 1458.

O'Callaghan, A. and D. van Sinderen. 2016. Bifidobacteria and their role as members of the human gut microbiota. Front. Microbiol. 7: 925.

Pannaraj, P.S., F. Li, C. Cerini, J.M. Bender, S. Yang, A. Rollie et al. 2017. Association between breast milk bacterial communities and establishment and development of the infant gut microbiome. JAMA Pediatr. 171(7): 647-654.

Perez-Muñoz, M.E., M.-C. Arrieta, A.E. Ramer-Tait and J. Walter. 2017. A critical assessment of the "sterile womb" and "in utero colonization" hypotheses: Implications for research on the pioneer infant microbiome. Microbiome 5(1): 48.

Quévrain, E., M. Maubert, C. Michon, F. Chain, R. Marquant, J. Tailhades et al. 2016. Identification of an anti-inflammatory protein from *Faecalibacterium prausnitzii*, a commensal bacterium deficient in Crohn's disease. Gut. 65(3): 415-425.

Rajilić-Stojanović, M., H.G. Heilig, S. Tims, E.G. Zoetendal and W.M. de Vos. 2013. Long-term monitoring of the human intestinal microbiota composition. Environ. Microbiol. 15(4): 1146-1159.

Rinninella, E., P. Raoul, M. Cintoni, F. Franceschi, G.A.D. Miggiano, A. Gasbarrini et al. 2019. What is the healthy gut microbiota composition: A changing ecosystem across age, environment, diet, and diseases. Microorganisms 7(1): 14.

Rosario, D., R. Benfeitas, G. Bidkhori, C. Zhang, M. Uhlen, S. Shoaie et al. 2018. Understanding the representative gut microbiota dysbiosis in metformin-treated type 2 diabetes patients using genome-scale metabolic modelling. Front. Physiol. 9: 775.

Rowland, I., G. Gibson, A. Heinken, K. Scott, J. Swann, I. Thiele et al. 2018. Gut microbiota functions: Metabolism of nutrients and other food components. Eur. J. Nutr. 57(1): 1-24.

Scher, J.U., C. Ubeda, A. Artacho, M. Attur, S. Isaac, S.M. Reddy et al. 2015. Decreased bacterial diversity characterizes the altered gut microbiota in patients with psoriatic arthritis, resembling dysbiosis in inflammatory bowel disease. Arthritis Rheumatol. 67(1): 128-139.

Segal, A.W. 2016. Making sense of the cause of Crohn's – A new look at an old disease. F1000Res. 5.

Singh, R.K., H.-W. Chang, D. Yan, K.M. Lee, D. Ucmak et al. 2017. Influence of diet on the gut microbiome and implications for human health. J. Transl. Med. 15(1): 73.

Sokol, H. and P. Seksik. 2010. The intestinal microbiota in inflammatory bowel diseases: Time to connect with the host. Curr. Opin. Gastroenterol. 26(4): 327-331.

Sommer, F. and F. Bäckhed. 2013. The gut microbiota—Masters of host development and physiology. Nat. Rev. Microbiol. 11(4): 227-236.

Srinivasan, R. and A.K. Akobeng. 2009. Thalidomide and thalidomide analogues for induction of remission in Crohn's disease. Cochrane Database Syst. Rev. 2009. 2: CD007350.

Stewart, C.J., N.J. Ajami, J.L. O'Brien, D.S. Hutchinson, D.P. Smith, M.C. Wong et al. 2018. Temporal development of the gut microbiome in early childhood from the TEDDY study. Nature 562(7728): 583-588.

Tap, J., M. Derrien, H. Törnblom, R. Brazeilles, S. Cools-Portier, J. Doré et al. 2017. Identification of an intestinal microbiota signature associated with severity of irritable bowel syndrome. Gastroenterol. 152(1): 111-123, e8.

Telle-Hansen, V., K. Holven and S. Ulven. 2018. Impact of a healthy dietary pattern on gut microbiota and systemic inflammation in humans. Nutrients 10(11): 1783.

Toscano, M., R. De Grandi, D.G. Peroni, E. Grossi, V. Facchin, P. Comberiati et al. 2017. Impact of delivery mode on the colostrum microbiota composition. BMC Microbiol. 17(1): 205.

Vaiopoulou, A., G. Karamanolis, T. Psaltopoulou, G. Karatzias and M. Gazouli. 2014. Molecular basis of the irritable bowel syndrome. World J. Gastroenterol. 20(2): 376-383.

van der Ark, K.C., R.G. van Heck, V.A.M. Dos Santos, C. Belzer and W.M. de Vos et al. 2017. More than just a gut feeling: Constraint-based genome-scale metabolic models for predicting functions of human intestinal microbes. Microbiome 5: 78.

Van Nood, E., A. Vrieze, M. Nieuwdorp, S. Fuentes, E.G. Zoetendal, W.M. de Vos et al. 2013 . Duodenal infusion of donor feces for recurrent Clostridium difficile. N. Engl. J. Med. 368(5): 407-415.

Vandeputte, D., G. Kathagen, K. D'hoe, S. Vieira-Silva, M. Valles-Colomer, J. Sabino et al. 2017. Quantitative microbiome profiling links gut community variation to microbial load. Nature 551(7681): 507-511.

Wampach, L., A. Heintz-Buschart, A. Hogan, E.E. Muller, S. Narayanasamy, C.C. Laczny et al. 2017. Colonization and succession within the human gut microbiome by archaea, bacteria, and microeukaryotes during the first year of life. Front. Microbiol. 8: 738.

Wang, M., M. Li, S. Wu, C.B. Lebrilla, R.S. Chapkin, I. Ivanov et al. 2015. Faecal microbiota composition of breast-fed infants is correlated with human milk oligosaccharides consumed. J. Pediatr. Gastroenterol. Nutr. 60(6): 825-833.

Winter, S.E. and A.J. Bäumler. 2014. Dysbiosis in the inflamed intestine: Chance favors the prepared microbe. Gut Microbes 5(1): 71-73.

Wu, G.D., C. Compher, E.Z. Chen, S.A. Smith, R.D. Shah, K. Bittinger et al. 2016. Comparative metabolomics in vegans and omnivores reveal constraints on diet-dependent gut microbiota metabolite production. Gut 65(1): 63-72.

Yu, L.C.-H., J.-T. Wang, S.-C. Wei and Y.-H. Ni. 2012. Host-microbial interactions and regulation of intestinal epithelial barrier function: From physiology to pathology. World J. Gastrointest. Pathophysiol. 3(1): 27-43.

Zhang, J., Z. Guo, Z. Xue, Z. Sun, M. Zhang, L. Wang et al. 2015. A phylo-functional core of gut microbiota in healthy young Chinese cohorts across lifestyles, geography and ethnicities. ISME J. 9(9): 1979-1990.

Zhernakova, A., A. Kurilshikov, M.J. Bonder, E.F. Tigchelaar, M. Schirmer, T. Vatanen et al. 2016. Population-based metagenomics analysis reveals markers for gut microbiome composition and diversity. Science 352(6285): 565-569.

Zhong, H., J. Penders, Z. Shi, H. Ren, K. Cai, C. Fang et al. 2019. Impact of early events and lifestyle on the gut microbiota and metabolic phenotypes in young school-age children. Microbiome 7(1): 2.

Zhou, Q., W.W. Souba, C.M. Croce and G.N. Verne. 2010. MicroRNA-29a regulates intestinal membrane permeability in patients with irritable bowel syndrome. Gut 59(6): 775-784.

Zhou, Y. and F. Zhi. 2016. Lower level of bacteroides in the gut microbiota is associated with inflammatory bowel disease: A meta-analysis. BioMed Res. Int. 2016: 5828959.

Zmora, N., J. Suez and E. Elinav. 2018. You are what you eat: Diet, health and the gut microbiota. Nat. Rev. Gastroenterol. Hepatol. 16: 35-56.

3

Probiotics and GIT Diseases/Stomach Ulcer

Nehal El-Deeb[1,*] and Lamiaa Al-Madboly[2]

[1] Biopharmaceutical Products Research Department, Genetic Engineering and Biotechnology Research Institute, City of Scientific Research and Technology Applications, New Borg El-Arab City 21934, Alexandria, Egypt

[2] Pharmaceutical Microbiology Dept., Faculty of Pharmacy, Tanta University, Tanta, Egypt

1. Introduction

Gastric ulcer is one of the most common chronic gastrointestinal diseases that is mainly characterized by a significant decrease in the mucosal barrier. The exact mechanisms of gastric ulcer are unclear yet; *Helicobacter pylori* infection and the frequent long-term uses of non-steroidal anti-inflammatory drugs are major factors that induce the development of gastric ulcer. *Helicobacter pylori*, a commonly known Gram-negative human gastric pathogen, plays an important etiologic role in gastric diseases particularly in peptic and gastric ulcers, gastric adenocarcinoma and to a little extent in the lymphoma of the mucosa-associated lymphoid tissue. The pathogenicity of this bacterium is attributed to virulence factors including; urease, *cagA, vacA, iceA*, as well as bacterial adhesion and maintenance factors. Furthermore, the virulence factors *cagA, iceA* and *vacA* are responsible for the great genetic diversity in *H. pylori* relative to their ancestor strain, and host and environmental factors that collectively affect the severity of the disease. Emergence of drug resistance remains a major challenge in the eradication of this pathogen in spite of the efficiency of the triple-therapy regimen followed in the empirical management of *H. pylori* infections. In this respect, probiotics exhibit antagonistic activities against *H. pylori* and hence have great treating potential for infections caused by this microbe. Moreover, probiotics could improve the host defense mechanism. Therefore, the current chapter will explain the possible interaction between probiotics with bacterial pathogens, non-steroidal anti-inflammatory drugs consumption and the host.

*Corresponding author: nehalmohammed83@gmail.com

2. Probiotic Health Benefits

According to NIH, Probiotics generally are live microorganisms (in most cases, bacteria that belong to groups called *Lactobacillus* and *Bifidobacterium* and may also be some yeasts such as *Saccharomyces boulardii*) that are more likely to be similar to the beneficial gut microbiota. Also, Probiotics are defined by FAO/WHO as "Live microorganisms which when administered in adequate amounts confer a health benefit on the host" (Sgouras et al., 2015). These friendly bacteria are available to the consumers as:

- **Dietary supplements and Foods with probiotics such as**, some juices and soy drinks, fermented and unfermented milk, yogurt, buttermilk, some soft cheeses, miso, tempeh, kefir, kimchi, sauerkraut, and many pickles.
- **Nutraceutical Supplements**, probiotic supplements, which are available in the form of capsules, tablets, powders and liquid extracts
- **Other Products** that aren't used orally, such as skin creams and cleansers.

The health benefits of probiotics include; enhanced immune responses (De *Francesco* et al., 2006) and protection against pathogenic microbes (Hamilton-Miller, 2003, Lorca et al., 2001, Malfertheiner et al., 2002). *Lactobacillus* and *Bifidobacterium* are the common genera of probiotics (Malfertheiner et al., 2002). They are characterized by being acid tolerant and possess antimicrobial attributes due to the production of lactic acid, hydrogen peroxide and bacteriocins, and competitive interaction with pathogens for microbial adhesion sites, and also host immune response modulation (Coconnier et al., 1998, Malfertheiner et al., 2002). Moreover, they can be used as delivery vectors for medical therapies, such as vaccines because of their acid resistant property (Midolo et al., 1995). In addition, several studies showed that *Lactobacillus* possesses antagonistic effects toward *H. pylori in vitro*. Furthermore, clinical trials have presented the beneficial outcomes for probiotics used in children in some different clinical situations such as rotavirus infections, antibiotic associated diarrhea, and irritable bowel syndrome. Other examples of less commonly used probiotic microorganisms are strains of *Streptococcus*, *Escherichia coli*, and *Saccharomyces* (Aiba et al., 1998).

Numerous studies have indicated that probiotics can be used for the treatment of gastric diseases, especially, gastric ulcers. Generally, Peptic ulcers are described as open sores that develop on the inside lining of stomach and upper portions of the small intestine with chronic stomach pains. Gastric ulcer is considered as one of the most common and serious chronic diseases of the upper gastrointestinal tract. In general, peptic ulcer development occurs as a result of imbalance between the defense mechanisms and damaging stress factors at the luminal surface of the stomach (Tarnawski et al., 2013). The high frequency of peptic and gastric ulcers in developing countries is due to the misuse of nonsteroidal anti-inflammatory drugs (NSAIDs), the high prevalence of *H. pylori* infections, and cigarette smoking that represent the major risk factors involved in ulcer development (Tarnawski et al., 2013, Wang et al., 2009).

3 Gastric Ulcer and *Helicobacter pylori* Infection

3.1 *Helicobacter pylori*

Helicobacter pylori is a bacterium of gram-negative reaction, microaerophilic, and spiral-shaped that colonizes the mucosa of gastric epithelium causing gastrointestinal diseases. This pathogen is formerly called *Campylobacter pyloridis*, renamed as *Campylobacter pylori* and then later categorized as *Helicobacter pylori*. This reclassification to a new genus is due to the presence of differences in morphology, structure and genetic characteristics. The prevalence of this bacterium is worldwide infecting more than half of the world population. Probability of exposure is high and widespread, and infection occurs early in life under the age of five. About 79-83% of the population is infected with *H. pylori* during the first two decades of life. Surveys using serological tests suggest a sero-prevalence of 22-57% in infants under the age of five, increasing to 80-90% by the age of 20 and remains constant thereafter. According to the data of both developing and developed countries, there is a direct relationship between disease prevalence in developing countries and age but there is no gender preference detected in *H. pylori* infections reflecting the lower socioeconomic level of those afflicted areas. Infection with *H. pylori* is highly associated with atrophic gastritis, peptic ulcers and gastric carcinoma, and rarely, lymphoma of the mucosa-associated lymphoid tissue. Furthermore, this bacteria pathogen was classified by an arm of the World Health Organization in 1994, the International Agency for Cancer Research, as a potential human carcinogen (Benson et al. 2004., Forbes et al., 1994, Graham et al., 1991, Marshall, 1994).

3.2. Pathogenesis of *H. pylori*

H. pylori adhere to the gastric and duodenal epithelial lining due to the presence of a strong affinity, and subsequently damage microvilli as well as the tight junctions between adjacent cells resulting in erosion of the epithelial lining permitting acid and bacteria to get through it and hence establish pathogen in the mucous layer. The majority of *H. pylori* strains are urease producers that decompose urea forming ammonia, which is deleterious to the gastric mucosa (Tarnawski et al., 2013). Moreover, this reaction leads to low acidity which allows the bacteria to colonize the stomach. Urease enzyme is an important virulence factor *in H. pylori* infections, and it is composed of four subunits, A, B, C and D. Subunit B, it is a peptide formed of 569 amino acids and encoded by the *ureB* gene. It has a strong antigenicity and protection among all proteins of *H. pylori* (Wang et al., 2009). Other important enzymes secreted by *H. pylori* include proteases and phospholipases that destroy the mucosal lining. The pathogen lipopolysaccharide triggers the inflammatory cells of the mucosa. Among them, neutrophils release myeloperoxide in response to pathogen attack. Moreover, the bacterial platelet-activating factor promotes thrombotic occlusion of surface capillaries. Taken together, these responses destroy the protective mucosal layer in the gastro-duodenal region. Exposed epithelial cells are highly susceptible to the damaging effect of acid-peptic digestion and eventually, gastric inflammation occurs resulting in peptic ulceration (Tarnawski et al., 2013, Wang et al., 2009).

Plenty of studies conducted in several regions of the world on *H. pylori* populations suggested the involvement of different blood group antigen binding adhesion factors including; *cagA, iceA, vacA,* and *babA* (Hussain et al., 2004, Kersulyte et al., 2000, Nogueira et al., 2001), *hp0169* and *comB4* (Kavermann et al., 2003) virulence IV secretion system (Fig. 1; Table 1) that facilitate translocation of some bacterial factors such as lipopolysaccharides (LPS), urease, broken fragments of peptidoglycan and porins through the epithelial layer and subsequent stimulation of the gastric mucosa macrophages. Activated macrophages release IL-1β, IL-8, IL-12, TNF-α

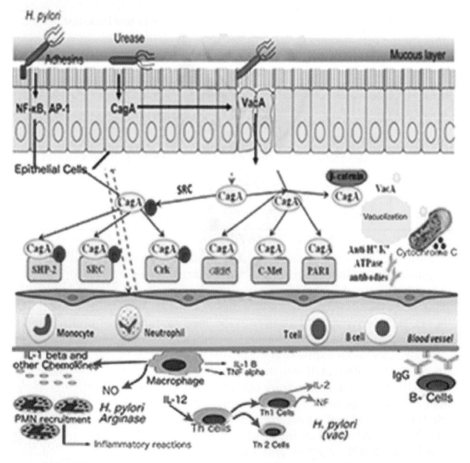

Fig. 1: Pathological changes in gastric epithelium up on *H. pylori* infection

Table 1: Virulence and maintenance factors of *H. pylori*

Factors	Function	References
Virulence factors		
VacA	Encodes a protein cytotoxin that induces vacuolation in eukaryotic cells	(Resta-Lenert and Barrett 2003)
cagA	Stimulates the production of interleukin 8; a part of it is also a code for type IV secretion system	(Jijon et al., 2004)
babA	Binds to Lewis antigen displayed on the surface of stomach epithelial cell	(Jijon et al., 2004)
iceA	Upregulated upon contact of *H. pylori* with gastric epithelium	(Jijon et al., 2004, Pretzer et al., 2005)
oicB	Induces IL-8 secretion by epithelial cells	(Kadowaki et al., 2001)
picB	Induces IL-8 expression in gastric epithelial cells	(Kadowaki et al., 2001)
urease	Neutralizes acid	(Yan and Polk, 2002)
Adhesion and maintenance factors		
Hp0169	Essential for *H. pylori* stomach colonization as it codes for a collagenase	(van Pijkeren et al., 2006)
Com84	Essential for colonization, as it codes a putative ATPase which is a part of DNA transformation associated type-IV transport system	(van Pijkeren et al., 2006)
rocF	Encodes arginase that facilitates production of ammonia and favors NO production in stimulated macrophages	(Van de Bovenkamp et al., 2003)
MUC4AC	Primary receptor for *H. pylori* in human stomach	(Van de Bovenkamp et al., 2003)

and induce expression of INF-γ by TH1 cells. Moreover, IL-8 and INF-γ disorganize the epithelial barrier function. Macrophages, antigen presenting cells, with displayed bacterial epitopes attached to class II MHC elicit clonal expansion in B-cells to produce antibodies against *H. pylori* and stimulate Tc cells via TH2 derived IL-2 that eventually lead to damage of infected cells. Apoptosis is stimulated when the effector molecules of macrophages, TNF-α and IL-1β, activate *fas* expression in sensitized cells. The presence of low GC content of the *cag* PAI relative to the remaining *Helicobacter* genomes indicates that this island was acquired from other bacterial species by horizontal gene transfer (Broutet et al., 2001). Infection with *cagA+* strains increases the tendency for duodenal ulcers and adeno-carcinoma development (Broutet et al., 2001). Production of interleukin 8 is stimulated by the *cag* PAI via intracellular NOD1 receptor and the nuclear factor κB pathway leading to increased gastric inflammation which subsequently results in disease development. Furthermore, host cytokine response potentiates pathogen clearance via initiation of apoptosis. Ultimately, all *H. pylori* strains carry *vacA* gene that encodes for a cytotoxin protein that triggers vacuolation in different eukaryotic cells. This gene is present in all strains, but it has a mosaicism in the terminal(s) and median(m) regions resulting in several alleles and consequently variation in the amounts of toxin produced. The allele *s1m1* corresponds to the highest production, followed by *s1m2*, while strains with the *s2m2* allele do not produce any toxin. Furthermore, two novel proteins; a secreted collagenase, encoded by *hp0169* and a putative ATPase encoded by *com*B4, being part of a DNA transformation-associated type IV transport system of *H. pylori* are absolutely necessary for colonization (Atherton et al., 1995, Crabtree et al., 1995, Kavermann et al., 2003, Xing et al., 2005).

Development of peptic ulcers or gastric carcinoma is more among patients infected with *cag*+ *H. pylori* strains than those infected with negative strains due to the stronger inflammatory response triggered in the stomach (Kusters et al., 2006). Figures 2 and 3 illustrate the pathological changes in the gastric epithelium induced by *H. pylori* infection. Following attachment of the pathogen to the gastric epithelium, expression of the type IV secretion system is initiated by the *cag* PAI which pushes its own peptidoglycan fragments acting as inflammation inducer into the gastric epithelial cells. Recognition of the injected peptidoglycan is mediated by the Nod1 receptor, which then activates the expression of chemokines promoting inflammation (Viala et al., 2004). Additionally, the bacteria also pushes *cagA* protein into the gastric epithelium, where it disorganizes the epithelial membrane barrier and different cellular activities (Backert and Selbach, 2008). When *cagA* protein goes inside the cell, it is phosphorylated on its tyrosine residues by tyrosine kinase present in association with the host cell membrane. Another important membrane protein associated with tyrosine kinase domain is the epidermal growth factor receptor (EGFR) that has been shown to be activated by the pathogen. The stimulated EGFR alters signal transduction and gene expression in the epithelial cells of the host that may contribute to pathogenesis. Moreover, C-terminal region of the *cagA* protein can regulate host cell gene transcription independent of protein tyrosine phosphorylation (Broutet et al., 2001, Baldwin et al., 2007).

Expression of inducible nitric oxide synthase (iNOS) and production of nitric oxide (NO) in macrophage are upregulated following the *H. pylori* infection both under *in vivo* and *in vitro* conditions. Additionally, urease positive strains (*ureA*+) activate iNOS and enhance NO production. Urease enzyme is considered as an essential survival factor for the pathogen during gastric colonization that leads to NO dependent damage and carcinogenesis (Gobert et al., 2002). Other virulence proteins including *vacA, cagPA*-1 and *picB* show a selective and significant decline in the activated iNOS mRNA, protein and NO2-production with the *ureA*-strain compared with wild type *H. pylori*. This bacterium also induces a weak immune response which fails to eliminate the pathogen. Translation of iNOS mRNA and NO production by *H. pylori* stimulated macrophages is also inhibited by the polyamine

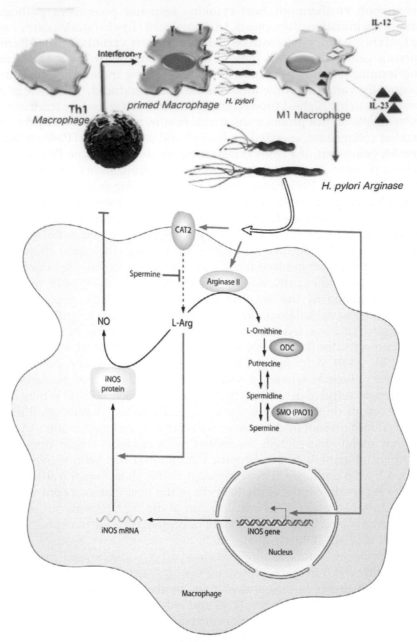

Fig. 2: Modulation of enzyme activities (Arginase, CT2 and ODC) and No production during acute and chronic gastritis.

spermine derived from ornithine decarboxylase (*ODC*) and is dependent on the availability of iNOS substrate i.e. L-arginine. siRNA knockdown studies of two inducible genes, cationic amino acid transporter (*CAT2*) and *ODC*

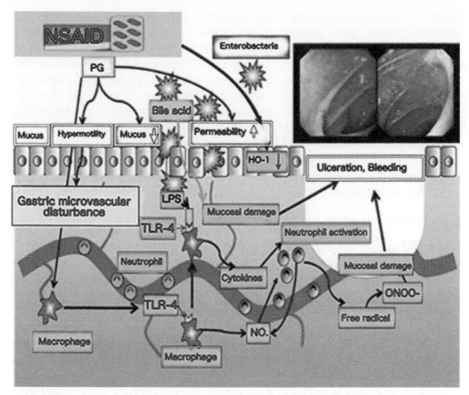

Fig. 3: The possible mechanisms of NSAID-induced gastric and intestine injuries.

in gastric macrophages, suggested that addition of spermine or knockdown of *CAT2* suppresses L-arginine uptake, decreases iNOS protein levels and NO production, whereas knockdown of *ODC* has the opposite effect. The high *ODC* activity of macrophages was mentioned in chronic gastritis which cause formation of polyamine spermine. Increased polyamine spermine concentration in turn decreases iNOS expression and NO generation in *H. pylori* stimulated macrophages that is essential for survival of pathogens during chronic diseases (Fig. 2). The constitutive expression of *rocF* arginase also facilitates the production of ammonia and also favors the production of nitric oxide in stimulated macrophages (Gobert et al., 2002). Taken together, these findings suggest that in case of chronic gastritis up regulation of *ODC* in gastric macrophages impairs host defense against *H. pylori* by inhibiting iNOS derived NO production (Chaturvedi et al., 2010, Gobert et al., 2002).

Gastrititis, peptic ulcers, and neoplasia are developed due to apoptosis induced during *Helicobacter pylori* infection suggesting it is a major etiology of these diseases. The infection induces cell death via stimulation of caspase-8, -9 and -3 along with the overexpression of the proapoptotic *Bcl-2* family proteins *Bad* and *Bid* in the human gastric epithelial cells (Shibayama et al., 2001). Surprisingly, just a membrane fraction of the pathogen is sufficient to stimulate the death pathway inspite of the presence of *cagA* or vacuolating toxin. In chronic cases, this pathogen causes peptic/gastric ulcers and adenocarcinoma. Two suggested mechanisms induce cancer by *H. pylori* infection. Regarding the first mechanism, there is improved production of free radicals near *H. pylori* infected sites and a concomitant enhanced rate of host cell mutation. For the second mechanism, the "perigenetic pathway" involves improvement of the transformed host cell phenotype through changes in cell proteins like adhesion (Tsuji et al., 2003). Additionally, inflammation is promoted through increased levels of TNF-α and IL-6 hence altering the gastric epithelial cell adhesion leading to disruption and migration of mutated epithelial cells without needing additional mutations in the brake pedal genes (tumor suppressor genes) such as those encoding cell adhesion proteins (Suganuma et al., 2008).

3.3 Treatment of *H. pylori* Infections

Triple therapy, a 2-week course, is the most effective treatment protocol. It involves administration of two antibiotics to eradicate the bacteria and either an acid inhibitor or a stomach-lining shielder. There are two types of acid-inhibiting drugs that might be used: H2 blockers such as Ranitidine, Cimetidine, Famotidine, and proton pump inhibitors like Lansoprazole, Omeprazole, Pantoprozole, Rabeprazole, Esomeprazole, Leminoprazole (Marshall, 1994). Additionally, Bismuth subsalicylate may be used to protect the gastric lining from acid. Another ulcer-preventing agent, sucralfate, is a basic aluminium sulphate sucrose complex and it has anti-pepsin and anti-acid properties. It interacts with gastric juice forming a sticky paste and hence protecting the mucosa by forming a protective coating, and also binds to ulcer-affected sites (Rafii et al., 2008).

3.4 Emergence of Drug Resistance in *H. pylori*

The prevalence of *H. pylori* related infections such as chronic gastritis, duodenal and gastric ulcers is quite high worldwide. The emergence of drug resistance in *H. pylori* leads to a bad situation. Various research studies highlight the emergence of drug resistance in *H. pylori* strains isolated from patients representing a major obstacle in the elimination of this gastro-duodenal bacterial pathogen. For instance, 90% of *H. pylori* strains isolated from patients in Calcutta showed resistance to Metronidazole and 7.5% strains to Tetracycline (Mukhopadhyay et al., 2000). Multidrug combinations showed efficiency in treatment of *H. pylori* infections. These combinations consisted of a proton pump inhibitor and two antibiotics i.e., Omeprazole, Clarithromycin and Amoxycillin (Chaudhuri et al., 2003). It was reported that bacterial *vacA m1* allele was the most common genotype detected among patients with therapeutic failures (68%) than in those with successful treatment (39%). However, there was no significant association of *vacAs1* (signal sequence allele) or *cag* pathogenicity island status with the persistence recorded. Persistent and recurrent infections following treatment is a great problem in *H. pylori* infection (Chaudhuri et al., 2003). Accordingly, current antibiotic-based triple therapies are not practical for global control because of the high cost, problems with patients' compliance, genotype variation in *H. pylori* strains, and the emergence of antibiotic-resistant strains (Michetti, 1997). However, the best approach to control *H. pylori* infection is vaccination via administration of oral bacterial antigens (Corthesy et al., 2005, Kleanthous et al., 1998). Therefore, alternative approaches are badly required to fight against *H. pylori* infections.

4. Probiotics, Gut Physiology and Health Benefits

The human gastro-intestinal tract has a complex microbiome of bacteria whether pathogenic or nonpathogenic such as probiotics. The latter supplements the community of gut microbes maintaining a healthy epithelial barrier function thus promoting a general immune homeostasis (Schaible and Kaufmann, 2005). Recently, there is a commercial focus on the importance of supplementing food with cultures of live probiotics in the form of fermented milk products containing a single strain (*B. longum* BB536, commercial name ProCult3, *L. acidophilus* La1, commercial name LC1; and others) or mixed cultures of various lactobacilli (*L. acidophilus, L. plantarum, L. rhamnosus, L. fermentum*), bifidobacteria (*B. infantis, B. bifidum, B. longum*), yeasts (*Saccharomyces* sp.) and other microbes (Streptococci). Interestingly, the probiotics stimulatory capacity has been assessed in farm animals and this evaluation revealed improvement in the overall health status of the animal, immune system functions, lowering the susceptibility to infection and enhancement in poultry and meat products yield (Reuter 2001). Many review articles have been published outlining the beneficial effects of probiotics and prebiotics in human health (Bengmark, 2003, Fedorak and Madsen, 2004, Steidler, 2003, Tuohy et al., 2003).

Probiotic microorganisms compete with harmful pathogens in the gastro-intestinal tract preventing them from adhesion, colonization and invasion. Synthesis of organic molecules by some microorganisms is an important mechanism required for their survival and maintenance. These molecules include, amino acids, nucleotides, short chain fatty acids, and enzyme cofactors that may be used directly or modified and metabolized from nutrients present in the host gut. Availability of such nutrients within the microenvironments of the host results in the loss of genes encoding their biosynthesis in many of these microorganisms (Schaible and Kaufmann, 2004, Schaible and Kaufmann 2005). Therefore, the dependency of these microbes on certain host nutrients enforces them to select distinct host habitats. An instructive example for nutritive host–pathogen competition is represented by the mutual requirement for iron. Iron is a necessary micronutrient for growth, basic metabolism and maintenance of the majority of living organisms. Additionally, probiotics synthesize and release interferon gamma (IFN-γ) that activates the immune system of the host by enhancing the function of phagocytes (Schaible and Kaufmann, 2005, Schiffrin et al., 1997).

Thus, sensitized macrophages suppress the growth of bacterial pathogens like Mycobacterium (Schaible and Kaufmann, 2004).

Probiotics are present in natural food products and hence known as GRAS (Generally Recognized as Safe). They are able to produce several health benefits beyond their basic nutrition. These beneficial microorganisms show a lot of antagonistic activities due to the production of numerous organic acids and other compounds like bacteriocins and antifungal compounds (Jack et al., 1995, Lavermicocca et al., 2000, Magnusson and Schnurer, 2001, Prema et al., 2010). A number of review articles addressed the applications of bacteriocin starter cultures in different food systems(Galvez et al., 2007). Various health claims were reported about probiotic strains including; normalization of the vaginal (Szajewska et al., 2006) and gastro-intestinal ecosystems (Galvez et al., 2007), enhancement of specific and non-specific immune responses (Schiffrin et al., 1997), detoxification of carcinogens and tumor suppression (Kulkarni and Reddy, 1994), normalization of blood pressure in patients with hypertension and cholesterol (Simons et al., 2006).

4.1 Mechanism of the Action of Probiotics

The growth of lactic acid bacteria in food cultures lowers the pH suppressing the growth and division of most food borne pathogens. Moreover, some biochemical conversions occur during the growth process improving the flavor, enhancing the organoleptics, as well as nutritional properties (Schillinger and Lucke, 1989). Several probiotic strains produce antagonistic compounds such as short chain fatty acids, bacteriocins, and hydrogen peroxide. The first microbiologist to discover the antimicrobial property of bacteria was Gratia. He reported that bacteria are able to synthesis proteinaceous toxins which are able to suppress the growth of closely related strains (Bhunia et al., 1988). Later on, many bacteriocin producing strains were identified and they are very important in food fermentations. The development of potential novel antimicrobial and therapeutic agents has increased following isolation of pediocins and their candidate producers like *Pediococcus acidilactici, P. damnosus,* and *P. pentosaceous* (Bhunia et al., 1988).

Some of the protective functions of probiotics may be enforced by modulating immune response and epithelial functions within the large and small intestines. Models studied *in vitro* assumed that both immune and epithelial cells distinguish between different bacterial species via stimulation of Toll-like receptors (Kadowaki et al., 2001). Resta-Lenert and Barrett reported that live *Streptococcus thermophilus* as well as *Lactobacillus acidophilus* were able to suppress the adhesion and invasion of enteroinvasive *E. coli* into human intestinal cell lines (Resta-Lenert and Barrett, 2003). Other studies also mentioned that *Lactobacillus rhamnosus* GG was able to inhibit cytokine-induced programmed cell death in the intestinal epithelial cell models via prevention of TNF-induced activation of the proapoptotic p38/mitogen-activated protein kinase (Yan and Polk, 2002). Interleukin-8 is released by normal epithelial cells in response to bacterial pathogens but not against probiotic strains. Bacterial DNA of pathogenic strains evoke phosphorylation of the extracellular signal-regulated kinase pathways and turn on activator protein-1 (Lammers et al., 2002), and of probiotic strains by modulating nuclear factor-κB (NF-κB) pathway in response to TNF-α as indicated in Fig. 1 (Llewellyn and Foey, 2017). Some selected probiotic strains can activate the regulatory functions of the host dendritic cells (DCs) by targeting specific pattern-recognition receptors and pathways and conferring protection against 2, 4, 6-trinitrobenzenesulfonic acid (TNBS)-induced colitis. The preventive effect of probiotic-pulsed DCs require a high local expression of the immunoregulatory enzyme indolamine 2, 3 dioxgenase, MyD88-, TLR2- and NOD2-dependent signaling and induction of CD4+, CD25+ regulatory cells in an IL-10-independent pathway. Probiotics have a vital role in counteracting stress-induced changes within the intestinal cells, visceral sensitivity and gut motility in a strain dependent manner. Effects may be due to the bacterial-host cell interaction and/or through soluble factors. Probiotics may trigger these beneficial responses via different mechanisms including; competition with pathogenic microbes for necessary nutrients, activation of epithelial heat-shock proteins, normalization of tight junction protein structure, mucin genes up-regulation, defensins secretion, and regulatory function of the NF-κB signaling pathway (Lammers et al., 2002, Llewellyn and Foey, 2017).

4.2 Macromolecules Supporting Probiotic Function

Several important models are used to study the probiotics adherence to Human epithelials such as cell CaCo-2 and HT-29 human-derived adenocarcinoma cells (Velez et al., 2007). Pretzer *et al.*, 2005 identified a glycoprotein called mannose-specific adhesin encoded by *lp_1229* gene (renamed as *msa*) in *L. plantarum* WCFS1 using genome-wide microarray-based genotyping. Therefore, it is possible to identify genes encoding for certain proteins that are responsible for facilitating bacterial adhesion to intestinal cells using *in silico* techniques for genome wide analysis that became available recently. Such approaches are less labor-intensive, time saving and highly precise. It is fast when used for initial screening and just wet lab experiments are enough for validation of the results. An *in-silico* study reported that adherence of *L. acidophilus* NCFM to Caco-2 cells is mediated by multiple factors including, a mucin-binding protein (*Mub*), fibronectin-binding protein (*FbpA*), and a surface layer protein (*SlpA*) and it also identified the encoding of genes for all of them. Another study mentioned the important adhesion factors of *L. salivarius* UCC118 that were sortase and sortase dependent proteins (SDPs) (van Pijkeren et al., 2006). Interestingly, the binding of *L. johnsonii* NCC533 to Caco-2 and HT-29 intestinal epithelial cells and mucins was maintained by two peculiar cytoplasmic proteins; Tu (EF-Tu) which is an elongation factor and a heat shock protein *GroEL* (Bergonzelli et al., 2006, Granato et al., 2004). Additionally, probiotics can modulate *H. pylori* adhesion and colonization to the gastric mucosa via the production of certain compounds such as lactic acid, bacteriocins or hydrogen peroxide. Various reports support this hypothesis and have proven the efficacy of probiotics in treating *H. pylori* infections as *L. salivarius* WB1004, *L. acidophilus* La1, *L. acidophilus, Bacillus clausii, L. casei* strain Shirota *L. rhamnosus* R0011 and *L. acidophilus* R0052 (Schillinger and Lucke 1989, Szajewska et al., 2006, Watanabe et al., 2009, Yan and Polk, 2002). Moreover, certain studies reported that *L. johnsonii* La1, *L. salivarius, L. acidophilus,* and *Weissella confusa* prevent the adhesion of *H. pylori* to intestinal HT-29 cells (Coconnier et al., 1998) or to MKN 45 gastric cell lines (Kabir et al., 1997). Probiotics are able to attenuate *H. pylori*-induced gastritis via interference in the attachment of the pathogen to the epithelial cells as well as the activation of the host immune system (Felley and Michetti, 2003, Ushiyama et al., 2003). Furthermore, a double-masked, randomized, controlled clinical trial, involving 326 school children from a low socioeconomic area of Santiago, Chile, suffering from *H. pylori* infection were administered live and heat-killed strains of *L. johnsonii, L. paracasei* and/or carrier once daily for 4 weeks. They were then subjected to the 13C-urea breath test that revealed a significant decline in *H. pylori* colonization among those children receiving only live *L. johnsonii* (Cruchet et al., 2003). To have a therapeutic effect on *H. pylori* infections, *Lactobacillus* must share MUC5AC glycolipid specificity as a pre-requisite for eradication of this pathogen because MUC5AC has been identified as the primary receptor for *H. pylori* in the human stomach (Van de Bovenkamp et al., 2003). From the previous studies, one can conclude the presence of a complementary effect of probiotics in the management of the *H. pylori* infection.

4.3 Gastric Ulcers, Cytological and Immunological Mechanisms

The mechanisms involved in the development of gastric ulcers or gastrointestinal bleeding are not well understood yet. To initiate the complications, the mucosal barrier and submucosal blood vessels must be damaged for the bleeding to occur that in turns erodes the vessel wall and interferes with blood coagulation (Lanas and Chan, 2017). This disruption could occur as a result of *H. pylori* infection and/or NSAID use or low doses of Aspirin use.

4.4. Nonsteroidal Anti-inflammatory Drugs (NSAIDs) and Gastric Ulcers

Nonsteroidal anti-inflammatory drugs (NSAIDs) are a class of drugs that reduce pain, decrease fever, prevent blood clots and, in higher doses, decrease inflammation.

The term nonsteroidal is used to distinguish these drugs from steroids, which while having a similar eicosanoid-depressing, anti-inflammatory action, have a broad range of other effects. These drugs were first used in 1960, and the term served to separate these medications from steroids, which were particularly stigmatized at the time due to the connotations with anabolic steroid abuse (Buer, 2014). Nonsteroidal anti-inflammatory drugs and low doses of Aspirin are the second cause of peptic and gastric ulcers (Lanas and Chan, 2017) with a high-risk in aging populations, and acute damage of the gastric mucosa occurs in up to 100% of the patients after short-term use.

Using high-doses or multiple-NSAID, in people with a history of previous ulcer bleeding, co-therapy with corticosteroids, and probably coexistence with *H. pylori* infection is common (Gabriel et al., 1991). The mechanisms of NSAIDs toxicity on the gastrointestinal tracts is explained by the reduction in prostaglandins production which in turn decrease epithelial cells mucus, bicarbonate secretion, mucosal perfusion, epithelial proliferation, and ultimately the mucosal resistance to injury (Laine et al., 1995). The development of gastric lesion in response to NSAIDs is completely inhibited by prior administrations of PEG2 (Takeuchi, 2012), parallel to the protective effects of PEG2 on gastric ulcer development, and administration of antisecretory drugs also prevents these lesions confirming the important roles of luminal acids in gastric lesions pathogenesis (Nishikawa et al., 2006).

The protection roles of Prostaglandins in topical gastric portions was explained by acid production inhibition and stimulation of mucous and bicarbonate sections (Robert et al., 1979). Also, the endogenous Prostaglandins showed additional cytoprotective roles against subsequent mild and severe gastric irritants that induce gastric damage (Konturek et al., 1982). During injury, Prostaglandin E2 (PGE2) is produced by two main enzymes; cyclooxygenase 1 (COX1; also known as PTGS1) and cyclooxygenase 2 (COX2; also known as PTGS2) (Peskar, 2001), whose work is catalyzed by their binding to their receptors EP1 and EP4 respectively (Ohnishi et al., 2001) to inhibit acid secretion and to stimulate mucus and bicarbonate secretions (Peskar, 2001).

The exact cytological mechanisms of gastric ulcerogenic responses to NSAIDS is more complicated than the researchers expected and included closely interacted pathways which resulted in gastric hypermotility, free radical accumulation, neutrophil migration (Asako et al., 1992). The exact actions of NSAIDS which clearly explained gastric ulcer induction, is the inhibition of gastric COX1 and not COX2 (Futaki et al., 1993). While, another study clarified that, by selectively using COX1 and COX2 inhibition models, NSAIDS gastric ulcerogenic effects are accomplished by the inhibition of both COX1 and COX2, and they showed that the inhibition of gastric COX1 upregulates the induction of COX2 that may help in mucous integrity in the event of Prostaglandins deficiency caused by COX1 inhibitions (Takeuchi 2012,Takeuchi et al., 2010,Tanaka et al., 2002).

Also, the pathogenesis effects of stomach hypermotility and mucosal folding in gastric ulcerogenic actions were explained by (Takeuchi , 2012), and it was reported that at ulcerogenic doses, all NSIDA (except Aspirin) showed abilities to increase gastric motility that in turn induces gastric lesions (Tanaka et al., 2002). It was reported that gastric hypermotility induces temporal restriction of blood flow in mucosa with disturbance in gastric microvascular architecture that leads to the interaction of neutrophils with the endothelium (Garrick et al., 1986), and all these consequences decreased gastric resistance to injury (Fig. 3). NSAIDs have the ability to diffuse into both synthetic and biological membranes; modify the membrane hydrophobicity by interacting with membrane phospholipid molecules that in turn alter the stability of membrane bilayers. These interactions, could alter the 'gatekeeping functions' of cells, causing an inevitable GI mucosal damage such as erosions, ulcers, and bleeding, consequent to membrane pore formation (Lichtenberger et al., 2006) causing penetration of bile acid, proteolytic enzymes, intestinal bacteria, or toxin.

NSAIDs resulted in localized decrease of PG, resulting in the reduction of intestinal mucus, and microcirculatory disturbances accompanying abnormally increased intestinal motility. Gastric hypermotility induces temporal restriction of blood flow with a disturbance in gastric microvascular bundles that leads to the interaction of neutrophils with the endothelium, with all these resulting in decreased gastric resistance to injury. Mucosal injuries resulted from penetration of bile acid, proteolytic enzymes, intestinal bacteria, or toxins into the endothelium (Higuchi et al., 2009).

4.5 Probiotics and NSAIDs; Interactions and Protective Roles

The most affected site of NSAIDs injuries is the lower gastrointestinal tract compared to the upper portion. It is presumable that these pathological features may be related to the alterations of the intestinal microbiota community, similar to that occurring with other intestinal disorders such as inflammatory bowel diseases. Several studies explained the possible relationship between intestinal microbiota community alterations and NSAIDs by the induction of overgrowth of Gram-negative and anaerobic bacterial species, which are able to exacerbate the NSAID-induced intestinal injury (Lanas and Scarpignato, 2006). The exact mechanism of bacterial imbalance increase in Gram-negative bacteria in NSAIDs treatments is not clearly understood as yet , and the possible theory is that it could be caused by bacterial penetrations into the mucosa (Fig. 3) as a result of enhanced mucosal permeability induced by the NSAIDS. Consequently, secretion of more cytokines and neutrophils migration due to bacterial lipopolysaccharide (LPS) recognition by the intestinal epithelial cells occurs (Hagiwara et al. 2004). To further explained the bacterial host interaction in NSAID-induced intestinal damage, (Watanabe et al., 2008) recently reported that only a decrease in Gram-negative, but not Gram-positive, bacteria populations could counteract the indomethacin-induced intestinal injury, explaining that the LPS/toll-like receptor 4 (TLR4)/MyD88-dependent signaling pathway plays a key role in the NSAID-induced intestinal injuries development. It is obviously impossible to make people "germ-free" to enhance better tolerance to NSAIDs, but it has been suggested that admiration probiotics could provide protection against NSAID-induced enteropathy, without any significant adverse effects Fig. 4 (Scarpignato, 2008). Moreover, Hagiwara et al. (2004) reported that germ-free rats and gnotobiotic rats associated solely

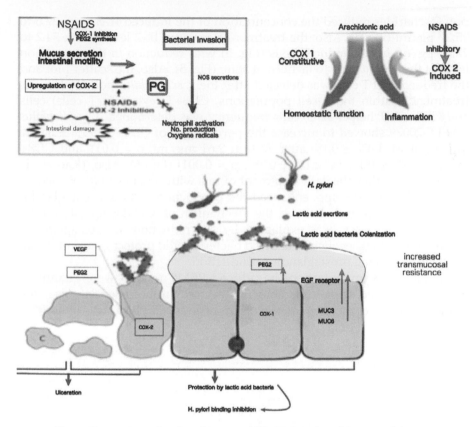

Fig. 4: Protective role of probiotic in NSAIDS -induced intestinal injury

with Bifidobacterium or Lactobacillus didn't show any intestinal ulcers after NSAIDs administration but, NSAIDs induced ileal ulcers in gnotobiotic rats infected with *Eubacterium limosum* or *E. coli*.

The decrease in Gram-negative, as opposed to Gram-positive, bacteria populations could counteract the indomethacin-induced intestinal injury, explaining that the LPS/toll-like receptor 4 (TLR4)/MyD88-dependent signaling pathway has a role to pay. The preventive role of probiotics has been reported in gastrointestinal tract disorders such as inflammatory bowel disease (Bibiloni et al., 2005) and irritable bowel syndrome (O'Mahony et al., 2005). *Lactobacillus casei* strain Shirota, the probiotic bacterium that is isolated from humans (Matsuzaki, 1998) was examined for its protective effects on indomethacin-induced small intestinal damage in rats (Watanabe et al., 2009). The obtained results showed that, the administration of *L. casei* for one week protects against indomethacin-induced small intestinal damage and suppresses neutrophil infiltration and regulates the expression of inflammatory cytokines, including TNF-α.

In addition, El-Deeb et al. (2018) explained the immunomodulatory effects of *L. acidophilus* polysaccharides, and reported that, *L. acidophilus* polysaccharides reduced the concentration of the induced IL-8 from 27.6 to 25.27 pg/ml (Fig. 5). Also, the treatment decreased IL-2 levels from 54.2 to 27.07 pg/ml in LPS-induced cells (Fig. 5) with reduction in the expression level of the TNF-α gene to its normal state (Fig. 5). Also, the scatter plot and the frequency of T cells was detected after the *L. acidiophilus* polysaccharides treatment. Within the T cell populations, CD8+T (cytotoxic T cells) cells and CD4+ T (T helper cells) cells frequency was determined. *L. acidiophilus* LAEPS-20079 showed to increase the percentages of CD4+ (0.95% ± 0.01% in control vs. 1.4% ± 0.1% after 72 h at 2.61 mg/ml, p = 0.01), and CD8+ T cells (1.5% ± 0.1% vs. 2.1% ± 0.3%, p = 0.001) (Fig. 6). Also, (Kao et al., 2018) reported that, the daily treatment of rats with beneficial *Bifidobacterium* spp. and *Lactobacillus* spp. enhanced the production rates of prostaglandin E_2 (PGE$_2$) so, as recorded above, the mechanism of NSAIDs toxicity on the gastrointestinal tracts was explained by the reduction in prostaglandins production so that beneficial lactic acid bacteria could protect gastrointestinal tracts by enhancing prostaglandins production.

    IL2 and IL8 concentrations in induced PBMC models were quantified using ELISA, and the results showed reduced induced levels. Also, the induced TNF-α gene expressions levels in the induced PBMC was quantified using RTqPCR.

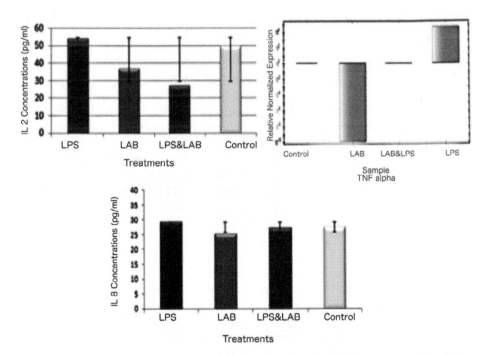

Fig. 5: The regulation of IL2, IL8 and TNF α in induced models using *L. acidiophilus* polysaccharides

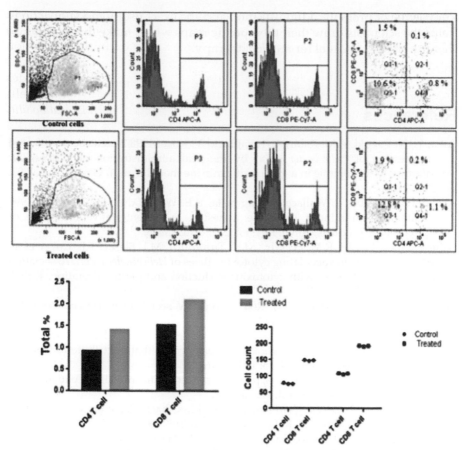

Fig. 6: The effects of *L. acidophilus* polysaccharides on CD8 and CD8 T cells. The CD8 and CD8 T cells populations were checked using flow cytometry.

Conclusion

Gut bacterial communities play important roles by expressing the host physiology throughout many functions including metabolic, barrier effect, trophic and immunological functions. Current misuse of antibiotics, immunosuppressive therapies and radiotherapies with other treatments may alter the gastrointestinal tract flora. So, probiotic therapy could be a promising option to reestablish the microbial balance and prevent disease. Proper evaluations should be taken into consideration before bringing probiotics into routine usage as probiotic quality and reliability are critical. Therefore, future well-designed placebo-controlled reports with validated results are obligatory for determining the exact health benefits of these probiotics therapies. In addition, probiotic specie selection for certain diseased conditions is very important in the treatment therapy protocol. All this is important keeping the future in mind to ensure that extensive knowledge of probiotic administration and the exact useful mode of actions helps the scientists draw their therapeutic maps which make this traditional therapy an effective tool for medical therapy.

References

Aiba, Y., N. Suzuki, A.M. Kabir, A. Takagi and Y. Koga. 1998 . Lactic acid-mediated suppression of *Helicobacter pylori* by the oral administration of *Lactobacillus salivarius* as a probiotic in a gnotobiotic murine model. Am. J. Gastroenterol. 93: 2097-2101.

Asako, H., P. Kubes, J. Wallace, T. Gaginella, R.E. Wolf, D.N. Granger et al. 1992 . Indomethacin-induced leukocyte adhesion in mesenteric venules: role of lipoxygenase products. Am. J. Physiol. 262: G903-G908.

Atherton, J.C., P. Cao, R.M. Peek Jr., M.K. Tummuru, M.J. Blaser, T.L. Cover et al. 1995. Mosaicism in vacuolating cytotoxin alleles of *Helicobacter pylori*. Association of specific vacA types with cytotoxin production and peptic ulceration. J. Biol. Chem. 270: 17771-17777.

Backert, S. and M. Selbach., 2008. Role of type IV secretion in *Helicobacter pylori* pathogenesis. Cell Microbiol. 10: 1573-1581.

Baldwin, D.N., B. Shepherd, P. Kraemer, M.K. Hall, L.K. Sycuro, D.M. Pinto-Santini et al. 2007. Identification of *Helicobacter pylori* genes that contribute to stomach colonization. Infect. Immunol. 75: 1005-1016.

Bengmark, S. 2003. Use of some pre-, pro- and synbiotics in critically ill patients. Best Pract. Res. Clin. Gastroenterol. 17: 833-848.

Benson, J.A., K.A. Fode-Vaughan and M.L. Collins. 2004. Detection of *Helicobacter pylori* in water by direct PCR. Lett. Appl. Microbiol. 39: 221-225.

Bergonzelli, G.E., D. Granato, R.D. Pridmore, L.F. Marvin-Guy, D. Donnicola and I.E. Corthesy-Theulaz et al. 2006. GroEL of *Lactobacillus johnsonii* La1 (NCC 533) is cell surface associated: potential role in interactions with the host and the gastric pathogen *Helicobacter pylori*. Infect. Immunol. 74: 425-434.

Bhunia, A.K., M.C. Johnson and B. Ray. 1988. Purification, characterization and antimicrobial spectrum of a bacteriocin produced by *Pediococcus acidilactici*. J. Appl. Bacteriol. 65: 261-268.

Bibiloni, R., R.N. Fedorak, G.W. Tannock, K.L. Madsen, P. Gionchetti, M. Campieri et al. 2005. VSL#3 probiotic-mixture induces remission in patients with active ulcerative colitis. Am. J. Gastroenterol. 100: 1539-1546.

Broutet, N., A. Marais, H. Lamouliatte, A. de Mascarel, R. Samoyeau, R. Salamon et al. 2001. cagA Status and eradication treatment outcome of anti-*Helicobacter pylori* triple therapies in patients with nonulcer dyspepsia. J. Clin. Microbiol. 39: 1319-1322.

Buer, J.K. 2014. Origins and impact of the term 'NSAID'. Inflammopharmacol. 22: 263-267.

Chaturvedi, R., M. Asim, S. Hoge, N.D. Lewis, K. Singh, D.P. Barry et al. 2010. Polyamines impair immunity to *Helicobacter pylori* by inhibiting L-Arginine uptake required for nitric oxide production. Gastroenterol. 139: 1686-1698, 1698.

Chaudhuri, S., A. Chowdhury, S. Datta, A.K. Mukhopadhyay, S. Chattopadhya, D.R. Saha et al. 2003. Anti-*Helicobacter pylori* therapy in India: differences in eradication efficiency associated with particular alleles of vacuolating cytotoxin (vacA) gene. J. Gastroenterol. Hepatol. 18: 190-195.

Coconnier, M.H., V. Lievin, E. Hemery and A.L. Servin. 1998. Antagonistic activity against Helicobacter infection *in vitro* and *in vivo* by the human *Lactobacillus acidophilus* strain LB. Appl. Environ. Microbiol. 64: 4573-4580.

Corthesy, B., S. Boris, P. Isler, C. Grangette and A. Mercenier. 2005. Oral immunization of mice with lactic acid bacteria producing *Helicobacter pylori* urease B subunit partially protects against challenge with *Helicobacter felis*. J. Infect. Dis. 192: 1441-1449.

Crabtree, J.E., A. Covacci, S.M. Farmery, Z. Xiang, D.S. Tompkins, S. Perry et al. 1995. *Helicobacter pylori* induced interleukin-8 expression in gastric epithelial cells is associated with CagA positive phenotype. J. Clin. Pathol. 48: 41-45.

Cruchet, S., M.C. Obregon, G. Salazar, E. Diaz and M. Gotteland. 2003. Effect of the ingestion of a dietary product containing *Lactobacillus johnsonii* La1 on *Helicobacter pylori* colonization in children. Nutrition. 19: 716-721.

De Francesco, V., M. Margiotta, A. Zullo, C. Hassan, L. Troiani, O. Burattini et al. 2006. Clarithromycin-resistant genotypes and eradication of *Helicobacter pylori*. Ann. Intern. Med. 144: 94-100.

El-Deeb, N.M., A.M. Yassin, L.A. Al-Madboly and A. El-Hawiet. 2018. A novel purified *Lactobacillus acidophilus* 20079 exopolysaccharide, LA-EPS-20079, molecularly regulates both apoptotic and NF-kappaB inflammatory pathways in human colon cancer. Microb. Cell Fact. 17: 29.

Fedorak, R.N. and K.L. Madsen. 2004. Probiotics and prebiotics in gastrointestinal disorders. Curr. Opin. Gastroenterol. 20: 146-155.

Felley, C. and P. Michetti. 2003. Probiotics and *Helicobacter pylori*. Best Pract. Res. Clin. Gastroenterol. 17: 785-791.

Forbes, G.M., M.E. Glaser, D.J. Cullen, J.R. Warren, K.J. Christiansen, B.J. Marshall et al. 1994. Duodenal ulcer treated with *Helicobacter pylori* eradication: seven-year follow-up. Lancet. 343: 258-260.

Futaki, N., K. Yoshikawa, Y. Hamasaka, I. Arai, S. Higuchi, H. Iizuka et al. 1993. NS-398, a novel non-steroidal anti-inflammatory drug with potent analgesic and antipyretic effects, which causes minimal stomach lesions. Gen. Pharmacol. 24: 105-110.

Gabriel, S.E., L. Jaakkimainen and C. Bombardier. 1991. Risk for serious gastrointestinal complications related to use of nonsteroidal anti-inflammatory drugs. A meta-analysis. Ann. Intern. Med. 115: 787-796.

Galvez, A., H. Abriouel, R.L. Lopez and N. Ben Omar. 2007. Bacteriocin-based strategies for food biopreservation. Int. J. Food Microbiol. 120: 51-70.

Garrick, T., F.W. Leung, S. Buack, K. Hirabayashi and P.H. Guth. 1986. Gastric motility is stimulated but overall blood flow is unaffected during cold restraint in the rat. Gastroenterol. 91: 141-148.

Gobert, A.P., B.D. Mersey, Y. Cheng, D.R. Blumberg, J.C. Newton, K.T. Wilson et al. 2002. Cutting edge: urease release by *Helicobacter pylori* stimulates macrophage inducible nitric oxide synthase. J. Immunol. 168: 6002-6006.

Graham, D.Y., E. Adam, G.T. Reddy, J.P. Agarwal, R. Agarwal, D.J. Evans et al. 1991. Seroepidemiology of *Helicobacter pylori* infection in India. Comparison of developing and developed countries. Dig. Dis. Sci. 36: 1084-1088.

Granato, D., G.E. Bergonzelli, R.D. Pridmore, L. Marvin, M. Rouvet, I.E. Corthesy-Theulaz et al. 2004. Cell surface-associated elongation factor Tu mediates the attachment of *Lactobacillus johnsonii* NCC533 (La1) to human intestinal cells and mucins. Infect. Immun. 72: 2160-2169.

Hagiwara, M., K. Kataoka, H. Arimochi, T. Kuwahara and Y. Ohnishi. 2004. Role of unbalanced growth of gram-negative bacteria in ileal ulcer formation in rats treated with a nonsteroidal anti-inflammatory drug. J. Med. Invest. 51: 43-51.

Hamilton-Miller, J. M. 2003. The role of probiotics in the treatment and prevention of *Helicobacter pylori* infection. Int. J. Antimicrob. Agents. 22: 360-366.

Higuchi, K., E. Umegaki, T. Watanabe, Y. Yoda, E. Morita, M. Murano et al. 2009. Present status and strategy of NSAIDs-induced small bowel injury. J. Gastroenterol. 44: 879-888.

Hussain, M.A., F. Kauser, A.A. Khan, S. Tiwari, C.M. Habibullah, N. Ahmed et al. 2004. Implications of molecular genotyping of *Helicobacter pylori* isolates from different human populations by genomic fingerprinting of enterobacterial repetitive intergenic consensus regions for strain identification and geographic evolution. J. Clin. Microbiol. 42: 2372-2378.

Jack, R.W., J.R. Tagg and B. Ray. 1995. Bacteriocins of gram-positive bacteria. Microbiol. Rev. 59: 171-200.

Jijon, H., J. Backer, H. Diaz, H. Yeung, D. Thiel, C. McKaigney et al. 2004 . DNA from probiotic bacteria modulates murine and human epithelial and immune function. Gastroenterol. 126: 1358-1373.

Kabir, A.M., Y. Aiba, A. Takagi, S. Kamiya, T. Miwa and Y. Koga et al. 1997. Prevention of Helicobacter pylori infection by lactobacilli in a gnotobiotic murine model. Gut. 41: 49-55.

Kadowaki, N., S. Ho, S. Antonenko, R.W. Malefyt, R.A. Kastelein, F. Bazan et al. 2001. Subsets of human dendritic cell precursors express different toll-like receptors and respond to different microbial antigens. J. Exp. Med. 194: 863-869.

Kao, L., T.H. Liu, T.Y. Tsai and T.M. Pan. 2018. Beneficial effects of the commercial lactic acid bacteria product, Vigiis 101, on gastric mucosa and intestinal bacterial flora in rats. J. Microbiol. Immunol. Infect. 53: 266-273.

Kavermann, H., B.P. Burns, K. Angermuller, S. Odenbreit, W. Fischer, K. Melchers et al. 2003. Identification and characterization of *Helicobacter pylori* genes essential for gastric colonization. J. Exp. Med. 197: 813-822.

Kersulyte, D., A.K. Mukhopadhyay, B. Velapatino, W. Su, Z. Pan, C. Garcia et al. 2000. Differences in genotypes of *Helicobacter pylori* from different human populations. J. Bacteriol. 182: 3210-3218.

Kleanthous, H., G.A. Myers, K.M. Georgakopoulos, T.J. Tibbitts, J.W. Ingrassia, H.L. Gray et al. 1998. Rectal and intranasal immunizations with recombinant urease induce distinct local and serum immune responses in mice and protect against *Helicobacter pylori* infection. Infect. Immunol. 66: 2879-2886.

Konturek, S.J., T. Brzozowski, I. Piastucki, T. Radecki, A. Dembinski, A. Dembinska-Kiec et al. 1982. Role of locally generated prostaglandins in adaptive gastric cytoprotection. Dig. Dis. Sci. 27: 967-971.

Kulkarni, N. and B.S. Reddy. 1994. Inhibitory effect of Bifidobacterium longum cultures on the azoxymethane-induced aberrant crypt foci formation and fecal bacterial beta-glucuronidase. Proc. Soc. Exp. Biol. Med. 207: 278-283.

Kusters, J.G., A.H. van Vliet and E.J. Kuipers. 2006. Pathogenesis of *Helicobacter pylori* infection. Clin. Microbiol. Rev. 19: 449-490.

Laine, L., R. Sloane, M. Ferretti and F. Cominelli. 1995. A randomized double-blind comparison of placebo, etodolac, and naproxen on gastrointestinal injury and prostaglandin production. Gastrointest. Endosc. 42: 428-433.

Lammers, K.M., U. Helwig, E. Swennen, F. Rizzello, A. Venturi, E. Caramelli et al. 2002. Effect of probiotic strains on interleukin 8 production by HT29/19A cells. Am. J. Gastroenterol. 97: 1182-1186.

Lanas, A. and F.K.L. Chan. 2017. Peptic ulcer disease. Lancet. 390: 613-624.

Lanas, A. and C. Scarpignato. 2006. Microbial flora in NSAID-induced intestinal damage: A role for antibiotics. Digestion. 73 Suppl 1: 136-150.

Lavermicocca, P., F. Valerio, A. Evidente, S. Lazzaroni, A. Corsetti, M. Gobbetti et al. 2000. Purification and characterization of novel antifungal compounds from the sourdough *Lactobacillus plantarum* strain 21B. Appl. Environ. Microbiol. 66: 4084-4090.

Lichtenberger, L.M., Y. Zhou, E.J. Dial and R.M. Raphael. 2006. NSAID injury to the gastrointestinal tract: Evidence that NSAIDs interact with phospholipids to weaken the hydrophobic surface barrier and induce the formation of unstable pores in membranes. J. Pharm. Pharmacol. 58: 1421-1428.

Llewellyn, A. and A. Foey. 2017. Probiotic modulation of innate cell pathogen sensing and signaling events. Nutrients 9: 1156.

Lorca, G.L., T. Wadstrom, G.F. Valdez and A. Ljungh. 2001. *Lactobacillus acidophilus* autolysins inhibit *Helicobacter pylori in vitro*. Curr. Microbiol. 42: 39-44.

Magnusson, J. and J. Schnurer. 2001. *Lactobacillus coryniformis* subsp. coryniformis strain Si3 produces a broad-spectrum proteinaceous antifungal compound. Appl. Environ. Microbiol. 67: 1-5.

Malfertheiner, P., F. Megraud, C. O'Morain, A.P. Hungin, R. Jones, A. Axon et al. 2002. Current concepts in the management of *Helicobacter pylori* infection – The Maastricht 2-2000 Consensus Report. Aliment. Pharmacol. Ther. 16: 167-180.

Marshall, B.J. 1994. *Helicobacter pylori*. Am. J. Gastroenterol. 89: S116-S128.

Matsuzaki, T. 1998. Immunomodulation by treatment with *Lactobacillus casei* strain Shirota. Int. J. Food Microbiol. 41: 133-140.

Michetti, P. 1997. Vaccine against *Helicobacter pylori*: Fact or fiction? Gut 41: 728-730.

Midolo, P.D., J.R. Lambert, R. Hull, F. Luo and M.L. Grayson. 1995. *In vitro* inhibition of *Helicobacter pylori* NCTC 11637 by organic acids and lactic acid bacteria. J. Appl. Bacteriol. 79: 475-479.

Mukhopadhyay, A.K., D. Kersulyte, J.Y. Jeong, S. Datta, Y. Ito, A. Chowdhury et al. 2000. Distinctiveness of genotypes of *Helicobacter pylori* in Calcutta, India. J. Bacteriol. 182: 3219-3227.

Nishikawa, K., A. Yokota, Y. Mashita, M. Taniguchi and K. Takeuchi. 2006. Oral but not parenteral aspirin upregulates COX-2 expression in rat stomachs: A causal relationship between COX-2 expression and prostaglandin deficiency. Gastroenterology 130: A407.

Nogueira, C., C. Figueiredo, F. Carneiro, A.T. Gomes, R. Barreira, P. Figueira et al. 2001. *Helicobacter pylori* genotypes may determine gastric histopathology. Am. J. Pathol. 158: 647-654.

O'Mahony, L., J. McCarthy, P. Kelly, G. Hurley, F. Luo, K. Chen et al. 2005. *Lactobacillus* and *bifidobacterium* in irritable bowel syndrome: Symptom responses and relationship to cytokine profiles. Gastroenterology 128: 541-551.

Peskar, B.M. 2001. Role of cyclooxygenase isoforms in gastric mucosal defence. J. Physiol. Paris 95: 3-9.

Prema, P., D. Smila, A. Palavesam and G. Immanuel. 2010. Production and characterization of an antifungal compound (3-Phenyllactic Acid) produced by *Lactobacillus plantarum* strain. Food Bioproc. Technol. 3: 379-386.

Pretzer, G., J. Snel, D. Molenaar, A. Wiersma, P.A. Bron, J. Lambert et al. 2005. Biodiversity-based identification and functional characterization of the mannose-specific adhesin of *Lactobacillus plantarum*. J. Bacteriol. 187: 6128-6136.

Rafii, F., J.B. Sutherland and C.E. Cerniglia. 2008. Effects of treatment with antimicrobial agents on the human colonic microflora. Ther. Clin. Risk Manag. 4: 1343-1358.

Resta-Lenert, S. and K.E. Barrett. 2003. Live probiotics protect intestinal epithelial cells from the effects of infection with enteroinvasive *Escherichia coli* (EIEC). Gut 52: 988-997.

Reuter, G. 2001. Probiotics – possibilities and limitations of their application in food, animal feed, and in pharmaceutical preparations for men and animals. Berl. Munch Tierarzt. Wochenschr. 114: 410-419.

Robert, A., J.E. Nezamis, C. Lancaster and A.J. Hanchar. 1979. Cytoprotection by prostaglandins in rats. Prevention of gastric necrosis produced by alcohol, HCl, NaOH, hypertonic NaCl, and thermal injury. Gastroenterology. 77: 433-443.

Scarpignato, C. 2008. NSAID-induced intestinal damage: Are luminal bacteria the therapeutic target? Gut 57: 145-148.

Schaible, U.E. and S.H. Kaufmann. 2004. Iron and microbial infection. Nat. Rev. Microbiol. 2: 946-953.

Schaible, U.E. and S.H. Kaufmann. 2005. A nutritive view on the host-pathogen interplay. Trends Microbiol. 13: 373-380.

Schiffrin, E.J., D. Brassart, A.L. Servin, F. Rochat and A. Donnet-Hughes. 1997. Immune modulation of blood leukocytes in humans by lactic acid bacteria: Criteria for strain selection. Am. J. Clin. Nutr. 66: 515S-520S.

Schillinger, U. and F.K. Lucke. 1989. Antibacterial activity of *Lactobacillus sake* isolated from meat. Appl. Environ. Microbiol. 55: 1901-1906.

Sgouras, D.N., T.T. Trang and Y. Yamaoka. 2015. Pathogenesis of *Helicobacter pylori* infection. Helicobacter 20(Suppl 1): 8-16.

Shibayama, K., Y. Doi, N. Shibata, T. Yagi, T. Nada, Y. Iinuma et al. 2001. Apoptotic signaling pathway activated by *Helicobacter pylori* infection and increase of apoptosis-inducing activity under serum-starved conditions. Infect Immun. 69: 3181-3189.

Simons, L.A., S.G. Amansec and P. Conway. 2006. Effect of *Lactobacillus fermentum* on serum lipids in subjects with elevated serum cholesterol. Nutr. Metab. Cardiovasc. Dis. 16: 531-535.

Steidler, L. 2003. Genetically engineered probiotics. Best. Pract. Res. Clin. Gastroenterol. 17: 861-876.

Suganuma, M., K. Yamaguchi, Y. Ono, H. Matsumoto, T. Hayashi, T. Ogawa et al. 2008. TNF-alpha-inducing protein, a carcinogenic factor secreted from *H. pylori*, enters gastric cancer cells. Int. J. Cancer 123: 117-122.

Szajewska, H., M. Setty, J. Mrukowicz and S. Guandalini. 2006. Probiotics in gastrointestinal diseases in children: Hard and not-so-hard evidence of efficacy. J. Pediatr. Gastroenterol. Nutr. 42: 454-475.

Takeuchi, K. 2012. Pathogenesis of NSAID-induced gastric damage: Importance of cyclooxygenase inhibition and gastric hypermotility. World J. Gastroenterol. 18: 2147-2160.

Takeuchi, K., A. Tanaka, S. Kato, K. Amagase and H. Satoh. 2010. Roles of COX inhibition in pathogenesis of NSAID-induced small intestinal damage. Clin. Chim. Acta. 411: 459-466.

Tanaka, A., S. Hase, T. Miyazawa and K. Takeuchi. 2002. Up-regulation of cyclooxygenase-2 by inhibition of cyclooxygenase-1: A key to nonsteroidal anti-inflammatory drug-induced intestinal damage. J. Pharmacol. Exp. Ther. 300: 754-761.

Tarnawski, A., A. Ahluwalia and M.K. Jones. 2013. Gastric cytoprotection beyond prostaglandins: Cellular and molecular mechanisms of gastroprotective and ulcer healing actions of antacids. Curr. Pharm. Des. 19: 126-132.

Tsuji, S., N. Kawai, M. Tsujii, S. Kawano and M. Hori. 2003. Review article: Inflammation-related promotion of gastrointestinal carcinogenesis – A perigenetic pathway. Aliment Pharmacol. Ther. 18(Suppl 1): 82-89.

Tuohy, K.M., H.M. Probert, C.W. Smejkal and G.R. Gibson. 2003. Using probiotics and prebiotics to improve gut health. Drug Discov. Today 8: 692-700.

Ushiyama, A., K. Tanaka, Y. Aiba, T. Shiba, A. Takagi, T. Mine et al. 2003. *Lactobacillus gasseri* OLL2716 as a probiotic in clarithromycin-resistant *Helicobacter pylori* infection. J. Gastroenterol. Hepatol. 18: 986-991.

Van de Bovenkamp, J.H., J. Mahdavi, A.M. Korteland-Van Male, H.A. Buller, A.W. Einerhand, T. Boren et al. 2003. The MUC5AC glycoprotein is the primary receptor for *Helicobacter pylori* in the human stomach. Helicobacter 8: 521-532.

van Pijkeren, J.P., C. Canchaya, K.A. Ryan, Y. Li, M.J. Claesson, B. Sheil et al. 2006. Comparative and functional analysis of sortase-dependent proteins in the predicted secretome of *Lactobacillus salivarius* UCC118. Appl. Environ. Microbiol. 72: 4143-4153.

Velez, M.P., S.C. De Keersmaecker and J. Vanderleyden. 2007. Adherence factors of *Lactobacillus* in the human gastrointestinal tract. FEMS Microbiol. Lett. 276: 140-148.

Viala, J., C. Chaput, I.G. Boneca, A. Cardona, S.E. Girardin, A.P. Moran et al. 2004. Nod1 responds to peptidoglycan delivered by the *Helicobacter pylori* cag pathogenicity island. Nat. Immunol. 5: 1166-1174.

Wang, X.Q., P.D. Terry and H. Yan. 2009. Review of salt consumption and stomach cancer risk: Epidemiological and biological evidence. World J. Gastroenterol. 15: 2204-2213.

Watanabe, T., K. Higuchi, A. Kobata, H. Nishio, T. Tanigawa, M. Shiba et al. 2008. Non-steroidal anti-inflammatory drug-induced small intestinal damage is toll-like receptor 4 dependent. Gut 57: 181-187.

Watanabe, T., H. Nishio, T. Tanigawa, H. Yamagami, H. Okazaki, K. Watanabe et al. 2009. Probiotic *Lactobacillus casei* strain Shirota prevents indomethacin-induced small intestinal injury: Involvement of lactic acid. Am. J. Physiol. Gastrointest. Liver Physiol. 297: G506-G513.

Xing, J.Z., C. Clarke, L. Zhu and S. Gabos. 2005. Development of a microelectronic chip array for high-throughput genotyping of *Helicobacter* species and screening for antimicrobial resistance. J. Biomol. Screen 10: 235-245.

Yan, F. and D.B. Polk. 2002. Probiotic bacterium prevents cytokine-induced apoptosis in intestinal epithelial cells. J. Biol. Chem. 277: 50959-50965.

Research Progress on the Probiotic Function and Application of *Clostridium butyricum*

Wenxiu Zheng[1,2], Xian Xu[1], Qing Xu[1], He Huang[1] and Ling Jiang[3]*

[1] College of Pharmaceutical Sciences, Nanjing Tech University, Nanjing 210009, People's Republic of China
[2] School of Food and Biological Engineering, Hefei University of Technology, Hefei, 230009, People's Republic of China
[3] College of Food Science and Light Industry, Nanjing Tech University, Nanjing 210009, People's Republic of China

1. Introduction

Probiotics (directly-fed microbials) are defined as live microorganisms which are beneficial for the host animals when administered in adequate amounts (Pan et al., 2008b). They act by killing pathogenic microorganisms, stimulating the intestinal immune response, improving growth performance and nutrient digestibility, and promoting the balance of intestinal microorganisms (Yang et al., 2012). Probiotics have been considered as potential substitute for antibiotics in the prevention and treatment of animal diseases, reducing mortality and improving the growth performance of livestock and poultry (Sumon et al., 2016). *Clostridium butyricum (C. butyricum)*, is a butyrate-producing Gram-positive, obligately anaerobic and endospore-forming probiotic organism, which can provide nutrients for the regeneration and repair of the intestinal epithelium, as well as contributes to the regulation of a healthy intestinal micro-ecological environment (Duan et al., 2017). Some strains have been widely used as important probiotics for the prevention and treatment of both human and veterinary intestinal diseases in East Asian countries, such as Japan, Korea and China (Kamiya et al., 1997, Oka et al., 2018).

    C. butyricum is a strictly anaerobic spore-forming bacilli which is found in the soil and intestines of healthy animals and humans (Hosny et al., 2019,

*Corresponding author: jiangling@njtech.edu.cn

Miyaoka et al., 2019, Pan et al., 2008b). It can also be found in natural yoghurt, cheese and plant leaves. In China, *C. butyricum* was approved as a new feed additive in 2009, and later in 2015, *C. butyricum* was approved as a national second-class new veterinary drug. In Europe, the European Commission had issued several decisions on the approval of *C. butyricum* (strain FERM-BP 2789 and MIYAIRI588) as a novel ingredient and feed additive for use in food supplements on the European market, and stipulated its usage specifications and indicators. The detailed policies in regard to *C. butyricum* of China and Europe are shown in Table 1. Since it can produce endospores, butyrate as well as lactic acid and other short-chain fatty acids (SCFAs) in the colon, thus it is able to survive in media with low pH and relatively high bile concentrations (Zhang et al., 2013). Furthermore, due to it has already been used in food, medical, breeding and industrial segments, there is no doubt that it has a bright future for further development and utilization.

In this review, we illustrate the characteristics of *C. butyricum*, possible mechanisms of its probiotic function, and its utilization in food, medical, industrial and animal feed applications, as well as its future prospects.

Table 1: The polities of *C. butyricum* in China and Europe

Year	Area	Strain	Policies
2015	China	*C. butyricum*	As National second-class new veterinary drug
2009	China	*C. butyricum*	As a newly approved feed additive
2014	European Union	*C. butyricum* (CBM 588)	As a novel food ingredient
2014	European Union	*C. butyricum* (FERM BP-2789)	As a feed additive for turkeys for fattening and rearing turkeys for breeding
2013	European Union	*C. butyricum* (FERM BP-2789)	As a feed additive for chickens for fattening and minor avian species (excluding laying birds)
2011	European Union	*C. butyricum* (FERM-BP 2789)	As a feed additive for minor avian species except laying birds, weaned piglets and minor porcine species (weaned)
2009	European Union	*C. butyricum* MIYAIRI 588 (FERM-P 1467)	As a feed additive for chickens for fattening
2009	European Union	*C. butyricum* (FERM-BP 2789)	As a feed additive for chickens for fattening

2. Characteristics of *C. butyricum*

2.1 Morphological Characteristics

C. butyricum is an anaerobic Gram-positive bacilli. When observed under a microscope, the cell bodies are present as straight or slightly curved rods, with both ends blunt and the middle part slightly swollen with a diameter of 0.5-1.7 and a length of 2.4-7.6 μm. The cells can be single, in pairs, or partially filamentous. The body has flagella all around which give it motility. When growing on agar plates, the cells form small white or light grey irregular round colonies which are opaque, with a slightly blunt surface and the characteristic odor of butyrate.

2.2 Possible Mechanism of Probiotic Function

Although *C. butyricum* has been widely used for the treatment of intestinal diseases of humans and animals or as a feed additive to improve animals' immunity, the specific mechanisms of action by which *C. butyricum* exerts its health-promoting effect on the intestine is still obscure and has not yet been fully uncovered. Nevertheless, some scholars have found different possible mechanisms of action in different fields of study (Liu et al., 2017, Sun et al., 2016, Cassir et al., 2015, Wydau-Dematteis et al., 2015).

For example, Kuroiwa et al. (1990) demonstrated that *C. butyricum* promotes the growth of intestinal microbiota because it produces amylase which can hydrolyze starch to produce oligosaccharides, which in turn can be reused by beneficial intestinal bacteria such as *Bifidobacterium* spp., *Lactobacillus* spp. and *Enterococcus faecalis*, and as a consequence facilitating the digestion and absorption of protein and fat in the feed. In addition, *C. butyricum* cannot decompose protein so no harmful substances such as ammonia, hydrogen sulfide, and hydrazine can generate, and no symptoms of poisoning or pathological changes of organs are caused (Kuroiwa et al., 1990).

Furthermore, *C. butyricum* can produce large amounts of short-chain fatty acids such as butyrate, acetic, and lactic acid by fermentation of sugars, and these short-chain fatty acids can reduce the pH of the intestine to maintain an acidic environment, and thus prevent the abnormal proliferation of pathogenic- and spoilage bacteria in the intestine and reduce the enterotoxin burden. Furthermore, butyrate has a proliferative effect on mucosal cells in the intestine and has a certain therapeutic efficacy against inflammatory bowel disease (Okamoto et al., 2000). It has been reported that *C. butyricum* has obvious antagonistic inhibitory effects on harmful microbes such as *Candida albicans*, *Clostridium difficile*, enterotoxigenic *E. coli*, *Klebsiella* spp., *Salmonella* spp., *Vibrio* spp. and *Helicobacter pylori* (Takahashi et al., 2004].

In addition, *C. butyricum* can also produce hydrogen. H_2 metabolism reflects the balance between H_2-producing (hydrogenogenic) and H_2-utilizing (hydrogenotrophic) microbes, which has a primary influence on the final composition of colonic gases (Carbonero et al., 2012, Park et al., 2019). Molecular H_2 has beneficial effects in different systems as an optimal antioxidant agent by selectively scavenging free hydroxyl radicals (-OH), and it is an important physiological regulatory factor with antioxidant, anti-inflammatory, and antiapoptotic protective effects on cells and organs (Liao et al., 2015a). The particular production of *C. butyricum* and their probiotic function can be seen from the Table 2.

Table 2: The products of *C. butyricum* and their probiotic function

Production	Examples	Function
Enzymes	Cellulases Glycosidases Amylases Proteases Lysase Pectin Methylase	Help the body digest, provide necessary growth substances for intestinal beneficial microorganisms, and promote the growth and reproduction of beneficial microorganisms in the intestine
Vitamins	B vitamins Vitamin K Niacin	Improve feed utilization and animal production performance
Short chain Fatty acids	Butyrate Acetic acid Propionic acid Formic acid	Change the intestinal pH, inhibit the activity of some inflammatory cytokines, stimulate intestinal peristalsis, improve intestinal microenvironment, regulate microecological balance and treat related diseases
Amino acids		Promote the absorption of VE and supplement essential nutrients
Butyricin		Inhibits and kills certain *Clostridium* species

3. Applications of *C. butyricum*

3.1 In Food

Even though probiotics have a long history of applications in food fermentation, because these were traditional, empirical techniques, *C. butyricum* in foods was not discovered until recently. *C. butyricum* can be made into broth, capsules, granules and other health foods. The applications of *C. butyricum* in the food and beverage industries are widespread because the pure acid it produces can be applied in the dairy industry, or in the form of esters as a food additive to increase fruit-like flavors (He et al., 2005). *C. butyricum* is naturally present in cheese and the product butyrate promotes the formation of special flavors. The long-term screening of *C. butyricum* in jiuqu and cellar mud during the wine making process makes it highly resistant to adverse effects, and the fermentation products enrich the flavor of the wine. At the same time, its probiotic effect also inhibits contamination by harmful bacteria and ensures the quality of the wine. Nagase invented a *C. butyricum* chewing gum, in which live bacteria and spores were added to chewing gum or lozenges. When chewing, *C. butyricum* enters into the gaps of gingiva and proliferates with saliva, maintaining the balance of oral flora, making the mouth clean and odorless as well as preventing periodontal disease and dental caries (Nagase, 2002).

3.2 In Medicine

C. butyricum has great potential in medical applications. According to a previous study, the strain MIYAIRI 588, which displays a specific phenotype of *C. butyricum*, has been approved as a probiotic for human clinical use in Japan since 1968 (Kato et al., 2018, Takahashi et al., 2004). It has been investigated in the prevention and treatment of inflammatory bowel disease (IBD) (Kong et al., 2011, Sun et al., 2018) and non-antimicrobial-induced diarrhea as well as antimicrobial-associated diarrhea in humans and animals (Seki et al., 2013, Pan et al., 2008b, Peeters et al., 2019). Oral administration of *C. butyricum* MIYAIRI 588 may enhance immunity in the respiratory organs of mice and the spores it forms can germinate and grow in the intestinal tract of mice (Murayama et al., 1995). Takahashi et al. (2004) proved that *C. butyricum* MIYAIRI 588 has prophylactic and therapeutic effects against enterohemorrhagic *Escherichia coli* O157:H7 infection in gnotobiotic mice (Takahashi et al., 2004). In another study, Seo et al. (2013) suggested that *C. butyricum* MIYAIRI 588 successfully improves non-alcoholic fatty liver disease and it may have a potential for decreasing lipid accumulation in the liver (Seo et al., 2013). In another study, the researchers examined the safety and efficacy of *C. butyricum* MIYAIRI treatment in the prevention of pouchitis, which suggested that the probiotic was beneficial for preventing it in patients with ulcerative colitis (UC) who have undergone ileal pouch anal anastomosis (IPAA) (Yasueda et al., 2016).

C. *butyricum* CGMCC0313.1 (CB0313.1), another butyrate-producing strain, is also used as a probiotic to treat and prevent non-antimicrobial-induced diarrhea and irritable bowel syndrome (Shang et al., 2016) and it has good effects on maintaining intestinal mucosa as well as mitigating gut-associated ailments like ulcerative colitis and Crohn's disease in clinical practice (Jia et al., 2017, Zhang et al., 2009). Shang et al. (2016) have indicated that CB0313.1 is a promising probiotic agent for the treatment of obesity and associated metabolic disorders which can counteract the detrimental effects of a high fat diet (HFD) on body weight gain and insulin resistance, and helps maintain gut immunity and barrier homeostasis, thereby contributing to the modulation of systemic adipose inflammation (Shang et al., 2016). Jia et al. (2017) reported a novel use of CB0313.1 in the prevention and treatment of hyperglycemia and associated metabolic dysfunctions and delineated the potential mechanism of mitigating hyperglycemia in diabetic host mice (Jia et al., 2017). The same year, Zhang et al. (2017) investigated the preventive and therapeutic effects of CB0313.1 on food allergies for the first time. They found that CB0313.1 has anti-inflammatory and anti-anaphylactic activity, which can significantly modulate the mucosal immune response, with positive effects on patients with food allergies (Zhang et al., 2017).

In addition to the protective effects on the digestive tract, other *C. butyricum* strains also have shown efficacy against non-alcoholic, high-fat diet-induced liver disease (NAFLD) and acute liver injury (ALI) (Liu et al., 2017, Zhou et al., 2017). Furthermore, *C. butyricum* was effective in the treatment of vascular dementia (VaD) by regulating the gut-brain axis, which is considered as a new therapeutic strategy against VaD (Liu et al., 2015).

C. *butyricum* is also used in combination with other strains and treatments in addition to isolated utilization. Live *C. butyricum* and

Bifidobacterium sp. combined powder (LCBBCP) is a new mixed intestinal probiotic independently developed in China, mainly used for the clinical treatment of premature infants with diarrhea, indigestion, constipation and other diseases (Ling et al., 2015). Zhang et al. (2016b) found that LCBBCP can help to boost the development and maturation of immunity in late preterm infants by improving the balance of $CD4^+/CD8^+$ T cells to drastically increase the percentage of $CD4^+$ T cells. Moreover, excessive proliferation and activation was prevented at the same time by maintaining high B and T lymphocyte attenuator (BTLA) expression on the immune cells' surface (Zhang et al., 2016b). Shi et al. (2015) reported that *C. butyricum* can markedly increase the efficacy of specific immunotherapy (SIT) against antigen specific allergic inflammations in the intestine. The effect is attributed to the butyrate derived from *C. butyricum* regulating the signal transduction pathway of IgE production in antigen-specific B cells and inducing IL-10 expression (Shi et al., 2015). Liao et al. (2016) also used the same approach and discovered that it can dramatically enhance the therapeutic effect on asthma by inducing antigen-specific regulatory B-cells (Liao et al., 2016).

3.3 In Chemical Industry

C. butyricum is also widely used in industry to produce biohydrogen and organic solvents such as ethanol, acetone, butyrate, butanol and 1,3-propanediol (Bruce et al., 2016, Dolejš et al., 2019). It has been reported that some pure cultures of *Clostridium* sp. can use carbohydrates to produce hydrogen. Especially *C. butyricum* was found to produce the highest H_2 yields per mole of hexose via dark fermentation, which is a promising method for hydrogen production due to its convenience, high efficiency and low cost (Hawkes et al., 2002, Junghare et al., 2012, Ortigueira et al., 2019). Rafieenia and Chaganti studied the metabolic flux distribution in the metabolism of monosaccharides and disaccharides via metabolic pathway analysis in *C. butyricum*, and their results showed that disaccharides (trehalose and sucrose) were better carbon sources for H_2 production by *C. butyricum* than monosaccharides (Rafieenia and Chaganti, 2015). In addition, *C. butyricum* is also an effective H_2 producer in co-cultures with other strains (Aly et al., 2018, Laurinavichene and Tsygankov, 2015). Finally, the butyrate produced by *C. butyricum* has many potential applications in industry. *C. butyricum* can ferment many different and readily available carbon sources such as glucose, lactose from whey, sucrose from molasses as well as starch, potato wastes, wheat flour, cellulose or dextrose to produce butyrate (He et al., 2005). The metabolic pathway for hexose and pentose fermentation by *C. butyricum* is shown in Figure 1, and we can clearly know the way *C. butyricum helps* metabolism. *C. butyricum* has several end products containing acetate, butyrate, lactate, CO_2 and H_2 and the metabolic pathway from hexose (like glucose) to pyruvate is different from the pathway from pentose (like xylose) to pyruvate in that hexose is metabolized to pyruvate via the Embden-Meyerhof-Parnas (EMP) pathway while pentoses are metabolized by the pentose-phosphate pathway when they are transported into the cytoplasm, and the detailed explanations of the metabolic pathways and reactions can be found in a previous study (Zhu and Yang, 2004).

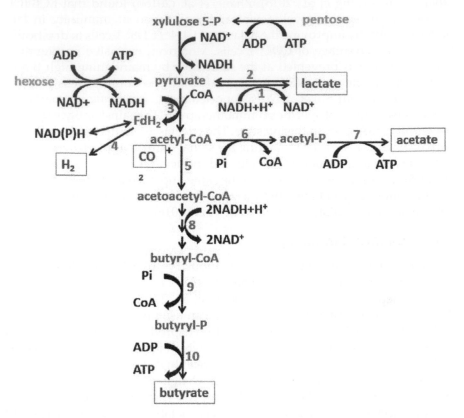

Fig. 1: The metabolic pathway for hexose and pentose fermentation by *C. butyricum*. The number indicates enzymes as follows: (1) lactate dehydrogenase; (2) NAD-independent lactate dehydrogenase; (3) pyruvate-ferredoxin oxido-reductase; (4) hydrogenase; (5) acetyl CoA acetyltransferase (or thiolase); (6) phosphotransacetylase; (7) acetate kinase; (8) three enzymes producing butyryl-CoA; (9) phosphotransbutyrylase; (10) butyrate kinase

3.4 In Animal Feed

As a potential substitute for antibiotics, the application of probiotics in the poultry industry is rapidly being accepted by people (Dawood and Koshio, 2016). *C. butyricum* is added to animal feeds as live microbial feed supplements which can improve the animals' intestinal microbial balance, control specific enteric pathogens, and thereby promote the vitality, egg production and immunity, as well as reduce the feed ratio. Among the utilized strains, *C. butyricum* MIYAIRI 588 is not only applied in the medical field, but also currently authorized in the European Union as a feed additive for turkeys, chickens and related minor avian species, weaned piglets, and as a novel ingredient for use in food supplements (Isa et al., 2016).

In poultry farming, several studies investigated the effects of *C. butyricum* on broiler chickens. On one hand, *C. butyricum* improves the growth performance, benefits the immune function and the balance of the intestinal microflora in broiler chickens (Yang et al., 2012). On the other hand, when broiler chickens are challenged with *E. coli* K88, it can reverse low body weight (BW) and average day gain (ADG), decreasing intestinal barrier functions and digestive enzyme activities and promoting the immune response (Zhang et al., 2013, Zhang et al., 2016a). Furthermore, *C. butyricum* might increase the concentration of polyunsaturated fatty acids (PUFAs) in meat by improving the physiological antioxidant activity in broiler chicks (Liao et al., 2015b).

C. butyricum can be administered not only as a feed supplement but also as an additive to water in aquaculture, which has been proved to be effective in a wide range of species for the promotion of growth, enhanced nutrition, feed efficiency, digestibility, immunity and survival rate (De et al., 2014). This administration route is possible because *C. butyricum* has a high adaptive capacity to low pH and high temperature environments as well as resistance to many antibiotics (Duan et al., 2017). Additionally, live *C. butyricum* and lipoteichoic acid extracted from *C. butyricum* has been shown to be effective on suppressing the growth of pathogens (Gao et al., 2011, Gao et al., 2012). It has been reported that *C. butyricum* has preventive and therapeutic effects against pathogen infection in farmed fish (Pan et al., 2008b, Gao et al., 2013), and *C. butyricum*-incorporating diets were found to be beneficial for *Macrobrachium rosenbergii* cultures in terms of hindering the growth of pathogenic bacteria and increasing the growth rate as well as the protease and amylase activities of the cultured prawn (Sumon et al., 2016). Interestingly, even heat-killed *C. butyricum* retained notable immunomodulating properties in *Miichthys miiuy* (Pan et al., 2008a). The results of these studies will help the improvement and development of more environmentally friendly aquacultures.

4. Conclusions and Future Prospects

In recent years, much attention has been paid to microecological agents such as *Bifidobacterium* spp., *Lactobacillus* spp., and *Saccharomyces* spp., which have been widely used in aquaculture and achieved great economic and social benefits. Currently, the research on *C. butyricum* is in the initial stages in our country, but it has enormous prospects for development and application. *C. butyricum* as a novel probiotic is green, safe, non-toxic, non-polluting, residue-free and has no side effects. Moreover, the probiotic agent stores easily and the number of live bacteria is much more stable than in non-sporulating probiotics. Due to its unique morphological and biochemical characteristics, *C. butyricum* has become a promising animal health product and antibiotic substitute, which can also be used in combination with antibiotics. The metabolism of *C. butyricum* can produce a variety of substances such as enzymes, vitamins, fatty acids and the bacteriocin butyricin. As a health food supplement and drug, it can adjust the intestinal microecological balance, reduce the incidence of intestinal inflammation and improve immunity. As a feed additive, it can promote animal growth, enhance animal immunity, increase the feed conversion rate, maintain the intestinal microecological environment and treat intestinal diseases. If used in industrial fermentation, it can produce hydrogen, butyrate and 1,3-propylene glycol, which promote the development of the green industry.

However, some issues are also worth noting. In animal feed applications, in order to ensure economic benefits, to increase the product quality and reduce the use of antibiotics, the activity of *C. butyricum* in the gastrointestinal tract should be further improved and its adaptation to a variety of external conditions in the feed production process should be strengthened. The engineering of tolerance to the conditions in the feed preparation process should pay attention to symbiotic fermentation with other bacteria in order to improve feed palatability, nutrition, conversion rate and cost reduction. At the same time, the synergy of *C. butyricum* and other drugs requires further study in order to effectively treat animal diseases and reduce mortality. Furthermore, different doses of *C. butyricum* probiotic are added for different animals in different growth stages to ensure the greatest positive effect in animal husbandry, poultry industry, and fishery. In terms of clinical applications, *C. butyricum* and its metabolites have important physiological functions, and exert various effects on the health of humans and animals. However, many of these beneficial effects on health are not decisive. The mechanisms of action and probiotic mechanisms of metabolites need to be studied further, especially in regard to maintaining microecological balance, treating inflammation and cancer, and providing the necessary theoretical basis for the application of *C. butyricum* probiotics. Furthermore, different probiotics have their own effects. *C. butyricum* has a strong resistance to various antibiotics, and its synergy with other probiotics and antibiotics in vitro or in vivo remains to be discovered. In addition, other effects of *C. butyricum* and its combination with new technologies also need more research.

C. butyricum has strong adaptability and good stability which meets the expectations of the consumers. It is a promising probiotic product and will become a very valuable species with good prospects for development in the future. The further development of feed additives such as *C. butyricum* and its metabolites will help reduce the current abuse of antibiotic products in feed, reduce the residues of drugs in meat products, reduce the resistance of animal-borne bacteria, and protect the health of animals. As a result, it will have a significant impact on green and healthy animal husbandry and the safe production of animal foods. Therefore, its application value and prospects require further exploration.

Acknowledgements

This work was supported by the National Science Foundation of China (31922070, U1603112), and the Natural Science Foundation of Jiangsu Province (BK20180038, BK20171461), the Six Talent Peaks Project in Jiangsu Province (2015-JY-009), and the Jiangsu Synergetic Innovation Center for Advanced Bio-Manufacture (XTE1838).

References

Aly, S.S., T. Imai, M.S. Hassouna, D.M.K. Nguyen, T. Nguyen, A. Kanno et al. 2018. Identification of factors that accelerate hydrogen production by *Clostridium butyricum* RAK25832 using casamino acids as a nitrogen source. Int. J. Hydrogen Eng. 43(10): 5300-5313.

Bruce, T., F.G. Leite, M. Miranda, C.C. Thompson, N. Pereira, M. Faber et al. 2016. Insights from genome of *Clostridium butyricum* INCQS635 reveal mechanisms to convert complex sugars for biofuel production. Arch. Microbiol. 198(2): 115-127.

Carbonero, F., A.C. Benefiel and H.R. Gaskins. 2012. Contributions of the microbial hydrogen economy to colonic homeostasis. Nat. Rev. Gastro. Hepat. 9(9): 504-518.

Cassir, N., S. Benamar, J.B. Khalil, O. Croce, M. Saint-Faust, A. Jacquot et al. 2015. *Clostridium butyricum* strains and dysbiosis linked to necrotizing enterocolitis in preterm neonates. Clin. Infect. Dis. 61(7): 1107-1115.

Dawood, M.A. and S. Koshio. 2016. Recent advances in the role of probiotics and prebiotics in carp aquaculture: A review. Aquaculture 454: 243-251.

De, B.C., D.K. Meena, B.K. Behera, P. Das, P.D. Mohapatra, A. P. Sharma et al. 2014. Probiotics in fish and shellfish culture: Immunomodulatory and ecophysiological responses. Fish Physiol. Biochem. 40(3): 921-971.

Dolejš, I., M. Líšková, V. Líšková, K. Markošová, M. Markošová, F. Markošová et al. 2019. Production of 1,3-Propanediol from pure and crude glycerol using immobilized *Clostridium butyricum*. Catalysts 9(4): 317.

Duan, Y., Y. Zhang, H. Dong, Y. Wang and J. Zhang. 2017. Effect of the dietary probiotic *Clostridium butyricum* on growth, intestine antioxidant capacity and resistance to high temperature stress in kuruma shrimp *Marsupenaeus japonicus*. J. Therm. Biol. 66: 93-100.

Gao, Q.X., T.X. Wu, J.B. Wang and Q.C. Zhuang. 2011. Inhibition of bacterial adhesion to HT-29 cells by lipoteichoic acid extracted from *Clostridium butyricum*. Afr. J. Biotechnol. 10(39): 7633-7639.

Gao, Q., L. Qi, T. Wu and J. Wang. 2012. Ability of *Clostridium butyricum* to inhibit *Escherichia coli*-induced apoptosis in chicken embryo intestinal cells. Vet. Microbiol. 160(3-4): 395-402.

Gao, Q., Y. Xiao, P. Sun, S. Peng, F. Yin, X. Ma et al. 2013. *In vitro* protective efficacy of *Clostridium butyricum* against fish pathogen infections. Indian J. Microbiol. 53(4): 453-459.

Hawkes, F.R., R. Dinsdale, D.L. Hawkes and I. Hussy. 2002. Sustainable fermentative hydrogen production: Challenges for process optimization. Int. J. Hydrogen Eng. 27(11-12): 1339-1347.

He, G.Q., Q. Kong, Q.H. Chen and H. Ruan. 2005. Batch and fed-batch production of butyric acid by *Clostridium butyricum* ZJUCB. J. Zhejiang Univ-SC B. 6(11): 1076-1080.

Hosny, M., J.Y.B. Khalil, A. Caputo, R.A. Abdallah, A. Levasseur et al. 2019. Multidisciplinary evaluation of *Clostridium butyricum* clonality isolated from preterm neonates with necrotizing enterocolitis in South France between 2009 and 2017. Sci. Rep-UK 9(1): 2077.

Isa, K., K. Oka, N. Beauchamp, M. Sato, K. Wada, K. Ohtani et al. 2016. Safety assessment of the *Clostridium butyricum* MIYAIRI 588® probiotic strain including evaluation of antimicrobial sensitivity and presence of *Clostridium toxin* genes *in vitro* and teratogenicity *in vivo*. Hum. Exp. Toxicol. 35(8): 818-832.

Jia, L., D. Li, N. Feng, M. Shamoon, Z. Sun, L. Ding et al. 2017. Anti-diabetic effects of *Clostridium butyricum* CGMCC0313.1 through promoting the growth of gut butyrate-producing bacteria in type 2 diabetic mice. Sci. Rep-UK. 7(1): 7046.

Junghare, M., S. Subudhi and B. Lal. 2012. Improvement of hydrogen production under decreased partial pressure by newly isolated alkaline tolerant anaerobe, *Clostridium butyricum* TM-9A: Optimization of process parameters. Int. J. Hydrogen Eng. 37(4): 3160-3168.

Kamiya, S., H. Taguchi, H. Yamaguchi, T. Osaki, M. Takahashi and S. Nakamura et al. 1997. Bacterioprophylaxis using *Clostridium butyricum* for lethal caecitis by *Clostridium difficile* in gnotobiotic mice. Rev. Med. Microbiol. 8: S57-S59.

Kato, M., Y. Hamazaki, S. Sun, Y. Nishikawa and E. Kage-Nakadai. 2018. *Clostridium butyricum* MIYAIRI 588 increases the lifespan and multiple-stress resistance of *Caenorhabditis elegans*. Nutrients 10(12): 1921.

Kong, Q., G.Q. He, J.L. Jia, Q.L. Zhu and H. Ruan. 2011. Oral administration of *Clostridium butyricum* for modulating gastrointestinal microflora in mice. Curr. Microbiol. 62(2): 512-517.

Kuroiwa, T., M. Iwanaga, K. Kobari, A. Higashionna, F. Kinjyo, A. Saito et al. 1990. Preventive effect of *Clostridium butyricum* M588 against the proliferation of *Clostridium difficile* during antimicrobial therapy. *Kansenshogaku zasshi*. J. Jpn. Assoc. Infect. Dis. 64(11): 1425-1432.

Laurinavichene, T. and A. Tsygankov. 2015. Hydrogen photoproduction by co-culture *Clostridium butyricum* and *Rhodobacter sphaeroides*. Int. J. Hydrogen Eng. 40(41): 14116-14123.

Liao, H.Y., L. Tao, J. Zhao, J. Qin, G.C. Zeng, S.W. Cai et al. 2016. *Clostridium butyricum* in combination with specific immunotherapy converts antigen-specific B cells to regulatory B cells in asthmatic patients. Sci. Rep-UK. 6: 20481.

Liao, X.D., G. Ma, J. Cai, Y. Fu, X.Y. Yan. 2015a. Effects of *Clostridium butyricum* on growth performance, antioxidation, and immune function of broilers. Poultry Sci. 94(4): 662-667.

Liao, X., R. Wu, G. Ma, L. Zhao, Z. Zheng, R. Zhang et al. 2015b. Effects of *Clostridium butyricum* on antioxidant properties, meat quality and fatty acid composition of broiler birds. Lipids Health Dis. 14(1): 36-45.

Ling, Z., X. Liu, Y. Cheng, Y. Luo, L. Yuan, L. Li et al. 2015. *Clostridium butyricum* combined with *Bifidobacterium* infantis probiotic mixture restores fecal microbiota and attenuates systemic inflammation in mice with antibiotic-associated diarrhea. BioMed Res. Int. 2015(2015): 582048.

Liu, J., Y. Fu, H. Zhang, J. Wang, J. Zhu, Y. Wang et al. 2017. The hepatoprotective effect of the probiotic *Clostridium butyricum* against carbon tetrachloride-induced acute liver damage in mice. Food Funct. 8(11): 4042-4052.

Liu, J., J. Sun, F. Wang, X. Yu, Z. Ling, H. Li et al. 2015. Neuroprotective effects of *Clostridium butyricum* against vascular dementia in mice via metabolic butyrate. BioMed Res. Int. 2015(2015): 412946.

Miyaoka, T., M. Kanayama, R. Wake, S. Hashioka, M. Hayashida, M. Nagahama et al. 2018. *Clostridium butyricum* MIYAIRI 588 as adjunctive therapy for treatment-resistant major depressive disorder: A prospective open-label trial. Clin. Neuropharmacol. 41(5): 151-155.

Murayama, T.I., N. Mita, M. Tanaka, T. Kitajo, T. Asano, K. Mizuochi et al. 1995. Effects of orally administered *Clostridium butyricum* MIYAIRI 588 on mucosal immunity in mice. Vet. Immunol. Immunop. 48(3-4): 333-342.

Nagase, M. 2002. Chew Gum Including Butyric Acid Bacteria. Japan, JP20020383182. 2002-12-11.

Oka, K., T. Osaki, T. Hanawa, S. Kurata, E. Sugiyama, M. Takahashi et al. 2018. Establishment of an endogenous *Clostridium difficile* rat infection model and evaluation of the effects of *Clostridium butyricum* MIYAIRI 588 probiotic strain. Front. Microbiol. 9: 1264.

Okamoto, T., M. Sasaki, T. Tsujikawa, Y. Fujiyama, T. Bamba, M. Kusunoki et al. 2000. Preventive efficacy of butyrate enemas and oral administration of *Clostridium butyricum* M588 in dextran sodium sulfate-induced colitis in rats. J. Gastroenterol. 35(5): 341-346.

Ortigueira, J., L. Martins, M. Martins, C. Martins and P. Moura. 2019. Improving the non-sterile food waste bioconversion to hydrogen by microwave pretreatment and bioaugmentation with *Clostridium butyricum*. Waste Mang. 88: 226-235.

Pan, X., T. Wu, Z. Song, H. Tang and Z. Zhao. 2008a. Immune responses and enhanced disease resistance in Chinese drum, *Miichthys miiuy* (Basilewsky), after oral administration of live or dead cells of *Clostridium butyrium* CB2. J. Fish Dis. 31(9): 679-686.

Pan, X., T. Wu, L. Zhang, Z. Song, H. Tang, Z. Zhao et al. 2008b. *In vitro* evaluation on adherence and antimicrobial properties of a candidate probiotic *Clostridium butyricum* CB2 for farmed fish. J. Appl. Microbiol. 105(5): 1623-1629.

Park, J.H., D.H. Kim, H.S. Kim, G.F. Wells and H.D. Park. 2019. Granular activated carbon supplementation alters the metabolic flux of *Clostridium butyricum* for enhanced biohydrogen production. Biores. Technol. 281: 318-325.

Peeters, L., L. Mostin, P. Wattiau, F. Boyen, J. Dewulf, D. Maes et al. 2019. Efficacy of *Clostridium butyricum* as probiotic feed additive against experimental *Salmonella typhimurium* infection in pigs. Livest. Sci. 221: 82-85.

Rafieenia, R. and S.R. Chaganti. 2015. Flux balance analysis of different carbon source fermentation with hydrogen producing *Clostridium butyricum* using cell net analyzer. Biores. Technol. 175: 613-618.

Seki, H., M. Shiohara, T. Matsumura, N. Miyagawa, M. Tanaka, A. Komiyama et al. 2003. Prevention of antibiotic-associated diarrhea in children by *Clostridium butyricum* MIYAIRI. Pediatr. Int. 45(1): 86-90.

Seo, M., I. Inoue, M. Tanaka, N. Matsuda, T. Nakano, T. Awata et al. 2013. *Clostridium butyricum* MIYAIRI 588 improves high-fat diet-induced non-alcoholic fatty liver disease in rats. Digest. Dis. Sci. 58(12): 3534-3544.

Shang, H., J. Sun and Y.Q. Chen. 2016. *Clostridium butyricum* CGMCC0313.1 modulates lipid profile, insulin resistance and colon homeostasis in obese mice. PLoS One 11(4): e0154373.

Shi, Y., L.Z. Xu, K. Peng, W. Wu, R. Wu, Z.Q. Liu et al. 2015. Specific immunotherapy in combination with *Clostridium butyricum* inhibits allergic inflammation in the mouse intestine. Sci. Rep-UK. 5: 17651.

Sumon, M.S., F. Ahmmed, S.S. Khushi, M.K. Ahmmed, M.A. Rouf, M.A.H. Chisty et al. 2016. Growth performance, digestive enzyme activity and immune response of *Macrobrachium rosenbergii* fed with probiotic *Clostridium butyricum* incorporated diets. J. King Saud Univ. Sci. 30(1): 21-28.

Sun, J., F. Wang, Z. Ling, X. Yu, W. Chen, H. Li et al. 2016. *Clostridium butyricum* attenuates cerebral ischemia/reperfusion injury in diabetic mice via modulation of gut microbiota. Brain Res. 1642: 180-188.

Sun, Y.Y., M. Li, Y.Y. Li, L.X. Li, W.Z. Zhai, P. Wang et al. 2018. The effect of *Clostridium butyricum* on symptoms and fecal microbiota in diarrhea-dominant irritable bowel syndrome: A randomized, double-blind, placebo-controlled trial. Sci. Rep-UK. 8(1): 2964.

Takahashi, M., H. Taguchi, H. Yamaguchi, T. Osaki, A. Komatsu, S. Kamiya et al. 2004. The effect of probiotic treatment with *Clostridium butyricum* on enterohemorrhagic *Escherichia coli* O157: H7 infection in mice. FEMS Immunolo. Med. Mic. 41(3): 219-226.

Wydau-Dematteis, S., M. Louis, N. Zahr, R. Lai-Kuen, B. Saubaméa, M.J. Butel et al. 2015. The functional dlt operon of *Clostridium butyricum* controls the D-alanylation of cell wall components and influences cell septation and vancomycin-induced lysis. Anaerobe 35: 105-114.

Yang, C.M., G.T. Cao, P.R. Ferket, T.T. Liu, L. Zhou, L. Zhang et al. 2012. Effects of probiotic, *Clostridium butyricum*, on growth performance, immune function, and cecal microflora in broiler chickens. Poult. Sci. 91(9): 2121-2129.

Yasueda, A., T. Mizushima, R. Nezu, R. Sumi, M. Tanaka, J. Nishimura et al. 2016. The effect of *Clostridium butyricum* MIYAIRI on the prevention of pouchitis and alteration of the microbiota profile in patients with ulcerative colitis. Surg. Today 46(8): 939-949.

Zhang, H.Q., T.T. Ding, J.S. Zhao, X. Yang, H.X. Zhang et al. 2009. Therapeutic effects of *Clostridium butyricum* on experimental colitis induced by oxazolone in rats. World J. Gastroenterol. 15(15): 1821-1828.

Zhang, L., G.T. Cao, X.F. Zeng, L. Zhou, P.R. Ferket, Y.P. Xiao et al. 2013. Effects of *Clostridium butyricum* on growth performance, immune function, and cecal microflora in broiler chickens challenged with *Escherichia coli* K88. Poultry Sci. 93(1): 46-53.

Zhang, L., L. Zhang, X. Zeng, L. Zhou, G. Cao, C. Yang et al. 2016a. Effects of dietary supplementation of probiotic, *Clostridium butyricum*, on growth performance, immune response, intestinal barrier function, and digestive enzyme activity in broiler chickens challenged with *Escherichia coli* K88. J. Anim. Sci. Biotechnol. 7(1): 1-9.

Zhang, S.F., Z.S. Tang, L. Tong, X.X. Tao, Q.F. Suo, X.M. Xu et al. 2016b. Effects of *Clostridium butyricum* and *Bifidobacterium* on BTLA expression on CD4+ T cells and lymphocyte differentiation in late preterm infants. Microb. Pathogenesis. 100: 112-118.

Zhang, J., H. Su, Q. Li, H. Wu, M. Liu, J. Huang et al. 2017. Oral administration of *Clostridium butyricum* CGMCC0313.1 inhibits β-lactoglobulin-induced intestinal anaphylaxis in a mouse model of food allergy. Gut Pathog. 9(1): 11-21.

Zhou, D., Q. Pan, X.L. Liu, R.X. Yang, Y.W. Chen, C. Liu et al. 2017. *Clostridium butyricum* B1 alleviates high-fat diet-induced steatohepatitis in mice via enterohepatic immunoregulation. J. Gastroen. Hepatol. 32(9): 1640-1648.

Zhu, Y. and S.T. Yang. 2004. Effect of pH on metabolic pathway shift in fermentation of xylose by *Clostridium tyrobutyricum*. J. Biotechnol. 110(2): 143-157.

CHAPTER
5

Technology and Health Claim Evaluation of Probiotic Dairy Products

Nuray Yazihan[1,2] and Barbaros Özer[3*]

[1] Ankara University, Faculty of Medicine, Internal Medicine, Pathophysiology Department, Ankara, Turkey
[2] Ankara University, Institute of Health Sciences, Interdisciplinary Food, Metabolism and Clinical Nutrition Department, Ankara, Turkey
[3] Ankara University Faculty of Agriculture Department of Dairy Technology, Ankara, Turkey

1. Introduction

Development of new food, food ingredients and food production technologies have a dynamic nature. Major driving forces of this dynamic structure are globalization, increasing ethnic diversity and consumers' desire for more nutritious foods/food ingredients. Undoubtedly, novel foods and functional foods are conceptually different. Since we understand that foods and food ingredients may have a far more positive impact on our health than we think, novel foods are largely expected to have functional characteristics too. Definition, regulations and acceptance of novel foods and functional foods including probiotics and prebiotics differ from one country to another. Novel foods are defined as *"foods that had not been consumed to a significant degree by humans in the EU before 15 May 1997, when the first Regulation on novel food came into force"* (https://ec.europa.eu/food/safety/novel_food_en).

According to EU, functional foods must have particular effects on the human body and are intended to be consumed for a particular nutritional purpose (PARNUT) (https://ec.europa.eu/food/safety/novel_food_en). A similar definition is in place in the Japanese FOSHU (Food for Specialized Health Uses) and FNFC (Food with Nutrient Functional Claims) regulations. According to FOSHU; functional foods must exert a beneficial effect on nutrition, enhance the physiological state of the body and all these functional characteristics must be proven by scientific and clinical studies (https://

*Corresponding author: adabarbaros@gmail.com

gain.fas.usda.gov/Recent%20GAIN%20Publications/Functional%20
Food%20Report_Tokyo%20ATO_Japan_8-13-2014.pdf,https://www.fas.
usda.gov/data/japan-japan-s-new-health-claims-labeling-system-creates-
opportunities).

The concept of functional foods was first introduced by Japanese
researchers in 1984 who discovered the correlation between nutrition, sensory
quality and physiological system modulation (Tufarelli and Laudadio, 2016).
Since then many functional foods have been introduced into global markets
and they have become an important component of global trade. In 2017, the
revenue generated by global functional foods market was USD 29.932 billion
and is forecasted to reach USD 44.156 billion in 2022 (https://www.statista.
com/statistics/ 252803/global-functional-food-sales/).

Among functional foods, probiotic, prebiotic and synbiotic products
occupy a significant place. In 2016, the global probiotics market was assessed
at USD 35.9 billion and is expected to reach USD 96 billion by 2020 (https://
www.transparencymarketresearch.com/pressrelease/probiotics-market.
htm). Increasing health concerns globally coupled with the efficacy of
probiotics are the key factors for the growth of probiotics market. Although
there is no precise definition of probiotics in many countries, FAO/WHO
defined probiotics in 2002 as *"live microorganisms which when administered in
adequate amounts confer a health benefit on the host"* (FAO/WHO, 2002). This
definition was later revised grammatically in 2014 as follows: *"probiotics are
live microorganisms that, when administered in adequate amounts, confer a health
benefit on the host"* (Hill et al., 2014). Another definition developed by World
Gastroenterology Organization (WGO) says *"probiotics are live microorganisms
that confer a health benefit on the host when administered"*. According to FAO/
WHO definition, probiotic microorganisms are primarily expected to
pass gastric environments and replicate themselves in the large intestine
(Michalak and Chojancka, 2016). Additionally, probiotics are expected to be
barriers to colonization of pathogens, as stated in WGO Practice Guideline-
Probiotics and Prebiotics (2017). Gut environments have the capacity to
prevent colonization of ingested microorganisms. This effect results from
joint actions of gastric and pancreatic juices and enzymes, bile salts and host
microbiota as well as mucosal immunity and dietary changes.

During the last two decades, efforts to improve health and well-being of
people by modulating gut microflora have been intensified (Lockyer 2017).
Microbial content of the gastrointestinal system differs from the oesophagus
to cecum. It is estimated that more than 40 trillion microorganisms are found
in the colon of an adult and more than 10 million genes are found in gut.
Viruses and bacteriophages are also found in the intestinal microenvironment
(Jandhyala et al., 2015, D'Argenio and Salvatore, 2015). Gut microflora of
individuals are established by vertical transmission at birth and differ from
person to person. Gut microflora composition is modulated by a number of
factors including genetics, diet, stress, age and medical treatments (David
et al., 2014). Both potentially harmful and beneficial microorganisms are
present in the gut system. Beneficial bacteria act as suppressors or inhibitors

of the harmful bacteria as well as aiding digestion and absorption of nutrients, modulating/stimulating the immune system and synthesizing some beneficial nutrients, e.g. vitamins (Wallace et al., 2011). The theory of improving the well-being and health of people by unbalancing gut microflora in favor of beneficial bacteria has been known since Nobel Prize winner Elie Metchnikoff had released his cult book "Prolongation of Life: Optimistic Studies" in 1907. Roughly six decades later, Lily and Stillwell (1965) set the concept of "probiotics" after they showed that beneficial bacteria were able to produce growth-promoters.

In the EU, probiotics are used in a wide range of product categories including food (food supplements, foods for specific medical purposes), medical devices, pharmaceuticals/ cosmetics and feed supplements. These products are regulated based on the legal requirements of the category into which they fall. According to the Regulation (EC) No 1924/2006, consumers must be protected from misleading claims. The EU has tightened rules on using the term "probiotic" in 2012 based on Nutritional Health Claim Regulation (NHCR). This *de facto* ban is based on the interpretation of the phrase "contains probiotics" to be a health claim instead of a nutrition claim. Within Europe, scientific claims of a food are assessed by EFSA (European Food Safety Authority) based on the following criteria:

- Satisfactory scientific evidence.
- Strong scientific evidence based on human clinical studies, *in vivo* animal studies and/or *in vitro* cell culture studies.
- Combination of human efficacy studies and supportive studies.

Although the gut microflora is modulated by ingesting beneficial bacteria (so called "probiotic effect"), satisfactory direct evidence of a cause-effect relationship between increased number of beneficial microorganims in the gut and positive health impact is yet to be established (Plaza-Diáz et al., 2015). Debates over the health claims of probiotics have been ongoing between the probiotic industry and the EU. Based on the above approach, EFSA had evaluated many requests from manufacturers including probiotic strains of combinations with a number of beneficial health effects and rejected a majority of these requests. Since this rule was put into effect within EU, many products have been withdrawn from the markets and many products are still awaiting approval. Among the EFSA's main reasons of rejections are:

- Insufficient characterization.
- Unvalidated/unproven claims.
- Insufficient beneficial impact on nutrition and/or physiology of the human body.
- Insufficient number of placebo controlled and/or double-blind clinical studies.
- Poorly designed trials/insufficient scientific evidence.

Given the reasons for rejection by the EFSA for probiotic applications, assessment of health effects of probiotics by appropriate methodological approaches is essential. In many countries, in order to convert scientific findings into clinically relevant information, a number of clinical guides have been released. These clinical guides are primarily considered in the decision-making stage for a specific health claim. Such a guide which has been in effect in the USA since 2008, has been updated on a regular basis (http://usprobioticguide.com/). According to this guide, three stages of clinical assessment of probiotic microorganisms should be followed successively: *Level 1*- randomized clinical trials, *Level 2*-properly designed controlled, cohort or case control clinical trials and *Level 3*-systemic analysis and expert reports.

In the USA and Canada, manufacturers must provide satisfactory published clinical evidence for the strains(s) used in each probiotic food. In case of using multiple strains, evidence must address the specific combination rather than extrapolating the evidence for separate strains.

Based on the PUBMED search, since 1950s, over 20,000 scientific articles/ reports related to probiotics have been published in various journals and roughly 2200 of these studies include clinical trials. The subject of about 9000 articles was *Lactobacillus* spp. which is by far the most studied probiotic species, and 1265 of these studies were based on clinical trials. This was followed by *Bifidobacterium* spp. with 625 clinical trials and *Streptococcus*, *Bacillus* and *Saccharomyces* spp. were used in 180, 110 and 100 clinical studies, respectively. In these trials, the immune system, anti-microbial activity and microbial resistance were the most studied subjects.

The probiotic bacteria most commonly used for commercial applications belong to genera *Lactobacillus* and *Bifidobacteria* (Yadav and Shukla, 2017). These groups of bacteria have long been taken by the human body through traditional fermented foods including yogurt and other fermented dairy products (Zoumpopolou et al., 2017). Apart from some strains of *Lactobacillus* spp. and *Bifidobacterium* spp., microorganisms belonging to genus *Saccharomyces* (i.e., *S. boulardii*), *Enterococcus* (i.e., *E. faecium, E. faecalis*), *Propionibacterium* (i.e., *P. freudenreichii* ssp. *shermanii*), *Streptococcus* (i.e., *Str. thermophilus*) and *Bacillus* (*B. subtilis, B. clausii, B. pumilus, B. polymyxa, B. cereus*) are used in the manufacture of various probiotic products (Shah, 2007). Especially, probiotic *Bacillus* spores which are heat-and acid-stable, provide a number of advantages to food industry (Cutting, 2011).

2. Criteria for Selection of Probiotic Strains

Up until now, a great number of researches have been conducted on screening and selection of probiotic microorganisms obtained from human sources. Majority of these studies have followed a multi-step approach for isolation, identification and characterization of potential probiotic microorganisms by employing a diverse range of molecular, microbiological, biochemical and technological tools. This multi-step approach includes:

(i) Isolation of strains from a suitable source (those intended for human consumption must be of human origin),

(ii) Phenotypic and genotypic identification of strains (for common methodologies refer to Table 1),
(iii) Safety assessment of identified strains,
(iv) Conformity of potential probiotic strains to preliminary selection criteria,
(v) Justification of specific health claims associated with probiotics by a verifiable and repeatable *in vitro, in vivo* and *ex vivo* test model ,
(vi) Conformity of potential probiotic strains to technological applications (i.e. fermentation, spray or freeze drying, competitiveness with other microflora of the food in question etc.)

WGO defines the requirements of commercial probiotics and content of the criteria as follows:

• Identification and nomenculature of genus and species,
• Designation of strain,
• Viability of each strain throughout shelf-life,
• Optimized storage conditions,
• Conditions for safe use,
• Recommended level of ingestion for intended health effect(s),
• Description of the physiological effect as far as is allowable by legislations,
• Communication channels for post-market surveillance.

Probiotic strains intended for human and animal consumption must have QPS (Qualified Presumption of Safety) or GRAS (Generally Recognized as Safe) status according to EFSA Official Journal 2014; 12(3): 3593 and under sections 201(s) and 409 of the Federal Food, Drug, and Cosmetic Act by FDA, respectively. Certain species/strains of lactic acid bacteria (LAB) and bifidobacteria have been granted GRAS or QPS status since LAB and bifidobacteria have been used safely for a long time with no report of problems associated with consumption of foods containing these bacteria.

Today, methodologies for isolation and phenotypic/genotypic identification of the pure strains are well-established. Likewise, *in vitro* methods for checking the conformity of pure isolates to preliminary selection criteria are employed with almost no difficulty. A probiotic strain defined with its safety is primarily expected to survive during gastrointestinal transit (resistance to low pH and bile salts) and adhere to the intestinal mucosa or cell lines as an indicator of temporary gut colonization. Many health effects attributed to the probiotics are directly related with the degree of adhesion capacity of probiotics to the intestinal mucosa. Adhesion of probiotics to the intestinal mucosa may be specific (through interaction with the mucus layer of the gastrointestinal tract or with the proteins of the extracellular matrix through the cell envelope and secreted proteins) or non-specific (depending on the type of sticky compounds such as proteinaceous compounds, teichoic and lipoteichoic acids, peptidoglycans, and exopolysaccharides etc.). However, a clear correlation between *in vitro* and *in vivo* assays employed in selection of probiotic strains from native species/strains of bacteria is still

Table 1: Phenotypic and genotypic methods commonly employed in identification of probiotic microorganisms

Identification method	Tools	Specificity level	Remarks	References
Phenotypic	Morphological examination Growth temperature (15 °C vs. 45 °C)	Species Species	Sensitive, inexpensive, lengthy and time-consuming requiring 24–48 h or longer	
	Gram-staining Sugar fermentation Salt resistance Acidity resistance Hyphal formation	Species Species Species Species Species		
Genotypic	16S-18S-23S-28S rRNA sequencing	Species-specific	Rapid identification of probiotic bacteria using primers targeting 16S rRNA, yeasts by targeting 23S rRNA, fungi by targeting 28S rRNA, all targeting the ITS region of the rRNA genes	Luo et al., 2012 Hallen-Adams and Suhr, 2017 Suhr et al., 2016 Boix-Amorós et al., 2017
	Denaturing gradient gel electrophoresis (DGGE)	Strain-specific	No prior culturing of cells is required. Recommended to use in combination with 16S rRNA sequencing of the V3 region	Yadav and Shukla, 2017 Liu et al., 2012
	Pulsed Field Gel Electrophoresis (PFGE)	Strain-specific	Allowing the separation of DNA fragments larger than 10 kb. Successfully discriminating *B. animalis* and *B. longum* as well as *Lb. casei* and *Lb. rhamnosus*	Tynkkynen et al., 1999
	Amplified rDNA Restriction Analysis (ARDRA)	Species-specific	Quick and accurate discrimination between different species and strains	Srutkova et al., 2011

Amplified Fragment Length Polymorphism (AFLP)	Strain-specific	Allowing accurate identification of strains of *Lactobacilli* and *Bifidobacterium longum*	Makino et al., 2011
Random Amplified Polymorphic DNA (RAPD)-PCR	Strain-specific	Effective in distinguishing various species of *Lactobacillus* at inter- and intra-species level as well as successfully distinguishing bifidobacteria and *Lactobacillus acidophilus* but not between *Lb. gasseri* and *Lb. fermentum*.	Gosiewski et al., 2012 Saxami et al., 2016
Ribotyping	Strain-specific	High discrimination of *Lactobacillus* at species, subspecies and strain levels	Rodtong and Tannock, 1993
Fluorescent activated cell sorting (FACS)	Strain-specific	Effective in sorting heterogeneous biological cell mixtures based on fluorescent characteristics and light scattering	Chen et al., 2012
Fluorescent in situ Hybridization (FISH)	Strain-specific	Direct imaging and visual enumeration of live cells. Recommended to use in combination with DNA/rRNA hybridization method. Replacement of DNA probe by peptide nucleic acid (PNA) probes is more effective in quantification of viable cells.	Stender et al., 2002
Metagenomic analysis	Strain-specific	Evaluation of complex microbial environment, common usage in genomic analysis of micro-organisms by direct extraction and subsequent cloning of DNA from an assemblage of microorganisms from a particular environment. Gives information about functionality	Kalyuzhnaya et al., 2011 Vietes et al., 2010 Ferrer et al., 2010

obscure. When a strain is adapted to specific environments such as nutritional changes, it could cause a decoy in the genome size. Phylogenetic arrangement and diversity are important for functions and effects of bacteria. For example, the genome size and GC content of vaginal *Lactobacillus* spp. was reported to be reduced compared to intestinal *Lactobacilli* due to microbial interactions in the vaginal flora, possibly. Based on the comparative functional genomics studies, it is hypothesized that overrepresented genes in the core genome are primarily responsible for the probiotic characteristics of strains.

Shortly, major challenges in probiotic studies are viability and survivability of probiotic strains during and after technological applications, passage through gastrointestinal tract (GIT) and justification of health claims with proper *in vivo*, *in vitro* and *ex vivo* models. Breast milk is one of the major sources of many probiotic strains commercially available in the markets. It is known that breast milk contains many species or strains of *Lactobacilli* and *Bifidobacteria*, however; not all strains isolated from breast milk can survive under harsh environmental conditions such as freeze-drying, gastric digestion, long term storage etc. At this point, the EFSA wants to ensure that *"any health claim a product makes is true and based on sound, substantiated science"* in order to avoid misguidance of consumers. EFSA's major argument is that microorganisms are living organisms, and their growth and metabolic activity patterns may well change as a function of time. Therefore, correct identification of the microbial species and strains is critically important, as the observed effects are specific to species and strains. During the last three decades, a lot of research has been conducted on the isolation, identification and putative functional properties of potentially probiotic candidates. These studies have been evaluated through several meta-analyses. The common outcome of these meta-analyses is that probiotic lactobacilli and bifidobacteria are effective against various microbiota-associated diseases in a strain-dependent manner. Novel species/strains of probiotic microorganisms must meet certain safety criteria before being considered for further evaluations. Among them, resistance to antibiotics and infective capacity are the major criteria for safety assessment. According to the EFSA, the origin and metabolisms of novel probiotic strain/species should be clearly defined. In addition, non-toxic and non-pathogenic properties of the potentially probiotic microorganisms should be demonstrated clearly. Potential probiotic microorganisms which do not belong to the groups previously given QPS (Qualified Presumption of Safety) status by EFSA should be evaluated extensively. Other selection parameters of probiotics are given below:

(a) Acid and bile salt resistance,
(b) Resistance to antibiotics,
(c) Resistance to gastric juice,
(d) Cholesterol assimilation,
(e) Survivability, proliferation and colonization in gastrointestinal tract,
(f) Adherence to gut epithelial cells,

(g) Antimicrobial activity,
(h) Genotypic and phenotypic stability,
 (i) Biocompatibility.

3. Technology of Probiotic Dairy Products

Food industry—especially the dairy industry—has taken quick action to exploit the market potential driven by the possible beneficial health impacts of probiotic microorganisms. This motivation has triggered efforts to incorporate probiotics into dairy matrices such as yogurt, cheese or dairy-based beverages. A major challenge for incorporation of probiotics into dairy foods is to maintain the number of probiotic microorganisms higher than the required level for "therapeutic minimum" at the time of consumption. This threshold level varies depending on countries' regulations but is generally not accepted below 6-7 log cfu/g or mL. Gastrointestinal survival and adhesion ability of probiotics to Caco-2 cells and tolerance to digestion *in vitro* cell culture are largely determined by the type of carrier food matrix. Solid food matrices offer better protection of probiotics than liquid foods. Recently, Lollo et al. (2013) showed that probiotic yogurt had improved neutrophils and lymphocytes (immune response indicators), cytokines responses (TNF-α and IL-1β), and improved exhaustive exercise caused immune depression mainly in monocytes and neutrophils, and various standard health parameters compared to probiotic whey-based beverages. On the other hand, dairy foods generally have an acidic nature and some dairy products contain starter and non-starter bacteria which may suppress the growth of probiotics. Therefore, probiotic microorganisms must cope with relatively harsh environments in dairy products. Up until now many strategies have been developed to increase viability and survivability of probiotics in dairy foods. These approaches and basic manufacturing steps of selected probiotic dairy foods are discussed below.

3.1 Probiotic Yogurt (Bio-Yogurt) and Fermented Dairy Products

Yogurt is one of the most appropriate delivery medium for probiotics. Basically, probiotic yogurt is produced in the same way as standard yogurt. Overall, heat-treated milk (85-95 °C for 5-20 min) is cooled to 43 °C and yogurt culture (*Lactobacillus delbrueckii* subsp. *bulgaricus* and *Streptococcus thermophilus*) and probiotic strain(s) are inoculated into the milk simultaneously. Fermentation is allowed until the pH of fermenting milk reaches 4.6-4.7. During fermentation, probiotic bacteria interact with starter microorganisms and this interaction may have positive or negative impacts on product characteristics. Therefore, selection of probiotic strains and their culturing conditions should be maintained properly in order to obtain yogurt of an acceptable quality. The tolerance of probiotic strains to environmental stress conditions in yogurt has a strain-dependent characteristic. Jayamanne and Adams (2006) evaluated stress tolerance levels of various *Bifidobacterium* spp. in bio-yogurt and showed that *Bifidobacterium animalis* subsp. *lactis* had superior survival abilities and oxidative stress tolerance compared with *B. longum*, *B. breve*, *B. bifidum*, *B. adolescentis* and *B. longum* biotype *infantis*.

Some metabolites produced by probiotic microorganisms may adversely affect the growth of yogurt starter bacteria and therefore may extend the fermentation period. Especially, acidophilin LA-1 produced by *Lactobacillus acidophilus* (*L. acidophilus*) partly suppresses the growth of *L. bulgaricus*. In reverse, hydrogen peroxide produced by *L. bulgaricus* limits the growth of *L. acidophilus*. Majority of *Bifidobacterium* spp. strains do not have proteolytic enzyme systems. Therefore, nitrogenous compounds that *Bifidobacterium* spp. require for growth are generally generated by *L. bulgaricus*. *Str. thermophilus* is able to consume oxygen in yogurt and therefore, creates a suitable environment for the growth of *Bifidobacterium* spp. which are strictly anaerobic. Shortly, the yogurt matrix is very complex and dynamic, and proper design of this matrix is essential to convert it to a suitable medium for the growth of probiotic microorganisms.

A number of factors including acidity, dissolved oxygen level, redox potential, competition with yogurt starter bacteria for nutrients and hydrogen peroxide concentration affect the viability of probiotic bacteria in yogurt during production or storage. Especially dissolved oxygen and acidity are the major parameters affecting the growth of bifidobacteria. A majority of the probiotic strains demonstrate weak growth under acidic conditions. The growth of *Bifidobacterium* spp. especially slows down under a pH of 5.0. Therefore, efforts have been intensified to develop efficient solutions to stimulate the growth and survivability of probiotic microorganisms in yogurt and other fermented dairy foods. Common strategies developed for stimulating the growth or survivability of probiotic microorganisms are:

- Addition of prebiotics (synbiotic products),
- Microencapsulation,
- Packaging yogurt with materials having restricted oxygen permeability or incorporated oxygen scavenging agent(s),
- Double-stage fermentation,
- Use of strains resistant to acid and/or salt,
- Inclusion of fruit matrices,
- Supplementation of yogurt with growth promoting substances (e.g. cysteine, tryptone, whey protein concentrates and casein hydrolysates).

Prebiotics are non-digestible specific dietary components that are used to stimulate the growth and viability of probiotic strains. Among the prebiotics, fructooligosaccahrides (FOS), inulin, galactooligosaccharides (GOS), lactulose and isomaltooligosaccharides are widely used in the manufacture of yogurt-type probiotic dairy products. The stimulating effect of prebiotics on growth of probiotic cells depends on the type and concentration of prebiotics used. Prebiotics also affect the physical and sensory characteristics of the final products. Therefore, the concentration and type of prebiotics should be decided properly. Bitaraf et al. (2012) employed response surface methodology to optimize the independent variables of probiotic yogurt production. Authors showed that addition of a high level of inulin softens the yogurt matrix and a low level of inoculum is recommended for a stronger gel structure in the end product. Similar results were obtained by Başyiğit-Kılıç and Akpınar-Kankaya (2016), who found that while β-glucan improved the growth and viability of probiotic lactobacilli, the degree of syneresis was higher in β-glucan-treated samples compared with the yogurts containing no prebiotic. β-glucan derived from oat or barley was recommended for use in probiotic yogurt at 0.24% (w/w) or less to obtain acceptable physical properties in the end product as well as high probiotic counts. Shakerian et al. (2014) showed that inulin (at a level of 1.0%, w/v) did not stimulate the growth of *L. acidophilus* but promoted the growth of *B. animalis* subsp. *lactis*. Similarly, Özer et al. (2005) showed that lactulose was a far more effective growth-promoter for *L. acidophilus* La-5 than inulin. In contrast, Rezaei et al. (2014) demonstrated that inulin (at a level of 2.0%, w/v) promoted the growth of *L. acidophilus* and *B. animalis* subsp. *lactis* in frozen probiotic yogurt as well as improving the overrun, flow and melting properties of the end product. The stability and prebiotic properties of inulin are affected by degree of polymerization (DP). Long chain inulin (average DP of 23 and 25 units) are more stable against thermal treatments than short-chain inulin. On the other hand, short-chain inulin were reported to stimulate the growth of *Lactobacillus rhamnosus* (*L. rhamnosus*) in probiotic yogurt more efficiently than long-chain inulin. Recently, sea buckthorn has been proposed as a new prebiotic source for functional foods and nutraceutical applications by Gunenc et al. (2016). Addition of sea buckthorn to yogurt at a level of 2% (w/w) resulted in higher probiotic counts, antioxidant capacity and oxygen scavenging activity (Gunenc et al., 2016). Readers may refer to Figueroa-Gonzales et al. (2011) for more details about type and mode of action of prebiotics in dairy foods.

Microencapsulation is a promising technology to protect probiotic cells from detrimental environmental conditions in a food matrix. A wide range of commercial wall and coating materials are available for the microencapsulation of microbial cells. Riberio et al. (2014) evaluated the viability of encapsulated *L. acidophilus* La-5 in yogurt. The authors employed ionic gelation and complex coacervation techniques in microencapsulation by using pectin and whey proteins as the wall and coating materials, respectively. The rates of viability of encapsulated probiotic cells both in yogurt and during simulated gastrointestinal digestion processes were significantly higher than that of the non-encapsulated cells. However, yogurt with encapsulated probiotic bacteria received less sensory scores than the control yogurt containing free probiotic cells. Dual coating is an emerging innovative technology in the fields of probiotics. This fourth-generation coating technology allows target microorganisms to pass through the gastrointestinal tract without being damaged and in numbers sufficient enough to confer beneficial health impacts on the host. The first layer consists of protein and protects bacteria from gastric acid and bile salts. The second layer is made up of polysaccharide which protects bacteria from environmental conditions including humidity, temperature, pressure and digestion. Dual-coating technology is yet to be employed in dairy industry.

Cell immobilization is another method for improving probiotic cell viability in yogurt. Cereal dietary fibers can be used as prebiotic immobilization carriers for probiotic bacteria (Terpou et al., 2017a). Terpou et al. (2017b) showed that wheat bran acting as a carrier for the immobilization of *Lactobacillus casei* ATCC393 effectively protected the cells against simulated gastrointestinal digestion conditions (pH 3.0). In the meantime, the formation and release of volatile compounds in yogurt was influenced in a positive manner. Owing to their structural and physicochemical properties, milk proteins are considered as efficient natural vehicles for incorporating probiotic cells into a food matrix (Livney, 2010). Sidira et al. (2017) immobilized *Lactobacillus plantarum* 2035 on whey proteins and inserted immobilized cells into yogurt. The counts of probiotic bacteria in yogurt were above the therapeutic minimum (>6 log cfu/g) at the end of the storage period and the sensory properties of the end product were acceptable.

Since *L. acidophilus* is microaerophilic and *Bifidobacterium* spp. is strictly anaerobic, oxygen toxicity is one of the most limiting factors for the viability of probiotic cells in yogurt (Dave and Shah, 1997a). Active packages or packages with improved oxygen barriers with the incorporation of oxygen absorbers into plastic films are suitable for probiotic foods (Miller et al., 2003). Glass packaging is the best for probiotic foods. However, due to the limitations in the use of glass as packaging material (i.e., cost, storage, recycling etc.), packages with high-barrier plastic laminates offer certain advantages for the probiotic industry (Korkebandi et al., 2011). Although state-of-the-art of packaging technology allows the production of packaging materials with significantly reduced oxygen permeability, a complete prevention of gas transfer from outer sources into probiotic food is not possible. It is possible to reduce the permeability of polymeric materials by means of crystallization. On the other hand, there is no clear evidence that the degree of crystallization of polymeric materials affects the number of probiotic bacteria (Jansson et al., 2002).

The packaging industry has developed a new-generation of packaging materials with limited oxygen permeability (trade name Nupak™) or combined with oxygen scavenging agents (trade name $Zero_2$™), allowing probiotics to survive better (Miller et al., 2003). Nupak™ consists of a multi-layered structure of HIPS/tie/EVOH/tie/PE (Karaman et al., 2015). High-impact polystyrene (HIPS) is effectively used to strengthen the outer layer of the packages against physical damaging. EVOH has a limiting effect on the oxygen level in probiotic yogurts. As mentioned above, the incorporation of oxygen scavenging agents to packaging materials is an effective way of reducing oxygen in a package. However, sometimes more than one packaging technology or approach like packaging material with reduced permeability along with oxygen absorbents may be needed to improve the viability of probiotics in products (Tripathi and Gri, 2014) or microencapsulation of probiotic bacteria, and storage at <4 °C (Hisiao et al., 2003, Rathore et al., 2013, Riaz and Masud, 2014). Another suitable alternative for stimulating the growth of probiotic bacteria in yogurt is to add ascorbic acid into yogurt as an oxygen scavenging agent (Dave and Shah, 1997b). Glucose oxidase is capable of reducing oxidative stress in probiotic yogurts (Cruz et al., 2012). Addition of glucose oxidase and glucose in probiotic yogurt remarkably increases the viability of probiotic bacteria (*B. longum* BL 05) in cold stored stirred yogurt for 15 days, as reported by Cruz et al. (2012).

Double-stage fermentation is a practical option to stimulate the growth of probiotic bacteria in yogurt. Yogurt bacteria grow faster than probiotic bacteria and partly limit the growth of probiotics. Lactic acid produced by yogurt starter bacteria also negatively affects the growth and viability of probiotic strains. Therefore, probiotics are added to yogurt milk about two hours before yogurt starter bacteria are inoculated. This pre-fermentation period allows probiotic bacteria to proliferate without being affected by metabolites such as hydrogen peroxide produced by yogurt starter bacteria (Lankaputhra and Shah, 1996).

Supplementation of probiotic yogurt with fruit pulps/flours or cereals have gained popularity in recent years. These supplements act as growth-promoters for probiotic cells as well as improving physical and sensory properties of the end product. Table 2 summarizes the recent researches on probiotic yogurts supplemented with various fruit- or cereal-based ingredients.

Anti-oxidant and anti-mutagenic properties of probiotic bacteria have long been known. Efforts have been intensified to search for novel bioactive peptides from food sources since some synthetic peptides have been associated with health risks. Bioactive peptide production by probiotic bacteria has a strain-dependent nature. For example, *L. acidophilus* (ATCC®4356™), *L. casei* (ATCC®393™) and *L. paracasei* subsp. *paracasei* (ATCC®BAA52™) were shown to liberate peptides in yogurt with high radical scavenging activity and strong anti-mutagenicity (Sah et al., 2014).

In recent years, yogurt type fermented products from milk of small ruminants as an alternative to fermented cow's milk has gained popularity (Zhang et al., 2015). Especially goat's milk products offer a solution to cow's milk allergies and gastrointestinal intolerances (Haenlein, 2004). Pinto et al. (2017) compared the physicochemical and biological properties of probiotic fermented milk made from ewe's milk or goat's milk. Probiotic co-cultures (*L. acidophilus* LAFTI®L10, *L. acidophilus* Ki and *B. animalis* Bo) grew better in products of ewe milk compared to those of goat milk and inulin had no effect on the growth of probiotic cells but improved the storage stability of yogurts.

Novel food processing technologies such as high hydrostatic pressure (HHP) can also be employed in the manufacture of probiotic yogurt (Jankowska et al., 2012). Jankowska et al. (2005) processed yogurt milk with HHP under 550 MPa/15 min/18 °C conditions. Although the number

Table 2: Probiotic yogurt supplemented with prebiotics, cereal-, vegetable- or fruit-based ingredients

Product	Culture	Results	Source
Probiotic yogurt supplemented with fruit flours (1.0%)	*Str. thermophilus, Lb.acidophilus* La-5 and *B. animalis* subsp. *lactis* Bb-12	Banana, apple or grape flours improved *Lb. acidophilus* tolerance to simulated gastrointestinal conditions. Only banana flour showed the protective effect on *B. animalis* subsp. *lactis* during 28 days of storage	Casarotti and Pena, 2015
Stirred probiotic yogurt supplemented with açai pulp	*Str. thermophilus, Lb. acidophilus* L10, *B. animalis* subsp. *lactis* B104, *B. longum* B105, *B. animalis* subsp. *lactis* B94	Açai pulp stimulated the growth of *Lb. acidophilus* L10, *B. animalis* ssp. *lactis* Bl04 and *B. longum* Bl05 and the end product contained higher level of mono and polyunsaturated fatty acids	Santo et al., 2010
Probiotic goat yogurt supplemented with water soluble soybean extract	Yogurt culture plus *B.lactis*	Physical properties of the yogurts containing soy extracts improved	da Silva et al., 2012
Probiotic fermented milk supplemented with chesnut flour	*Lb. acidophilus, B. animalis* subsp. *lactis, Lb. rhamnosus*	Viability and antioxidant capacity of chesnut flour enriched fermented milks increased significantly	Özcan et al., 2016
Apricot probiotic drinking yogurt supplemented with inulin (0.5-1.0-2.0%) and oat fiber	*Str. thermophilus, Lb. acidophilus* La-5 and *B. animalis* subsp. *lactis* BB-12	Incorporation of oat fibers stimulated the growth of probiotic bacteria as well as improved the physical properites of the end products	Güler-Akın et al., 2016
Synbiotic yogurt added with FOS	*Lb. plantarum* CFR2194, *Lb. fermentum* CFR2192	DPPH radical scavenging activity of synbiotic yogurt containing *Lactobacillus plantarum* CFR2194 aand FOS was higher than that of the control sample	Madhu et al., 2012

Product	Culture	Findings	Reference
Probiotic fermented goat milk supplemented with polymerised whey protein isolate (0.3 %) as thickening agent	Kefir Mild 01 (*Lc. lactis* subsp. *lactis*, *Lc. lactis* subsp. *cremoris*, *Lc. lactis* subsp. *lactis* biovar *diacetylactis*, *Leuconostoc mesenteroides* subsp. *mesenteroides*, *Str. thermophilus*, *Lb. acidophilus*)	PWP (0.3%) and pectin (0.2%) mixture yielded a low syneresis, high viscosity and higher scores of flavour and taste in the formulated probiotic fermented milk. The number of *Lb. acidophilus* was above the threshold level for therapeutic effect	Wang et al., 2017
Probiotic goat milk yogurt enriched with polymerised whey protein isolate (PWP) and pectin	*Lb. acidophilus*, *Lb. casei*, *Bifidobacterium* spp.	The counts of *Lb. casei* and *Bifidobacterium* spp. Remained almost unchanged during 4 weeks of storage. *Lb. acidophilus* completely died off after 4 weeks. PWP had a potential of being used as protein-based thickening agent in goat milk yogurt	Wang et al., 2012
Probiotic yogurt with added spices (cardamom, cinnamon, nutmeg)	Yogurt culture plus *Bifidobacterium animalis* subsp. *lactis* BB-12, *Lactobacillus acidophilus* La-5	The presence of spice oleoresins did not affect the growth and viability of probiotic strains and the antioxidant capacity of the probiotic yogurts did not changed over the storage period of 4 weeks. The probiotic yogurt with added cardamom had the highest sensory scores	Illupapalayam et al., 2014
Probiotic goat's milk yogurt with added Cupuassu fruit pulp	Yogurt culture plus *Lb. acidophilus* La-5	Cupuassu fruit pulp improved the texture of probiotic goat's milk yogurt without adversely affecting the probiotic counts throughout cold storage of 28 days	Costa et al., 2015
Probiotic yogurt with added organic green banana flour (1.0 to 5.0%)		Addition of green banana flour to yogurt milk at a level of 3% resulted in better physical properties and lower atherogenic and thrombogenic indices with acceptable sensory properties	Batista et al., 2017

(Contd.)

Table 2: (*Contd.*)

Product	Culture	Results	Source
Probiotic yogurt supplemented with green banana pulp (3.0% to 10.0%)	Yogurt culture plus *B.animalis* subsp. *lactis* BB-12, *Lb. acidophilus* La-5	The green banana pulp added to the yogurt stimulated the multiplication of *Lb. acidophilus* after the first day of fermentation and *B. bifidum* after seven days in cold storage. Green banana pulp had a prebiotic potential with no detrimental effect on physico-chemical and sensory properties of probiotic yogurt	da Costa et al., 2017
Probiotic yogurt supplemented with fruit pulp (10%) of apricot, plum, raspberries and jamun as antioxidant sources	Yogurt culture plus *Lb.rhamnosus* in free or microencapsulated form	Alginate microencapsulated probiotic culture was more stable than carragenan encapsulated or free probiotic cultures. The antioxidant capacity of the probiotic yogurts decreased during storage up to 15 days at 4 °C.	Kumar and Kumar, 2016
Fermented milk supplemented with *Aloe vera*	*Lb.casei* NCDC 19	*Aloe vera* powder stimulated the proteolytic capacity and percent ACE inhibitory activity of *Lactobacillus casei* NCDC 19. Also, the growth of probiotic bacteria was stimulated by *Aloe vera*.	Basannavar et al., 2014
Probiotic stirred yogurt enriched with *Aloa vera* foliar gel	*B. animalis* subsp. *lactis* BB-12, *Lb. acidophilus* La-5	*Aloe vera* leaf gel had negative effect on probiotic viability. Physical properties of the end product were also affected adversely by *Aloe vera*.	Azari-Anpar et al., 2017
Yogurt with added pineapple waste (oven and freeze-dried peel and pomace powder)	Yogurt culture plus *Lb. acidophilus* (ATCC® 4356™), *Lb. casei* (ATCC® 393™) and *Lb. paracasei* spp. *paracasei* (ATCC®BAA52™)	Pineapple waste showed a prebiotic effect on probiotic bacteria and led to increase in probiotic population by 0.3 to 1.4 log cycle in yogurt. Yogurt containing pineapple waste powder had high antioxidant and antimutagenic capacities.	Sah et al., 2016a

Probiotic yogurt with highbush blueberry fruit	Yogurt culture plus *B. animalis* subsp. *lactis* Bb-12, *Lb. acidophilus* La-5, *Lb. paracasei* subsp. *paracasei* (LCP)	The anthocyanine stability of yogurt made with LCP culture was lower than BB-12 or La-5. The malvidin- predominant anthocyanine-degradation in the former yogurt was more pronounced during cold storage.	Ścibisz et al., 2012
Probiotic yogurt with added pulses	Yogurt cultures plus *Lb. rhamnosus, Lb. acidophilus*	Pulse ingredients had no effect on acidification trends of yogurts	Zare et al. 2012
Probiotic yogurt with added microalgae (*Chlorella vulgaris* and *Arthrospira platensis*)	Yogurt culture plus *B.animalis* subsp. *lactis* BB-12, *Lb. acidophilus* La-5	Addition of microalgae significantly promoted the growt of lactobacilli and bifidobacteria. *A. platensis* extended the fermentation time of yogurt	Behestipour et al., 2012
Yogurt enriched with probiotic-cultured banana puree	Yogurt culture plus *Lb.paracasei* LCP37)	Probiotic cultured bana puree was a suitable medium for the incorporation of probiotics into fermented milk. 2 g of polydextrose/100 mL was the most appropriate concentration regarding physical and sensory properties of yogurt	Srisuvor et al., 2013
Probiotic fermented milk supplemented with quinoa flour (1.0 to 3.0%)	*B. animalis* subsp. *lactis* BB-12, *Lb. acidophilus* La-5	Addition of quinoa flour did not protect probiotic cells during simulated gastrointestinal digestion process and had no positive effect on adhesion of probiotic bacteria to Caco-2 cells	Casarotti et al., 2014a
Probiotic non-fat yogurt supplemented with dietary fibers from banana, passion fruit or apple	Yogurt culture plus *B.animalis* subsp. *lactis*BL04, HN019, B94, *Lb. acidophilus* L10	Fibers from banana and apple stimulated the viability of probiotic bacteria throughout shelf life. A synergestic effect of fiber type and probiotic strain on conjugated linoleic acid was evident. Fruit fiber improved the fatty acid profile of pobiotic yogurts	Santo et al., 2012
Probiotic yoghurt supplemented with grape extract	Yoghurt culture plus *Lb. acidophilus, B. animalis* Bb-12	The addition of grape extract resulted in pseudoplastic behaviour in yoghurt with decreased yield stress and increased k value	da Silva et al., 2017

(Contd.)

Table 2: (Contd.)

Product	Culture	Results	Source
Probiotic yogurt enriched with barberry extract (4.0-5.0%, w/v)	Yogurt culture plus B. animalis subsp. lactis, Lb. acidophilus	The viable counts of Lb. acidophilus and B. bifidum in the set and stirred probiotic yogurts containing barberry extract were significantly higher than the control yogurt during the storage period	Hassani et al., 2016
Probiotic goat's milk yogurt supplemented with honey	Yogurt culture plus Lb. acidophilus La-5	Supplementation of honey resulted in an increase in the count of Lb. acidophilus ca. 1 log cycle	Machado et al., 2017
Probiotic soy-yogurt supplemented with soybean by-product okara flour and inulin	Lb. acidophilus La-5, B.animalis Bb-12, Str.thermophilus	Okara flour may well be used in the manufacture of probiotic soy yogurt with improved nutritional and sensory properties	Bedani et al., 2014
Probiotic yogurt enriched with green, white and black tea	Lb. acidophilus LA-5, B.animalis subsp. lactis Bb-12, Lb. casei LC-01, Str. thermophilus Th-4, Lb. delbrueckii subsp. bulgaricus	The antioxidant activity of all tea supplemented yogurts remained almost unchanged throughout 21 days of storage	Muniandy et al., 2016
Probiotic yogurt supplemented with whole ground green lentils	Yogurt cultures plus Lb. acidophilus, B. animalis subsp. lactis	Lentil polysaccharides stimulated the growth of probiotics as well as showing a strong antioxidant activity	Agil et al., 2013
Probiotic yogurt supplemented with fiber-rich pineapple peel powder (PPP)	Yogurt culture plus Lb. acidophilus (ATCC® 4356™), Lb. casei (ATCC® 393™), Lb. paracasei subsp. paracasei (ATCC® BAA52™)	PPP supplementation at 1.0% significantly reduced the fermentation time of probiotic yogurt. Syneresis of PPP-supplemented yogurt was comparable to the yogurt with added inulin	Sah et al., 2016b

Probiotic non fat yogurt supplemented with commerical plant extract (prepared from garlic, olive, onion and citrus extracts with sodium acetate as carrier)	Yogurt culture plus *Lb. acidophilus* ATCC 4356, *B. animalis* subsp. *lactis* ATCC 25527	Extract supplementation led to a greater buffering capacity in yogurt, thus promoted the growth of probiotics	Michael et al., 2015
Oat-based probiotic yogurt	Yogurt culture plus *Lb. brevis* SBP49, *Lb. acidophilus* SBP55	Colony counts of probiotic bacteria was high (>8 log cfu/g) and the resistance to artifical digestive juices and the adherence to ephitelial cells of these probiotics were also high	Lim, 2018
Probiotic yogurt with added fresh apple pieces, dried raisin or wheat grains enriched with *Lactobacillus casei*	Yogurt culture plus *Lb. casei*	Novel yogurts had stable texture and high probiotic counts throughout storage period	Bosnea et al., 2017
Probiotic Greek Dahi supplemented with pomegranate pulp (PP) and flaxseed powder (FP)	*Lb. acidophilus* NCDC 195 *Lb. casei* NDRI RTS *Lb. plantarum* NDRI 184	Optimum conditions for probiotic Greek Dahi was 15% PP, 2% FP and 12 h incubation at 37°C	Kumar et al., 2018

of probiotic bacteria declined continuously during the 4-week storage at 4 °C, the counts of probiotics were still above the therapeutic minimum (>10⁷ cfu/g). Penna et al. (2006) proposed a combination of HHP (676 MPa for 5 min) and heat treatment (85 °C for 20 min) in the manufacture of low fat probiotic yogurt with improved rheological and textural properties and creamy and thick consistency. At relatively milder high-pressure conditions (i.e. 100 MPa) microorganisms are not totally inactivated. When the high pressure is released, bacteria return to normal metabolic activity mode (Mota et al., 2015). Therefore, it may be possible to use mild high pressure as an on/off switch to stop/start fermentation, similar to refrigeration, as proposed by Mota et al. (2015).

Probiotic yogurt may be used in the manufacture of yogurt powder which is a high value-added dairy product. During drying, the survival rates of starter bacteria may be diminished. This phenomenon is dependent on the drying method and conditions (i.e. outlet temperature and yogurt pH). At an outlet temperature of 70-75 °C, a satisfactory level of survival of *L. bulgaricus* (13.7-15.8%) and *Str. thermophilus* (51.6-54.7%) were reported by Bielecka and Majkowska (2000). However, when the outlet temperature was increased to 80 °C, a considerable decrease in the survivability rates of both bacteria were noticed. Kearney et al. (2009) showed that *L. paracasei* NFBC 338 remained stable in spray dried probiotic yogurt at 4 °C and 15 °C for up to 42 days of storage. Izadi et al. (2014) optimized the drying conditions of probiotic yogurt using response surface methodology and recommended the following conditions: inlet air temperature of 150 °C, drying air flow rate of 478 m³/h, feed flow rate of 2 L/h and outlet temperature of 63.3 °C, respectively. In order to increase the survivability rates of probiotic bacteria during drying of yogurt, a number of technological solutions have been developed. Among such solutions are adding cryoprotectants and prebiotics to yogurt milk or microencapsulation of probiotic cells. Capela et al. (2006) demonstrated that the use of a commercial cryoprotectant-Unipectine™ RS 150-yielded a 7% approximate increase in survivability of *L. casei* 1520 in freeze-dried yoghurt. Same authors also showed that addition of prebiotic Raftilose® P95 at a level of 1.5% (w/v) improved the viability of the same probiotic bacteria by 1.42 log during 4 weeks of cold storage. However, the effectiveness of microencapsulation of probiotic cells was found to be rather limited compared with adding cryoprotectants or prebiotics. Addition of maltodextrin to yogurt at a level of 10% (w/w) may also be an option to preserve probiotic bacteria during drying of probiotic yogurt (de Mederios et al., 2014).

3.2 Probiotic Cheese

Although yogurt and fermented dairy products are suitable vehicles for the delivery of probiotics, probiotic bacteria may show rather low viability in such products due to their acidic nature (Vinderola et al., 2000a). On the other hand, since cheese usually has higher pH and fat content and more solid consistency than fermented milks, it can be a more suitable probiotic carrier medium to the human body (Ong and Shah, 2009, Özer et al., 2009). The cheese matrix affords additional protection to the acid-sensitive probiotic cells both in the product and in the GIT (Corbo et al., 2001, Vinderola et al., 2002b). Since cheese has a longer shelf-life than fresh fermented milk products, a stringent selection of probiotic strains for use in probiotic cheese-making is required to secure the viability of the cells throughout the ripening period (Phillips et al., 2006). In most cheese varieties, the pH varies between 4.8 and 5.2 during ripening period, which is favored by acid-sensitive probiotic bacteria. Cheese is essentially an anaerobic system with oxidation-reduction potential of *ca.* –250 mV, allowing facultatively or obligatory anaerobic bacteria growth (Fox et al., 2000). The redox potential (E_h) of the inner and outer parts of cheese may vary, resulting in different patterns of bacterial growth in different layers of a cheese block. Abraham et al. (2007) demonstrated that while the E_h of the outer part of Camembert cheese ranged between +330 mV and +360 mV, the E_h of the inner part was between –300 mV and –360 mV. This difference was due to the differences in the metabolic activities of cheese microflora in the inner and outer surfaces of the cheese. Oxygen in the cheese matrix is usually consumed by cheese microflora within a few weeks depending on the type and manufacturing practices. This is favored by probiotic bacteria-especially bifidobacteria (van den Tempel et al., 2002). Majority of cheese varieties available in the global markets are high cooked or ripened in dense brine solutions. All these technological applications negatively affect the growth and viability of most probiotic strains. Therefore, probiotic cheese manufacturers should find suitable solutions to keep the probiotic counts in cheese high enough for a therapeutic effect without adversely affecting the quality of the end product. This requires some modifications in the cheese production lines. Some practical options for such modifications are lowering salt level in the brine, vacuum packaging, pre-incubation of milk with probiotic bacteria, addition of growth stimulators (i.e. inulin, oligofructose) and protection of cells against undesired environmental conditions (i.e. microencapsulation, cell incubation under sub-lethal conditions or cell propagation in an immobilized biofilm).

Since fresh cheeses do not undergo ripening and have a shorter shelf-life, they are considered as more suitable mediums for the delivery of probiotics (Fritzen-Freire et al., 2010). Jeon et al. (2016), for example, demonstrated that Cottage cheese was a good delivery medium for *L. plantarum* Lb41 isolated from Kimchi. On the contrary, hard and semi-hard cheeses offer extra advantages for probiotics due to higher pH and a more compact texture that they have (Sabikhi et al., 2014). Metabolic products produced by cheese microflora during long term ripening may positively or negatively affect the viability of probiotic bacteria in cheese. Therefore, probiotic strains used in probiotic cheese making should be selected based on the cheese type and production conditions. *Lactobacillus* GG, for example, was demonstrated to remain viable in a 5-year old hard cheese sample at a level of 2×10^7 cfu/g

(Jatila and Matilainen, 2008). Probiotic bacteria can be incorporated into cheese milk prior to or simultaneously with the cheese starter culture or into the curd after partly or completely removing cheese whey. The stage of addition of probiotic bacteria may affect the quality of cheese. Although pre-incubation of milk with probiotic bacteria presents some advantages such as increases in the counts of lactobacilli in the initial inoculum; a crumbly body and lower pH were evident in Pategras cheese made this way (Bergamini et al., 2005, 2010). In contrast, Daigle et al. (1999) made a Cheddar-like probiotic cheese from microfiltered milk standardized with cream enriched with native phosphocaseinate retentate. This method stimulated the growth of probiotic bacteria with no adverse effect on the cheese quality during 12 weeks of ripening. Cheddar cheese was demonstrated to be a suitable medium for the growth and viability of *L. paracasei* (Ross et al., 2002) and *B. infantis* (Daigle et al., 1999). Successful incorporation of *Bifidobacterium* spp. into Edam cheese was reported by Sabikhi et al. (2014). Since the growth and viability patterns of probiotics are affected by many factors including cheese variety, manufacturing methods of cheese, strain or species of probiotic bacteria and symbiosis of probiotics with cheese microflora, it is not easy to generalize a growth trend for probiotics. As a striking example, while *L. gasseri* K7 (a bacteriocin-gassericin K7-producing strain) remained viable at a level of 10^6 cfu/g in a 90-day old semi-hard cheese sample, the other probiotic strain of *L. gasseri* which was tested lost its viability within a shorter period (Zehnter, 2008). Combination of *L. gasseri* K7 with *Str. thermophilus* yielded better results regarding probiotic counts and cheese quality (Matijasic, 2008). This may be linked with a high oxygen consumption ability of *Str. thermophilus* which creates a microaerophilic environment at which probiotics grow better. A commercial strain of *L. acidophilus* La-5 grew poorly in 32 week-old Cheddar cheese sample (Phillips et al., 2006); however, this strain showed better survivability in white brined cheese (Yilmaztekin et al., 2004). During the last two decades, evaluations of the probiotic potential of *Enterococcus* spp. have been intensified (Gardiner et al., 1998) and some strains of *Enterococcus* spp. are now classified as probiotic (Özer and Kırmacı, 2011). Apart from the protective effect of the cheese matrix on probiotic cells during ripening, it may also affect the fate of probiotic bacteria in GIT. Human origin strains of *L. acidophilus* and *L. gasseri* delivered to the human body through fresh cheese were well-protected against bile salts and gastric conditions (Masuda et al., 2005). In order to investigate the influence of cheese matrix on the survivability of *L. rhamnosus* HN001 and *L. acidophilus* NCFM during simulated human digestion, Makelainen et al. (2009) used three different models including human upper gastrointestinal tract, human colon and colonocytes in cell culture. Probiotics in cheese were able to survive in the simulated upper GIT model, and counts of both probiotic strains were increased during the simulated colonic fermentation of the probiotic cheese. Viable cell counts of *L. acidophilus* (La-5), *L. casei* subsp. *paracasei*, (*L. casei* 01) and *B. animalis* subsp. *lactis* (Bb-12) used as adjunct cultures in the manufacture of semi-hard goat cheese (Coalho type) during simulated GIT passage was log 5.5-6.0 cfu/g as reported by de Oliveira et al. (2014). Meira et al. (2015) showed that goat ricotta was an efficacious food matrix for maintaining the viability of *L. acidophilus* La-5 and *B. animalis* Bb-12 under the conditions of simulated gastric digestion. The survival of *L. casei* LAFTI®L26, *L. acidophilus* LAFTI® L10 or *B. animalis* Bo incorporated into soft-textured whey cheese was monitored during artificial digestion stages of the mouth, oesophagus-stomach, duodenum and ileum (Madureira et al., 2011). Mouth conditions had almost no effect on the viability of the probiotic strains but the numbers of *L. casei* and *L. acidophilus* decreased slightly under oesophagus-stomach, duodenum and ileum conditions. An *in vitro* human colon simulation study showed that *L. rhamnosus* HN001 or *L. acidophilus* HCFM used as adjunct probiotics in Gouda cheese survived the simulated passage through the upper GIT and the counts of these probiotics increased during colonic fermentation simulation (Ouwehand et al., 2010). According to a human intervention trial, consuming 15 g of probiotic Gouda cheese (equals to 10^9 cfu/day) increased both natural killer (NK) cells and phagocytic activities (Ouwehand et al., 2010).

Bacterial heteropolysaccharides may protect cells against heat, acid, bile and salt stress conditions *in vitro*. *Lactobacillus mucosae* DPC 6426 is a well-known exopolysaccharide producing probiotic strain with a potential hypocholesterolemic effect. *L. mucosae* DPC 6426 has a high GIT adhesion capability. Theoretically, this adhesion ability is expected to help in competitive exclusion of attaching and effacing enteric pathogens *in vivo* and therefore reducing the opportunity of infections (Ryan et al., 2015). Low-fat Cheddar and Swiss cheeses were reported to be suitable matrices for the delivery of *L. mucosae* DPC 6426 to the human body (Ryan et al., 2015). Pino et al. (2017) monitored the survivability of *L. rhamnosus* H25 and *L. paracasei* N24-potential probiotic strains- during passage through the GIT of healthy volunteers by using rep-PCR technique. The probiotic strains retrieved from faecal samples of healthy volunteers after 15 days of consumption of the experimental cheeses had a high level of survivability. This indicated that survivability of these strains during passage into the GIT was high. The fate of probiotics during passage through GIT has a strain-dependent characteristic. Pisano et al. (2008), for example, showed that 6 lactobacillus strains out of 14 strains demonstrated great resistance to GIT conditions. In Boursin-type goat cheese, *B. animalis* subsp. *lactis* showed higher resistance to the artificial gastric and enteric juices than *L. rhamnosus*, with average decreases in the initial populations of 0.2 and 4.0 log cfu/g after 35 days of storage, respectively (Martins et al., 2018). Almost all of the probiotic cheese studies presented in the literature are at laboratory or pilot scale. Most recently, Blaiotta et al. (2017) developed a commercially standardized method for the manufacture of probiotic Italico cheese using *L. rhamnosus* LbGG and SP1. *L. rhamnosus* LbGG and SP1 showed a fairly high tolerance to acid-gastric and duodenal stresses, up to 40 days of ripening at 4 °C, maintaining a viability level higher than 10^8 cfu/g.

Probiotic yeasts are widely used in the manufacture of many fermented foods including wine, beer, bread, cider, kefir and so on. Their probiotic potential has long been known and especially *Saccharomyces boulardii* is associated with the general immune stimulatory, anti-inflammatory, non-specific prebiotic effects and elimination of bacterial toxins and pathogen binding (Arevalo-Villena et al., 2017). Probiotic strains of yeasts show high hydrophobicity, auto-aggregation ability, biofilm formation (e.g. adhesion to glass sides), adhesion to cell lines (e.g. Caco-2, Mucin, TC7), antibiotic resistance and antimicrobial effects (Arevalo-Villena et al., 2017). *S. boulardii* is usually taken to the human body through oral administration in the form of tablets, pills etc. *S. boulardii* is known for its tolerance to gastric acidity, bile salts and pancreatic fluids (Czerucka et al., 2007). Incorporation of probiotic strains of yeasts to dairy products is not common due to the detrimental effect of CO_2 and other metabolites of yeasts on physical and sensory properties of dairy foods such as yogurt and cheese. Karaolis et al. (2013) incorporated *S. boulardii* into non-fat goat milk yogurt successfully. The counts of *S. boulardii* were high enough for a probiotic effect during 4 weeks of storage at 6 °C and no organoleptic defects in the samples were noted during that period. Potent anti-microbial properties of some probiotic yeasts may have a potential of being used in cheese manufacture. *Kluyveromyces marxianus* S-2-05 and *Kluyveromyces lactis* S-3-05 which were new strains isolated from French cheese Tomme d'Orchies showed an anti-Salmonella activity and down regulation of the virulance *sopD* gene of *Salmonella enterica* subsp. *enterica* serovar. *typhimurium*. Apart from this successful antagonism against *Salmonella* spp., these yeasts were able to survive under simulated gastrointestinal conditions and also displayed a high hydrophilic cell surface and capacity to adhere to intestinal Caco-2 cells. All these properties imply high probiotic potential of these yeast strains (Ceugniez et al., 2017).

Viability of most bifidobacteria strains in cheese is affected by the salt level in the cheese matrix and exceeding the threshold level of 4-5% NaCl may cause a dramatic decrease in the colony counts of probiotic strains (Gobbetti et al., 1998). It was reported that *L. acidophilus* was susceptible to salt concentrations above 6% and a maximum 3.5% (w/w) salt in goat cheese provided counts of *B. animalis* subsp. *lactis* and *L. acidophilus* high enough for a probiotic effect (Gomes and Malcata, 1998). The brined-type cheeses have a relatively higher level of NaCl than pasta-filata type cheeses. Slight modifications in the production technology of brined-type cheeses may effectively increase the counts of probiotic bacteria to a level of $>10^6$ cfu/g (Samona and Robinson, 1994, Gomes et al., 1995). One of the most applicable solutions is to increase the initial level of probiotic inoculation. Gomes et al. (1995) recommended an inoculum level of 3.5-7.0% (corresponding to average levels of 2.0×10^9 cfu/g and $3\text{-}4\times10^9$ cfu/g, respectively) of probiotic bacteria to obtain a proper acidification rate in Gouda cheese. Partial replacement of NaCl with KCl and addition of arginine to cheese milk may also yield a cheese (i.e. Minas cheese) with a high sensory acceptability and probiotic counts (Felicio et al., 2016).

Yilmaztekin et al. (2004) compared two inoculum levels of *L. acidophilus* La-5 and *B. bifidum* Bb-02 (2.5% and 5.0% corresponding to average levels of 1.0-1.3x10⁹ cfu/g and 2.0-2.1 x10⁹ cfu/g, respectively) in white brined Turkish cheese. Although the growth trend of probiotic bacteria was the same in both the cheeses, the sample with higher inoculum had higher cell counts for both probiotic bacteria throughout 90-day ripening. Similar findings were published for white brined cheese (Ghoddushi and Robinson, 1996), cottage cheese (Blanchette et al., 1996) and fresh cheese (Roy et al., 1997) as well. In brined type cheese, salt penetration from the brine into the cheese block normally ends within a few weeks, depending on the salt concentration of brine, brine temperature, size and texture of the cheese block and their pH. Therefore, it is fair to expect that the loss of probiotic bacteria in cheese occurs at the early stage of ripening in brine. Daigle et al. (1999) showed rapid decline in the counts of *B. bifidum* in Cheddar cheese during the first week of ripening (the loss of 0.5 log cycle) and the colony counts remained almost stable during the rest of the ripening period (11 weeks). Salt is the major but not the sole criteria that influences the growth and viability of probiotic cells. The type of salt in brine solution also affects the growth pattern and metabolic activities of probiotics-especially *Bifidobacterium* spp. It was reported that the morphology, acid-producing ability and other growth characteristics of bifidobacteria were affected by calcium and sodium ions in brine (Misra and Kuila, 1990, Modler et al., 1990, Samona and Robinson, 1991).

Probiotic bacteria have low proteolytic activity and weak growth patterns in milk (Ziarno et al., 2010). Therefore, it is not quite possible to make cheese by using probiotic bacteria alone. Probiotic adjunct cultures should be used in combination with proper mesophilic or thermophilic lactic acid bacteria in the manufacture of probiotic cheese. No antagonism should be present between probiotic adjunct culture and cheese starter bacteria. In order to stimulate the growth of probiotic bacteria which was used in combination with a mesophilic starter culture, the ripening temperature was proposed to be increased from 6 °C to 14 °C (Ziarno et al., 2010). On the contrary, Gobbetti et al. (1998) showed that the lactic acid produced by thermophilic *Str. thermophilus* caused a decrease in the counts of *B. infantis* within 14 days of storage in Crescenze cheese, while the numbers of *B. bifidum* and *B. longum* increased 1 to 2 log cycles under the same conditions. Souza and Saad (2009) investigated the symbiosis between thermophilic yogurt starter bacteria and *L. acidophilus* La-5 with thermophilic yogurt starters in Minas fresh cheese. The authors found that thermophilic yogurt bacteria did not affect the growth of *L. acidophilus* La-5 but the overall quality of cheese improved, except for a slight post-acidification problem (Souza et al., 2008, Souza and Saad, 2009). It is possible to manufacture probiotic fresh cheese with a high sensory quality by using a combination of probiotic ABT cultures (*L. acidophilus* La-5, *B. animalis* subsp. *lactis* Bb-12 and *Str. thermophilus*) and mesophilic cheese starters (Marcatti et al., 2009, Fritzen-Freire et al., 2010). Buritti et al. (2007) proposed to replace ABT cultures with classical O-type cultures. Gürsoy and Kınık (2010) stated that *L. paracasei* subsp. *paracasei, Enterococcus faecium* and *B. bifidum* showed no antagonistic effects against lactococcal cheese starters in white brined Turkish cheese. Although a majority of probiotic strains show rather weak proteolytic activity, whey cheese made with *B. animalis* subsp. *lactis*, *L. acidophilus* and *L. casei* as probiotic adjunct has a high ACE-I inhibitory peptide activity, which is more pronounced in cheese with *B. animalis* subsp. *lactis* and *L. casei* (Madureira et al., 2012). Specifically peptides having a molecular weight in a range <3 kDa exhibited higher ACE inhibitory properties.

In order to protect probiotic bacteria against undesirable environmental conditions in a cheese matrix, a number of strategies/methods have been developed. Each of these techniques have advantages and disadvantages. Among these methods (some of them are discussed in section 3.1 of this chapter) microencapsulation seems more promising (Krasaekoopt et al., 2003). Özer et al. (2009) demonstrated that the number of encapsulated probiotic bacteria in white brined cheese decreased by about 1 log cycle after 90 days. The decrease in the number of control cheese samples inoculated with probiotics in free state was 3 log cycles. In contrast, it was claimed that the loss of microencapsulated probiotic bacteria in Feta cheese was higher, probably due to restricted interaction of encapsulated probiotics with the environment for survival or limited disposal of cell metabolites that may be accumulated within the capsules resulting in cell death (Godward and Kailasapathy, 2003). Feta cheese has an open texture, allowing salt penetration from brine to cheese faster and more efficiently. Disintegration of probiotic microcapsules in brine solution and higher salt uptake are more likely (Kailasapathy and Masondole, 2005, Özer and Kırmacı, 2011). Mobility of metabolites in and out of the microcapsules is widely affected by the texture of the cheese matrix. Restricted exchange of cell metabolites from and to microcapsules in Cheddar cheese was reported by Godward (2000). The degree of protection of probiotic cells by microencapsulation depends not only on cheese texture but also on microcapsule matrix design which affects the disintegration rate of capsules. A microcapsule is made up of a semipermeable, spherical, thin and strong membranous wall (Kailasapathy and Masondole, 2005). Nutrients and metabolites move in and out of the capsules through this semipermeable membrane. Emulsion, extrusion, interfacial polymerization and/or coacervation are the most common microencapsulation techniques. The viability of probiotic bacteria could be extended up to 80-95% by emulsion or extrusion techniques (Doleyres and Lacroix, 2005). κ-carrageenan or calcium-alginate are the most common support materials used in the preparation of biopolymer gels in which cells are entrapped. It was reported that while the viability of *L. rhamnosus* colonies microencapsulated in alginate matrix was maintained for 48 h at pH 2.0, cells in free state were inactivated completely under the same conditions (Goderska et al., 2003). Some examples of use of microencapsulated and immobilized probiotic bacteria in cheese production are presented in Table 3.

Table 3: Some examples of microencapsulation and immobilization of probiotic cells for cheese production

Cheese type	Probiotic bacteria	Method	Supporting material/ immobilization medium	Source
Cheddar	*Lb. paracasei*	Milk fat	-	Stanton et al., 1998
White Brined Turkish cheese Kasar cheese	*Lb.acidophilus +B. bifidum* *Lb.acidophilus +B. bifidum*	Extrusion Extrusion	Ca-alginate Ca-alginate	Özer et al., 2009 Özer et al., 2008
White Brined Turkish cheese Kasar cheese	*Lb.acidophilus +B. bifidum* *Lb.acidophilus +B. bifidum*	Emulsion Emulsion	κ-carrageenan κ-carrageenan	Özer et al., 2009 Özer et al., 2008
Cheddar	*Ent. faecium*	Milk fat	-	Gardiner et al., 1998
White brined cheese	*B. bifidum,* *B. adolescentis*	Cream		Ghoddusi and Robinson, 1996
Crescenza	*B. bifidum* *B. infantis* *B. longum*		Ca-alginate	Gobbetti et al., 1997
Feta	*Lb. acidophilus* *B. lactis*	Emulsion	Ca-alginate	Kailasapathy and Masondole, 2005
Cheddar	*Lb. acidophilus* *B. infantis*	Emulsion	Ca-alginate	Godward and Kailasapathy, 2003
Part-skim Mozzarella cheese	*Lactobacillus paracasei* ssp. *paracasei* LBC-1	Emulsion	Ca-alginate	Ortakci et al., 2012
Feta	*Lb. casei*	Immobilization	Fruit pieces	Kourkoutas et al., 2006
Cheddar	*B. bifidum*	Immobilization	κ-carrageenan	Dinakar and Mistry, 1994

(Contd.)

Table 3: (*Contd.*)

Cheese type	Probiotic bacteria	Method	Supporting material/ immobilization medium	Source
Feta	Lb. casei ATCC 393	Immobilization	Sea buckthorn berries (superfood)	Terpeu et al., 2017c
Whey cheese	Lb. casei ATCC 393 Lb. delbrueckii subsp. bulgaricus ATCC 11842	Immobilization	Casein	Dimitrellou et al., 2017

The use of prebiotics in probiotic cheese-making is not as common as probiotic yogurt manufacturing. The combination of inulin, oligofructose and oligosaccharide from honey was reported to stimulate the growth of *B. animalis* subsp. *lactis* and *L. acidophilus* in synbiotic Petit-Suisse cheese without affecting the sensory quality of the end product (Cardarelli et al., 2007, 2008). Similarly, inulin HPX effectively promoted the growth of *L. plantarum* in soft cheese (Modzelewska-Kapitula et al., 2007). On the other hand, FOS was found to be ineffective in stimulating the growth of bifidobacteria in Edam cheese (Hayes et al., 1996). Fructans added to cheese milk are not degraded both by probiotics and cheese starter bacteria (Buriti et al., 2007).

Biopreservative compounds may stimulate the growth and viability of probiotic microorganisms in cheese. Some metabolic products produced by *Propionibacterium thoenii* P-127 demonstrated promotion of growth of *B. bifidum* in Domiati cheese (El-Kholy et al., 2006). On the other hand, probiotic bacteria have a more inhibitory capacity on lactic acid bacteria and pathogens than *vice versa* (Vinderola et al., 2002a). Biopreservative compounds generated by probiotic bacteria such as hydrogen peroxide, alcoholic compounds, diacetyl and bacteriocins may show hostile effects on pathogenic and, in some cases, non-pathogenic lactic acid bacteria (Buriti et al., 2007). For example, *L. paracasei* ssp. *paracasei*, *L. rhamnosus*, *L. acidophilus*, *B. animalis* subsp. *lactis* and *Propionibacterium* spp. were demonstrated antimicrobial effects on *Aspergillus niger*, *Penicillium roqueforti*, *Candida albicans* and *Saccharomyces cerevisiae* in cheese-based dips (Tharmaraj and Shah, 2009). Similarly, *L. helveticus* CU631 in cream cheese had a strong antagonistic effect against *Helicobacter pylori* (Song et al., 2001). *L. gasseri* LF221 and K7 were able to inhibit the growth of *Clostridium* spp. and hence partly eliminated the risk of late blowing in cheese during ripening (Perko et al., 2002).

Improving cold and acid tolerances of probiotic bacteria by exposing them to sub-lethal levels of the given stress is another strategy to increase the number of viable bacteria (Roy, 2005). Collado and Sanz (2006) showed that the recovered strains of *B. breve* and *B. adolescentis* obtained after prolonged exposure to human faeces and homologous lethal stress conditions were resistant to high concentrations of bile salts and sodium chloride, making these strains suitable for cheese-making. Combination of sub-lethal conditions may be more effective in enhancing the viability of probiotic cells. Maus and Ingham (2003) demonstrated that the exposure of early stationary phase bacterial cells to combinations of reduced temperature (~6 °C), reduced pH (improved resistance at a pH of 3.5 when the growth medium pH is decreased from 6.0 to 5.2) and starvation (30 to 60 min under reduced pH and temperature conditions) could enhance the viability of some strains of *B. longum* and *B. animalis*. These strains are considered to be suitable for the manufacture of fresh cheeses and other varieties with low pH. Thermo-tolerance of probiotic bacteria can be improved by means of pressure pre-treatment. Pressure-treated *L. rhamnosus* GG remained viable after exposure to heat treatment at 60 °C, as reported by Ananta and Knorr (2004).

Choice of packaging materials for probiotic cheese is critical for the viability of probiotic bacteria. For cultured foods, high impact polystyrene packaging materials with a wall thickness of 300-350 µm are suitable (Miller et al., 2002, Sarkar, 2010). Wall thickness of 300-350 µm allows an oxygen diffusion rate of 1.0 to 5.0 $cm^2/kg/d$. For probiotic cheeses, vacuum packaging with low oxygen permeability is ideal. It was reported that the survival of *L. acidophilus* in white brined cheese ripened in a vacuum package at 4°C for 12 weeks was satisfactory enough (i.e. >10^7 cfu/g) (Kasımoğlu et al., 2004). A brief summary of research outcomes of various probiotic cheeses is given in Table 4.

Presence of probiotic bacteria in foods is usually determined with culture-dependent methods which are time-consuming and generally imprecise (Desfossés-Foucault et al., 2012). Viable but non-cultivable bacteria cannot be detected by traditional culturing techniques and it is not always possible to differentiate closely related lactic acid bacteria in the cheese microbial community (Ndoye et al., 2011). Molecular-based techniques such as quantitative-polymerase chain reaction (qPCR) offer alternatives to quantify bacterial cells more precisely. Since DNA can be detected after cell death, RNA-seq can be a better option to monitor the cell viability. RNA-seq is based on obtaining a digital expression from high throughput sequencing and has greater sensitivity, greater specificity and nucleotide level resolution than the cDNA microarray technique (Wang et al., 2009). RNA-Seq offers a deeper understanding of how the yogurt environment may affect probiotic organisms during and after fermentation and facilitates an increase in the probiotic effects of cultured dairy products (Bisanz et al., 2014). However, RNA is fairly fragile and requires an additional step of retrotranscription to cDNA prior to further analysis (Ndoye et al., 2011, Desfossés-Foucault et al., 2012). All these methods require extra care and the possibility of analytical errors is high. To avoid such problems, a novel method which is based on blocking amplification of dead cells by chemical means such as propidium monoazide (PMA-qPCA) has been developed. PMA is able to penetrate only cell membranes of dead cells, and with exposure to light it binds to DNA covalently and inhibits amplification (Nocker and Camper, 2009). Ganesan et al. (2014) used PMA followed by species-specific qPCR to identify probiotic lactobacilli and bifidobacteria in Cheddar cheese. They found that probiotic adjunct cultures at levels 100-fold below that of cheese starter bacteria, could modify starter and non-starter bacterial levels in commercial Cheddar cheese. Similar to PMA, a DNA cross-linking agent and eukaryotic topoisomerase II poison ethidium monoazide (EMA) is used for discrimination of live and dead microorganisms in pasteurized milk. EMA + Light + T-poisons inhibited the PCR product of dead *Listeria monocytogenes* cells (Soejima et al., 2008). Molecular assessments of microorganisms are done by different nucleic acid-based methods. Due to the diversity of the death microorganisms, different methods are needed to define and differentiate results of nucleic acid methods varying from PCR to metagenomics (Cangelosi and Meschke, 2014). The discrimination of live and dead microorganisms is important for

Table 4: Summary of research outcomes of probiotic cheeses

Product	Probiotic bacteria included	Remarks	Source
Soft and semi-soft cheeses			
Probiotic fresh cheese	Human origin *Lactobacillus salivarius* strains CECT5713 and PS2 (8 $\log_{10}$ cfu/g)	Probiotic strains were added after cooking at 38 °C for 40 min and whey drainage. Up to 21 days, the viability of probiotics were high enough and the cheeses had acceptable sensory properties.	Cárdenas et al., 2014
Cream cheese with added dried tomatoes	*Lb. paracasei* Lpc-37	The population of *Lb. paracasei* remained greater than 10^7 cfu/g after 21 days of storage at 4 °C. Sensory properties of the end product were acceptable.	Santini et al., 2012
Reduced sodium probiotic Cottage cheese	*B. lactis*, *Lb. acidophilus*	Both probiotic bacteria effectively suppressed the growth of *Listeria monocytogenes* in cheese	Jesus et al., 2016
Minas Frescal cheese	*Lb. casei* Zhang	Cheeses presented a viscous-like rheological behaviour with time-dependent decrease in rigidity. Organoleptically the probiotic cheeses were disliked	Dantas et al., 2016
Prato cheese with reduced salt content	*Lb. casei* 01	Sodium reduction (substitution of NaCl with KCl) and supplementation with flavour enhancer did not affect the counts of probiotics in cheese. Flavour enhancer did not constitute an obstacle to the organoleptical properties of the end product.	Silva et al., 2017
Pont-L'Eveque cheese	*Lb. plantarum* UCMA 3037	The strain which was isolated from Camembert cheese and characterized as acid- and salt resistant, grew well and survived in sufficient numbers in Pont-L'Eveque cheese during ripening period of 75 days. The strain did not affect the sensory quality of the cheese	Coeuret et al., 2004

(Contd.)

Table 4: (*Contd.*)

Product	Probiotic bacteria included	Remarks	Source
Camembert	Lb. paracasei ssp.	Lb. paracasei sp. have the ability to grow more than 2 log units during cheese ripening and lower inoculum level of this probiotic bacteria than 10^6 cfu/mL to cheese milk could be posible.	van de Casteele et al., 2003
Semi-soft cheese	Lb. acidophilus NCFM, Lb. rhamnosus HN001	Lb. acidophilus NCFM and Lb. rhamnosus HN001 tolerated the simulated GIT conditions and their numbers were increased in the colonic fermentaton simulations. Cyclooxygenase-gene expression of colonocytes in a cell culture model was positively affected by the cheese matrix	Makelainen et al., 2009
Sweet whey cheese added with sugar alone or combinations of sugar-aloe vera, sugar-chocolate, sugar-jam	Lb. paracasei LAFTI®L26	Number of the probiotic strain was satisfactorily high at the end of storage period of 21 days at 7 °C. Lactose in the cheese matrix was partly hydrolysed by Lb. paracasei. Additives enhanced the sensory properties of the cheese but not affected the colony counts	Madureira et al., 2008
Cottage cheese	Lb. acidophilus UFVH2b20	After storage of cheese for 3 months at 5 °C, the numbers of Lb. acidophilus UFVH2b20 was over therapeutic minimum (>10^8 cfu/g)	Araujo et al., 2009
Semi-hard cheeses			
Brazilian semi-hard goat cheese (coalho)	Lb.acidophilus, Lb. paracasei or B. lactis or mixture of three probiotic strains	No marginal difference between cheeses was noted regarding proteolytic profiles. Probiotic counts were between 6-7 log cfu/g at day 7. Combination of cheese startes plus three probiotic mixtures yielded better organoleptical properties in cheese.	Garcia et al., 2012

Dutch-type cheese	Lb. rhamnosus HN001	Lb. rhamnosus HN001 increased the availability of calcium and magnesium from Dutch-type cheese significantly	Aljewicz and Cichosz, 2015
Semi-hard rennet cheese	6 strains of probiotic bacteria (3 of Lb. acidophilus and 3 of Lb. casei group)	Lb. casei group contributed to the cheese peptidolysis at lower degrees. Lb. acidophilus group was more effective on secondary proteolysis in cheese and therfore could be more suitable for probiotic cheese-making	Bergamini et al., 2009
Semi-hard rennet cheese	Lb. acidophilus, Lb. paracasei	Pre-incubation of cheese milk with probiotic cultures caused a crumbly body and low pH as well as high numbers of lactobacilli. Simultaneous inoculation of probiotics with cheese starters could be a better option regarding cheese quality.	Bergamini et al., 2005
Ras cheese	Lb. reuteri, Lb. casei, Lb. gasseri	The probiotic strains showed a potent antimycotic effect in cheese. The numbers of both probiotic strains were higher than 10^6 cfu/g at the end of ripening	Dabiza and Fathi, 2006
Hard cheeses			
Cheddar cheese	Lb. helveticus R0052, B. longum R0175 B. lactis BB-12, B. longum 15708, B. infantis 15697	Inoculation before cheddaring was more effective regarding viability of probiotics. Salting caused about 13% loss in viability of probiotics and B. longum 15708 was the more sensitive strain to cheese environment.	Fortin et al., 2011
Cheddar cheese	Lb. plantarum K25	Cheese with Lb. plantarum K25 showed that the levels of serum total cholesterol, low density lipoprotein cholesterol and triglycerides decreased significantly, and the level of serum high-density lipoprotein cholesterol increased in mice fed with the probiotic cheese.	Zhang et al., 2013

(Contd.)

Table 4: (*Contd.*)

Product	Probiotic bacteria included	Remarks	Source
Cheddar cheese	*B. longum* 1941, *B. animalis* ssp. *lactis* LAFTI◊B94, *Lb. casei* 279, *Lb. casei* LAFTI◊ L26, *Lb.acidophilus* 4962, *Lb.acidophilus* LAFTI◊L10	The numbers of probiotic bacteria in cheeses ripened at 4 °C or 8 °C for 24 weeks declined from 9.10-9.74 $\log_{10}$ to 6.81-7.53 $\log_{10}$ cfu/g. Cheese made with *Bifidobacterium* spp. or *Lb. casei* 279 had bitter-acidic and vinegar tastes.	Ong and Shah, 2009
Swiss- and Dutch-type cheeses	*Lb. acidophilus* NCFM or *Lb. rhamnosus* HN001 or *Lb. paracasei* LPC-37	*Lactobacillus* sp. used in the study are characterized by good viability and have no negative influence on starter cultures in cheese production	Aljewicz and Cichosz, 2016
Pecorino cheese	*Lb. acidophilus*, *B. longum*, *B. lactis*	Cheese inoculated with lamb rennet paste containing *Lb.acidophilus* had high levels of conjugated linoleic acid. Cheese made with lamb paste containing bifidobacterium group had high levels of free linoleic acids.	Santillo et al., 2009
Kashar cheese	*Lb. acidophilus* La5, *B. animalis* subsp. *lactis* Bb-12 (in microencapsulated form)	Microencapsulated probiotic strains survived scalding and remained viable at high levels ($>10^6$ cfu/g) during storage period in Kashar cheese	Özer et al., 2008
Kefalotyri-type hard cheese (from caprine milk)	*Lb.rhamnosus* LC705, *Lb.paracasei* ssp. *paracasei* DC412	Degrees of hydrolyzation of α_s-CN and β-CN and lipolysis in the probiotic cheeses were higher than the control cheese. Probiotic cheese received higher sensory scores than the control sample	Kalavrouzioti et al., 2005
Briend cheese			
White brined cheese supplemented with *Cuminum cyminum* L.	*Lb. acidophilus* DK2970	Combination of *Cuminum cyminum* L. essential oils (15 mL/100 mL) and *Lb. acidophilus* (0.5%) showed a remarkable anti-microbial effect on *Staphylococcus aureus* in Iraninan White brined cheese.	Sadeghi et al., 2013

Product	Probiotic	Description	Reference
Feta cheese	*P. freudenreichii* subsp. *shermanii* LMG 16424	The presence of *P. freudenreichii* subsp. *shermanii* was monitored by 16S rRNA during 60 days of storage. Feta cheese contained high number of probiotic adjunct and the sensory quality of the end product was acceptable.	Angelopoulou et al., 2017
Turkish white cheese	*Lb. acidophilus* La-5, *B. animalis* Bb12	Cheese inoculated with mixed probiotic strains at a level of 5% (v/v) had higher ripening coefficient than the cheese inoculated with 22.5% (v/v) probiotic adjunct culture. After 90 days, the cheese had probiotic colony counts over 10^6 cfu/g.	Özer et al., 2009 Yilmaztekin et al., 2004
Turkish white cheese	*Lb. acidophilus* La-5, *B. animalis* subsp. *lactis* Bb-12 (in microencapsulated form)	The decrease in the numbers of microencapsulated probiotic cells was limited (1 log cycle after 90 days). Medium- and long-chain free fatty acid contents of the cheeses with immobilized probiotics were much higher than that of the control cheese.	Özer et al.,2009
Turkish white cheese	*Lb. fermentum* AB5-18, *Lb. fermentum* AK4-120, *Lb. plantarum* AB16-65, *Lb. plantarum* AC18-42	Probiotic culture mix did not affect the quality parameters of the cheese and the counts of the probiotic bacteria was high enough for probiotic effect.	Kiliç et al., 2009
Turkish white cheese	*E. faecium*, *Lb. paracasei* subsp. *paracasei*, *B. bifidum*	Probiotic bacteria did not cause any detrimental effect on the quality of cheese and white cheese matrix was suitable medium for delivery of especially *B. bifidum*.	Gürsoy and Kınık, 2010
Iranian White cheese	*B. bifidum*, *B. adolescentis*	After 60 days of ripening at 6 °C, the numbers of bifidobacteria were over 10^7 cfu/g.	Ghoddusi and Robinson, 1996

determination of the source of biological activity of all the microorganisms and metabolites. Dead microorganisms have the capacity to stimulate human biological systems and can be used in vaccines (Sander et al., 2011).

3.3 Probiotic Milk- and Whey-based Beverages

Among the functional beverages, probiotic dairy drinks were the first commercialized products and, at present, their consumption rates are higher than other functional beverages. Although yoghurt and cheese are considered as suitable matrices for the delivery of probiotic bacteria into human body, probiotic beverages also offer certain advantages to both manufacturers and consumers. Various commercial fermented and non-fermented dairy-based probiotic beverages are available in the global dairy markets (Özer and Kırmacı, 2010). Some of these products are produced using a single probiotic strain and some are made with mixed probiotic strains (*see* Table 5). *B. animalis* subsp *lactis* Bb-12 (previously *B. bifidum* Bb-12), *L. acidophilus* La-5 and *L. rhamnosus* GG (LGG, acquired by Chr-Hansen A/S, Denmark) which have well-established probiotic backgrounds, are the most commonly used commercial strains in various dairy products including beverages. Due to the acidic nature of fermented probiotic dairy beverages, strains used in the manufacture of such products should be acid-tolerant.

Lactobacillus GG, for example, is known for its acid and bile-tolerant properties and has been used in the manufacture of Gefilus® (Endo et al., 2014, Kumar et al., 2015). *L. acidophilus* and *B. animalis* subsp. *lactis* are another species which are widely used in the production of well-known probiotic dairy beverages alone (i.e. acidophilus milk, bifidus milk, Bifigurt®) or in combination with other lactic acid bacteria (i.e. sweet acidophilus milk, acidophilin, acidophilus-yeast milk, Biomild, Cultura®, Nu-Trish, Yakult Miru-Miru, Biomild, Cultura®, Diphilus milk). In general, dairy-based fermented probiotic beverages are incubated at 37 °C and since most of the probiotic strains are slow acid-producers, the fermentation period may take about 14-16 h. While the fermentation of acidophilus milk is ended at a pH of 5.5, some fermented probiotic beverages including A/B milk, bifidus milk, acidophilus-bifidus milk, Bifighurt are fermented until a pH of 4.4-4.6 is reached (Akal et al., 2019).

L. casei Shirota has long been used as a fermenting bacterium in the manufacture of Japanese fermented probiotic beverages, namely Yakult, Yakult Miru-Miru and Mil-Mil. Since Yakult contains high level of sugar (~14%), high heat treatment (i.e. UHT treatment) to milk triggers Maillard reaction between added glucose and milk proteins, leading to light coffee brown color in the product (Surono and Hosono, 2002). Addition of flavoring agents to Yakult (i.e. tomato, celery, cabbage, carrot *etc.*) is a common practice. The production practices of Yakult Miru-Miru and Mil-Mil are fairly similar to each other and fermentation of both products is achieved by a mixed culture of *B. bifidum* (or *B. breve* in Mil-Mil case and *L. paracasei* subsp. *paracasei* in Miru-Miru case) and *L. acidophilus*. The viability of *L. acidophilus*

Table 5: Some examples of commerical probiotic dairy drinks (Based on Gürakan et al., 2010 and Özer and Kırmacı, 2011)

Product	Lb. acidophilus	Lb. casei Immunitus™	Lb.rhamnosus GG	Lb. johnsonii	Lb. helveticus	Lb. casei	Lb. plantarum	Lb. rhamnosus	Lb. casei Shirota	B. bifidum[7]	B. longum	B. breve	Bifidobacterium spp.
Acidophilus milk	+												
Sweet acidophilus millk	+												
Acidophilus buttermilk[1]	+												
Acidophilus-yeast milk[2]	+												
Acidophilin[3]	+												
A-38 fermented milk	+												
Actimel		+											
Aktifit, Biola, BioAktiv, YOMO, LGG+, Yoplait 360°, Kaiku Actif			+										
Bifidus mik										+	or +		
Bifighurt®											+		
Biomild	+											+	
Cultura® or A/B milk	+											+	
CHAMYTO				+	+								
Diphilus milk	+											+	
Gaio						+[8]							
Nu-Trish A/B	+												+
Nu-Trish plus ABC[4]	+												+

Table 5: (Contd.)

Product	Lb. acidophilus	Lb. casei Immunitus[2a]	Lb.rhamnosus GG	Lb. johnsonii	Lb. helveticus	Lb. casei	Lb. plantarum	Lb. rhamnosus	Lb. casei Shirota	B. bifidum[7]	B. longum	B. breve	Bifidobacterium spp.
Onaka He GG, Gefilus[5]			+										
Procult drink[5]											+		
ProViva							+[9]						
Verum								+[10]					
Vitagen	+												
Yakult									+				
Yakult Miru-Miru[6]	+									+		or+	

Note: [1]Plus Lc.lactis subsp. lactis, subsp. cremoris, subsp lactis biovar diacetylactis, [2]Plus Saccharomyces lactis, [3]Lc. lactis subsp. lactis, kefir grains, [4]Plus calcium, [5]Plus yoghurt cultures, [6]In alternative productions B. bifidum is replaced by Lb. paracasei subsp. paracasei, [7]B. bifidum Bb-12 was redefined as B. animalis subsp. lactis Bb-12, [8]Strain F9, [9]Strain 299v, [10]Strain LB21 in carbonated AT milk (made from milk inoculated with *L. acidophilus* and *Str. thermophilus*) was not affected by CO_2 treatment but the joint effect of CO_2 and *B. bifidum* led to a decrease in the counts of *L. acidophilus* in carbonated ABT milk (made from milk inoculated with *B. bifidum*, *L. acidophilus* and *Str. thermophilus*) (Vinderola et al., 2000b). Walsh et al. (2014) reported that neither the bacterial growth nor the sensory properties of carbonated synbiotic milk-based beverages was affected adversely by carbonation.

Similar to probiotic yogurt production, inulin or corn fibers may be incorporated to probiotic dairy-based beverages as growth-enhancers for probiotic microorganisms. The stimulatory effect of prebiotic substances may vary depending on the type of prebiotics and food structure. For example, inulin or corn fiber did not promote the growth of *B. animalis* Bb-12 and *L. acidophilus* La-5 in a synbiotic dairy beverage but polydextrose had positively affected the viability of same strains under cold storage conditions (Algeyer et al., 2010). Synbiotic dairy-based beverages supplemented with raffinose (Martinez-Vilalluenga et al., 2006), oligofructose (de Castro et al., 2009, Oliveira et al. 2009) and inulin (Varga et al., 2006) were reported. More recently, Coman et al. (2013) showed that supplementation of synbiotic fermented milk with buckwheat flour and oat bran helped to protect the viability of *L. rhamnosus* IMC501, *L. paracasei* IMC502 and SYMBIO®. Addition of carrot juice to probiotic dairy beverages stimulated the viability of *L. acidophilus* La-5, *L. rhamnosus* GG, *L. plantarum* and *B. animalis* subsp. *lactis* Bb-12 to between 88% and 98% after 3 weeks of storage (Daneshi et al., 2013). Cyanobacteria or blue-green microalgae (e.g. *Spirulina platensis*) also demonstrated promotion of the viability of mesophilic dairy starter cultures (Molnar et al., 2005). The growth stimulatory effect of cyanobacterial biomass stems from adenine, hypoxanthine and free amino acids (Beheshtipour et al., 2013). Supplementation of dry biomass from *Spirulina platensis* stimulates the viability of *Lactobacillus* spp. but its growth-stimulatory effect on *B. bifidum* or *B. animalis* is rather limited, if any (Varga et al., 1999). *S. platensis, Chlorella vulgaris* affected the growth of *L. acidophilus* La-5 and *B. animalis* Bb-12 to a great extent (Beheshtipour et al., 2012). Addition of grape pomace extract to probiotic fermented goat's milk increased the total phenolic content of the beverage and contributed to the viability of *L. rhamnosus* HN001 (dos Santos et al., 2017).

Probiotics may offer protection against pathogenic bacteria such as *Salmonella enteritidis* serovar. *thypimurium*. de Leblanc et al. (2010) investigated the adjuvant effect of probiotic fermented milk containing *L. casei* DN-114-001 in the protection against *S. enteritidis* serovar Thypimurium using mouse models. In the case of administration of probiotic fermented milk to adult mice (before and after infection) helped maintaining the intestinal barrier against *Salmonella* Thypimurium, diminishing the spread of this pathogenic organism into deeper tissues. Similarly, administration of fermented beverage to the offspring after weaning from the mother during suckling period had a protective effect against *Salmonella* Thypimurium as well. The growth suppression effect of *B. animalis* Bb-12, *L. acidophilus* La-5 and *L. casei*

HN001 on survival of *Mycobacterium avium* subsp. *paratuberculosis* (MAP) in probiotic fermented milks was also demonstrated by van Brandt et al. (2011).

Cheese whey is another potentially suitable food matrix for the growth and viability of probiotics (Pescuma et al., 2010, Castro et al., 2013, Buriti et al., 2014). In recent years, research have been intensified to develop whey-based probiotic beverage formulations. A successful formulation from deproteinized whey fermented with *L. rhamnosus* NCDO 243, *B. bifidum* NCDO 2715 and *P. freudenreichii* subsp. *shermanii* with high sensory acceptability was developed by Maity et al. (2008). Other examples of fermented whey-based beverage formulations include UF whey-retentate (Pereira et al., 2015), fresh acid whey supplemented with buttermilk powder or sweet whey powder (Skryplonek and Jasińska, 2015), flavored beverage from mixture of UHT goat's milk and goat cheese whey (da Silveira et al., 2015), Channa-whey based beverages (Priti et al., 2017), fresh cheese whey supplemented with pineapple juice (Shukla et al., 2013) and whey-based goat milk beverage with added guava or soursop pulps (Buriti et al. 2014). The ratio of cheese whey in the probiotic beverage formulations is the limiting factor for the sensory acceptance of the product. Shukla et al. (2013) reported that the consumers' demand for beverages containing cheese whey in the formulation at a ratio higher than 65% decreased remarkably. Novel probiotic beverages should deliver a reliable organoleptic quality beyond their safety and functional properties. Various optimization models and methodologies are employed to produce foods with acceptable sensory properties. Faisal et al. (2017) used fuzzy logic analysis to evaluate sensory properties of three different probiotic whey beverages formulated from orange powder and flavor. One of the formulated beverages (contained 1 L of whey, 0.7 g of stabilizer, 8 g of sugar, 1% of orange powder and 0.4 mL of flavor) outperformed standard commercial whey beverages and other formulated beverages. Fuzzy logic study revealed that color and flavor were more decisive quality parameters than texture and taste.

Lollo et al. (2013) compared probiotic yogurt and probiotic whey beverages in terms of their immune-protection capacities using an exhausting physical-exercise protocol with Wistar rats. The probiotic samples were prepared by using yogurt starter cultures and *L. acidophilus* La-14 and *B. longum* Bl-05. It was shown that yogurt outperformed the whey samples in blood-cell indicators (neutrophils and lymphocytes ratios) and inflammatory and oxidative status (TNF-α, IL-1β, IL-6 and oxidative stress parameters).

3.4 Probiotic Ice Cream

Comparing to yogurt, cheese and beverages, ice cream is a rather new and innovative carrier for probiotics, prebiotics and synbiotics (Cruz et al., 2009). The survival rate of probiotic bacteria in ice cream without any prebiotic support is generally low (Turgut and Cakmakci, 2009), but supplementation of the ice cream mix with FOS or inulin helps maintain the viability of the probiotic bacteria (Akin et al., 2007). Apart from acting as a growth-promoter, inulin may be replaced partially with stabilizers in the ice cream mix (di Criscio et al., 2010). Microencapsulation may be an effective solution for keeping the viability of probiotic bacteria in ice cream but in order to overcome organoleptic defects in the final product, microcapsules should be embedded into a proper matrix (i.e. chocolate pieces) (El-Sayed et al., 2014, Champagne et al., 2015). On the other hand, Spigno et al. (2015) showed that spray drying of microcapsules made from inulin and sodium alginate gave 50.8% yield and 82% mortality in ice cream. Probiotic ice cream may also be manufactured from vegetable milks (soy, coconut) or composite milk (cow or coconut milk with soy milk) (Aboulfazli et al., 2016). The tolerance of *B. animalis* subsp. *lactis* Bb-12 and *L. acidophilus* La-5 to gastrointestinal conditions was improved in the composite milk ice cream (cow's milk + soy milk), as reported by Aboufazli and Baba (2015). Human trials to test the probiotic efficiency of ice cream are scarce. Ashwin et al. (2015) investigated the impact of probiotic ice cream consumption on levels of Salivary Mutans Streptococci (SMS) during and after the trial. During treatment, the SMS level declined but after 6 months the SMS levels of the treated group was similar to the baseline.

References

Aboufazli, F. and A.S. Baba. 2015. Effect of vegetable milk on survival of probiotics in fermented ice cream under gastrointestinal conditions. Food Sci. Technol. Res. 21(3): 391-397.

Aboufazli, F., A.B. Shori and A.S. Baba. 2016. Effects of the replacement of cow milk with vegetable milk on probiotics and nutritional profile of fermented ice cream. LWT-Food Sci. Technol. 70: 261-270.

Abraham, S., R. Cachon, B. Colas, G. Feron and J. de Coninck. 2007. Eh and pH gradients in Camembert cheese during ripening: Measurements using microelectrodes and correlations with texture. Int. Dairy J. 17: 954-960.

Agil, R., A. Gaget, J. Gliwa, T.J. Avis, W.G. Wilmore, F. Hosseinian et al. 2013. Lentils enhance probiotic growth in yogurt and provide added benefit of antioxidant protection. LWT-Food Sci. Technol. 50: 45-49.

Akal, C., N. Turkmen and B. Özer. 2019. Technology of dairy-based beverages. pp. 331-372. *In*: A.M. Grumezescu and A.M. Holban (eds.). Milk-Based Beverages, Chapter 10. Elsevier, CA, USA.

Akin, M.B., M.S. Akin and Z. Kirmaci. 2007. Effects of inulin and sugar levels on the viability of yogurt and probiotic bacteria and the physical and sensory characteristics in probiotic ice cream. Food Chem. 104(1): 93-99.

Algeyer, L.C., M.J. Miller and S.Y. Lee. 2010. Drivers of liking yogurt drinks with prebiotics and probiotics. J. Food Sci. 75: S212-S219.

Aljewicz, M. and G. Cichosz. 2015. The effect of probiotic *Lactobacillus rhamnosus* HN001 on the *in vitro* availability of minerals from cheeses and cheese-like products. LWT-Food Sci. Technol. 60: 841-847.

Aljewicz, M. and G. Cichosz. 2016. Influence of probiotic (*Lactobacillus acidophilus* NCFM, *L. paracasei* LPC37, and *L. rhamnosus* HN001) strains on starter cultures and secondary microflora in Swiss- and Dutch-type cheeses. J. Food Proces. Preserv. Article no: 41:e13253.

Alvarez-Calatayud, G. and A. Margolles. 2016. Dual-coated lactic acid bacteria: An emerging innovative technology in the fields of probiotics. Future Microbiol. 11(3): 467-475.

Ananta, E. and D. Knorr. 2004. Evidence on the role of protein biosynthesis in the induction of heat tolerance of *Lactobacillus rhamnosus* GG by pressure pre-treatment. Int. J. Food Microbiol. 96(3): 307-313.

Angelopoulou, A., V. Alexandraki, M. Georgalaki, R. Anastasiou, E. Manolopoulou, E. Tsakalidou et al. 2017. Producton of probiotic Feta cheese using *Propionibacterium freudenreichii* subsp. *shermanii* as adjunct. Int. Dairy J. 66: 135-139.

Araujo, E.A., A.F. de Carvalho, E.S. Leandro, M.N. Furtado and C.A. de Moraes. 2009. Production of cottage-like symbiotic cheese and study of probiotic cells survival when exposed to different stress levels. Pesquisa Agropecuaria Tropical, 39(2): 111-118.

Arevalo-Villena, A., A. Briones-Perez, M.R. Corbo, M. Sinigaglia and A. Bevilacqua. 2017. Biotechnological application of yeasts in food science: Starter cultures, probiotics and enzyme production. J. Appl. Microbiol. 123: 1360-1372.

Ashwin, D., V. Ke, M. Taranath, N. Ramagoni, A. Nara, M. Sarpangala et al. 2015. Effect of probiotic containing ice-cream on Salivary Mutans Streptococci (SMS) levels in children of 6-12 years of age: A randomized controlled double blind study with six months follow up. J. Clin. Diagn. Res. 9(2): ZC06-9.

Azari-Anpar, M., H. Payeinmahali, A.D. Garmakhany and A.S. Mahounak. 2017. Physicochemical, microbial, antioxidant, and sensory properties of probiotic stirred yoghurt enriched with *Aloe vera* foliar gel. J. Food Proces. Preserv. 41: e13209

Basannavar, S., R. Pothuraju and R.K. Sharma. 2014. Effect of *Aloe vera* (*Aloe barbadensis* Miller) on survivability, extent of proteolysis and ACE inhibition of potential probiotic cultures in fermented milk. J. Sci. Food Agric. 94: 2712-2717.

Başyiğit-Kılıç, G. and D. Akpınar-Kankaya. 2016. Assessment of technological characteristics of non-fat yoghurt manufactured with prebiotics and probiotic strains. J. Food Sci. Technol. 53(11): 864-871.

Batista, A.L.D., R. Silva., L.P. Cappato, M.V.S. Ferreira, K.O. Nascimento, M. Schmiele et al. 2017. Developing a synbiotic fermented milk using probiotic bacteria and organic green banana flour. J. Function. Foods 38: 242-250.

Bedani, R., M.M.S. Campos, I.A. Castro, E.A. Rossi and S.M.I. Saad. 2014. Incorporation of soybean by-product okara and inulin in a probiotic soy yoghurt: Texture profile and sensory acceptance. J. Sci. Food Agric. 94: 119-125.

Beheshtipour, H., A.M. Mortazavian, P. Haratian and K.K. Darani. 2012. Effects of *Chlorella vulgaris* and *Arthrospira platensis* addition on viability of probiotic bacteria in yogurt and its biochemical properties. Eur. Food Res. Technol. 235: 719-728.

Beheshtipour, H., A.M. Mortazavian, R. Mohammadi, S. Sohrabvandi and K. Khosravi-Darani. 2013. Supplementation of *Spirulina platensis* and *Chlorella vulgaris* algae into probiotic fermented milks. Comprehen. Rev. Food Sci. Food Safety. 12: 144-154.

Bergamini, C.V., E.R. Hynes, A. Quiberoni, V.B. Suarez and C.A. Zalazar. 2005. Probiotic bacteria as adjunct starters: Influence of the addition methodolgy on their survival in a semi-hard Argentinian cheese. Food Res. Int. 38(5): 597-604.

Bergamini, C.V., E.R. Hynes, M.C. Candioti and C.A. Zalazar. 2009. Multivariate analysis of proteolysis patterns differentiated the impact of six strains of probiotic bacteria on a semi-hard cheese. J. Dairy Sci. 92(6): 2455-2467.

Bergamini, C.V., E. Hynes, C. Meinardi, V. Suarez, A. Quiberoni, C. Zalazar et al. 2010. Pategras cheese as a suitable carrier for six probiotic cultures. J. Dairy Res. 77(3): 265-272.

Bielecka, M. and A. Majkowska. 2000. Effect of spray drying temperature of yoghurt on the survival of starter cultures, moisture content and sensoric properties of yogurt powder. Nahrung. 44: 257-260.

Bisanz, J.E., J.M. Macklaim, B.G. Gloor and G. Reid. 2014. Bacterial metatranscriptome analysis of a probiotic yogurt using an RNA-Seq approach. In. Dairy J. 39: 284-292.

Bitaraf, M.S., F. Khodaiyan, M.A. Mohammadifar and S.M. Mousavi. 2012. Application of response surface methodology to improve fermentation time and rheological properties of probiotic yogurt containing *Lactobacillus reuteri*. Food Bioprocess Technol. 5: 1394-1401.

Blaiotta, G., N. Murru, A. di Cerbo, M. Succi, R. Coppola and M. Aponte. 2017. Commercially standardized process for probiotic "Italico" cheese production. LWT-Food Sci. Technol. 79: 601-608.

Blanchette, L., D. Roy, G. Belanger and S.F. Gauthier. 1996. Production of cottage cheese using dressing fermented by bifidobacteria. J. Food Sci. 79(1): 8-15.

Boix-Amorós, A., C. Martinez-Costa, A. Querol, M.C. Collado and A. Mira. 2017. Multiple approaches detect the presence of fungi in human breastmilk samples from healthy mothers. Sci. Rep. 7(1): 13016.

Bosnea, L.A., N. Kopsahelis, V. Kokkali, A. Terpou and M. Kanellaki. 2017. Production of a novel probiotic yogurt by incorporation of *L. casei* enriched fresh apple pieces, dried raisins and wheat grains. Food Bioproducts Proces. 102: 62-71.

Buriti, F.C.A., T.Y. Okazaki, J.H.A. Alegro and S.M.I. Saad. 2007. Effect of a probiotic mixed culture on texture profile and sensory performance of Minas fresh cheese in comparison with the traditional products. Arch. Latinoamericanos Nutr. 57: 179-185.

Buriti, F.C.A., S.C. Freitas, A.S. Egito and K.M.O. dos Santos. 2014. Effects of tropical fruits pulps and partially hydrolysed galactomannan from *Caesalpinia pulcherrima* seeds on the dietary fibre content, probiotic viability, texture and sensory features of goat dairy beverages. LWT-Food Sci. Technol. 59: 196-203.

Canbulat, Z. and T. Özcan. 2015. Effects of short-chain and long-chain inulin on the quality of probiotic yogurt containing *Lactobacillus rhamnosus*. J. Food Proces. Preserv. 39: 1251-1260.

Cangelosi, G.A. and J.S. Meschke. 2014. Dead or alive: Molecular assessment of microbial viability. Appl. Environ. Microbiol. 80(19): 5884-5891.

Capela, P., T.K.C. Hay and N.P. Shah. 2006. Effect of cryoprotectants, prebiotics and microencapsulation on survival of probiotic organisms in yoghurt and freeze-dried yoghurt. Food Res. Int. 39: 203-211.

Cardarelli, H.R., S.M.I. Saad, G.R. Gibson and J. Vulevic. 2007. Functional *petit-suisse* cheese: Measure of the prebiotic effect. Anaerobe. 13(5-6): 200-207.

Cardarelli, H.R., F.C.A. Buriti, I.A. Castro and S.M.I. Saad. 2008. Inulin and oligofructose improve sensory quality and increase the probiotic viable count in potentially synbiotic *petit-suisse* cheese. LWT-Food Sci. Technol. 41(6): 1037-1046.

Cárdenas, N., J. Calzada, A. Peirotén, E. Jiménez, R. Escudero, J.M. Rodríguez et al. 2014. Development of a potential probiotic fresh cheese using two *Lactobacillus salivarius* strains isolated from human milk. Biomed Res. Int. Article ID 801918.

Casarotti, S.N., B.M. Carniro and A.L.B. Penna. 2014. Evaluation of the effect of supplementing fermented milk with quinoa flour on probiotic activity. J. Dairy Sci. 97: 6027-6035.

Casarotti, S.N. and A.L.B. Penna. 2015. Acidification profile, probiotic in vitro gastrointestinal tolerance and viability in fermented milk with fruit flour. Int. Dairy J. 41: 1-6.

Castro, W.F., A.G. Cruz, M.S. Bisinotto, L.M.R. Guerreiro, J.A.F. Faria, H.M.A. Bolini et al. 2013. Development of probiotic dairy beverages: Rheological properties and application of mathematical models in sensory evaluation. J. Dairy Sci. 96: 16-25.

Ceugniez, A., F. Coucheney, P. Jacques, G. Daube, V. Delcenserie, D. Drider et al. 2017. Anti-*Salmonella* activity and probiotic trends of *Kluyveromyces marxianus* S-2-05 and *Kluyveromyces lactis* S-3-05 isolated from a French cheese, Tomme d'Orchies. Res. Microbiol. 168: 575-582.

Champagne, C., Y. Raymond, N. Guertin and G. Bélanger. 2015. Effects of storage conditions, microencapsulation and inclusion in chocolate particles on the stability of probiotic bacteria in ice cream. Int. Dairy J. 47: 109-117.

Chen, S., Y. Cao, L.R. Ferguson, Q. Shu and S. Garg. 2012. Flow cytometric assessment of the protectants for enhanced in vitro survival of probiotic lactic acid bacteria through simulated human gastro-intestinal stresses. Appl. Microbiol. Biotechnol. 95: 345-356.

Coeuret, V., M. Gueguen, and J.P. Vernoux. 2004. *In vitro* screening of potential probiotic activities of selected lactobacilli isolated from unpasteurized milk products for incorporation into soft cheese. J. Dairy Res. 71(4): 451-460.

Collado, M.C. and Y. Sanz. 2006. Method for direct selection of potentially probiotic *Bifidobacterium* strains from human faeces based on their acid-adaptation ability. J. Microbiol. Methods 66: 560-563.

Coman, M.M., M.C. Verdenelli, C. Cecchini, S. Silvi, A. Vasile, G.E. Bahrim et al. 2013. Effect of buckwheat flour and oat bran on growth and cell viability of the probiotic strains *Lactobacillus rhamnosus* IMC 501®, *Lactobacillus paracasei* IMC 502® and their combination SYNBIO®, in synbiotic fermented milk. Int. J. Food Microbiol. 167: 2681-2688.

Corbo, M.R., M. Albenzio, M. de Angelis, A. Sevi and M. Gobbetti. 2001. Microbiological and biochemical properties of Canestrato Pugliese hard cheese supplemented with bifidobacteria. J. Dairy Sci. 84: 551-561.

Costa, M.P., B.S. Frasao, A.C.O. Silva, M.Q. Freitas, R.M. Franco, C.A. Conte-Junior et al. 2015. Cupuassu (*Theobroma grandiflorum*) pulp, probiotic, and prebiotic: Influence on color, apparent viscosity, and texture of goat milk yogurts. J. Dairy Sci. 98: 5995-6003.

Cruz, A.G., A. Antunes, A. Sousa, J. Faria and S. Saad. 2009. Ice-cream as a probiotic food carrier. Food Res. Int. 42(9): 1233-1239.

Cruz, A.G., F.W. Castro, J.A.F. Faria, S. Jr. Bogusz, D. Granato, R.M.S. Celeguini et al. 2012. Glucose oxidase: A potential option to decrease the oxidative stress in stirred probiotic yogurt. LWT-Food Sci. Technol. 47: 212-215.

Cutting, S.M. 2011. Bacillus probiotics. Food Microbiol. 28: 214-220.

Czerucka, D., T. Piche and P. Rampal. 2007. Review article: Yeast as probiotics - *Saccharomyces boulardii*. Aliment. Pharma. Therap. 26: 767-778.

Dabiza, N.M.A. and F.A. Fathi. 2006. Characteristics and chemical composition of probiotic Ras cheese. Deutsche Lebensmitt.-Runds. 102(12): 561-566.

da Costa, E.L., N.M.M. Alencar, B.G.S. Rullo and R.L. Taralo. 2017. Effect of green banana pulp on physicochemical and sensory properties of probiotic yoghurt. Food Sci. Technol. Campinas. 37(3): 363-368.

Daigle, A., D. Roy, G. Belanger and J.C. Vuillemard. 1999. Production of probiotic cheese (Cheddar-like cheese) using enriched cream fermented by *Bifidobacterium infantis*. J. Dairy Sci. 82(6): 1081-1091.

Daneshi, M., M.R. Ehsani, S.H. Razavi and V. Labbafi. 2013. Effect of refrigerated storage on the probiotic survival and sensory properties of milk/carrot juice mix drink. Elect. J. Biotechnol. 16(5): 1-12.

Dantas, A B., V.F. Jesus, R. Silva, C.N. Almada, E.A. Esmerino, L.P. Cappato et al. 2016. Manufacture of probiotic Minas Frescal cheese with *Lactobacillus casei* Zhang. J. Dairy Sci. 99: 18-30.

D'Argenio, V. and F. Salvatore. 2015. The role of the gut microbiome in the healthy adult status. Clinica Chimica Acta 451(Pt A): 97-102.

da Silva, D.C.G., L.R. de Abreu and G.M.P. Assumpção. 2012. Addition of water-soluble soy extract and probiotic culture, viscosity, water retention capacity and syneresis characteristics of goat milk yogurt. Ciencia Rural, Santa Maria 42(3): 545-550.

da Silva, D.F., N.N. Tenorio-Junior, R.G. Gomes, M.S.S. Pozza, M. Britten, P.T. Matumoto-Pintro et al. 2017. Physical, microbiological and rheological properties of probiotic yogurt supplemented with grape extract. J. Food Sci. Technol. 54(6): 1608-1615.

da Silveira, O.E., J.H.L. Neto, L.A. da Silva, A.E.S. Raposo, M. Magnani, H.R.Cardarelli et al. 2015. The effects of inulin combined with oligofructose and goat cheese whey on the physicochemical properties and sensory acceptance of a probiotic chocolate goat dairy beverage. LWT-Food Sci. Technol. 62: 445-451.

Dave, R.I. and N.P. Shah. 1997a. Effect of cysteine on the viability of yoghurt and probiotic bacteria in yoghurts made with commercial starter cultures. Int. Dairy J. 7: 537-545.

Dave, R.I. and N.P. Shah. 1997b. Effectiveness of ascorbic acid as an oxygen scavenger in improving viability of probiotic bacteria in yoghurts made with commercial starter cultures. Int. Dairy J. 7: 435-443.

David, L.A., C.F. Maurice, R.N. Carmody, D.B. Gootenberg, J.E. Button, B.E. Wolfe et al. 2014. Diet rapidly and reproducibly alters the human gut microbiome. Nature 505: 559-563.

de Castro, F.P., T.M. Cunha, P.L.M. Barreto, R.D.M.C. de Amboni and E.S. Prudêncio. 2009. Effect of oligofructose incorporation on the properties of fermented probiotic lactic beverages. Int. J. Dairy Technol. 62: 68-74.

de Leblanc, A.M., C.M. Galdeano, C.A. Dogp, E. Carmuega, R. Weill, G. Perdigonj et al. 2010. Adjuvant effect of a probiotic fermented milk in the protection against *Salmonella enteritidis* serovar Typhimurium infection: Mechanisms involved. Int. J. Immunpathol. Pharma. 23(4): 1235-1244.

de Mederios, A.C.L., M. Thomazini, A. Urbano, R.T.P. Correia and C.S. Favaro-Trindade. 2014. Structural characterisation and cell viability of a spray dried probiotic yoghurt produced with goats' milk and *Bifidobacterium animalis* subsp. *lactis* (BI-07). Int. Dairy J. 39: 71-77.

de Oliveira, M.E.G., E.F. Gracia, C.E.V. de Oliveira, A.M.P. Gomes, M.M.E. Pinadp, A.R.M.F. Madureira et al. 2014. Addition of probiotic bacteria in a semi-hard goat cheese (coalho): Survival to simulated gastrointestinal conditions and inhibitory effect against pathogenic bacteria. Food Res. Int. 64: 241-247.

Desfossés-Foucault, E., V. Dussault-Lepage, C. Le Boucher, P. Savard, G. LaPointe, D. Roy et al. 2012. Assessment of probiotic viability during Cheddar cheese manufacture and ripening using propidium monoazide-PCR quantification. Front. Microbiol. 3: 350.

di Criscio, T., A. Fratianni, R. Mignogna, L. Cinquanta, R. Coppola, E. Sorrentino et al. 2010. Production of functional probiotic, prebiotic, and symbiotic ice creams. J. Dairy Sci. 93(10): 4555-4564.

Dimitrellou, D., P. Kandyllis, Y. Kourkoutas and M. Kanellaki. 2017. Novel probiotic whey cheese with immobilized lactobacili on casein. LWT-Food Sci. Technol. 86: 627-634.

Dinakar, P. and V.V. Mistry. 1994. Growth and viability of *Bifidobacterium bifidum* in Cheddar cheese. J. Dairy Sci. 77: 2854-2864.

Doleyres, Y. and C. Lacroix. 2005. Technologies with free and immobilised cells for probiotic bifidobacteria production and protection. Int. Dairy J. 15(10): 973-988.

dos Santos, K.M.O., I.C. de Oliveira, M.A.C. Lopes, A.P.G. Cruz, F.C.A. Buriti, L.M. Cabrala et al. 2017. Addition of grape pomace extract to probiotic fermented goat milk: The effect on phenolic content, probiotic viability and sensory acceptability. J. Sci. Food Agric. 97: 1108-1115.

El-Kholy, W.I., B.A. Effat, N.F. Tawfik and O.M. Sharaf. 2006. Improving soft cheese quality using growth stimulator for *Bifidobacterium bifidum*. Deut. Lebensmit.-Rundschau. 102(3): 101-109.

El-Sayed, H.S., H.H. Salama and S.M. El-Sayed. 2014. Production of synbiotic ice cream. Int. J. ChemTech Res. 7(1): 138-147.

Endo, A., J. Terasjarvi and S. Salminen. 2014. Food matrices and cell conditions influence survival of *Lactobacillus rhamnosus* GG under heat stresses and during storage. Int. J. Food Microbiol. 174: 110-112.

Faisal, S., S. Chakraborty, W.E. Devi, M.K. Hazarika and V. Puranik. 2017. Sensory evaluation of probiotic whey beverages formulated from orange powder and flavour using fuzzy logic. Int. Food Res. J. 24(2): 703-710.

FAO/WHO. 2002. Report of a Joint FAO/WHO Working Group on Drafting Guidelines for the Evaluation of Probiotics in Food. London, ON, Canada: FAO/WHO.

Felicio, T.L., E.A. Esmerino, V.A.S. Vidal, L.P. Cappato, R.K.A. Garcia, R.N. Cavalcanti et al. 2016. Physico-chemical changes during storage and sensory acceptance of low sodium probiotic Minas cheese added with arginine. Food Chem. 196: 628-637.

Ferrer, M., A. Beloqui and P.N. Golyshin. 2010. Screening metagenomic libraries for laccase activities. Methods Molecular Biol. 668: 189-202.

Figueroa-Gonzales, I., G. Quijano, G. Ramirez and A. Cruz-Guerrero. 2011. Probiotics and prebiotics: Perspectives and challenges. J. Sci. Food Agric. 91: 1341-1348.

Fortin, M.H., C.P. Champagne, D. St-Gelais, M. Britten, P. Fustier, M. Lacroix et al. 2011. Effect of time of inoculation, starter addition, oxygen level and salting on the viability of probiotic cultures during Cheddar cheese production. Int. Dairy J. 21: 75-82.

Fox, P.F., T.P. Guinee, T.M. Cogan and P.L.H. McSweeney. 2000. Fundamentals of Cheese Sciences. Aspen Publication, Maryland, NJ, USA.

Fritzen-Freire, C.B., C.M.O. Mueller, J.B. Laurindo and E.S. Prudencio. 2010. The influence of *Bifidobacterium* Bb-12 and lactic acid incorporation on the properties of Minas Frescal cheese. J. Food Eng. 96(4): 621-627.

Ganesan, B., B.C. Weimer, J. Pinzon, N. Dao Kong, G. Rompato, C. Brothersen et al. 2014. Probiotic bacteria survive in Cheddar cheese and modify populations of other lactic acid bacteria. J. Appl. Microbiol. 116: 1642-1656.

Garcia, E.F., M.E.G. de Oliveira, R.C.R.E. Queiroga, T.A.D. Machado and E.L. de Souza. 2012. Development and quality of a Brazilian semi-hard goat cheese (coalho) with added probiotic lactic acid bacteria. Int. J. Food Sci. Nutr. 63(8): 947-956.

Gardiner, G., R.P. Ross, J.K. Collins, G. Fitzgerald and C. Stanton. 1998. Development of a probiotic cheddar cheese containing human-derived *Lactobacillus paracasei* strains. Appl. Environ. Microbiol. 64(6): 2192-2199.

Ghoddushi, H.B. and R.K. Robinson. 1996. The test of time. Dairy Ind. Int. 61(7): 25-28.

Gobbetti, M., A. Corsetti, E. Smacchi, M. de Angelis and J. Rossi. 1997. Microbiology and chemistry of Pecorino Umbro cheese during ripening. Italian Food Sci. 2: 111-126.

Gobbetti, M., A. Corsetti, E. Smacchi, A. Zocchetti and M. de Angelis. 1998. Production of Crescenza cheese by incorporation of bifidobacteria. J. Dairy Sci. 81: 37-47.

Goderska, K., M. Zybala and Z. Czarnecki. 2003. Characterization of microcapsulated *Lactobacillus rhamnosus* LR7 strain. Polish J. Food Nutr. 12: 237-238.

Godward, G. 2000. Studies on enhancing the viability and survival of probiotic bacteria in dairy foods through strain selection and microencapsulation. MSc. thesis, University of Western Sydney, Australia.

Godward, G. and K. Kailasapathy. 2003. Viability and survival of free and encapsulated probiotic bacteria in Cheddar cheese. Milchwiss. 58(11-12): 624-627.

Gomes, A.M.P., F.X. Malcata, F.A.M. Klaver and H.J. Grande. 1995. Incorporation and survival of *Bifidobacterium* sp. strain Bo and *Lactobacillus acidophilus* strain Ki in a cheese product. Neth. Milk Dairy J. 49(2-3): 71-95.

Gomes, A.M.P. and F.X. Malcata. 1998. Development of probiotic cheese manufactured from goat milk: Response surface analysis via technological manipulation. J. Dairy Sci. 81(6): 1492-1507.

Gosiewski, T., A. Chmielarczyk, M. Strus, W.M. Brzychczy and P.B. Heczko. 2012. The application of genetics methods to differentiation of three *Lactobacillus* species of human origin. Ann. Microbiol. 62: 1437-1445.

Gunenc, A., C. Khoury, C. Legault, H. Mirrashed, J. Rijke, F. Hosseinian et al. 2016. Seabuckthorn as a novel prebiotic source improves probiotic viability in yogurt. LWT-Food Sci. Technol. 66: 490-495.

Güler-Akın, M.B., I. Ferliarslan and M.S. Akın. 2016. Apricot probiotic drinking yoghurt supplied with inulin and oat fiber. Adv. Microbiol. 6: 999-1009.

Gürakan, C., A. Cebeci and B. Özer. 2010. Probiotic dairy beverages. pp. 165-195 In: F. Yildiz (ed.). Development and Manufacture of Yogurt and Other Functional Dairy Products. CRC Press. Boca Raton, FL, USA.

Gürsoy, O. and O. Kınık. 2010. Incorporation of adjunct cultures of *Enterococcus faecium*, *Lactobacillus paracasei* subsp *paracasei* and *Bifidobacterium bifidum* into white cheese. J. Food Agric. Environ. 8(2): 107-112.

Haenlein, G.F.W. 2004. Goat milk in human nutrition. Small Rum. Res. 51: 155-163.

Hallen-Adams, H.E. and M.J. Suhr. 2017. Fungi in the healthy human gastrointestinal tract. Virulence 8(3): 352-358.

Hassani, M., A. Sharifi, A.M. Sani and B. Hassani. 2016. Growth and survival of *Lactobacillus acidophilus* and *Bifidobacterium bifidum* in probiotic yogurts enriched by barberry extract. J. Food Safety 36: 503-507.

Hayes, W.W., S.C. Feijoo and J.H. Martin. 1996. Effect of lactulose and oligosaccharides on survival of *Bifidobacterium longum* in Edam cheese. J. Dairy Sci. 79(suppl. 1): 118.

Hill, C., F. Guarner, G. Reid, G.R. Gibson, D.J. Merenstein, B. Pot et al. 2014. Expert consensus document. The International Scientific Association for Probiotics and

Prebiotics consensus statement on the scope and appropriate use of the term probiotic. Nature Rev. Gastro. Hepato. 11(8): 506-514.

Hisiao, H.C., W.C. Lian and C.C. Chou. 2004. Effect of packaging conditions and temperature on viability of microencapsulated *Bifidobacteria* during storage. J. Sci. Food Agric. 52: 134-139.

Illupapalayam, V.V., S.C. Smith and S. Gamlath. 2014. Consumer acceptability and antioxidant potential of probiotic yoğurt with spice. LWT-Food Sci. Technol. 55: 255-262.

Izadi, M., M.H. Eskandari, M. Niakousari, S. Shekarforoush, M.A. Hanifpour, Z. Izadi et al. 2014. Optimisation of a pilot-scale spray drying process for probiotic yoghurt, using response surface methodology. Int. J. Dairy Technol. 67(2): 211-219.

Jandhyala, S.M., R. Talukdar, C. Subramanyam, H. Vuyyuru, M. Sasikala, R.D. Nageshwar et al. 2015. Role of the normal gut microbiota. World J. Gastro. 21(29): 8787-8803.

Jankowska, A., K. Wisniewska and A. Reps. 2005. Application of probiotic bacteria in production of yoghurt preserved under high pressure. High Pres. Res. 25(1): 57-62.

Jankowska, A., K. Wisniewska, A. Grzeskiewicz and A. Reps. 2012. Examining the possibilities of applying high pressure to preserve yoghurt supplemented with probiotic bacteria. High Pres. Res. 32(3): 339-346.

Jansson, S.E.A., G. Gallet, T. Hefti, S. Karlsson, U.W. Gedde, M. Hendenqvist et al. 2002. Packing materials for fermented milk, Part 2: Solute-induced changes and effects of material polarity and thickness on food quality. Packag. Technol. Sci. 15(6): 287-300.

Jatila, H. and K. Matilainen. 2008. Probiotic cheese quality. Proceedings of 5th IDF Symposium on Cheese Ripening, IDF, 9-13 March, Bern, Switzerland, p. 86.

Jayamanne, V.S. and M.R. Adams. 2005. Determination of survival, identity and stress resistance of probiotic bifidobacteria in bio-yoghurts. Lett. Appl. Microbiol. 42: 189-194.

Jeon, E.B., S.-H. Son, R.K.C. Jeewanthi, N.-K. Lee and H.-D. Paik. 2016. Characterization of *Lactobacillus plantarum* Lb41, an isolate from kimchi and its application as a probiotic in cottage cheese. Food Sci. Biotechnol. 25(4): 1129-1133.

Jesus, A.L.T., M.S. Fernandes, B.A. Kamimura, L. Prado-Silva, R. Silva, E.A. Esmerino et al. 2016. Growth potential of *Listeria monocytogenes* in probiotic cottage cheese formulations with reduced sodium content. Food Res. Int. 81: 180-187.

Kailasapathy, K. and L. Masondole. 2005. Survival of free and microencapsulated *Lactobacillus acidophilus* and *Bifidobacterium lactis* and their effect on texture of feta cheese. Aus. J. Dairy Technol. 60(3): 252-258.

Kalavrouzioti, I., M. Hatzikamari, E. Litopoulou-Tzanetaki and N. Tzanetakis. 2005. Production of hard cheese from caprine milk by the use of two types of probiotic cultures as adjuncts. Int. J. Dairy Technol. 58(1): 30-38.

Kalyuzhnaya, M.G., D.A. Beck and L. Chistoserdova. 2011. Functional metagenomics of methylotrophs. Methods Enzym. 495: 81-98.

Karaman, A.D., B. Özer, M.A. Pascall and V. Alvarez. 2015. Recent advances in dairy packaging. Food Rev. Int. 31: 295-318.

Karaolis, C., G. Botsaris, I. Pantelides and D. Tsalta. 2013. Potential application of *Saccharomyces boulardii* as a probiotic in goat's yoghurt: Survival and organoleptic effects. Int. J. Food Sci. Technol. 48: 1445-1452.

Kasımoğlu, A., M. Göncüoğlu and S. Akgün. 2004. Probiotic white cheese with *Lactobacillus acidophilus*. Int. Dairy J. 14(12): 1067-1073.

Kearney, N., X.C. Meng, C. Stanton, J. Kelly, G.F. Fitzgerald, R.P. Ross et al. 2009. Development of a spray dried probiotic yogurt containing *Lactobacillus paracasei* NFBC 338. Int. Dairy J. 19: 684-689.

Kılıç, G.B., H. Kuleaşan, I. Eralp and A.G. Karahan. 2009. Manufacture of Turkish Beyaz cheese added with probiotic strains. LWT-Food Sci. Technol. 42(5): 1003-1008.

Klaver, F.A.M., F. Kingma and A.C. Bolle. 1990. Growth relationships between bifidobacteria and lactobacilli in milk. Voedingsmiddelen Technol. 23(9): 13-16.

Korbekandi, H., A.M. Mortazavian and S. Iravani. 2011. Technology and stability of probiotic in fermented milks. pp. 131-169. *In*: N. Shah, A.G. Cruz, J.A.F. Fariai (eds.). Probiotic and Prebiotic Foods: Technology, Stability and Benefits to the Human Health. Nova Science Publishers, New York, USA.

Kourkoutas, Y., L. Bosnea, S. Taboukos, C. Baras, D. Lambrou, M. Kanellaki et al. 2006. Probiotic cheese production using *Lactobacillus casei* cells immobilized on fruit pieces. J. Dairy Sci. 89(5): 1439-1451.

Krasaekoopt, W., B. Bhandari and H. Deeth. 2003. Evaluation of encapsulation techniques of probiotic yogurt. Int. Dairy J. 13: 3-13.

Kumar, B.V., S.V.N. Vijayendra and O.V.S. Reddy. 2015. Trends in dairy and non-dairy probiotic products: A review. J. Food Sci. Technol. 52: 6112-6124.

Kumar, A. and D. Kumar. 2016. Development of antioxidant rich fruit supplemented probiotic yogurts using free and microencapsulated *Lactobacillus rhamnosus* culture. J. Food Sci. Technol. 53(1): 667-675.

Kumar, S.R., R. Rasane and R. Nimmanapalli. 2018. Optimization of a process for production of pomegranate pulp and flaxseed powder fortified Greek Dahi. Int. J. Dairy Technol. 71(3): 753-763.

Lankaputhra, W.E.V. and N.P. Shah. 1996. A simple method for selective enumeration of *Lactobacillus acidophilus* in yoghurt supplemented with *L. acidophilus* and *Bifidobacterium* spp. Milchwiss. 51: 446-451.

Larsen, N., F.K. Vogensen, R.J. Gøbel, K.F. Michaelsen, S.D. Forssten, S.J. Lahtinen et al. 2013. Effect of *Lactobacillus salivarius* Ls-33 on fecal microbiota in obese adolescents. Clinical Nutrition 32(6): 935-940.

Lily, D.M. and R.H. Stillwell. 1965. Probiotics: Growth-promoting factors produced by microorganisms. Science 147(3659): 747-748.

Lim, E.-S. 2018. Preparation and functional properties of probiotic and oat-based synbiotic yogurts fermented with lactic acid bacteria. Appl. Biol. Chem. 61(1): 25-37.

Liu, W.J., Q.H. Baq, Jirimutu, M.J. Qing, Siriguleng, X. Chen et al. 2012. Isolation and identification of lactic acid bacteria from Tarag in Eastern Inner Mongolia of China by 16S rRNA sequences and DGGE analysis. Microbiol. Res. 167: 110-115.

Livney, Y.D. 2010. Milk proteins as vehicles for bioactives. Curr. Opin. in Coll. and Interface Sci. 15: 73-83.

Lockyer, S. 2017. Are probiotics useful for the average consumer? British Nutr. Found. Nutr. Bull. 42: 42-48.

Lollo, P.C.B., C.S. de Moura, P.N. Morato, A.G. Cruz, W.F. Castro, C.B. Betim et al. 2013. Probiotic yogurt offers higher immune-protection than probiotic whey beverage. Food Res. Int. 54: 118-124.

Lourens-Hattingh, A. and B.C. Viljoen. 2001. Yogurt as probiotic carrier food. Int. Dairy J. 11: 1-17.

Luo, Y., B.C. Ma, L.K. Zou, J.G. Cheng, Y.H. Cai, J.P. Kang et al. 2012. Identification and characterization of lactic acid bacteria from forest musk deer feces. African J. Microbiol. Res. 6: 5871-5881.

Machado, T.A.D.G., M.E.G. de Oliveira, M.I.F. Campos, P.O.A. de Assis, E.L. de Souza, M.S. Madruga et al. 2017. Impact of honey on quality characteristics of goat yogurt containing probiotic *Lactobacillus acidophilus*. LWT-Food Sci. Technol. 80: 221-229.

Madureira, A.R., J.C. Soares, M.M.E. Pintado, A.P. Gomes, A.C. Freitas, F.X. Malcata et al. 2008. Sweet whey cheese matrices inoculated with the probiotic strain *Lactobacillus paracasei* LAFTI⁰L26. Dairy Sci. Technol. 88(6): 649-665.

Madureira, A.R., M. Amorim, A.M. Gomes, M.M.E. Pintado and F.X. Malcata. 2011. Protective effect of whey cheese matrix on probiotic strains exposed to simulated gastrointestinal conditions. Food Res. Int. 44: 465-470.

Madureira, A.R., J.C. Soares, M. Amorim, T. Tavares, A.M. Gomes, M.M. et al. 2012. Bioactivity of probiotic whey cheese: Characterization of the content of peptides and organic acids. J. Sci. Food Agric. 93: 1458-1465.

Madhu, A.N., N. Amrutha and S.G. Prapulla. 2012. Characterization and antioxidant property of probiotic and synbiotic yogurts. Probio. Antimic. Proteins. 4: 90-97.

Maity, T.K., R. Kumar and A.K. Misra. 2008. Development of healthy whey drink with *Lactobacillus rhamnosus, Bifidobacterium bifidum* and *Propionibacterium freudenreichii* subsp. *shermanii*. Mljekarstvo. 58(4): 315-325.

Makelainen, H., S. Fossten, K. Olli, L. Granlund, N. Rautonen, A.C. Ouwehand et al. 2009. Probiotic lactobacillli in a semi-soft cheese survive in the simulated human gastrointestinal tract. Int. Dairy J. 19(11): 675-683.

Makino, H., A. Kushiro, E. Ishikawa, D. Muylaert, H. Kubota, T. Sakai et al. 2011. Transmission of intestinal *Bifidobacterium longum* subsp. *longum* strains from mother to infant, determined by multilocus sequencing typing and amplified fragment length polymorphism. Appl. Environ. Microbiol. 77: 678-893.

Marcatti, B., A.M.Q.B. Habitante, P.J.A. Sobral and C.S. Favaro-Trindade. 2009. Minas-type fresh cheese developed from buffalo milk with addition of *L. acidophilus*. Sci. Agricola. 66(4): 481-485.

Martinez-Villaluenga, C., J. Frias, R. Gomez and C. Vdal-Valverde. 2006. Influence of addition of raffinose family oligosaccharides on survival in fermented milk during refrigerated storage. Int. Dairy J. 16: 768-774.

Martins, I.B.A., R. Deliza, K.M.O. dos Santos, E.H.M. Walter, J.M. Martins and A. Rosenthal et al. 2018. Viability of probiotics in goat cheese during storage and under simulated gastrointestinal conditions. Food Bioprocess Technol. 11: 853-863.

Masuda, T., R. Yamanari and T. Itoh. 2005. The trial for production of fresh cheese incorporated probiotic *Lactobacillus acidophilus* group lactic acid bacteria. Milchwiss. 60(2): 167-171.

Matijasic, B.B. 2008. Application of probiotic *Lactobacillus gasseri* K7 in cheese production. Proceedings of 5th IDF Symposium on Cheese Ripening Symposium. IDF, 9-13 March, Bern, Switzerland, p. 21.

Maus, J.E. and S.C. Ingham. 2003. Employment of stressful conditions during culture production to enhance subsequent cold and acid tolerance of bifidobacteria. J. Appl. Microbiol. 95(1): 146-154.

Meira, Q.G.S., M. Magnani, Jr. F.C. de Mederios, R.C.R.E. Queiroga, M.S. Madruga, B. Gullon et al. 2015. Effects of added *Lactobacillus acidophilus* and *Bifidobacterium lactis* probiotics on the quality characteristics of goat ricotta and their survival under simulated gastrointestinal conditions. Food Res. Int. 76: 828-838.

Mendes-Soares, H., H. Suzuki, R.J. Hickey and L.J. Forney. 2014. Comparative functional genomics of *Lactobacillus* spp. reveals possible mechanisms for specialization of vaginal lactobacilli to their environment. J. Bacteriol. 196(7): 1458-1470.

Michael, M., R.K. Phebus and K.A. Schmidt. 2015. Plant extract enhances the viability of *Lactobacillus delbrueckii* subsp. bulgaricus and *Lactobacillus acidophilus* in probiotic nonfat yogurt. Food Sci. Nutr. 3(1): 48-55.

Michalak, I. and K. Chojnacka. 2016. Functional fermented food and feed from seaweed. pp. 231-247. *In*: D. Montet and R.C. Ray (eds.). Fermented Foods: Biochemistry and Biotechnology. CRC Press, Boca Raton, FL, USA.

Miller, C.W., M.H. Nguyen, M. Rooney and K. Kailasapathy. 2002. The influence of packaging materials on the dissolved oxygen content of probiotic yogurt. Pack. Technol. Sci. 15: 133-138.

Miller, C.W., M.H. Nguyen, M. Rooney and K. Kailasapathy. 2003. The control of dissolved oxygen content in probiotic yogurts by alternative packaging methods. Pack. Technol. Sci. 16: 61-67.

Misra, A.K. and R.K. Kuila. 1990. Cultural and biochemical activities of *Bifidobacterium bifidum*. Milchwiss. 45: 155-158.

Modler, H.W., R.C. McKeller and M. Yaguchi. 1990. Bifidobacteria and bifidogenic factors. Canadian Inst. of Food Sci. Technol. J. 23: 29-41.

Modzelewska-Kapitula, M., L. Klebukowska and K. Kornacki. 2007. Influence of inulin and potentially probiotic *Lactobacillus plantarum* strain on microbiological quality and sensory properties of soft cheese. Polish J. Food Nutr. Sci. 57(2): 143-146.

Molnar, N., B. Gyenis and L. Varga. 2005. Influence of a powdered Spiriluna platensis biomass on lactococci in milk. Milchwiss. 60(4): 380-382.

Mota, M.J., R.P. Lopes, I. Delgadillo and J.A. Saravia. 2015. Probiotic yogurt production under high pressure and the possible use of pressure as an on/off switch to stop/start fermentation. Process Biochem. 50: 906-911.

Muniandy, P., A.B. Shori and A.S. Baba. 2016. Influence of green, white and black tea addition on the antioxidant activity of probiotic yogurt during refrigerated storage. Food Pack. Shelf Life 8: 1-8.

Ndoye, B., E.A. Rasolofo, G. La Pointe and D. Roy. 2011. Are view of the molecular approaches to investigate the diversity and activity of cheese microbiota. Dairy Sci. Technol. 91: 495-524.

Nocker, A. and A.K. Camper. 2009. Novel approaches toward preferential detection of viable cells using nucleic acid amplification techniques. FEMS Microbiol. Lett. 291: 137-142.

Oliveira, R.P.S., A.C.R. Florence, R.C. Silva, P. Perego, A. Converti, L.A. Goielli et al. 2009. Effect of different prebiotics on the fermentation kinetics, probiotic survival and fatty acid profiles in nonfat synbiotic fermented milk. Int. J. Food Microbiol. 128(3): 467-472.

Ong, L. and N.P. Shah. 2009. Probiotic Cheddar cheese: Influence of ripening temperatures on survival of probiotic microorganisms, cheese composition and organic acid profiles. LWT-Food Sci. Technol. 42(7): 1260-1268.

Ortakci, F., J.R. Broadbent, W.R. McManus and D.J. McMahon. 2012. Survival of microencapsulated probiotic *Lactobacillus paracasei* LBC-1e during manufacture of Mozzarella cheese and simulated gastric digestion. J. Dairy Sci. 95: 6274-6281.

Ouwehand, A.C., T. Suomalainen, S. Tölkkö and S. Salminen. 2002. *In vitro* adhesion of propionic acid bacteria to human intestinal mucus. Lait. 82: 123-130.

Ouwehand, A.C., I. Fandi and S.D. Forssten. 2010. Cheese as a carrier food for probiotics: *In vitro* and human studies. Aus. J. Dairy Technol. 65(3): 165-169.

Özcan, T., L. Yılmaz-Ersan, A. Akpınar-Bayizit and B. Delikanlı. 2016. Antioxidant properties of probiotic fermented milk supplemented with chestnut flour (*Castanea sativa* Mill). J. Food Proces. Preserv. 41: e13156.

Özer, D., M.S. Akın and B. Özer. 2005. Effect of inulin and lactulose on survival of *Lactobacillus acidophilus* LA-5 and *Bifidobacterium bifidum* BB-02 in acidophilus-bifidus yoghurt. Food Sci. Technol. Int. 11(1): 19-26.

Özer, B., Y.S. Uzun and H.A. Kırmacı. 2008. Effect of microencapsulation on viability of *Lactobacillus acidophilus* LA-5 and *Bifidobacterium bifidum* BB-12 during Kasar cheese ripening. Int. J. Dairy Technol. 61(3): 237-244.

Özer, B., H.A. Kırmacı, E. Şenel, M. Atamer and A.A. Hayaloğlu. 2009. Improving the viability of *Bifidobacterium bifidum* BB-12 and *Lactobacillus acidophilus* LA-5 in white-brined cheese by microencapsulation. Int. Dairy J. 19(1): 22-29.

Özer, B. and H.A. Kirmaci. 2010. Functional milks and dairy beverages. Int. J. Dairy Technol. 63(1): 1-15.

Özer, B. and H.A. Kirmaci. 2011. Technological and health aspects of probiotic cheese. pp. 1-42. *In*: R.D. Foster (ed.). Cheese: Types, Nutrition and Consumption. Nova Science Publisher, NY, USA.

Pavunc, A.L., J. Beganović, B. Kos, A. Buneta, S. Beluhan, J. Šušković et al. 2010. Influence of microencapsulation and transglutaminase on viability of probiotic strain *Lactobacillus helveticus* M92 and consistency of set yoghurt. Int. J. Dairy Technol. 64(2): 254-261.

Penna, A.L.B., S. Gurraam and G.V. Barbosa-Canovas. 2006. Effect of high hydrostatic pressure processing on rheological and textural properties of probiotic low-fat yogurt fermented by different starter cultures. J. Food Proces. Eng. 28: 447-461.

Pereira, C., M. Henriques, D. Gomes, A. Gomez-Zavaglia and G. de Antoni. 2015. Novel functional whey-based drinks with great potential in the dairy industry. Food Technol. Biotechnol. 53: 307-314.

Perko, B., B. Matijasic and I. Rogelj. 2002. Production of probiotic cheese with addition of *Lactobacillus gasseri* LF221(Rif^r) and K7(Rif^r). Research reports of the Biotechnical Faculty University of Ljubljana, Agriculture (Zootechny) 80(1): 61-70.

Pescuma, M., E.M. Hébert, F. Mozzi and G.C. de Valdez. 2010. Functional fermented whey-based beverage using lactic acid bacteria. Int. J. Food Microbiol. 141: 73-81.

Phillips, M., K. Kailasapathy and L. Tran. 2006. Viability of commercial probiotic cultures (*L. acidophilus, Bifidobacterium* sp., *L. casei, L. paracasei* and *L. rhamnosus*) in cheddar cheese. Int. J. Food Microbiol. 108(2): 276-280.

Picard, C., J. Fioramonti, A. Francois, T. Robinson, F. Neant, C. Matuchansky et al. 2005. Bifidobacteria as probiotic agents - physiological effects and clinical benefits. Aliment. Pharma. Therapeutics 22: 495-512.

Pino, A., K. van Hoorde, I. Pitino, N. Russo, S. Carpino, C. Caggia et al. 2017. Survival of potential probiotic lactobacilli used as adjunct cultures on Pecorino Siciliano cheese ripening and passage through the gastrointestinal tract of healthy volunteers. Int. J. Food Microbiol. 252: 42-52.

Pinto, J.M.S., S. Sousa, D.M. Rodrigues, X.F. Malcata, A.C. Duarte, T.A.P. Rocha-Santos et al. 2017. Effect of probiotic co-cultures on physico-chemical and biochemical properties of small ruminants' fermented milk. Int. Dairy J. 72: 29-35.

Priti, S., P.R. Ray, P.K. Ghatak, S.K. Bag and T. Hazra. 2017. Physico-chemical quality and storage stability of fermented Channa whey beverages. Ind. J. Dairy Sci. 70: 398-403.

Pisano, M.B., M. Casula, A. Corda, M.E. Fadda, M. Deplano, S. Cosentino et al. 2008.

In vitro probiotic characteristics of *Lactobacillus* strains isolated from Fiore Sardo cheese. Italian J. Food Sci. 20(4): 505-516.

Plaza-Diáz, J., J.A. Fernándéz-Caballero, N. Chueca, F. Garcia, C. Gòmez-Llorante, M.J. Sáez-Lara et al. 2015. Pyrosequencing analysis reveals changes in intestinal microbiota of healthy adults who received a daily dose of immunomodulatory probiotic strains. Nutrients 7: 3999-4015.

Ranadheera, S.C., C.A. Evans, M.C. Adams and S.K. Baines. 2012. *In vitro* analysis of gastrointestinal tolerance and intestinal cell adhesion of probiotics in goat's milk ice cream and yogurt. Food Res. Int. 49: 619-625.

Rezaei, R., M. Khomeiri, A. Aalami and M. Kashaninejad. 2014. Effect of inulin on the physicochemical properties, flow behavior and probiotic survival of frozen yogurt. J. Food Sci. Technol. 51(10): 2809-2814.

Rathore, S., M.P. Desai, C.V. Liew, L.W. Chan and P.W.S. Heng. 2013. Microcapsulation of microbial cells. J. Food Eng. 116: 369-381.

Riaz, Q.A. and T. Masud. 2014. Recent trends and applications of encapsulatig materials for probiotic stability. Critical Rev. Food Sci. Nutr. 53: 231-244.

Riberio, M.C.E., K.S. Chaves, C. Gebara, F.N.S. Infante, C.R.F. Grosso et al. 2014. Effect of microencapsulation of *Lactobacillus acidophilus* LA-5 on physicochemical, sensory and microbiological characteristics of stirred probiotic yogurt. Food Res. Int. 66: 424-431.

Rodtong, S. and G.W. Tannock. 1993. Differentiation of *Lactobacillus* strains by ribotyping. Appl. Environ. Microbiol. 59: 3480-3484.

Ross, R.P., G. Fitzgerald, K. Collins and C. Stanton. 2002. Cheese delivering biocultures-probiotic cheese. Aus. J. Dairy Technol. 57(2): 71-78.

Roy, D., I. Mainville and F. Mondou. 1997. Selective enumeration and survival of bifidobacteria in fresh cheese. Int. Dairy J. 7(12): 785-793.

Roy, D. 2005. Technological aspects related to the use of bifidobacteria in dairy products. Lait. 85(1-2): 39-56.

Ryan, P.M., Z. Burdíková, T. Beresford, M.A.E. Auty, G.F. Fitzgerald, R.P. Ross et al. 2015. Reduced-fat Cheddar and Swiss-type cheeses harboring exopolysaccharide-producing probiotic *Lactobacillus mucosae* DPC 6426. J. Dairy Sci. 98: 8531-8544.

Sabikhi, L., M.H.S. Kumar and B.N. Mathur. 2014. *Bifidobacterium bifidum* in probiotic Edam cheese: Influence on cheese ripening. Int. J. Food Sci. Technol. 51(12): 3902-3909.

Sadeghi, E., A.A. Basti, N. Noori, A. Khanjari and R. Partovi. 2013. Effect of *Cuminum cyminum* L. essential oil and *Lactobacillus acidophilus* (a probiotic) on *Staphylococcus aureus* during the manufacture, ripening and storage of white brined cheese. J. Food Proces. Preserv. 37: 449-455.

Sah, B.N.P., T. Vasiljevic, S. McKechnie and O.N. Donkor. 2014. Effect of probiotics on antioxidant and antimutagenic activities of crude peptide extarct from yogurt. Food Chem. 156: 264-270.

Sah, B.N.P., T. Vasiljevic, S. McKechnie and O.N. Donkor. 2016a. Effect of pineapple waste powder on probiotic growth, antioxidant and antimutagenic activities of yogurt. J. Food Sci. Technol. 53(3): 1698-1708.

Sah, B.N.P., T. Vasiljevic, S. McKechnie and O.N. Donkor. 2016b. Physicochemical, textural and rheological properties of probiotic yogurt fortified with fibre-rich pineapple peel powder during refrigerated storage. LWT-Food Sci. Technol. 65: 978-986.

Samona, A. and R.K. Robinson. 1991. Enumeration of bifidobacteia in dairy products. J. Soc. Dairy Technol. 44: 64-66.

Samona, A. and R.K. Robinson. 1994. Effects of yogurt cultures on the survival of bifidobacteria in fermented milks. J. Soc. Dairy Technol. 47: 58-60.

Sander, L.E., M.J. Davis, M.V. Boekschoten, D. Amsen, C.C. Dascher, B. Ryffel et al. 2011. Detection of prokaryotic mRNA signifies microbial viability and promotes immunity. Nature 474(7351): 385-390.

Santillo, A., M. Albenzio, M. Quinto, M. Caroprese, R. Marino, R. Sevi et al. 2009. Probiotic in lamb rennet paste enhances rennet lipolytic activity, and conjugated linoleic acid and linoleic acid content in Pecorino cheese. J. Dairy Sci. 92(4): 1330-1337.

Santini, M.S.S., E.C. Koga, D.C. Aragon, E.H.W. Santana, M.R. Costa, G.N.Costa et al. 2012. Dried tomato-flavored probiotic cream cheese with *Lactobacillus paracasei*. J. Food Sci. 77(11): M604-M608.

Santo, A.P.E., R.C. Silva, F.A.S.M. Soares, D. Anjos, L.A. Gioielli, M.N. Oliveria et al. 2010. Açai pulp addition improves fatty acid profile and probiotic viability in yoghurt. Int. Dairy J. 20: 415-422.

Santo, A.P.E., N.S. Cartolano, T.F. Silva, F.A.S.M. Soares, L.A. Gioielli, P. Perego et al. 2012. Fibers from fruit by-products enhance probiotic viability and fatty acid profile and increase CLA content in yoghurts. Int. J. Food Microbiol. 154: 135-144.

Sarkar, S. 2010. Approaches for enhancing the viability of probiotics: A review. British Food J. 112(4): 329-349.

Saxami, G., O.S. Papadopoulou, N. Chorianopoulos, Y. Kourkoutas, C.C. Tassou, A. Galanis. et al. 2016. Molecular detection of two potential probiotic *Lactobacilli* strains and evaluation of their performance as starter adjuncts in yogurt production. Int. J. Molecul. Sci. 17: 668-680.

Ścibisz, I., M. Ziarno, M. Mitek and D. Zareba. 2012. Effect of probiotic cultures on the stability of anthocyanins in blueberry yoghurts. LWT-Food Sci. Technol. 49: 208-212.

Shah, N. 2007. Funtional cultures and health benefits. Int. Dairy J. 17: 1262-1277.

Sharp, M.D., D.J. McMahon and J.R. Broadbent. 2008. Comparative evaluation of yogurt and low-fat Cheddar cheese as delivery media for probiotic *Lactobacillus casei*. J. Food Sci. 73(7): M375-M377.

Shakerian, M., S.H. Razavi, F. Khodaiyan, S.A. Ziai, M.S. Yarmand, A. Moayedi et al. 2014. Effect of different levels of fat and inulin on the microbial growth and metabolites in probiotic yogurt containing nonviable bacteria. Int. J. Food Sci. Technol. 49: 261-268.

Shukla, M., Y. Jha and S. Admassu. 2013. Development of probiotic beverage from whey and pineapple juice. J. Food Proces. Technol. 4: 2 doi: 10.4172/2157-7110.1000206

Siciliano, R.A. and M.F. Mazzeo. 2012. Molecular mechanisms of probiotic action: A proteomic perspective. Curr. Opin. Microbiol. 15: 390-396.

Sidira, M., V. Santarmaki, M. Kiourtzidis, A.A. Argyri, O.S. Papadopoulou, N. Chorianopoulos et al. 2017. Evaluation of immobilized *Lactobacillus plantarum* 2035 on whey protein as adjunct probiotic culture in yogurt production. LWT-Food Sci. Technol. 75: 137-146.

Silva, H.L.A., C.F. Balthazara, E.A. Esmerino, A.H. Vieira, L.P. Cappato, L.P.C. Neto et al. 2017. Effect of sodium reduction and flavor enhancer addition on probiotic Prato cheese processing. Food Res. Int. 99: 247-255.

Skryplonek, K. and M. Jasińska. 2015. Fermented probiotic beverages based on acid whey. Acta Sci. Polonorum Technol. Aliment. 14(4): 397-405.

Soejima, T., K. Iida, T. Qin, H. Taniai, M. Seki, S. Yoshida et al. 2008. Method to detect only live bacteria during PCR amplification. J. Clinical Microbiol. 46(7): 2305-2313.

Song, E.H., B.R. Won and Y.H. Yoon. 2001. Production of probiotic cream cheese by utilizing *Lactobacillus helveticus* CU 631. J. Animal Sci. Technol. 43(6): 919-930.

Souza, C.H.B., F.C.A. Buriti, J.H. Behrens and S.M.I. Saad. 2008. Sensory evaluation of probiotic Minas fresh cheese with *Lactobacillus acidophilus* added solely or in co-culture with a thermophilic starter culture. Int. J. Food Sci. Technol. 43(5): 871-877.

Souza, C.H.B. and S.M.I. Saad. 2009. Viability of *Lactobacillus acidophilus* La-5 added solely or in co-culture with a yogurt starter culture and implications on physico-chemical and related properties of Minas fresh cheese during storage. LWT-Food Sci. Technol. 42(2): 633-640.

Spigno, G., G.D. Garrido, E. Guidesi and M. Elli. 2015. Spray-drying encapsulation of probiotics for ice-cream application. Chem. Eng. Trans. 43: 49-54.

Srisuvor, N., N. Chinprahast, C. Prakitchaiwattana and S. Subhimaros. 2013. Effect of inulin and polydextrose on physiochemical and sensory properties of low-fat set yoghurt with probiotic cultured banana puree. LWT-Food Sci. Technol. 51: 30-36.

Srutkova, D., A. Spanova, M. Spano, V. Drab, M. Schwarzer, H. Kozakova et al. 2011. Efficiency of PCR-based methods in discriminating *Bifidobacterium longum* ssp. *longum* and *Bifidobacterium longum* ssp. *infantis* strains of human origin. J. Microbiol. Methods 87: 10-16.

Stanton, C., G. Gardiner, P.B. Lynch, J.K. Collins, G. Fitzgerald, R.P. Ross et al. 1998. Probiotic cheese. Int. Dairy J. 8(5-6): 491-496.

Stender, H., M. Fiandaca, J. Hyldig-Nielsen and J. Coull. 2002. PNA for rapid microbiology. J. Microbiol. Methods 48: 1-17.

Suhr, M.J., N. Banjara and H.E. Hallen-Adams. 2016. Sequence-based methdos for detecting and evaluating the human gut mycobiome. Lett. Appl. Microbiol. 62(3): 209-215.

Sun, Z., H.M. Harris, A. McCann, C. Guo, S. Argimòn, W. Zhang et al. 2015. Expanding the biotechnology potential of lactobacilli through comparative genomics of 213 strains and associated genera. Nature Commun. 6: 8322. DOI: 10.1038/ncomms9322

Surono, I.S. and A. Hosono. 2002. Fermented milks: Types and standards of identity. pp. 108-123. *In*: H. Roginski, J. Fuquay, P.F. Fox (eds.). Encyclopedia of Dairy Sciences. Elsevier Applied Sciences. Amsterdam, the Netherlands.

Talwalkar, A. and K. Kailasapathy. 2004. A review of oxygen toxicity in probiotic yogurts: Influence on the survival of probiotic bacteria and protective techniques. Comprehen. Rev. Food Sci. Food Safety. 3: 17-124.

Terpou, A., A.-I. Gialleli, A. Bekatorou, D. Dimitrellou, V. Ganatsios, E. Barouni et al. 2017a. Sour milk production by wheat bran supported probiotic biocatalyst as starter culture. Food Bioprod. Bioproc. 101: 184-192.

Terpou, A., A. Bekatorou, M. Kanellaki and A.A. Koutinas. 2017b. Enhanced probiotic viability and aromatic profile of yogurts produced using wheat bran (*Triticum aestivum*) as cell immobilization carrier. Process. Biochem. 55: 1-10.

Terpou, A., A.I. Gialleli, L. Bosnea, M. Kkanellaki, A.A. Koutinas, G.R. Castro et al. 2017c. Novel cheese production by incorporation of sea buckthorn berries (*Hippophae rhamnoides* L.) supported probiotic cells. LWT-Food Sci.Technol. 79: 616-624.

Tharmaraj, N. and N.P. Shah. 2009. Antimicrobial effects of probiotic bacteria against selected species of yeasts and moulds in cheese-based dips. Int. J. Food Sci. Technol. 44(10): 1916-1926.

Tripathi, M.K. and S.K. Gri. 2014. Probiotic functional foods: Survival of probiotics during processing and storage. J. Funct. Foods. 9: 225-241.

Tufarelli, V. and V. Laudadio. 2016. An overview on the functional food concept: Prospectives and applied researches in probiotics, prebiotics and synbiotics. J. Exp. Biology Agric. Sci. 4(3S): 273-278.

Turgut, T. and S. Cakmakci. 2009. Investigation of the possible use of probiotics in ice cream manufacture. Int. J. Dairy Technol. 62(3): 444-451.

Tynkkynen, S., R. Satokari, M. Saarela, T. Mattila-Sandholm and M. Saxelin. 1999. Comparison of ribotyping, randomly amplified polymorphic DNA analysis, and pulsed-field gel electrophoresis in typing of *Lactobacillus rhamnosus* and *L. casei* strains. Appl. Environ. Microbiol. 65: 3908-3914.

van Baarlen, P., F.J. Troost, S. van Hemert, C. van der Meer, W.M. de Vos, P.J. de Groot et al. 2009. Differential NF-κB pathways induction by *Lactobacillus plantarum* in the duodenum of healthy humans correlating with immune tolerance. PNAS. 106(7): 2371-2376.

van Brandt, L., K. Coudijzer, L. Herman, C. Michiels, M. Hendrickx, G. Vlaemynck et al. 2011. Survival of *Mycobacterium avium* ssp. *paratuberculosis* in yoghurt and in commercial fermented milk products containing probiotic cultures. J. Appl. Microbiol. 110: 1252-1261.

van de Casteele, S., T. Ruyssen, T. Vanheuverzwijn and P. van Assche. 2003. Production of Gouda cheese and Camembert with probiotic cultures: The suitability of some commercial probiotic cultures to be implemented in cheese. Commun. Agric. Appl. Biol. Sci. 68(2B): 539-542.

van den Tempel, T., J.K. Gundersen and M.S. Nielsen. 2002. The micro distribution of oxygen in Danablu cheese measured by a microsensor during ripening. Int. J. Food Microbiol. 75: 157-161.

Varga, L., J. Szigeti and V. Ördög. 1999. Effect of a *Spirulina platensis* biomass enriched with trace elements on combinations of starter culture strains employed in the dairy industry. Milchwis. 54(5): 247-248.

Varga, L., J. Szigeti and B. Gyenis. 2006. Influence of chicory inulin on the survival of microbiota of a probiotic fermented milk during refrigerated storage. Annal. Microbiol. 56(2): 139-141.

Vasiljevic, T., T. Kealy and V.K. Mishra. 2007. Effects of β-Glucan addition to a probiotic containing yogurt. J. Food Sci. 72(7): C405-C411.

Vieites, J.M., M.E. Guazzaroni, A. Beloqui, P.N. Golyshin and M. Ferrer. 2010. Molecular methods to study complex microbial communities. Method Molecul. Biol. 668: 1-37.

Vinderola, C.G., W. Prosello, D. Ghiberto and J.A. Reinheimer. 2000a. Viability of probiotic (*Bifidobacterium, Lactobacillus acidophilus* and *Lactobacillus casei*) and nonprobiotic microflora in Argentinian Fresco cheese. J. Dairy Sci. 83: 1905-1911.

Vinderola, C.G., M. Gueimonde, T. Delgado, J.A. Reinheimer and C.G. de los Reyes-Gavilan. 2000b. Characteristics of carbonated fermented milk and survival of probiotic bacteria. Int. Dairy J. 10: 213-220.

Vinderola, C.G., G.A. Costa, S. Regenhardt and J.A. Reinheimer. 2002a. Influence of compounds associated with fermented dairy products on the growth of lactic acid starter and probiotic bacteia. Int. Dairy J. 12: 579-589.

Vinderola, C.G., P. Mocchiutti and J.A. Reinheimer. 2002b. Interactions among lactic acid starter and probiotic bacteria used for fermented dairy products. J. Dairy Sci. 85: 721-729.

Vinderola, G., G. Gueimonde, C. Gomez-Gallego, L. Delfederico and S. Salminen. 2017. Correlation between in vitro and in vivo assays in selection of probiotics from traditional species of bacteria. Trends Food Sci. Technol. 68: 83-90.

Wallace, T.C., F. Guarner, K. Madsen, M.D. Cabana, G. Gibson, E. Hentges et al. 2011. Human gut microbiota and its relationship to health and disease. Nutr. Rev. 69: 392-403.

Walsh, H., J. Cheng and M. Guo. 2014. Effects of carbonation on probiotic survivability, physicochemical and sensory properties of milk-based symbiotic beverages. J. Food Sci. 79(4): M604-M613.

Wang, Z., M. Gerstein and M. Snyder. 2009. RNA-Seq: A revolutionary tool for transcriptomics. Nature Rev. Gene. 10: 57-63.

Wang, W., Y. Bao, G.M. Hendricks and M. Guo. 2012. Consistency, microstructure and probiotic survivability of goat's milk yoghurt using polymerized whey protein as a co-thickening agent. Int. Dairy J. 24: 113-119.

Wang, X., O. Wu, K. Deng, Q. Wei, P. Hu, J. He et al. 2015. A novel method for screening of potential probiotics for high adhesion capability. J. Dairy Sci. 98: 4310-4317.

Wang, H., C. Wang, M. Wang and M. Guo. 2017. Chemical, physicochemical, and microstructural properties, and probiotic survivability of fermented goat milk using polymerized whey protein and starter culture kefir Mild 01. J. Food Sci. 82(11): 2650-2658.

Yadav, R. and P. Shukla. 2017. An overview of advanced technologies for selection of probiotics and their expediency: A review. Critical Rev. Food Sci. Nutr. 57(15): 3233-3242.

Yilmaztekin, M., B.H. Özer and F. Atasoy. 2004. Survival of *Lactobacillus acidophilus* LA-5 and *Bifidobacterium bifidum* BB-02 in white-brined cheese. Int. J. Food Sci. Nutr. 55(1): 53-60.

Zare, F., C.P. Champagne, B.K. Simpson, V. Orsat and J.J. Boye. 2012. Effect of the addition of pulse ingredients to milk on acid production by probiotic and yogurt starter cultures. LWT-Food Sci. Technol. 45: 155-160.

Zehntner, U. 2008. Behaviour of the probiotic strain *Lactobacillus gasseri* K7 in ripened semi-hard cheese. Agrarforsch. 15(4): 194-197.

Zhang, L., X. Zhang, C. Liu, C. Li, S. Li, T. Li et al. 2013. Manufacture of Cheddar cheese using probiotic *Lactobacillus plantarum* K25 and its cholesterol-lowering effects in a mice model. World J. Microbiol. Biotechnol. 29: 127-135.

Zhang, T., J. McCarthy, G. Wang and M. Guo. 2015. Physicochemical properties, microstructure, and probiotic survivability of nonfat goats' milk yogurt using heat-treated whey protein concentrate as fat replacer. J. Food Sci. 80(4): M788-M794.

Ziarno, M., Z. Dorota and A. Bzducha-Wrobel. 2010. Study on dynamics of microflora growth in probiotic rennet cheese models. Polish J. Food Nutr. Sci. 60(2): 127-131.

Zoumpopolou, G., B. Pot, E. Tsakalidou and K. Papadimitriou. 2017. Dairy probiotics: Beyond the role of promoting gut and immune health. Int. Dairy J. 67: 46-60.

6

Milk Fermentation with Kefir Grains and Health Benefits

Siqing Liu

Research Scientist USDA-ARS-NCAUR, 1815 N University, Peoria, Illinois 61604, USA

1. Introduction

Kefir fermentation refers to the process of creating carbonated, acidic, and lightly alcoholic milk beverages from dairy milk by using kefir grains as starter cultures. Depending on the grains to milk ratio, kefir fermentation can be completed within 24 hours. Kefir grains are retrieved by sieving and can be used repeatedly to inoculate batches of fresh milk. Fermented kefir milk is stored at 4 °C for 24 hours or longer before consumption. The low temperature storage allows thickening and cooling of the beverage for a refreshing, smooth taste. Kefir grains are composed of living microorganisms and biopolymers including polysaccharides, proteins and lipids. The microflora of kefir grains contains lactic acid bacteria, acetic acid bacteria, yeasts, and other fungal species. The cohabitated microbial community in the grain can be activated and released into the surrounding liquid when milk or any other nutrient solution is available. The nutrients in milk are used for microbial growth and fermentation. The dynamic metabolic interactions of these microorganisms produce milk kefir beverages. Milk kefir has been used by our ancestors in many parts of the world. Kefir grains had originated from naturally occurring microorganisms clustered together. Initially, kefir was used to preserve milk from being spoiled, but it has become popular in modern times because of its reported health benefits. In this chapter, we will discuss the research progress of kefir fermentation and promote public awareness and acceptance for fermenting milk with kefir grains. This chapter serves as an important stepping stone for researchers to identify specific unique compounds and related products critical to improving human health and lifespan.

Corresponding author: siqing.Liu@ars.usda.gov

2. History of Kefir Fermentation

Radiocarbon dating enables researchers to determine the age of pottery & ceramic collections recovered by archaeologists from Neolithic sites. Gas chromatography–mass spectrometry (GC-MS), high-performance liquid chromatography–mass spectrometry (HPLC-MS) analyses of residues from pottery vessels from Jiahu, a Neolithic village in Henan, China, suggested the presence of mixed fermented beverages from rice, honey, grape and Chinese hawthorn traced back to 7000 B.C (McGovern et al., 2004). Proteomic analyses of dairy foods near Early Bronze Age Xiaohe tombs in Xinjian, China revealed symbiotic microorganisms and kefir-like milk fermentation that originated some 3800 years ago (Yang et al., 2014). While uncovering farming and domestic animal management practices at Neolithic sites near the Adriatic Sea, carbon-dated rhyta pottery vessels were found to be used as containers for cheese making and milk fermentation around 7200 years ago. GC-MS analyses confirmed fatty acid residues from these pottery vessels (McClure et al., 2018). It was believed that these potteries were used to store milk. Under stringent supplies of food, ancient human beings learned that milk from cows and sheep was more sustainable than consumption of meat. Without refrigeration, milk fermentation was a necessary practice to preserve harvested milk. When kept at room temperature, milk lactose spontaneously converted to organic acids through naturally available microbes. The resulting coagulated acidified milk did not contain lactose, and it was preferable to raw milk among the lactose-intolerant populations. Thus, milk fermentation became a preferred practice.

3. Kefir Grains

Kefir grains are a consortium of microorganisms embedded in bacterial surface polysaccharides intertwined with proteins and lipids. Kefir grains are white-yellow in color and around 0.2-5 cm in size with a cauliflower shape and have an irregular, twisted rough surface (Dobson et al., 2013). Kefir grains are initially built up by microbial self-aggregations of lactic and acetic acid bacteria, with significant biofilm formation properties. Yeasts are present at the granule surfaces. Upon continuous transfer and subculture, more microbes and biopolymers are produced, and grains grow and multiply (Wang et al., 2012). Kefir grains are robust and can be propagated indefinitely (Simova et al., 2002). Different grains contain unique combinations of microbial strains, resulting in various fermented products. The origin and property of kefir grains varies with geological areas. The most well-known traditional grains are from the Tibet region, the Caucasus mountains, Russia, Eastern Europe, and Asia (Bengoa, Iraporda, Garrote, & Abraham, 2018).

3.1 Kefir Milk

In addition to the source grains, other factors that contribute to the features of kefir milk include fermentation time, temperatures, and sources of raw milk and other added substrates. As such, there are a wide variety of kefir products based on acidity, texture, color, flavor, aroma and consistency.

Kefir differs when made from goat milk, sheep milk, cow milk and buffalo milk (Cais-Sokolinska et al., 2015). Differences in fermentation substrates are originated from a variety of animal feeds and available nutrient supplements from dairy milk. For example, feeding goats with false flax cake results in kefir milk with an increased production of bioactive polyunsaturated fatty acids, such as conjugated linoleic acid (CLA) and n-3 fatty acids (Cais-Sokolinska et al., 2015). There are kefir drinks made from other sources including water kefir and soy kefir that also provide health benefits (see below for more detailed descriptions).

3.2 Microorganisms Found in Kefir

Traditional studies of kefir grain microbes focus on culture-based isolation and phenotypical identification of single species with characteristic morphology, physiology and biochemical properties (Feofilova, 1958). The most abundant bacterial species include *Lactobacillus kefiri*, *Lactobacillus kefriranofaciens*, *Lactococcus lactis*, and yeast species include *Saccharomyces cerevisiae*, *Kluyveromyces* and *Candida*. These abundant species were readily identified using already available techniques at the time (La Riviere, 1969, La Riviere and Kooiman, 1967, Simova et al., 2002). With advanced molecular biology and DNA sequencing technology, previously undiscovered microbial species were identified in kefir by genomic DNA isolation of pure cultures and PCR based DNA sequence analyses of 16 S rRNA for bacteria and 26S rRNA for yeasts, respectively (Ninane et al., 2007, Vardjan et al., 2013). In recent years a sensitive culture-independent method known as PCR Denaturing Gradient Gel Electrophoresis (PCR-DGGE) allowed direct identification of less popular species in kefir. Researchers discovered that microorganisms present in kefir milk are different but less diversified when compared to microorganisms embedded within kefir grains. The most comprehensive identifications are attributed to the recent advance of single genome and metagenomic sequencing (Kim et al., 2017). Below is a summary of microbial species identified in kefir grains that have been described in recent literature.

Lactobacilli

Lactobacillus amylovorus	(Nalbantoglu et al., 2014)
Lactobacillus acidophilus	(Dobson et al., 2011)
Lactobacillus brevis	(Simova et al., 2002)
Lactobacillus buchneri	(Nalbantoglu et al., 2014)
Lactobacillus bulgaricus	(Simova et al., 2006)
Lactobacillus casei	(Simova et al., 2002)
Lactobaccilus crispatus	(Nalbantoglu et al., 2014)
Lactobacillus delbrueckii subsp. *bulgaricus*	(Simova et al., 2002)
Lactobacillus delbrueckii	(Nalbantoglu et al., 2014)
Lactobacillus diolivorans	(Nalbantoglu et al., 2014)

Lactobacillus gallinarum	(Nalbantoglu et al., 2014)
Lactobacillus gasseri	(Nalbantoglu et al., 2014)
Lactobacillus helveticus	(Simova et al., 2006)
Lactobacillus johnsonni	(Nalbantoglu et al., 2014)
Lactobacillus kefiranofaciens	(Nalbantoglu et al., 2014)
Lactobacillus kefiri	(Vardjan et al., 2013)
Lactobacillus kefirgranum	(Vardjan et al., 2013)
Lactobacillus otakiensis	(Nalbantoglu et al., 2014)
Lactobacillus parabuchneri	(Nalbantoglu et al., 2014)
Lactobacillus paracasei	(Nalbantoglu et al., 2014)
Lactobacillus parakefiri	(Vardjan et al., 2013)
Lactobacillus pentosus	(Walsh et al., 2016)
Lactobacillus plantarum	(Gao, Gu, Abdella, Ruan, & He, 2012)
Lactobacillus renteri	(Nalbantoglu et al., 2014)
Lactobacillus rhamnosus	(Nalbantoglu et al., 2014)
Lactobacillus rossiae	(Nalbantoglu et al., 2014)
Lactobacillus sakei	(Nalbantoglu et al., 2014)
Lactobacillus salivarius	(Walsh et al., 2016)
Lactobacillus sunkii	(Nalbantoglu et al., 2014)

Lactococci

Lactococcus lactis subsp. *Cremoris*	(Zhou, Liu, Jiang, & Dong, 2009)
Lactococcus lactis subsp. *Lactis*	(Ninane et al., 2007)
Lactococcus lactis subsp. *Lactis biovar.* Diacetylactis	(Bengoa et al., 2018)
Lactococcus garvieae	(Nalbantoglu et al., 2014)
Leuconostoc mesenteroides	(Nalbantoglu et al., 2014)
Leuconostoc camosum	(Walsh et al., 2016)
Leuconostoc citreum	(Walsh et al., 2016)
Leuconostoc gelidum	(Walsh et al., 2016)
Leuconostoc kimchi	(Walsh et al., 2016)

Streptococci

Streptococcus thermophilus	(Garofalo et al., 2015)
Streptococcus kefirresidentii	(Kim et al., 2017)

Acetic acid bacteria

Acetobacter sp.	(Gao et al., 2012)
Acetobacter fabarum	(Garofalo et al., 2015)
Acetobacter lovaniensis	(Garofalo et al., 2015)
Acetobacter orientalis	(Garofalo et al., 2015)
Acetobacter pasteurianus	(Walsh et al., 2016)
Gluconobacter sp.	(Bengoa et al., 2018)

Other bacteria

Anoxybacillus sp.	(Walsh et al., 2016)
Bacillus subtilis	(Gao et al., 2012)
Bacillus kefirresidentii	(Kim et al., 2017)
Bifidobacterium choerinum	(Dobson et al., 2011)
Bifidobacterium longum	(Dobson et al., 2011)
Bifidobacterium pseudolongum	(Dobson et al., 2011)

Butyribibrio sp.	(Walsh et al., 2016)
Clostridium sp.	(Walsh et al., 2016)
Dysgonomonas sp.	(Gao et al., 2012)
Enterobacter sp.	(Walsh et al., 2016)
Enterococcus sp.	(Garofalo et al., 2015)
Geobacillus sp.	(Walsh et al., 2016)
Gluconacetobacter diazotrophicus	(Walsh et al., 2016)
Gluconobacter oxydans	(Walsh et al., 2016)
Oenococcus oeni	(Nalbantoglu et al., 2014)
Pediococcus sp.	(Bengoa et al., 2018)
Pseudomonas putida	(Zhou et al., 2009)
Rothia kefirresidentii	(Y. Kim et al., 2017)
Serratia sp.	(Walsh et al., 2016)
Tetragenococcus halophilus	(Nalbantoglu et al., 2014)
Thermus sp.	(Walsh et al., 2016)
Turicibacter sp.	(Walsh et al., 2016)

Fungi

Candida holmii	(Bengoa et al., 2018)
Candida inconspicua	(Simova et al., 2002)
Candida kefyr	(Bengoa et al., 2018)
Candida krusei	(Bengoa et al., 2018)
Candida lambica	(Bengoa et al., 2018)
Candida lipolytica	(Bengoa et al., 2018)
Candida maris	(Simova et al., 2002)
Dekkera anomala	(Garofalo et al., 2015)
Kazachstania aerobia	(Zhou et al., 2009)
Kazachstania exigua	(Vardjan et al., 2013)
Kazachstania unispora	(Vardjan et al., 2013)
Kluyveromyces lactis	(Zhou et al., 2009)
Kluyveromyces marxianus	(Gethins et al., 2016)
Pelomonas sp.	(Gao et al., 2012)
Pichia fermentans	(Wang et al., 2012)
Pichia guilliermondii	(Gao et al., 2012)
Pichia kudriavzevii	(Gao et al., 2012)
Pichia Propionibacterium	(Walsh et al., 2016)
Saccharomyces cerevisiae	(Vardjan et al., 2013)
Saccharomyces martiniae	(Zhou et al., 2009)
Saccharomyces turicensis	(S. Y. Wang et al., 2012)
Saccharomyces unisporus	(Zhou et al., 2009)
Torulospora delbrueckii	(Vardjan et al., 2013)
Weissella sp.	(Gao et al., 2012)

This list includes a total of 92 species. There is a wide range of microbial species, as well as dominant populations of microbial species based on geographic origin. Kefir grains from Argentina, Belgium, Brazil, Ireland, Italy, South Africa, Taiwan and Tibet all have unique combinations of microbes (Marsh et al., 2013). Even within a single kefir source, some strains dominate the early stage of fermentation and others are more prevalent in later stages based on nutrient availability, fermentation conditions, and interactions among the microbial community (Bengoa et al., 2018, Walsh et al., 2016).

Microbes rely on each other in kefir. The growth of individual strains in kefir can be either competitive or dependent in a symbiotic manner. Some pure bacterial cultures could not survive in milk unless yeast extract was added. Yeasts in kefir provide amino acids and vitamins for bacteria. The utilization of lactate as an energy source by yeasts helps maintain a less acidic environment for some bacteria (Hsieh et al., 2012, Marsh et al., 2013, Walsh et al., 2016).

4. Biochemical Compounds Found in Kefir Milk and Kefir Grains

Kefir grain is a matrix of biopolymers embedded with microbiota. While using electron microscopy to examine a kefir grain, the surface is rough with irregular holes, and the interior of the grain appears as a threadlike microfiber comprised of proteins, lipids and polysaccharides (Ismaiel et al., 2011).

4.1 Kefiran

The main component of the kefir grain is the water soluble polysaccharide originally found to be produced by *Lactobacillus brevis* (La Riviere and Kooiman, 1967). Kefiran is a carbohydrate polymer consisting of equal amounts of galactose and glucose residues through β-D-(1→6) linkages. With a generally recognized as safe (GRAS) status, Kefiran has been widely used in pharmaceutical industries and has become a desired exopolysaccharide because of its properties as a stabilizer, thickener, emulsifier and gelling agent in the food industry. Fermentation parameters affecting kefiran production in large-scale bioreactors by using *Lactobacillus kefiranofaciens* were evaluated (Enikeev, 2012). A recent study revealed that a mixed culture of *Lactobacillus kefiranofaciens* with *Saccharomyces cerevisiae* can increase kefiran production significantly when compared with kefiran produced from a pure *L. kefiranofaciens* culture. It was suggested that yeast in the mixed cultures contributed to reduction of lactic acid and promoting bacterial cell growth (Cheirsilp et al., 2003). Using low-cost whey lactose and spent yeast cells hydrolysate in combinations with fed-batch and controlled pH fermentation, the production of kefiran reached the highest level around 2514 mg per liter (Cheirsilp et al., 2018).

4.2 Proteins

Milk proteins are composed of caseins, including kappa-, alpha-, and beta-caseins (CN), and whey proteins, contain alpha-lactalbumin (alpha-LA) and beta-lactoglobulin (beta-LG). During kefir fermentation, caseins are hydrolyzed into peptides and amino acids by proteolytic enzymes produced by kefir microbiota. Digestion of 27 milk proteins by Kefir microbial proteases resulted in detection of 1500 peptides as revealed by Peptidomics analyses (Dallas et al., 2016). When the proteolytic products of major milk proteins were analyzed by reverse phase -HPLC, the rates of alpha-LA, alpha and beta CN and alpha-LA hydrolysis were found to be proportional to the amount of kefir added and the duration of fermentation (Ferreira et al., 2010). In a separate study, a total of 236 peptides were detected from kefir fermentation, and 16 were known as bioactive peptides with health promoting antioxidant, antimicrobial, and immunomodulating properties (Ebner et al., 2015). It has been suggested that most of the kefir peptides are released from kefir fermentation and are not available in raw milk (Ebner et al., 2015).

4.3 Lipids

Dairy milk lipids are mainly (96%) fatty acid esters with glycerol, commonly referred to as triglycerides with a glycerol backbone and fatty acids on the three side chains. Milk fat varies based on the fatty acid side chains, with C12-18 referred to as long chain and C4-10 as short chain fatty acids. About 65% of milk fatty acids consist of saturated fatty acids, such as lauric, myristic, palmitic and stearic, and 32% are unsaturated fatty acids such as oleic acid, and 3% are polyunsaturated fatty acids linoleic and linolenic (Vieira et al., 2015). There are more than 400 different fatty acids identified in milk fat (Jensen, 2002). A recent study documented the milk fatty acid profile changes after fermentation with several types of kefir grains. Using gas chromatography, fatty acids were quantitatively analysed before and after kefir fermentation. After fermentation, observations included an increase in palmitic acid (16:0), oleic acid (18:1n9) and saturated fatty acids and a decrease in monounsaturated fatty acids. However, fatty acid composition varied with the source of kefir grain and the types of milk. Kefir grain AV used in this study produced higher monounsaturated fatty acids due to higher desaturase activities, and Kefir grain AD produced higher lipase activities (Vieira et al., 2015).

An earlier study also documented the reduction of milk cholesterol after kefir fermentation (Vujicic et al., 1992). Another important finding was that after kefir fermentation, a higher concentration of conjugated linoleic acid (CLA) was reported (Guzel-Seydim et al., 2006).

5. Other Products Produced by Kefir Fermentation

A comprehensive analysis of kefir milk after fermentation suggested the presence of microbial metabolites and fermentation end products. Detectable compounds include carboxylic acids (lactic acid, butyric acid, citric acid and acetic acid), alcohols, CO_2 and other flavor agents such as aldehydes, esters, ketones and lactones (Dimitrellou et al. 2015, Walsh et al., 2016). Accumulation of free amino acids in kefir cultures was also detected (Simova et al., 2006). Specific amino acids such as glutamic acid, tyrosine, serine, histidine, alanine, methionine, and lysine were present in fermented milk (Gul et al., 2015). Some amino acids were synthesized by lactic acid bacteria including *Lactobacillus delbrueckii* and *Lactobacillus helveticus*, while others might have resulted from proteolytic degradation of milk proteins (Gul et al., 2015, Simova et al., 2006, Zheng et al., 2012).

6. Type of Kefir Beverages

When we discuss the characteristics of kefir beverages, one of the most desired features that contributes to the flavor of kefir drinks is the carbon dioxide produced from microbial fermentation. The presence of CO_2 renders a drink a sparkling texture. The importance of CO_2 concentration was evidenced as a mathematical model and was recently developed to mimic CO_2 release during kefir milk fermentation (Gorsek et al., 2018). Although kefir beverages are identified based on fermentation substrates, the natural production of CO_2 is viewed as a common signature trait of all types of kefir beverages.

6.1 Milk Kefir

Milk kefir is acidic and creamy in consistency, and is the most studied kefir drink so far. Some have an exotic taste due to a minor ethanol content. This type of beverage usually exhibits the distinguished flavors and sensory properties developed through the process of fermenting milk, with variations depending on the source of kefir grains and the conditions of the fermentation process. There are milk kefir drinks made from cow and buffalo milk (Gul et al., 2015), from camel, goat, ewe milk (Alimi et al., 2018) and from sheep's milk (de Lima et al., 2018). A detailed composition study of buffalo milk kefir indicated higher calcium, higher ethanol, higher amino acids, and less cholesterol when compared with cow milk kefir (Gul et al., 2015).

6.2 Water Kefir

Water kefir is a kefir drink produced from water with dried or fresh fruits and added sugar, which is then fermented with water kefir grains under anaerobic conditions (Laureys & De Vuyst, 2017). Water kefir is a beverage that is a yellowish, translucent, slightly alcoholic liquid, with a fruity taste. Just like milk kefir grains, water kefir grains contain specific strains of lactic acid bacteria and yeasts within the grains. Water kefir grains are thought to originate from the natural fermentation of the sap from Opuntia cactus plants, which appear crystal and brittle in nature (Laureys and De Vuyst, 2014). Dextran is the major polymer found in water kefir grains, which is formed by alpha (1→ 6) linkages of glucose molecules (Laureys and De Vuyst, 2014). The predominant microbes available after water kefir fermentation are lactic acid bacteria species *Lactobacillus hilgardii*, *Lactobacillus nagelii*, *Lactobacillus paracasei*, *Bifidobacterium* and yeast *Saccharomyces cerevisiae*. These microbes exist as a community and their dynamic metabolic activities result in a mixture of lactic acid, acetic acid, glycerol, ethanol, and mannitol plus aroma compounds that add flavor. The presence of isoamyl-alcohol, ethyl acetate, 2-methyl-1-propanol, ethyl octanoate have been reported and low levels of ethyl decanoate, isoamyl acetate, and ethyl hexanoate were also detected in water kefir drinks (Laureys and De Vuyst, 2017).

6.3 Soy Kefir

Soy milk contains no cholesterol, less fat, less calcium, very low levels of vitamin B12 and no sulfur-containing amino acids when compared with cow milk. Soy milk has become a preferred nutrient drink substitute for cow milk for lactose intolerant or vegan individuals, due to the presence of fiber and absence of lactose. Soy milk has more iron, magnesium, copper, and manganese than cow milk (Hajirostamloo, 2009). The major carbohydrates of soybean milk are sucrose, raffinose, stachyose, pentanal and n-hexanal. Fermentation of raffinose and stachyose by microbes from kefir grains is desired because these carbohydrates are indigestible by the human body due to the absence of α-galactosidase. The breakdown of unsaturated acids of soy milk to n-hexanal and pentanal often leads to an unpleasant taste which can be eliminated by using *Bifidobacterium breve* MB233 (Scalabrini et al., 1998). Isoflavones are another major component of soy milk. Studies have shown that β-glucosidase produced by kefir microbes can convert isoflavones into their corresponding bioactive aglycones, a type of flavonoids that have potential anti-inflammatory and anticancer effects (Bau et al., 2015). Consumers making kefir drinks at home have been using other milk substitutes such as coconut milk and almond milk, either alone or in combination with dairy milk. One recent paper described fermentation of almond milk with *Lactobacillus reuteri* and *Streptococcus thermophilus* cultures that resulted in positive functional assessments *in vitro* with potential probiotic benefits (Bernat et al., 2015). As for individual consumption, there are no specific guidelines for choosing a particular kefir beverage. In general, a choice of drinks might just be a personal preference based on diet or restrictions on dairy products due to allergic reactions, lactose intolerance or other health concerns. Because kefir has been a household fermentation product for thousands of years, individual preferences might be a heritage or a tradition independent of specific health benefits.

7. Health Benefits of Kefir Beverages

The health promoting advantages of kefir milk consumption have been documented long ago but more comprehensive studies have just resurfaced in recent years. Oral intake and consumption of kefir beverages has been studied in human volunteers with health issues and by using animal models with an induced illness. The known benefits include antimicrobial properties, antioxidant, immunity-modulating and anticancer functions. A summary of some recent scientific findings is presented below.

7.1 Probiotic

Nobel Laureate Elie Metchnikoff first introduced the probiotic concept in 1908, when he attributed the long life of Bulgarian peasants to their consumption of fermented milk (Otles et al., 2003). Yogurt and kefir drinks with beneficial microbes are viewed as the pioneering class of functional foods. Kefir's probiotic nature comes from cultures of kefir grains that form an ecosystem of various species among *Lactobacilli, Lactococci, Leuconostoc* spp., acetic acid bacteria, and yeasts including *Saccharomyces, Candida* and *Kazachstania* (Simova et al., 2002). The initial definition of probiotics refers to living microorganisms that can fight off pathogenic microbes and provide health benefits to human beings, with the ability to restore immune intestinal health. With time, the definition of probiotics has evolved and become more specific to include biochemical traits related to intestinal functions and the overall health of host bodies. For example, acid and bile salt tolerances have been used as indicators of probiotic properties, because after passing through gastrointestinal tract, the bacteria that survive and colonize in the human gut must have resistance mechanisms to cope with bile salts synthesized in the liver. A high extracellular polysaccharide producing strain *Lactobacillus kefiranofaciens* XL10 from a Tibetan kefir grain was described probiotic. This strain can tolerate acid and bile salt. It is also capable of colonization in the gut and modulating the gut microbiota in a mouse model (Xing et al., 2017). Another recent study of several *Lactococcus lactis* cultures isolated from kefir grain showed better bile salts deconjugation and antimicrobial activity than similar strains isolated from raw milk (Yerlikaya, 2018). Moreover, it was demonstrated that pure *L. lactis* culture could not produce the anticipated probiotic protections (Yerlikaya, 2018), thus stable microbiota found in kefir milk might be the best choice for probiotic applications.

7.2 Antimicrobial

The antimicrobial abilities of kefir can be attributed to a combination of factors from the fermentation process, including low acidity, organic acids, rapid acid production, bacteriocins, carbon dioxide, hydrogen peroxide, ethanol, diacetyl, low redox potential and nutrient depletion (Silva et al., 2009). Kefir fermented broth produced from brown sugar was found to inhibit the growth of harmful bacteria such as *Salmonella typhi, Shigella sonnei, Staphylococcus aureus,* and *Escherichia coli* (Silva et al., 2009). Kefir and fermented probiotics milk products also were reported to have an antibacterial effect on *Helicobacter pylori* infection. In addition, kefir consumption has prevented gastric ulcers in rats (Batinkov, 1971, Orhan et al., 2012). In a recent study, fermentation of kefir milk was reported to be effective in controlling and eliminating *Mycobacterium bovis*, a pathogen that causes Bovine tuberculosis. Significant reduction of *M. bovis* was observed when the fermentation time extended. After 60 hours of fermentation, *M. bovis* was eliminated completely. The main factors for controlling the growth of *M. bovis* are the accumulation of metabolites and decrease of pH

(Macuamule et al. 2016). Kefir cultures were found to significantly reduce counts of enterobacteria, coliforms, and staphylococci during the cheese making process. This is attributable to the metabolic activity of kefir cultures that deplete sugars, lower pH, and remove oxygen (Dimitrellou et al., 2015). Kefir milk was also reported as having an inhibitory role against *Escherichia coli 3* (Garrote, Abraham, & De Antoni, 2000). Both kefir and kefiran showed antibacterial activity against *Staphylococcus aureus, Listeria monocytogenes, Pseudomonas aeruginosa, Salmonella typhimurium* and *Candida albicans*. The highest activity was against *Streptococcus pyogenes* (Kim et al., 2016; Rodrigues et al., 2005). In another recent study, kefir culture broth was found inhibitory to food pathogens and spoilage bacteria *Bacillus cereus, Staphylococcus aureus, Listeria monocytogenes, Enterococcus faecalis, Escherichia coli, Salmonella Enteritidis, Pseudomonas aeruginosa,* and *Cronobacter sakazakii* (Kim et al., 2016). However, the antibacterial activity found in kefir is not always consistent. Kefirs with various origins have different antimicrobial spectra, and the antibacterial activity also varies with fermentation time. It is possible that the antibacterial activity might be derived from key compounds at different fermentation stages, such as specific organic acids and bacteriocins and not simply due to lowered pH (Kim et al., 2016). A separate report showed that milk fermented products from Tibetan kefir grains can inhibit the growth of opportunistic bacterial pathogens *Staphylococcus aureus, Bacillus mesentericus, Mycobacterium luteum, Proteus vulgaris* and fungus *Aspergillus niger* (Kukhtyn et al., 2018).

Lactic acid bacteria have been shown to produce unique fermentation products with antifungal activity (Hassan and Bullerman, 2008, Magnusson et al., 2003, Sjogren et al., 2003; Strom et al., 2002). It is not surprising to find antifungal properties after kefir fermentation, due to the presence of a collection of lactic acid bacteria. In addition to the reported inhibition of *A. niger* (Kukhtyn et al., 2018), kefir cell-free supernatants demonstrated higher antifungal activity against *Aspergillus flavus* when compared with a control of artificially acidified milk with organic acids. This kefir product was used to protect foods made from corn against fungi spoilage. The antifungal property could be applied to protect commercial corn products and used as a food preservative (Gamba et al., 2016). Kefir fermented supernatant was also used to reduce aflatoxin produced by *Aspergillus parasiticus* from infected pistachio nut (Ansari et al., 2015).

7.3 Anti-inflammatory

The exopolysaccharide from kefir grain has been described to delay tumor growth *in vivo* when administered orally and enhances T-cell mediated immune responses in a mouse model (Murofushi et al., 1986). Kefir derived extracellular vesicles were found to have the ability to inhibit the production of inflammatory cytokines in intestinal cells and alleviated inflammatory bowel disease in a mouse model (Seo et al., 2018). The products such as peptides and exopolysaccharides from milk fermentation by kefir microflora enhanced the health of mice by positively manipulating the microbiota of the intestines (Vinderola, Perdigon, Duarte, Farnworth, & Matar, 2006). Some bioactive peptides and proteins have been reported to stimulate immune responses including production of immunoglobulin and lymphocyte proliferation (Bengoa et al., 2018, Moller et al., 2008, Walther and Sieber, 2011). *Lactobacillus kefiranofaciens* M1 from Taiwanese milk kefir grain was shown to induce cytokine production in intestinal epithelial cells and a probiotic induced anticolitis effect in mice (Chen et al., 2012). The oral administration of kefir fermented milk appeared to be a promising therapy for intestinal disorder (Chen et al., 2012, Franco et al., 2013).

7.4 Anticancer

The antitumor activity of water-soluble polysaccharides from kefir grain was first reported in studies with mice. The growth of induced tumors was inhibited significantly by either oral administration or with intraperitoneal injections (Murofushi et al., 1983, Shiomi et al., 1982). A recent study compared donkey milk and donkey milk kefir for their antimutagenic and anti-tumor effects in mice. Donkey milk kefir was shown to induce apoptosis and inhibited tumor growth, making it a better choice for treatment of breast cancer over the donkey milk control (Esener et al., 2018). Oral administration of milk kefir and soy milk kefir to a mouse model with induced tumors resulted in 64.8% and 70.9% inhibition of tumor growth because of kefir-induced apoptotic tumor cell lysis. Those authors reported that total immunoglobulin A from the wall of the small intestine of tumor-bearing mice was significantly higher after feeding on kefir for a month (Liu et al., 2002). Kefir drinks with added safflower oil enhanced antioxidant and anti-cancerous effects (Farsad-Naeimi et al., 2015). In recent years, as one of the functional food components, conjugated linoleic acid (CLA) production by using milk kefir has drawn some attention since it has been recognized with antimutagenic and anticarcinogenic properties (Farsad-Naeimi et al., 2015, Guzel-Seydim et al., 2011).

7.5 Improve Blood Lipid Profile

Kefir grains can metabolize cholesterol present in dairy milk. A study reported that cholesterol levels in milk were significantly reduced after kefir fermentation (Vujicic et al., 1992). Consumption of kefir drinks increases microorganisms from the Kefir microbiota in the guts of humans and mice (Wang et al., 2009). These microorganisms have been reported to produce bile salts hydrolase that deconjugate salts into acids that are excreted out in animals. To maintain a normal physiological level of bile acid, new bile acid needs to be synthesized from serum cholesterol which leads to reduced cholesterol (St-Onge et al., 2000). Meanwhile, microbial production of propionic acid from kefir fermentation can inhibit cholesterol biosynthesis and result in a redistribution of cholesterol in clinical studies (Fathi et al., 2017, St-Onge et al., 2000). Studies also found that drinking kefir resulted in a significant improvement in serum lipid profile for overweight premenopausal women compared to a control group (Fathi et al., 2016). When diabetic rats were fed with a combination of goat milk kefir and soy milk kefir for 35 days, their plasma glucose levels were reduced, and serum triglyceride was maintained compared to diabetic control rats (Nurliyani et al., 2015). In a 12-week animal trial, researchers discovered that kefir milk modulated the gut microbiota of high-fat diet-induced obese mice by promoting fatty acid oxidation as compared to a control group fed with regular milk (Kim et al., 2017).

Kefir beverages are loaded with probiotic microorganisms that have been shown to improve the gut microbiota balance of the hosts and help the immune system defense against pathogens. Other beneficial effects of these probiotic microorganisms include stimulation of the immune system, reduction of blood cholesterol levels, synthesis of vitamins particularly the vitamin B group (Hugenholtz, 2013), and anticarcinogenic and antimicrobial activities.

8. Conclusions and Perspectives

The market and interests of naturally fermented kefir beverages has been growing rapidly in recent years as is evidenced by increased publications (Figure 1) and increased consumers with a stronger desire for functional foods that improve health and reduce the risk of illness. As we know that

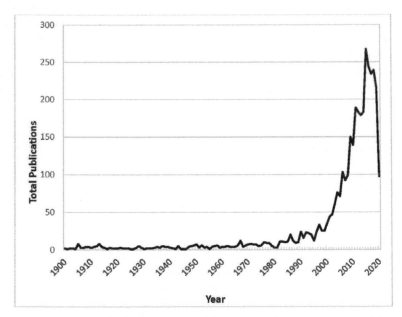

Fig. 1: The total number of publications of kefir related studies per year. The data were collected from 1900 to February 13, 2020 from a search of the CAS database (https://scifinder.cas.org/scifinder).

kefir products originated thousands of years ago and have been passed from household to household. Currently one can choose to purchase kefir grains and make preferred drinks at home or obtain various kefir products in grocery stores. However, Lifeway remains the only manufacturer of kefir products in US (http://lifewaykefir.com) ever since 1986. The commercial kefir products started to become globally popular with a market worth of $80.3 million (Lifeway, 2018). More research and education of the benefits of kefir beverages are needed to promote awareness and improve public perception. With the available metagenome sequencing technology, strains of low abundance in kefir have been identified and this technology will enable quick detection of the microbiota and real time monitor of the fermentation process to increase production efficiency. It is also possible to introduce known probiotics into the kefir grain for better products with preferred flavor and health benefits.

References

Alimi, D., M. Rekik and H. Akkari. 2018. Comparative *in vitro* efficacy of kefir produced from camel, goat, ewe and cow milk on Haemonchus contortus. J. Helminthol. 93: 440-446.

Ansari, F., F. Khodaiyan, K. Rezaei and A. Rahmani. 2015. Modelling of aflatoxin G1 reduction by kefir grain using response surface methodology. J. Environ. Health Sci. Eng. 13: 40.

Batinkov, E.L. 1971. Use of milk and kefir in peptic ulcer of the stomach and duodenum. Vopr. Pitan. 30(4): 89-91.

Bau, T.R., S. Garcia and E.I. Ida. 2015. Changes in soymilk during fermentation with kefir culture: Oligosaccharides hydrolysis and isoflavone aglycone production. Int. J. Food Sci. Nutr. 66(8): 845-850.

Bengoa, A.A., C. Iraporda, G.L. Garrote and A.G. Abraham. 2019. Kefir micro-organis: Their role in grain assembly and health properties of fermented milk. J. Appl. Microbiol. 126(3): 686-700.

Bernat, N., M. Chafer, A. Chiralt and C. Gonzalez-Martinez. 2015. Development of a non-dairy probiotic fermented product based on almond milk and inulin. Food Sci. Technol. Int. 21(6): 440-453.

Cais-Sokolinska, D., J. Pikul, J. Wojtowski, R. Dankow, J. Teichert, G. Czyzak-Runowska et al. 2015. Evaluation of quality of kefir from milk obtained from goats supplemented with a diet rich in bioactive compounds. J. Sci. Food Agric. 95(6): 1343-1349.

Cais-Sokolinska, D., J. Wojtowski, J. Pikul, R. Dankow, M. Majcher, J. Teichert et al. 2015. Formation of volatile compounds in kefir made of goat and sheep milk with high polyunsaturated fatty acid content. J. Dairy Sci. 98(10): 6692-6705.

Cheirsilp, B., H. Shimizu and S. Shioya. 2003. Enhanced kefiran production by mixed culture of *Lactobacillus kefiranofaciens* and *Saccharomyces cerevisiae*. J. Biotechnol. 100(1): 43-53.

Cheirsilp, B., S. Suksawang, J. Yeesang and P. Boonsawang. 2018. Co-production of functional exopolysaccharides and lactic acid by *Lactobacillus kefiranofaciens* originated from fermented milk, kefir. J. Food Sci. Technol. 55(1): 331-340.

Chen, Y.P., P.J. Hsiao, W.S. Hong, T.Y. Dai and M.J. Chen. 2012. *Lactobacillus kefiranofaciens* M1 isolated from milk kefir grains ameliorates experimental colitis in vitro and *in vivo*. J. Dairy Sci. 95(1): 63-74.

Dallas, D.C., Citerne F., T. Tian, V.L.M. Silva, K.M. Kalanetra, S.A. Frese et al. 2016. Peptidomic analysis reveals proteolytic activity of kefir microorganisms on bovine milk proteins. Food Chem. 197(Pt A): 273-284.

De Lima, M.D.S., R.A. da Silva, M.F. da Silva, P.A.B. da Silva, R.M.P.B. Costa, J.A.C. Teixeira et al. 2018. Brazilian kefir-fermented sheep's milk, a source of antimicrobial and antioxidant peptides. Probiot. Antimicrob. Prot. 10(3): 446-455.

Dimitrellou, D., P. Kandylis, Y. Kourkoutas, A.A. Koutinas and M. Kanellaki. 2015. Cheese production using kefir culture entrapped in milk proteins. Appl. Biochem. Biotechnol. 176(1): 213-230.

Dobson, A., O. O'Sullivan, P.D. Cotter, P. Ross and C. Hill. 2011. High-throughput sequence-based analysis of the bacterial composition of kefir and an associated kefir grain. FEMS Microbiol. Lett. 320(1): 56-62.

Ebner, J., A. Arslan, M. Fedorova, R. Hoffmann, A. Kucukcetin, M. Pischetsrieder et al. 2015. Peptide profiling of bovine kefir reveals 236 unique peptides released from caseins during its production by starter culture or kefir grains. J. Proteom. 117: 41-57.

Enikeev, R. 2012. Development of a new method for determination of exopolysaccharide quantity in fermented milk products and its application in technology of kefir production. Food Chem. 134(4): 2437-2441.

Esener, O., B.M. Balkan, E.I. Armutak, A. Uvez, G. Yildiz, M. Hafizoglu et al. 2018. Donkey milk kefir induces apoptosis and suppresses proliferation of Ehrlich ascites carcinoma by decreasing iNOS in mice. Biotech. Histochem. 93(6): 424-431.

Farsad-Naeimi, A., S. Imani, S.R. Arefhosseini and M. Alizadeh. 2015. Effect of safflower oil on concentration of conjugated linoleic acid of kefir prepared by low-fat milk. Recent Pat. Food Nutr. Agric. 7(2): 128-133.

Fathi, Y., S. Faghih, M.J. Zibaeenezhad and S.H. Tabatabaei. 2016. Kefir drink leads to a similar weight loss, compared with milk, in a dairy-rich non-energy-restricted diet in overweight or obese premenopausal women: A randomized controlled trial. Eur. J. Nutr. 55(1): 295-304.

Fathi, Y., N. Ghodrati, M.J. Zibaeenezhad and S. Faghih. 2017. Kefir drink causes a significant yet similar improvement in serum lipid profile, compared with low-fat milk, in a dairy-rich diet in overweight or obese premenopausal women: A randomized controlled trial. J. Clin. Lipidol. 11(1): 136-146.

Feofilova, E.P. 1958. Microflora of kefir grain. Mikrobiologiia 27(2): 229-234.

Ferreira, I.M.P.L.V.O., O. Pinho, D. Monteiro, S. Faria, S. Cruz, A. Perreira et al. 2010. Short communication: Effect of kefir grains on proteolysis of major milk proteins. J. Dairy Sci. 93(1): 27-31.

Franco, M.C., M.A. Golowczyc, G.L. De Antoni, P.F. Perez, M. Humen, M. de L. Serradell et al. 2013. Administration of kefir-fermented milk protects mice against *Giardia intestinalis* infection. J. Med. Microbiol. 62: 1815-1822.

Gamba, R.R., C.A. Caro, O.L. Martinez, A.F. Moretti, L. Giannuzzi, G.L. De Antoni et al. 2016. Antifungal effect of kefir fermented milk and shelf life improvement of corn arepas. Int. J. Food Microbiol. 235: 85-92.

Gao, J., F. Gu, N.H. Abdella, H. Ruan and G. He. 2012. Investigation on culturable microflora in Tibetan kefir grains from different areas of China. J. Food. Sci. 77(8): M425-M433.

Garofalo, C., A. Osimani, V. Milanović, L. Aquilanti, F. De Filippis, G. Stellato et al. 2015. Bacteria and yeast microbiota in milk kefir grains from different Italian regions. Food Microbiol. 49: 123-133.

Garrote, G.L., A.G. Abraham and G.L. De Antoni. 2000. Inhibitory power of kefir: The role of organic acids. J. Food Prot. 63(3): 364-369.

Gethins, L., M.C. Rea, C. Stanton, R.P. Ross, K. Kilcawley, M. O'Sullivan et al. 2016. Acquisition of the yeast *Kluyveromyces marxianus* from unpasteurised milk by a kefir grain enhances kefir quality. FEMS Microbiol. Lett. 363(16): fnw165.

Gorsek, A., J. Ritonja and D. Pecar. 2018. Mathematical model of CO_2 release during milk fermentation using natural kefir grains. J. Sci. Food Agric. 98(12): 4680-4684.

Gul, O., M. Mortas, I. Atalar, M. Dervisoglu and T. Kahyaoglu. 2015. Manufacture and characterization of kefir made from cow and buffalo milk, using kefir grain and starter culture. J. Dairy Sci. 98(3): 1517-1525.

Guzel-Seydim, Z.B., A.C. Seydim, A.K. Greene and T. Tas. 2006. Determination of antimutagenic properties of acetone extracted fermented milks and changes in their total fatty acid profiles including conjugated linoleic acids. Int. J. Dairy Technol. 59: 209-215.

Guzel-Seydim, Z.B., T. Kok-Tas, A.K. Greene and A.C. Seydim. 2011. Review: Functional properties of kefir. Crit. Rev. Food Sci. Nutr. 51(3): 261-268.

Hajirostamloo, B. 2009. Comparison of nutritional and chemical parameters of soymilk and cow milk. World Acad. Sci., Eng. and Technol., Int. J. Nutr. And Food Eng. 3(9): 455-457.

Hassan, Y.I. and L.B. Bullerman. 2008. Antifungal activity of *Lactobacillus paracasei* ssp. tolerans isolated from a sourdough bread culture. Int. J. Food Microbiol. 121(1): 112-115.

Hsieh, H.H., S.Y. Wang, T.L. Chen, Y.L. Huang and M.J. Chen. 2012. Effects of cow's and goat's milk as fermentation media on the microbial ecology of sugary kefir grains. Int. J. Food Microbiol. 157(1): 73-81.

Hugenholtz, J. 2013. Traditional biotechnology for new foods and beverages. Curr. Opin. Biotechnol. 24(2): 155-159.

Ismaiel, A.A., M.F. Ghaly and A.K. El-Naggar. 2011. Milk kefir: Ultrastructure, antimicrobial activity and efficacy on aflatoxin B1 production by *Aspergillus flavus*. Curr. Microbiol. 62(5): 1602-1609.

Jensen, R.G. 2002. The composition of bovine milk lipids: January 1995 to December 2000. J. Dairy Sci. 85(2): 295-350.

Kim, D.-H., D. Jeong, H. Kim, I.-B. Kang, J.-W. Chon, K.-Y. Song et al. 2016. Antimicrobial activity of kefir against various food pathogens and spoilage bacteria. Korean J. Food Sci. of Anim. Res. 36(6): 787-790.

Kim, D.-H., H. Kim, D. Jeong, I.-B. Kang, J.-W. Chon, H.-S. Kim et al. 2017. Kefir alleviates obesity and hepatic steatosis in high-fat diet-fed mice by modulation of gut microbiota and mycobiota: Targeted and untargeted community analysis with correlation of biomarkers. J. Nutr. Biochem. 44: 35-43.

Kim, Y., S. Blasche and K.R. Patil. 2017. Draft genome sequences of three novel low-abundance species strains isolated from kefir grain. Genome Announc. 5(39): e00869-17.

Kukhtyn, M., O. Vichko, Y. Horyuk, O. Shved and V. Novikov. 2018. Some probiotic characteristics of a fermented milk product based on microbiota of "Tibetan kefir grains" cultivated in Ukrainian household. J. Food Sci. Technol. 55(1): 252-257.

La Riviere, J.W. 1969. Ecology of yeasts in the kefir grain. Antonie Van Leeuwen 35(Suppl): D15-D16.

La Riviere, J.W. and P. Kooiman. 1967. Kefiran, a novel polysaccharide produced in the kefir grain by *Lactobacillus brevis*. Arch. Mikrobiol. 59(1): 269-278.

Laureys, D. and L. De Vuyst. 2014. Microbial species diversity, community dynamics, and metabolite kinetics of water kefir fermentation. Appl. Environ. Microbiol. 80(8): 2564-2572.

Laureys, D. and L. De Vuyst. 2017. The water kefir grain inoculum determines the characteristics of the resulting water kefir fermentation process. J. Appl. Microbiol. 122(3): 719-732.

Lifeway (2018) http://lifewaykefir.com/the-world-of-lifeway/

Liu, J.R., S.Y. Wang, Y.Y. Lin and C.W. Lin. 2002. Antitumor activity of milk kefir and soy milk kefir in tumor-bearing mice. Nutr. Cancer 44(2): 183-187.

Macuamule, C.L., I.J. Wiid, P.D. van Helden, M. Tanner and R.C. Witthuhn. 2016. Effect of milk fermentation by kefir grains and selected single strains of lactic acid bacteria on the survival of *Mycobacterium bovis* BCG. Int. J. Food Microbiol. 217: 170-176.

Magnusson, J., K. Strom, S. Roos, J. Sjogren and J. Schnurer. 2003. Broad and complex antifungal activity among environmental isolates of lactic acid bacteria. FEMS Microbiol. Lett. 219(1): 129-135.

Marsh, A.J., O. O'Sullivan, C. Hill, R.P. Ross and P.D. Cotter. 2013. Sequencing-based analysis of the bacterial and fungal composition of kefir grains and milks from multiple sources. PLoS One 8(7): e69371.

McClure, S.B., C. Magill, E. Podrug, A.M.T. Moore, T.K. Harper, B.J. Culleton et al. 2018. Fatty acid specific δ^{13}C values reveal earliest Mediterranean cheese production 7200 years ago. PLoS One 13(9): e0202807.

McGovern, P.E., J. Zhang, J. Tang, Z. Zhang, G.R. Hall, R.A. Moreau et al. 2004. Fermented beverages of pre- and proto-historic China. Proc. Natl. Acad. Sci. USA 101(51): 17593-17598.

Moller, N.P., K.E. Scholz-Ahrens, N. Roos and J. Schrezenmeir. 2008. Bioactive peptides and proteins from foods: Indication for health effects. Eur. J. Nutr. 47(4): 171-182.

Murofushi, M., M. Shiomi and K. Aibara. 1983. Effect of orally administered polysaccharide from kefir grain on delayed-type hypersensitivity and tumor growth in mice. Jpn. J. Med. Sci. Biol. 36(1): 49-53.

Murofushi, M., J. Mizuguchi, K. Aibara and T. Matuhasi. 1986. Immunopotentiative effect of polysaccharide from kefir grain, KGF-C, administered orally in mice. Immunopharmacol. 12(1): 29-35.

Nalbantoglu, U., A. Cakar, H. Dogan, N. Abaci, D. Ustek, K. Sayood et al. 2014. Metagenomic analysis of the microbial community in kefir grains. Food Microbiol. 41: 42-51.

Ninane, V., R. Mukandayambaje and G. Berben. 2007. Identification of lactic acid bacteria within the consortium of a kefir grain by sequencing 16S rDNA variable regions. J. AOAC. Int. 90(4): 1111-1117.

Nurliyani, E. Harmayani and Sunarti. 2015. Antidiabetic potential of kefir combination from goat milk and soy milk in rats induced with Streptozotocin-Nicotinamide. Korean J. Food Sci. Anim. Resour. 35(6): 847-858.

Orhan, Y.T., C. Karagozlu, S. Sarioglu, O. Yilmaz, N. Murat, S. Gidener et al. 2012. A study on the protective activity of kefir against gastric ulcer. Turk. J. Gastroenterol. 23(4): 333-338.

Otles, S., O. Cagindi and E. Akcicek. 2003. Probiotics and health. Asian Pac. J. Cancer Prev. 4(4): 369-372.

Pogacic, T., S. Sinko, S. Zamberlin and D. Samarzija. 2013. Microbiota of kefir grains. Mljekarstvo 63(1): 3-14.

Rodrigues, K.L., L.R. Caputo, J.C. Carvalho, J. Evangelista and J.M. Schneedorf. 2005. Antimicrobial and healing activity of kefir and kefiran extract. Int. J. Antimicrob. Agents 25(5): 404-408.

Scalabrini, P., M. Rossi, P. Spettoli and D. Matteuzzi. 1998. Characterization of bifidobacterium strains for use in soymilk fermentation. Int. J. Food Microbiol. 39(3): 213-219.

Seo, M.K., E.J. Park, S.Y. Ko, E.W. Choi and S. Kim. 2018. Therapeutic effects of kefir grain *Lactobacillus*-derived extracellular vesicles in mice with 2,4,6-trinitrobenzene sulfonic acid-induced inflammatory bowel disease. J. Dairy Sci. 101(10): 8662-8671.

Shiomi, M., K. Sasaki, M. Murofushi and K. Aibara. 1982. Antitumor activity in mice of orally administered polysaccharide from Kefir grain. Jpn. J. Med. Sci. Biol. 35(2): 75-80.

Silva, K.R., S.A. Rodrigues, L.X. Filho and A.S. Lima. 2009. Antimicrobial activity of broth fermented with kefir grains. Appl. Biochem. Biotechnol. 152(2): 316-325.

Simova, E., D. Beshkova, A. Angelov, T. Hristozova, G. Frengova, Z. Spasov et al. 2002. Lactic acid bacteria and yeasts in kefir grains and kefir made from them. J. Ind. Microbiol. Biotechnol. 28(1): 1-6.

Simova, E., Z. Simov, D. Beshkova, G. Frengova, Z. Dimitrov, Z. Spasov et al. 2006. Amino acid profiles of lactic acid bacteria, isolated from kefir grains and kefir starter made from them. Int. J. Food Microbiol. 107(2): 112-123.

Sjogren, J., J. Magnusson, A. Broberg, J. Schnurer and L. Kenne. 2003. Antifungal 3-hydroxy fatty acids from *Lactobacillus plantarum* MiLAB 14. Appl. Environ. Microbiol. 69(12): 7554-7557.

St-Onge, M.P., E.R. Farnworth and P.J. Jones. 2000. Consumption of fermented and nonfermented dairy products: Effects on cholesterol concentrations and metabolism. Am. J. Clin. Nutr. 71(3): 674-681.

Strom, K., J. Sjogren, A. Broberg and J. Schnurer. 2002. *Lactobacillus plantarum* MiLAB 393 produces the antifungal cyclic dipeptides cyclo(L-Phe-L-Pro) and cyclo(L-Phe-trans-4-OH-L-Pro) and 3-phenyllactic acid. Appl. Environ. Microbiol. 68(9): 4322-4327.

Vardjan, T., P. Mohar Lorbeg, I. Rogelj and A. Canzek Majhenic. 2013. Characterization and stability of lactobacilli and yeast microbiota in kefir grains. J. Dairy Sci. 96(5): 2729-2736.

Vieira, C.P., T.S. Alvares, L.S. Gomes, A.G. Torres, V.M. Paschoalin, C.A. Conte-Junior et al. 2015. Kefir grains change fatty acid profile of milk during fermentation and storage. PLoS One 10(10): e0139910.

Vinderola, G., G. Perdigon, J. Duarte, E. Farnworth and C. Matar. 2006. Effects of the oral administration of the products derived from milk fermentation by kefir microflora on immune stimulation. J. Dairy Res. 73(4): 472-479.

Vujicic, I.F., M. Vulic and T. Konyves. 1992. Assimilation of cholesterol in milk by kefir cultures. Biotechnol. Lett. 14(9): 847-850.

Walsh, A.M., F. Crispie, K. Kilcawley, O. O'Sullivan, M.G. O'Sullivan, M.J. Claesson et al. 2016. Microbial succession and flavor production in the fermented dairy beverage kefir. mSystems. 1(5): e00052-16.

Walther, B. and R. Sieber. 2011. Bioactive proteins and peptides in foods. Int. J. Vitam. Nutr. Res. 81(2-3): 181-192.

Wang, S.Y., K.N. Chen, Y.M. Lo, M.L. Chiang, H.C. Chen, J.R. Liu et al. 2012. Investigation of microorganisms involved in biosynthesis of the kefir grain. Food Microbiol. 32(2): 274-285.

Wang, Y., N. Xu, A. Xi, Z. Ahmed, B. Zhang, X. Bai et al. 2009. Effects of *Lactobacillus plantarum* MA2 isolated from Tibet kefir on lipid metabolism and intestinal microflora of rats fed on high-cholesterol diet. Appl. Microbiol. Biotechnol. 84(2): 341-347.

Xing, Z., W. Tang, W. Geng, Y. Zheng and Y. Wang. 2017. *In vitro* and *in vivo* evaluation of the probiotic attributes of *Lactobacillus kefiranofaciens* XL10 isolated from Tibetan kefir grain. Appl. Microbiol. Biotechnol. 101(6): 2467-2477.

Yang, Y., A. Shevchenko, A. Knaust, I. Abuduresule, W. Li, X. Hu et al. 2014. Proteomics evidence for kefir dairy in early bronze age China. J. Archaeol. Sci. 45: 178-186.

Yerlikaya, O. 2018. Probiotic potential and biochemical and technological properties of *Lactococcus lactis* ssp. *lactis* strains isolated from raw milk and kefir grains. J. Dairy Sci. 102: 124-134.

Zheng, H., E. Liu, P. Hao, T. Konno, M. Oda, Z.S. Ji et al. 2012. In silico analysis of amino acid biosynthesis and proteolysis in *Lactobacillus delbrueckii* subsp. *bulgaricus* 2038 and the implications for bovine milk fermentation. Biotechnol. Lett. 34(8): 1545-1551.

Zhou, J., X. Liu, H. Jiang and M. Dong. 2009. Analysis of the microflora in Tibetan kefir grains using denaturing gradient gel electrophoresis. Food Microbiol. 26(8): 770-775.

7

Engineering Probiotic Organisms

Jayashree Chakravarty[1] and Christopher Brigham[2]*

[1] Department of Bioengineering, University of Massachusetts Darmouth,
N. Darmouth, MA, USA

[2] Department of Interdisciplinary Engineering, Wentworth Institute of Technology,
Boston, MA, USA

1. Introduction

Probiotics are not necessarily an "invention." They have existed in traditional foods, such as beverages, salted fish, yogurt, cheeses, etc. for thousands of years (Amara, 2012). It is not very clear how or when food items containing probiotics were first used for medicinal applications. The word "probiotic" is derived from the Greek word meaning "for life." Possibly Ilya Ilyich Metchnikoff, winner of the Nobel Prize in Medicine in 1908, at the Pasteur Institute was the first to observe the effect of what is now known as "probiotic". Metchnikoff linked the health and longevity of Bulgarian peasants to the ingestion of bacteria present in yogurt (Metchnikoff, 1908, Metchnikoff, 2004). He had postulated that the bacteria involved in yogurt fermentation, *Lactobacillus bulgaricus* and *Streptococcus thermophilus*, suppress "putrefactive fermentation" of the intestinal flora and that consumption of these yogurts was important for the maintenance of health. In Japan, in the early 1930s, Minoru Shirota succeeded in isolating strains present as part of healthy individuals' intestinal microflora. He used these strains to develop fermented milk and tested its effects on patients. The first products were introduced into the market, sold under the name Shirota (later named *Lactobacillus casei* Shirota), produced by the Yakult Honsha Company (Amara and Shibi, 2013).

The definition of probiotics has been modified and changed many times. Probiotics as a term was first used by Lilly and Stillwell (Lilly and Stillwell,1965) to describe the "substances secreted by one microorganism that stimulate the growth of another." Parker (Parker, 1974), proposed that probiotics are "organisms and substances which contribute to intestinal

---

*Corresponding author: aatta75@yahoo.com

microbial balance." Marteau et al. (2002) defined them as "microbial cell preparations or components of microbial cells that have a beneficial effect on the health and well-being of human beings." After more than half a century the term "probiotics" was defined by FAO (Food and Agriculture Organization of the United Nations) and WHO (World Health Organization) to reflect Metchnikoff's idea, as "live strains of strictly selected microorganisms which, when administered in adequate amounts, confer a health benefit on the host" (FAO 2002). The definition was maintained by the International Scientific Association for Probiotics and Prebiotics (ISAPP) in 2013 (Hill et al., 2014).

The human gut microbiome is thought to comprise of up to 10^{14} organisms of over 2000 distinct bacterial, viral and eukaryotic species (Neish, 2009) and plays a crucial role in maintaining human health. It is affected by the health of its host, and it turn, shapes the host's metabolism to develop immunity. This interconnection makes the gut microbiome an attractive target for therapeutic applications. There is a fast-growing industry dedicated to the development and sale of probiotics. Consumption of probiotic organisms through food products is the most popular approach of administration. The global market for functional foods and beverages has grown from US$33 billion in 2000 to US $176.7 billion in 2013, accounting for 5% of the overall food market. It has been estimated that probiotic foods comprise between 60% to 70% of the total functional food market (Tripathi and Giri, 2014, Granato et al., 2010). According to Transparency Market Research (TMR), the global probiotics market accounted for around 6,762.2 million USD in

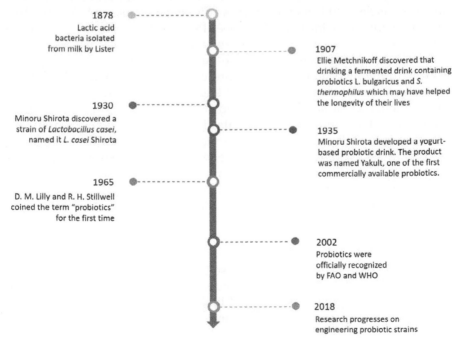

Fig. 1: Timeline of selected milestones in the history of probiotics

2018 and this is anticipated to increase to roughly about 12,753 million USD by 2026 (TMR, 2018).

Molecular and genetic engineering studies have helped to better understand the mechanistic basis for the beneficial activities of probiotics. Even after one hundred years, there are signs that Metchnikoff's hypothesis is being truly brought to life as many health benefits are indeed being conferred by the use of probiotics.

1.1 What Do Probiotics Do?

The use of probiotics has many beneficial effects. Lactic acid bacteria (LAB) have shown to increase folic acid content of yogurt, buttermilk, kefir and vitamin B12 and vitamin B6 in cheese (LeBlanc et al., 2007). In lactose intolerant individuals, milk lactose is hydrolyzed by probiotic strains, favoring calcium absorption (Patil and Reddy, 2006). Islam (2016) also suggested that a wide variety of anti-pathogenic compounds, like bacteriocins, ethanol, organic acids, diacetyl, acetaldehydes, hydrogen peroxide (H_2O_2) and peptides are produced by many probiotics. Probiotics may also show a beneficial effect on allergic reactions by improving the mucosal barrier function. Probiotics such as *Lactobacillus GG* may be helpful in alleviating some of the symptoms of food allergies such as those associated with milk protein (Wadher et al., 2010). *In vitro* studies of certain probiotics, such as *Lactobacillus plantarum* L67, have shown the potential to prevent allergy-associated disorders with the production of interleukin-12 and interferon-γ in their hosts (Song et al., 2016). Probiotic consumption has also proved to be useful in the treatment of many types of intestinal tract disorders like diarrhea, including antibiotic-associated diarrhea in adults, travelers' diarrhea, and diarrheal diseases in young children caused by rotaviruses (Jones, 2010, Galdeno and Perdigon, 2006). Probiotics have also been shown to preserve intestinal integrity and mitigate the effects of inflammatory bowel diseases, irritable bowel syndrome, and alcoholic liver disease (Wadher et al., 2010). Evidence from *in vitro* systems, animal models and humans suggest that probiotics can enhance both the specific and nonspecific immune response, possibly by activating macrophages, increasing levels of cytokines, increasing natural killer cell activity, and/or increasing levels of immunoglobulins (Sanders, 1999). In many animal and *in vitro* studies, LAB have been shown to reduce colon cancer risk by reducing the incidence and number of tumors. *In vitro* studies have demonstrated that probiotic strains, *Lactobacillus fermentum* NCIMB-5221 and -8829, are highly potent in suppressing colorectal cancer cells and promoting normal epithelial colon cell growth through the production of SCFAs (ferulic acid). This ability was also compared with other probiotics namely *L. acidophilus* ATCC 314 and *L. rhamnosus* ATCC 51303 both of which were previously characterized with–tumorigenic activity (Kahouli et al., 2015). One clinical study showed an increased recurrence-free period in subjects with bladder cancer (Saikali et al., 2004). Some preliminary evidence also suggests that food products derived from probiotics bacteria could possibly contribute to blood pressure control (Parvez et al., 2006). Other probiotic microbes such as *L. casei, Lactobacillus acidophilus* and *Bifidobacterium longum* have also been reported to have hypocholesterolemic effects (Karimi et al., 2015).

1.2 Natural Probiotic Strains

The microbes used as probiotics represent different genera and species of organism, such as bacteria, yeast or mold. LAB are important microorganisms in healthy human microbiota (Lenoir-Wijnkoop et al., 2007). Strains used as probiotics usually belong to species of the genera *Lactobacillus, Lactococcus, Enterococcus,* and *Bifidobacterium.* Indeed, several *Lactobacillus* species are beneficial microorganisms that have been associated with probiotic effects in humans and animals, such as reduction of acute diarrhea and allergy (Szajewska et al., 2001, Ouwehand, 2007), relief of inflammatory bowel disease (Ewaschuk and Dieleman, 2006, Limdi et al., 2006) and relief of antibiotic-associated gastrointestinal symptoms (Lenoir-Wijnkoop et al., 2007), as well as reduction of potentially pathogenic bacteria (Savard et al., 2011). However, the ways in which probiotic bacteria elicit their health effects are not fully understood. A list of commonly used probiotic organisms is presented in Table 1.

Table 1: Commonly used probiotic microorganisms

Microorganisms	References
Bacillus circulans PB7	Bandyopadhyay and Mohapatra, 2009
Bacillus subtilis	Zokaeifar et al., 2014, Larsen et al., 2014
Bacilus cereus	Trapecar et al., 2011
Bifidobacterium animalis subsp. *lactis*	Pinto et al., 2014
Bifidobacterium bifidum	Viljanen et al., 2005
Bifidobacterium breve	Bordoni et al., 2013
Bifidobacterium infantis	Wu et al., 2013
Bifidobacterium longum CMCC P0001	Yu et al., 2013
Enterococcus durans	Pieniz et al. 2013
Enterococcus faecium	Cao et al., 2013
Enterococcus faecium M-74	Mego et al., 2005
Enterococcus mundtii ST4SA	Botes et al., 2008
Escherichia coli	Niers et al., 2009, Soh et al., 2009, Lodinova-Zadnikova et al., 2003
Escherichia coli Nissle 1917	Boudeau et al., 2003
Lactobacillus acidophulus	Hawrelak, 2003, Abdin and Saeid, 2008; Phavichitr et al., 2013, Cortés-Zavaleta et al., 2014

Lactobacillus casei	Yamada et al., 2009, Ortiz et al., 2014
Lactobacillus delbrueckii subsp. *bulgaricus*	Moro-García et al., 2013
Lactobacillus fermentum RC-14	Homayouni et al., 2014
Lactobacillus GG	Hilton et al., 1977
Lactobacillus johnsonii LA1	Hawrelak, 2003, Fujimura et al., 2014
Lactobacillus plantarum	Michail and Abernathy, 2002
Lactobacillus plantarum DSMZ 12028	Cammarota et al., 2009
Lactobacillus reuteri RC-14	Martinez et al.,2009
Lactobacillus rhamnosus GR-1	Martinez et al., 2009
Lactobacillus rhamnosus HN001	Wickens et al., 2013
Lactobacillus salivarius	Aiba et al., 1998
Pediococcus acidilactici	Fernandez et al., 2013, Kaur et al., 2014
Saccharomyces boulardi	Guslandi et al., 2000, Choi et al., 2011
Streptococcus thermophilus	Li et al., 2013

2. Probiotic Mechanisms

The mode of action of probiotics is complex; however, there are a few common mechanisms that are evident in a wide variety of probiotic strains. Some mechanisms of action include adherence to the intestinal mucosal surface, which prevents colonization of pathogenic bacteria (Guarner and Malagelada, 2003) and stimulation of the intestinal immune system (Dieleman et al., 2003). Probiotics are also believed to function via the modulation of cell proliferation and apoptosis (Ichikawa et al., 1999, Yan and Polk, 2002). The mode of action of a given probiotic can differ based on the presence of other probiotics or enteric bacteria, and also eventual diseases to be treated (Soccol et al., 2014).

Studies at both molecular and genetic levels have shown that beneficial effects of probiotics involve four mechanisms:

1. Modulation of host immune system.
2. Inhibition of bacterial toxin production.
3. Inhibition through the production of anti-microbial substances.
4. Competitive exclusion of pathogenic strains for the adhesion to the epithelium.

The largest lymphoid tissue in the human body is the gut-associated lymphoid tissue (GALT). The earliest and largest exposure to microbial antigens occurs during birth, which depends on many factors such as the mode of birth, feeding, hygiene levels (Borchers et al., 2009). Establishment of the microbiota provides the host with the most substantial antigen challenge, with a strong stimulatory effect and is essential for the development of a fully functional and balanced immune system (Cebra, 1999, Gronlund et al., 2000). The intestinal epithelial cells are inherently designed to distinguish between pathogenic and non-pathogenic strains, the mechanism for which is still not fully understood, but is likely to be based on additional mucosal factors (Lan et al., 2007).

The bacteria of the human gut can be of two types: indigenous and transient. Most of the indigenous bacteria are part of the commensal microbiota. Microbial species possessing beneficial properties mostly belong to the genera *Bifidobacterium* and *Lactobacillus*, and some of these species exhibit powerful anti-inflammatory properties (Isolauri et al., 2002). Modulation of the host immune system by intestinal microbiota is based on initiation and continuation of a state of immunological tolerance to nutritional antigens and control of immunological reactions against pathogens (Isolauri et al., 2001). Immunological stimulation by probiotics can occur by the increased production of immunoglobulins, enhanced activity of macrophages and lymphocytes, and stimulation of γ-interferon production (Oelschlaeger, 2010). The host intestinal epithelial cells and immune cells are important in this context. Intestinal epithelial cells that have encountered certain bacteria participate in the immune response by producing chemokines and cytokines and upregulating adhesion molecules, thereby attracting and activating immune cells (Borchers et al., 2009). The adhesion of probiotic microorganisms to epithelial cells may trigger a signaling cascade, leading to modulation of the host immunological system. The release of some soluble components by the pathogenic strain may cause a direct or indirect (through epithelial cells) activation of immunological cells. This effect plays an important role in the prevention and treatment of contagious diseases, as well as in chronic inflammation of the alimentary tract or of a part thereof (Oelschlaeger, 2010). Probiotic bacteria can also stimulate the production of cytokines by immunocompetent cells of the gastrointestinal tract (Gill and Cross, 2002).

Probiotics have shown to inhibit production of bacterial toxins and help with the removal of toxins from the body. *Saccharomyces boulardii* has been shown to inhibit or neutralize the enterotoxicity of *Vibrio cholerae* toxin (Dias et al., 1995). The mechanism of this toxin neutralizing effect may be related to the ability of a yeast protein to bind to a receptor that in turn regulates intracellular adenylate cyclase levels (Brandao et al., 1998). In 2004, an *in vivo* study in mice showed that *Bifidobacterium breve* Yakult and *Bifidiobacterium pseudocatenulatum* DSM20439 could inhibit the expression of shiga toxin in *E. coli* (STEC) O157:H7 strains (Asahara et al., 2004). *S. boulardii* has shown to act against *C. difficile* toxin A in rat ileum. The *S. boulardii* could be interfering with the toxin A-induced inflammatory signal cascade that activates Erk1/2 and JNK/SAPK pathways (Chen et al., 2006). Live or killed *Lactobacillus rhamnosus* GG can bind deoxynivalenol (a mycotoxin that can cause gastroenteritis) and potentially restricts bioavailability of the toxin (Turner et al., 2008). Some strains can bind toxins to their cell wall and reduce their intestinal absorption (Schatzmayr et al., 2006, McCormick, 2013). However, detoxifying properties of probiotics is strictly a strain-related characteristic. The reduction of metabolic reactions leading to the production of toxins is also associated with the stimulation of pathways leading to the production of native enzymes, vitamins, and antimicrobial substances (Oelschlaeger, 2010).

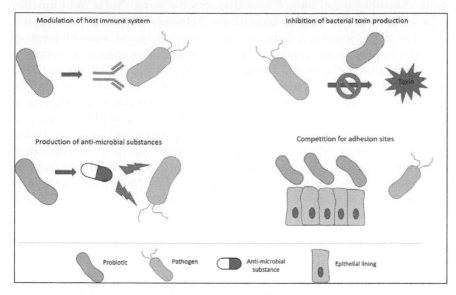

Fig. 2: Schematic drawings of how probiotic organisms can combat intestinal pathogens

Lactobacillus produce antimicrobial peptides like low-molecular-weight (LMWB) antibacterial peptides, as well as high-molecular-weight class III bacteriocins. *L. reuteri* strain ATC55730 produces a broad-spectrum of antimicrobial substances called reuterins, which are known to act against both gram positive and negative bacteria as well as yeast, fungi, protozoa and viruses (Oelschlaeger, 2010).

Deconjugation of bile acids is catalyzed by BSH enzymes. The ability of a probiotic strain to hydrolyze bile salts has been studied and several bile salt hydrolases (BSHs) have been identified. Probiotic bacteria (e.g., *Lactobacillus* and *Bifidobacterium*) may produce deconjugated bile acids (derivatives of bile acids), which have shown stronger antibacterial effects than the bile salts produced by their host. A probiotic strain showing BSH activity could maximize its rate of survival in the gastrointestinal environment, which would increase the overall beneficial effects associated with the strain (Begley et al., 2006).

Probiotic strains , as one of their mechanisms of action tend to co-aggregate, which may lead to the formation of a protective barrier preventing colonization of pathogenic bacteria on the epithelium (Schachtsiek et al., 2004). At the inter-colonic level, the commensal flora modulates the intestinal barrier function. For example, colonization of the excluded colonic loop in rats with *Escherichia coli* increases epithelial permeability, whereas a reduction in permeability is observed after colonization with *Lactobacillus brevis* (Garcia et al., 2001). These changes could be associated with the adhesion of probiotic bacteria to the intestinal mucosa. Other experiments have also shown that the adherence of enteropathogenic *E. coli* (EPEC) to the intestinal epithelial cell monolayers disrupts the paracellular tight junction (Spitz et al., 1995). These relationships between intestinal permeability (a contributor to intestinal distress) and bacteria suggest a positive role of probiotics in preventing alterations in paracellular permeability observed in several experimental models (Fioramonti et al., 2003). The passage of large molecules through the mucosa, however, does not always correlate with the alterations in mucosal permeability. In a neonatal rat model of necrotizing enterocolitis, severe lesions were found on the colonic wall in association with the passage of endotoxin in the plasma. A 3-day preventive treatment with *Bifidobacterium infantis* significantly reduced mortality, the number of colonic lesions and endotoxin passage but did not modify the lumen-to-blood permeability to dextran molecules (Caplan et al., 1999). In another study, the intestinal permeability was studied in a rat intestine placed in an Ussing chamber (an apparatus used for measuring epithelial membrane properties). The addition of *E. coli* to the mucosal part of the chamber strongly increased the passage of mannitol, the effect being abolished in rats previously treated with *L. plantarum* for one week (Mangell et al., 2002).

The luminal surface of the GI tract is covered by a layer of gel that acts as an important protective barrier against the harsh luminal environment. Gut pathogens must traverse this mucus layer before they adhere to, colonize and subsequently invade the epithelial cells. There exists a strong interaction between mucus and colonic bacteria (Fioramonti et al., 2003). About 45% of *Lactobacillus* GG and 30% of *B. lactis* administered orally to humans have been found to adhere to stool mucus (Kirjavainen et al.,1998). Another interesting property of probiotics is the inhibition of adhesion of enteropathogenic bacteria to mucus. For example, *Enterococcus faecium*, found in many probiotic preparations, inhibits the adhesion of enterotoxigenic *E. coli* K88 to porcine small intestine mucus (Jin et al., 2000). Some probiotic strains are unable to degrade gastrointestinal mucus. Using different techniques, this has been shown for *L. casei* strain GG, *L. acidophilus* and *B. bifidum* (Ruseler-Van et al., 1995), as well as for *L. rhamnosus* and *B. lactis* (Zhou et al., 2001).

The most commonly used probiotic strains are members of the LAB group, which have been in use for centuries for fermented food production (*e.g.*, yogurt, cheese, pickles). They are usually catalase-negative bacteria that grow under microaerophilic to strictly anaerobic conditions, and are non-spore forming (Stiles and Holzapfel, 1997). LAB such as *Lactococcus* and *Streptococcus* are important components of the endogenous microbiota in the human ileum, jejunum and colon (Hayashi et al., 2005).

In humans, the most studied aspect of probiotics with regards to digestion is their compensation for lactase insufficiency. Many studies have shown that lactose is better digested in lactose malabsorbers who have consumed yogurt rather than milk. This is likely due to parameters such as viscosity or pH, which are independent of the presence of bacteria (Fioramonti et al., 2003). Colonic microflora could also contribute to lactose degradation in lactose maldigesters (Marteau et al., 1990). Gut microbiota also plays a significant role in host metabolic processes like regulation of cholesterol absorption, blood pressure (BP), and glucose metabolism (Upadrasta and Madempuri, 2016, Ruan et al., 2015). Further research is required to evaluate the targeted and effective use of the wide variety of probiotic strains in various metabolic disorders to improve the overall health status of the host (Upadrasta and Madempudi, 2016). An important role in the action of probiotics is played by species- and strain-specific traits, such as: cellular structure, cell surface, size, metabolic properties, and substances secreted by microorganisms. The use of a combination of probiotics demonstrating various mechanisms of action may provide enhanced protection offered by a bio-therapeutic product (Lima-Filho et al., 2000).

3. Engineering Probiotic Strains

The gut microbiome is hugely individualized, and it varies between family members, across different ethnicity and cultures, and also changes during the course of a person's life. Since the gut microbiome is so important to human health and any imbalance to it is closely related to a diseased state, it is only natural for us to attempt to treat this state by rebalancing the microbiome through the administration of probiotics. Advances in genetic engineering and synthetic biology have made it possible to develop new strategies for tackling infectious diseases and fine-tune our control of probiotics. These bacteria can be genetically modified to act as sensors and have untapped potential as tools for sensing or detecting pathogens, enabling minimally invasive measurements of the gut microbiome conditions.

Over the past decade, several studies have been taking place to modify natural biological mechanisms to be used as signals and switches to detect a pathogen or modify a disease state. A few of them have been applied to the human gut microbiome. Researchers have successfully altered *E. coli* Nissle (*EcN*) to disrupt the virulence of *Vibrio cholerae* (Duan and March, 2010). Commensal species were also engineered to express anti-inflammatory compounds such as IL-10 (Steidler et al., 2000), KGF-2 (Hamady et al.,2010) to treat induced colitis in mouse models. The use of probiotics as a delivery mechanism to deliver GLP-1, a peptide that can decrease blood sugar level by enhancing the secretion of insulin, (Duan and Liu, 2015) shows that probiotics have a potential use in applying therapeutics that are otherwise difficult to use by traditional methods.

3.1 Taking Advantage of Quorum Sensing

Quorum sensing (QS) is a bacterial cell-to-cell communication mechanism that involves the production, detection, and response to autoinducers (extracellular signaling molecules). Quorum sensing allows bacteria to act in unison. As the bacterial population density increases, autoinducers (AIs) accumulate in the environment. Bacteria detect a threshold concentration of these AIs, which leads to altered gene expression. Using these signal-response systems, bacteria act as a multicellular entity, as opposed to individual cells all performing individual functions. Quorum sensing is involved in biofilm formation, secretion of virulence factors, sporulation, bioluminescence, competence, antibiotic production, etc. (Bai and Rai, 2011, Kaper and Sperandio 2005, Rutherford and Bassler, 2012). Table 2 lists some quorum sensing molecules that have been examined for pathogen detection by engineered probiotic organisms.

Pseudomonas aeruginosa is an opportunistic pathogen that colonizes human respiratory and gastrointestinal tracts and is among the leading causes of infection (Fujitani et al., 2011). This infection is usually treated with antibiotics, which however leads to unspecific killing of the human microbiome. In such cases, novel antimicrobial strategies that do not entirely rely on current antibiotics present an attractive potential solution. Saeidi and coworkers argued that bacteriocins might be more effective in such cases. They developed a synthetic genetic system which comprises of quorum sensing (QS), killing, and lysing devices, that enable non-pathogenic chassis like *E. coli* to sense and kill a pathogenic *P. aeruginosa*. The genetically modified cells could detect autoinducers, such as acyl homoserine lactones (AHLs) and respond by inducing its own lysin (driven by lysin E7, chosen here for its size and modularity), thus releasing an antibacterial toxin, pyocin S5 which kills the *P. aeruginosa*. The sensing device was based on the Type I quorum sensing mechanism *P. aeruginosa*. The *luxR* promoter is activated by LasR-30C_{12}HSL complex, leading to production of E7 lysis protein and S5 pyocin within the recombinant *E. coli*. After the E7 protein attains the threshold concentration that causes the chassis to lyse, the accumulated S5 is released into the exogenous environment and kills *P. aeruginosa*. They demonstrated that the engineered *E. coli* sensed and killed planktonic *P. aeruginosa*, as evidenced by a 99% reduction in the viable target cells. In co-culture, this system successfully inhibited the formation of *P. aeruginosa* biofilms by almost 90%, leading to much thinner biofilm matrices. This novel, synthetic biology-driven antimicrobial strategy could potentially be applied to fight *P. aeruginosa* and other infectious pathogens. The authors of this study claim that their developed genetic devices could potentially be transferred into a microbial chassis other than *E. coli*, such as primary residential microbes of the respiratory tract where the *P. aeruginosa* biofilm builds up in patients (Saeidi et al., 2011).

Enterohemorrhagic *E. coli* serotype O157:H7 (EHEC O157) is a clinically important pathogen. The attachment of enterohemorrhagic *Escherichia coli* O157:H7 (EHEC O157) to host intestinal epithelial cells can lead to the development of hemorrhagic colitis and hemolytic-uremic syndrome in humans. Probiotic *Lactobacillus acidophilus* bacteria interfere with the quorum sensing in *E. coli* O157:H7 and potentially regulate the genes of pathogenicity, colonization and inhibit toxicity. Transcription of important EHEC O157 virulence-related genes was studied by constructing promoter-reporter fusions and reverse transcriptase PCR. They found that when EHEC O157 was grown in the presence of *L. acidophilus* La-5 cell-free spent medium, there was a significant reduction of both extracellular Auto Inducer-2 (AI-2) concentrations and the expression of virulence-related genes *LEE*. The mechanism of action is not clear, but it is suggested that the possibility that *L. acidophilus* La-5 is somehow capable of quorum quenching EHEC O157. These results suggest that *L. acidophilus* La-5 secretes a molecule(s) that either acts as a quorum sensing signal inhibitor or that it directly interacts with bacterial transcriptional regulators, controlling the transcription of EHEC O157 genes involved in colonization (Medellin-Pena et al., 2007).

Vibrio cholerae is most widely known for its massive epidemics of the infectious disease cholera. A whole-cell based biosensor was developed in *E. coli*, based on quorum sensing mechanism in *V. cholerae* and CRISPRi technology. *V. cholerae* is known to control its infection cycle using a quorum sensing system, which involves established pathways like CqsS and LuxPQ which sense the *Vibrio cholerae* quorum-sensing autoinducer CAI-1 (cholera autoinducer-1) and AI-2 molecules, respectively. The QS based system in vibrios consists of the CqsA synthase/CqsA receptor pair. *V. cholerae* CqsA/S synthesizes and detects CAI-1. The membrane-bound 2 component sensor histidine kinase CqsS detects CAI-1, and the CqsS → LuxU → LuxO phosphorelay cascade converts the information provided by CAI-1 into the cell.

To make the *E. coli* sense *V. cholerae*, a genetic circuit comprising of CqsS, LuxU and LuxO proteins was used as the sensor, a genetic inverter and green fluorescence protein (GFP) as a reporter under a quorum-regulated, high expression promoter. The authors claim that the engineered *E. coli* system is able to sense the presence of *V. cholerae*, but there is only a 2.5-fold difference in reporter fluorescence intensity when cultures grown in *E. coli* and *V. cholerae* supernatant are compared. This sensor could be repurposed to sense other pathogens, produce anticholera infection molecules, etc. This sensor could also be a new *V. cholerae* treatment option based on highly sensitive, specific and easy-to-use whole cell biosensor (Holowko et al., 2016).

Many pathogens synthesize an extracellular signal named AI-2, such as *E. coli* O157:H7, *Clostridium perfringens*, and *C. difficile*. The inhibitory activity of AI-2, toxin production, and transcriptional levels for virulence-associated genes in *C. difficile* culture by *L. acidophilus* GP1B was studied by Yun et al., 2014.

The cell extract of *L. acidophilus* GP1B was capable of interfering with quorum sensing in *C. difficile* by decreasing AI-2 production. It was also shown that the addition of a cell extract of *L. acidophilus* GP1B resulted in downregulation of the gene associated with AI-2 production (*luxS*) and virulence genes *tcdA, tcdB, txeR* in *C. difficile* at the level of mRNA. *Lactobacillus acidophilus* GP1B also exhibited an inhibitory effect against the growth of *C. difficile* in the *C. difficile*-infection (CDI) mouse model. *L. acidophilus* GP1B

Table 2: Quorum sensing molecules used in pathogen detection by engineered probiotic strains

Organism	Signal molecule	Synthase (s)	References
Escherichia coli	AI-2	LuxS	Bansal et al., 2007
Pseudomonas aeruginosa	BHL	RhlI	Dubern and Diggle, 2008
	PQS	PqsR	Lee and Zhang, 2015
	OdDHL	LasI	Jimenez et al., 2012
Vibrio cholera	AI-2	LuxS	Milton, 2006
	CAI-1	CqsA	Ng and Bassler, 2009

Abbreviations: AI-2, autoinducer-2; BHL, N-butanoyl-L-homoserine lactone; PQS, Pseudomonas quinolone signal; OdDHL, N-(3-oxododecanoyl)-L-homoserine lactone; CAI-1, cholera autoinducer-1

protected mice from death, with 50% mortality observed with cell extract of *L. acidophilus* GP1B, as compared with 90% mortality in mice that did not receive any treatment. This may be related to a reduced pH resulting from organic acid production by the probiotic bacterium. Results in this study also indicated that the *L. acidophilus* GP1B used elicited a bactericidal effect on *C. difficile*. This study supports that probiotic preparations have potential for both prevention and treatment of *C. difficile*–associated diarrhea (Yun et al., 2014).

3.2 Probiotics "Sense and Kill" Pathogens via Their Quorum Sensing Molecules

Genetically modified *E. coli* was designed to specifically sense and destroy wild type *P. aeruginosa* (PAO1) via its quorum sensing molecule, N-3-oxododecanoyl homoserine lactone ($3OC_{12}HSL$). This engineered, or "sentinel" *E. coli* could detect the presence of QS molecule $3OC_{12}HSL$ and then secrete CoPy (a chimeric bacteriocin), into the extracellular medium using the flagellar secretion tag FlgM. Extracellular FlgM-CoPy is designed to kill PAO1 specifically. CoPy toxicity was demonstrated to be PAO1 specific, thus not affecting the sentinel *E. coli*. Quantitative analysis of the system showed that in liquid culture, ~10^5 sentinel *E. coli* could inhibit the growth of individual PAO1 cells. To test the integrated system *in vitro*, the sentinel *E. coli* and PAO1 cells were co-cultured on a semisolid agar plates system to get a rough approximation of how these would affect the bacterial environment in the GI tract. In the plate, *E. coli* cells were strongly induced because of the quorum sensing signals produced by the homogenously spread PAO1 cells and almost a 30-fold growth inhibition of PAO1 (>30-fold) by engineered *E. coli* sentinel cells was observed. Optical microscopy results show that the engineered *E. coli* sentinels successfully inhibit PAO1 growth (Gupta et al., 2013).

A novel genetic modification was used to reprogram *E. coli* to specifically recognize, migrate toward, and eradicate both dispersed and biofilm encased *P. aeruginosa* cells. To test the improvement of antimicrobial activity of the engineered *E. coli* using pathogen-specific localization, the cells were reprogrammed to "seek and kill" by the introduction of a quorum sensing (QS) device to detect the presence of N-Acyl homoserine lactone (AHL), a quorum sensing molecule produced and secreted by *P. aeruginosa*. This *E. coli* strain showed directed chemotaxis-guided motility. In the presence of *P. aeruginosa* AHL, the reprogrammed *E. coli* cells express the phosphatase CheZ (responsible for chemotaxis) and swim toward the pathogen. Further gene expression is initiated in *E. coli* to secrete an antibiofilm enzyme (DNaseI) and antimicrobial peptide (MccS) to degrade biofilms and kill planktonic or biofilm-residing cells which are released by the biofilm degradation. While most other studies rely on molecular diffusion of the antimicrobial compounds, this study was designed on motility-assisted localization of *E. coli* towards the pathogen (Hwang et al., 2014).

A "sense-and-kill" genetic system was engineered in *E. coli* to respond to the presence of *V. cholerae* by Jayaraman et al., 2017. The sensor was purposed to be able to deliver an antimicrobial payload upon *V. cholerae* detection. Artilysin (Art-085) is an endolysin-based antibacterial protein. This study showed the bactericidal activity of Art-085 against *V. cholerae* strains for the first time. To enable effective and rapid release of Art-085, the expression of the YebF secretion tag fused to the N-terminus of Art-085 was optimized. When the density of *V. cholerae* cells is high, they produce specific quorum sensing signal (CAI-1) at high concentrations. CAI-1 binding leads to the downregulation of guide RNAs (gRNAs), leading to the expression of the chimeric YebF-Art-085 protein. Thus, lysis proteins are produced in large quantities which rupture the *E. coli* host cell membrane. Upon lysis, the Art-085 killer proteins are released into the environment and subsequently kill the *V. cholerae* cells. To confirm the inhibitory activity of Art-085 on *V. cholerae* growth, both killer and control supernatants were exposed to a target *V. cholerae* strain. *V. cholerae* grown in the control supernatant lacking Art-085 showed normal growth, whereas the killer supernatant containing Art-085 showed high bactericidal activity against *V. cholerae*. It was also found that ≥50% dilution of the killer supernatants was enough to inhibit the growth of *V. cholerae*, whereas the control supernatant had no detrimental effect on the growth. Overall, this shows that Art-085 can be stably expressed inside the *E. coli* host chassis and is effective at inducing *V. cholerae* cell death. This is the first report that a sense-and-kill gene circuit has been developed against *V. cholerae* (Jayaraman et al., 2017).

In 2017, a laboratory strain of *E. coli* Nissle 1917 was genetically modified to include a gene encoding anti-biofilm enzyme Dispersin B (DspB), to promote the destabilization of mature biofilms. This 'Sense-Kill' construct comprises of *lasR*, pyocin S5 gene, E7 lysis protein gene and *dspB*. All of these are activated in an autonomous manner in response to autoinducers secreted by *P. aeruginosa*. To evaluate the modified strain of (*EcN*) with improved autoinducer sensitivity coupled with more robust downstream expression, the antimicrobial and anti-biofilm activity of the engineered *EcN* was tested against *P. aeruginosa*. When a *P. aeruginosa* strain constitutively expressing GFP was co-cultured with this modified *EcN* at various starting cell ratios, its cells showed a significant reduction in growth rates, as compared to its lysis control *EcN* E7. Further viability assays confirmed observed reduction in growth rate to be a result of efficient killing of the pathogen in this system.

In preventing biofilm formation and killing *P. aeruginosa* cells, the engineered *EcN* probiotic showed the greatest reduction in viable cells, resulting in decreased number of cells available to form microcolonies. This strain, expressing both S5 and DspB, was able to disassemble mature biofilm structures, resulting in the efficient killing of potential antibiotic-resistant cells encapsulated in the biofilm matrix (an 80% reduction compared to untreated controls). This engineered probiotic has been successfully tested on two animal models: *Caenorhabditis elegans* and mice. The *C. elegans* nematodes were infected with *P. aeruginosa* expressing GFP before treatment and then treated with the engineered *EcN* strain or with *EcN* controls. The greatest increase in *C. elegans* survival rate occurred in the engineered *EcN* treatment groups; the survival time increased >2-fold, and ~50% of the nematodes in the engineered *EcN* treatment groups remained alive after 96 hours post-infection, at which time all the nematodes in the infection control group had died. Enhanced clearance of *P. aeruginosa* from the digestive tract of the nematode in all likelihood contributed to the increased survival rate. Further evaluation of the engineered strain was conducted in a *P. aeruginosa*-infected murine model. When the mice infected with *P. aeruginosa* were fed with engineered *EcN*, a steady decline in the *P. aeruginosa* count in the feces was observed. Clearance of the bacterial load reached 77% as compared to the initial bacterial load prior to treatment. In both systems, the engineered *EcN* autonomously executed diagnostic and therapeutic activities that efficiently disassembled mature biofilm structures, resulting in efficient killing of potentially antibiotic-resistant cells encapsulated within the biofilm matrix (Hwang et al., 2017).

3.3 Other Probiotic Engineering Exploits – *E. coli* as Microbial Chassis

When selecting a microbial chassis to be used as a probiotic, the ability of the chassis to survive transit through the GI tract, ability to colonize the appropriate place in the GI tract and reach desired densities are some of the factors that should be considered. *E. coli* is one of the best studied prokaryotes and has highly tunable metabolic pathways. Hence, *E. coli* can be considered as a chassis strain to house the biological machinery for probiotic engineering. Multiple strains of *E. coli* including the laboratory strains NGF-1 (natural gut flora) and Nissle have been used as chassis for different purposes.

A therapeutic strategy to treat gastrointestinal infections like traveler's diarrhea was developed based on molecular mimicry of host receptors for bacterial toxins on the surface of harmless gut bacteria. Glycosyltransferase genes from *Neisseria meningitidis* or *Campylobacter jejuni* were transferred into *E. coli* CWG308. As a result, chimeric lipopolysaccharides were produced which are capable of binding heat-labile enterotoxins with stronger affinity. The majority (>94%) of the enterotoxin activity was neutralized in culture lysates of diverse pathogenic *E. coli* strains of both human and porcine origin. The interesting feature of this strategy is that it does not put any selective pressure for the evolution of resistance by the targeted pathogen. Thus, use of this toxin-receptor blockade strategy should not show any long-term side effects as a preemptive treatment (Paton et al., 2005). A similar approach was used to develop a probiotic for the treatment of cholera. The recombinant *E. coli* strain could produce a chimeric lipopolysaccharide terminating in a mimic of ganglioside GM_1. When administered on mice, this engineered probiotic could protect infant mice against *V. cholerae* infection, even in the case of delay in treatment after the infection was established (Forceta et al., 2006).

A *E. coli*-based probiotic was developed to deliver a therapeutic gene to the colonic mucosa to treat colitis. *In vitro* experiments showed that this novel bacterial vector could invade and transfer functional DNA to epithelial cells. This strain could efficiently deliver the therapeutic gene TGF-β1 to the intact intestinal mucosa in mice, which significantly reduced the severity of colitis. This tool could also be used to treat other intestinal disorders associated with the intestinal mucosa such as food allergies or lactose intolerance (Castagliuolo et al., 2005).

E. coli was also engineered to sense and record environmental stimuli in the gut. This engineered probiotic could "sense, remember, and report experiences" in the mammalian gut. The construct added to *E. coli* had two parts: a trigger element and a memory element, based on the bacteriophage λcl/cro gene switch. This work shows that *E. coli* can be modified to act as live diagnostic tool for nondestructive probing of the mammalian gut (Kotula et al., 2014). Based on the similar *cl/cro* system, *E. coli* strain NGF-1 was engineered to detect tetrathionate, which is produced during inflammation. The engineered probiotic strain could retain memory of exposure in the gut and detect tetrathionate in both infection-induced and genetic mouse models of inflammation for 6 months. The strain retained this exposure gut memory up to 7 days even after removal of signal. These strains show the potential of engineered bacteria as living diagnostics (Riglar et al., 2017).

Recently, a genetically engineered prototype probiotic was developed to inhibit opportunistic pathogen like *Salmonella* spp. The idea was to genetically engineer a probiotic strain to induce a specific microbiome correction during the onset or course of a disease. The class IIb sideromycin, microcin H47 (MccH47) originally isolated from *E. coli* strain H47, can inhibit *Salmonella* growth *in vitro*. A strain of *EcN* was engineered to contain a plasmid-based system carrying all genes necessary for microcin production, immunity, secretion and all operon genes that encode a *S. typhimurium*-inhibiting microcin MccH47 in response to environmental tetrathionate (a molecule resulting from the inflammation of the intestine, a hallmark of *Salmonella* infection). This strain could inhibit and outcompete *Salmonella*. Using *in vitro* assays, it was shown that *EcN* pttrMcH47 not only prevents *S. typhimurium* growth in static agar inhibition assays but also significantly reduces *S. typhimurium* fitness in pairwise ecological competition experiments. It was observed that MccH47-dependent inhibition increased in conditions limited by iron. This system was then transferred to a US FDA-approved probiotic *EcN* strain (Palmer et al., 2018).

The use of a versatile bacterial chassis holds much promise for the development of microbiota-based therapeutics. The systems discussed above are tools that will be more useful if combined into an engineered system with a more defined function. The next step is to apply these tools in the rational design of systems to tackle problems of gastrointestinal health. Care should be taken to see that these engineered species are not prematurely "flushed out" by other better-adapted microbes. With advances in genetic engineering techniques, researchers can work on improving the stability of the engineered strains during colonization of the gut. Another important topic of study is the risk involved with an engineered species dominating the gut microbiome. Significant work needs to be done to address these issues. The set of biosensors that have been successful *in vivo* is limited. The issues of metabolic load and yield are also important here. For the engineered strains to be versatile and effective therapeutically, they need to be able to detect and function in complex states. It is necessary to continue developing systems which can adapt their response to stimuli.

4. Summary and Outlook

4.1 In the Future – Better Than Fecal Transplants?

The hypothesis of the beneficial effects of ingested bacteria on human health has a long tradition in medicine. Experiences with fecal transplantation are somewhat limited and have had a much shorter time span. The first mention of fecal transplant goes back to around 60 years (Eiseman et al., 1958). In 1989, Borody et al. (1989) were the first to report on the therapeutic effects of fecal transplantation in patients with inflammatory bowel disease (IBD).

An investigation in 2010 has shown that the presence of an orally ingested probiotic in the stools of healthy subjects demonstrated the disappearance of *EcN* within some weeks (Joeres et al., 2010). A study of 10 healthy subjects found some index bacteria (i.e., indicators of fecal transplantation) to be present even after 24 weeks of the fecal transplantation treatment (Grehan et al., 2010). The difference in the colonization behavior of bacteria introduced by ingestion of probiotics and stool transplantation should be studied in more detail.

The use of probiotics has shown some promising results. They provide a viable alternative especially in the treatment and prevention of enteric diseases. Studies with specific probiotics offer interesting insights into their biological activity as well as their safety. In contrast, fecal transplantation has been less studied, the therapeutic and safety aspects of which are not as well understood. However, transplantation has shown some promising, but as-yet-unrefined, therapeutic benefits for many infectious and autoimmune GI disorders. Rationally designed combinations of microbial preparations may enable more efficient and effective treatment approaches tailored to diseases.

4.2 The GMO (Genetically Modified Organism) Question

Probiotics have been in use for over 100 years to treat a variety of infections of the gut, but the use of traditional treatments has diminished after the advent of modern antibiotics. With the rise of recent antibiotic-resistant strains of bacteria, probiotic agents are once again being focused on as alternatives to antibiotic treatments (D'Souza et al., 2002). Some species of LAB found in fermented food products like yogurt, kefir, etc. are not normally found in the human gut system. However, some LAB species are indeed found in both fermented foods and in the GI tract, albeit different strains of the same species. The US FDA classifies these LAB species as GRAS (Generally Recognized As Safe) organisms for human use (Teitelbaum and Walker, 2002). The most commonly used probiotics found in diary-based food belong to the indigenous human microflora and have a long history of safe use. The success of probiotic use has attracted the attention of genetic engineers, who want to "improve" on their successful application. The crosstalk between the human host and gut bacteria has evolved over millions of years. Its contribution to the health of the human host depend on an intricate network of bacteria-bacteria and bacteria-host interactions. The knowledge of the science behind the mode of action of probiotic organisms is still sparse. It is likely that the introduction of new probiotic strains that are produced by genetic modification will raise safety concerns. Social acceptance of genetically modified (GM) foods or ingredients is not uniform in developed countries (Campbell et al., 1998). The definition of a GMO also varies among countries. The main issue for GMOs is the evaluation of the risks to human health of uncontrolled product expression following transfer of the transgene into a commensal bacterium (Renault, 2002).

The prospect that genetic modification might "improve" probiotic microbes must be seriously balanced against the possibility of public rejection of these engineered organisms due to lingering mistrust in GMOs, however misguided or unfounded. With 11 case reports of *S. boulardii* septicemia (Beckly and Lewis, 2002), the acceptance of probiotics involves a certain amount of risk. Establishing the limit of the natural gene pool is not easy because gene shuffling occurs at different levels, such as chromosome, transposons, plasmids, and others (Ahmed, 2003). Strains that use the natural cellular machinery (e.g. competence) are emerging and these might not be considered GMOs.

A better understanding of the host pathogen interaction can lead to the development of bioengineered probiotics that can be used for the targeted elimination of pathogens. These engineered probiotics could overcome the short-half life and stability of other therapeutic alternatives and be a cost-effective alternative. However, there is a need to contain the modified organism to prevent its uninhibited spread. The biosafety and the ability of probiotics to cause allergy due to prolonged consumption should also be taken into consideration. New advances in science and technologies and further research will continue to provide novel bio-therapeutics for the treatment and prevention of various infections and diseases both in rich and economically challenged countries.

References

Abdin, A. A. and E.M. Saeid. 2008. An experimental study on ulcerative colitis as a potential target for probiotic therapy by *Lactobacillus acidophilus* with or without "olsalazine". J. Crohn's and Colitis 2: 296-303.

Ahmed, F.E. 2003. Genetically modified probiotics in food. Trends in Biotech. 21(11): 491-497.

Aiba, Y., N. Suzuki, A.M. Kabir, A. Takagi and Y. Koga. 1998. Lactic acid-mediated suppression of *Helicobacter pylori* by the oral administration of *Lactobacillus salivarius* as a probiotic in a gnotobiotic murine model. American J. Gastro. 93: 2097-2101.

Amara, A. 2012. Toward Healthy Genes. Amro Amara (ed.). Germany: Schüling Verlage.

Amara, A.A. and A. Shibl. 2013. Role of Probiotics in health improvement, infection control and disease treatment and management. Saudi Pharma. J. 23(2): 107-114.

Asahara, T., K. Shimizu, K. Nomoto, T. Hamabata, A. Ozawa,, Y. Takeda et al. 2004. Probiotic bifidobacterial protect mice from lethal infection with shiga toxin-producing *E. coli* O157:H7. Inf. Imm. 72: 2240-2247.

Bai, A.J. and V.R. Rai. 2011. Bacterial quorum sensing and food industry. Compre. Rev. in Food Sci. Food Safety 10: 183-193.

Bandyopadhyay, P. and P.K. Das Mohapatra. 2009. Effect of a Probiotic bacterium *Bacillus circulans* PB7 in the formulated diets: On growth, nutritional quality and immunity of *Catla catla* (Ham.). Fish Physio. Biochem. 35: 467-478.

Bansal, T., D. Englert, J. Lee, M. Hedge, T.K. Wood, A. Jayaraman et al. 2007. Differential effects of epinephrine, norepinephrine and indole on *Escherichia coli* O157:H7 chemotaxis, colonization, and gene expression. Infect. and Imm. 75: 4597-4607.

Beckly, J. and S. Lewis. 2002. Probiotics and antibiotic associated diarrhea: The case for probiotics remains unproved. British Med. J. 325: 901.

Begley, M., C. Hill and C.G. Gahan. 2006. Bile salt hydrolase activity in probiotics. Appl. and Env. Microbio. 72(3): 1729-1738.

Borchers, A.T., C. Selmi, F.J. Meyers, C.L. Keen and E. Gershwin. 2009. Probiotics and immunity. J. Gastroent. 44: 26-46.

Bordoni, A., A. Amaretti, A. Leonardi, E. Boschetti, F. Danesi, D. Matteuzzi et al. 2013. Cholesterol-lowering probiotics: *In vitro* selection and *in vivo* testing of *Bifidobacteria*. Appl. Microbio. Biotech. 9: 8273-8281.

Borody, T.J., L. George, P. Andrews, S. Brandl, S. Noonan, P. Cole et al. 1989. Bowel-flora alteration: A potential cure for inflammatory bowel disease and irritable bowel syndrome? Med. J. of Australia 150: 604.

Botes, M., C.A. van Reenen and L.M. Dicks. 2008. Evaluation of *Enterococcus mundtii* ST4SA and *Lactobacillus plantarum* 423 as probiotics by using a gastro-intestinal model with infant milk formulations as substrate. Int. J. of Food Microbio. 128(2): 362-370.

Boudeau, J., A.L. Glasser, S. Julien, J.F. Colombel and A. Darfeuille-Michaud. 2003. Inhibitory effect of probiotic *Escherichia coli* strain Nissle 1917 on adhesion to and invasion of intestinal epithelial cells by adherent-invasive *E. coli* strains isolated from patients with Crohn's disease. Aliment. Pharmacol. and Therap. 18: 45-56.

Brandao, R.L., I.M. Castro, E.A. Bambirra, S.C. Amaral, L.G. Fietto, M.J.M. Tropia et al. 1998. Intracellular signal triggered by cholera toxin in *Saccharomyces boulardii* and *Saccharomyces cerevisiae*. Appl. and Env. Microbio. 64: 564-568.

Cammarota, M., M. De Rosa, A. Stellavato, M. Lamberti, I. Marzaioli, M. Giuliano et al. 2009. *In vitro* evaluation of *Lactobacillus plantarum* DSMZ 12028 as a probiotic: Emphasis on innate immunity. Int. J. of Food Microbio. 135: 90-98.

Campbell, E.A., S.Y. Choi and H.R. Masure. 1998. A competence regulon in *Streptococcus pneumoniae* revealed by genomic analysis. Mol. Microbio. 27: 929-939.

Cao, G.T., X.F. Zeng, A.G. Chen, L. Zhou, L. Zhang, Y.P. Xiao et al. 2013. Effects of a probiotic, *Enterococcus faecium*, on growth performance, intestinal morphology, immune response, and cecal microflora in broiler chickens challenged with *Escherichia coli* K88. Poultry Sci. 92(11): 2949-2955.

Caplan, M.S., R. Miller-Catchpole, S. Kaup, T. Russell, M. Lickerman, M. Amer et al. 1999. *Bifidobacterial* supplementation reduces the incidence of necrotizing enterocolitis in a neonatal rat model. Gastroent. 117: 577-583.

Castagliuolo, I., E. Beggiao, P. Brun, L. Barzon, S. Goussard, R. Manganelli et al. 2005. Engineered *E. coli* delivers therapeutic genes to the colonic mucosa. Gene Therap. 12: 1070-1078.

Cebra, J.J. 1999. Influences of microbiota on intestinal immune system development. American J. of Clin. Nutrit. 69: 1046-1051.

Chen, X., E.G. Kokkotou, N. Mustafa, K.R. Bhaskar, S. Sougioultzis, M. O'Brien et al. 2006. *Saccharomyces boulardii* inhibis ERK1/2 nitrogen activated protein kinase activation both *in vitro* and *in vivo* and protects against *Clostridium difficile* toxin A-induced enteritis. J. of Bio. Chem. 281: 24449-24454.

Choi, C.H., S.Y. Jo, H.J. Park, S.K. Chang, J.S. Byeon, S.J. Myung et al. 2011. A randomized, double-blind, placebo-controlled multicenter trial of *Saccharomyces boulardii* in irritable bowel syndrome: Effect on quality of life. J. of Clin. Gastroent. 45(8): 679-683.

Cortés-Zavaleta, O., A. López-Malo, A. Hernández-Mendoza and H.S. García. 2014. Antifungal activity of *Lactobacilli* and its relationship with 3-phenyllactic acid production. Int. J. of Food Microbio. 173: 30-35.

Dias, R.S., E.A. Bambirra, M.E. Silva and J.R. Nicoli. 1995. Protective effect of *Saccharomyces boulardii* against the cholera toxin in rats. Brazilian J. of Med. and Bio. Res. 28: 323-325.

Dieleman, L.A., M.S. Goerres, A. Arends, D. Sprengers, C. Torrice, F. Hoentjen et al. 2003. *Lactobacillus* GG prevents recurrence of colitis in HLA-B27 transgenic rats after antibiotic treatment. Gut. 52: 370-376.

D'Souza, A.L., C. Rajkumar, J. Cooke and C.J. Bulpitt. 2002. Probiotic in prevention of antibiotic associated diarrhoea: Meta-analysis. British Med. J. 2002: 1361-1364.

Duan, F.F. and J.C. March. 2010. Engineered bacterial communication prevents *Vibrio cholerae* in an infant mouse model. Proceeds. of the Nat. Acad. of Sci. USA 107(25): 11260-11264.

Duan, F.F., J.H. Liu and J.C. March. 2015. Engineered commensal bacteria reprogram intestinal cells into glucose-responsive insulin-secreting cells for the treatment of diabetes. Diabetes 64 (5): 1794-1803.

Dubern, J.F. and S.P. Diggle. 2008. Quorum sensing by 2-alkyl-4-quinolones in *Pseudomonas aeruginosa* and other bacterial species. Mol. Biosys. 4: 882-888.

Eiseman, B., W. Silen, G.S. Bascom and A.J. Kauvar. 1958. Fecal enema as an adjunct in the treatment of pseudomembranous enterocolitis. Surgery 44: 854-859.

Ewaschuk, J.B. and L.A. Dieleman. 2006. Probiotics and prebiotics in chronic inflammatory bowel diseases. World J. of Gastroent. 12: 5941-5950.

FAO (Food and Agriculture Organization). 2002. Guidelines for the Evaluation of Probiotics in Food. Report of a Joint FAO/WHO Working Group on Drafting Guidelines for the Evaluation of Probiotics in Food.

Fernandez, B., R. Hammami, P. Savard, J. Jean and I. Fliss. 2013. *Pediococcus acidilactici* UL5 and *Lactococcus lactis* ATCC 11454 are able to survive and express their bacteriocin genes under simulated gastrointestinal conditions. J. of Appl. Microbio. 116(3): 677-688.

Fioramonti, J., V. Theodorou and L. Bueno. 2003. Probiotics: What are they? What are their effects on gut physiology? Best Prac. & Res. Clin. Gastroent. 17(5): 711-724.

Fujimura, K.E., T. Demoor, M. Rauch, A.A. Faruqi, S. Jang, C.C. Johnson et al. 2014. House dust exposure mediates gut microbiome *Lactobacillus* enrichment and airway immune defense against allergens and virus infection. Proceeds. of the Nat. Acad. of Sci. 111: 805-810.

Fujitani, S., H.Y. Sun, V.L. Yu. and J.A. Weingarten. 2011. Pneumonia due to *Pseudomonas aeruginosa*. Chest 139: 909-919.

Galdeano, C.M. and G. Perdigón. 2003. The probiotic bacterium *Lactobacillus casei* induces activation of the gut mucosal immune system through innate immunity. Clin. Vacc. Immun. 13(2): 219-226.

Garcia-Lafuente, A., M. Antolin, F. Guarner, E. Crespo and J. Malagelada. 2001. Modulation of colonic barrier function by the composition of the commensal flora in the rat. Gut. 48: 503-507.

Gill, H.S. and M.L. Cross. 2002. Probiotics and immune function. pp. 251-272. *In*: P.C. Calder, C.J. Field, H.S. Gill (eds.). Nutrit. and Imm. Func. CABI Publishing; Wallingford, UK.

Granato, D., G.F. Branco, F. Nazzaro, A.G. Cruz and J.A.F. Faria. 2010. Functional foods and nondairy probiotic food development: Trends, concepts, and products. Compre. Rev. in Food Sci. and Food Safety 9: 292-302.

Grehan, M.J., T.J. Borody, S.M. Leis, J. Campbell, H. Mitchell, A. Wettstein et al. 2010. Durable alteration of the colonic microbiota by the administration of donor fecal flora. J. of Clin. Gastroent. 44: 551-561.

Gronlund, M.M., H. Arvilommi, P. Kero, O.P. Lehtonen and E. Isolauri. 2000. Importance of intestinal colonization in the maturation of humoral immunity in early infancy: A prospective follow up study of healthy infants aged 0-6 months. Arch. of Diseas. in Child: Fetal and Neonat. Ed. 83(3): 186-192.

Guarner, F. and J.R. Malagelada. 2003. Gut flora in health and disease. Lancet 361: 512-519.

Gupta, S., E.E. Bram and R. Weiss. 2013. Genetically programmable pathogen sense and destroy. ACS Syn. Bio. 2: 715-723.

Guslandi, M., G. Mezzi, M. Sorghi and P.A. Testoni. 2000. *Saccharomyces boulardii* in maintenance treatment of Crohn's disease. Digest. Diseas. Sci. 45(7): 1462-1464.

Hamady, Z.Z.R., N. Scott, M.D. Farrar, J.P.A. Lodge, K.T. Holland, T. Whitehead et al. 2010. Xylan-regulated delivery of human keratinocyte growth factor-2 to the inflamed colon by the human anaerobic commensal bacterium *Bacteriodes ovatus*. Gut. 59(4): 461-469.

Hawrelak, J. 2003. Probiotics: Choosing the right one for your needs. J. of Australian Tradition. Med. Soc. 9 (2): 67-75.

Hayashi, H., R. Takahashi, T. Nishi, M. Sakamoto and Y. Benno. 2005. Molecular analysis of jejunal, ileal, caecal and recto-sigmoidal human colonic microbiota using 16S rRNA gene libraries and terminal restriction fragment length polymorphism. J. of Med. Microbio. 54: 1093-1101.

Hill, C., F. Guarner, G. Reid, G.R. Gibson, D.J. Merenstein, B. Pot et al. 2014. Expert consensus document: The International Scientific Association for Probiotics and Prebiotics consensus statement on the scope and appropriate use of the term probiotic. Natur. Rev. Gastroent. and Hepat. 11: 506-514.

Hilton, E., P. Kolakawaki, C. Singer and M. Smith. 1977. Efficacy of *Lactobacillus* GG as a diarrheal preventive in travelers. J. of Travel Med. 4: 41-43.

Holowko, M.B., H. Wang, P. Jayaraman and C.L. Poh. 2016. Biosensing *Vibrio cholerae* with Genetically Engineered *Escherichia coli*. ACS Syn. Bio. 5(11): 1275-1283.

Homayouni A., P. Bastani, S. Ziyadi, S. Mohammad-Alizadeh-Charandabi, M. Ghalibaf, A.M. Mortazavian et al. 2014. Effects of probiotics on the recurrence of bacterial vaginosis: A review. J. of Lower Genital Tract Diseas. 18: 79-86.

Hwang, I.Y., M.H. Tan, E. Koh, C.L. Ho, C.L. Poh, M.W. Chang et al. 2014. Reprogramming microbes to be pathogen-seeking killers. ACS Syn. Bio. 3: 228-237.

Hwang, I.Y., E. Koh, A. Wong, J.C. March, W.E. Bentley, Y.S. Lee et al. 2017. Engineered probiotic *Escherichia coli* can eliminate and prevent *Pseudomonas aeruginosa* gut infection in animal models. Nature Comm. 8: 15028.

Ichikawa, H., T. Kuroiwa, A. Inagaki, R. Shineha,T. Nishihira, S. Satomi et al. 1999. Probiotic bacteria stimulate gut epithelial cell proliferation in rat. Digest. Diseas. and Sci. 44: 2119-2123.

Islam, S.U. 2016. Clinical uses of probiotics. Med. (Baltimore). 95: 1-5.

Isolauri, E., Y. Sutas, P. Kankaanpaa, H. Arvilommi and S. Salminen. 2001. Probiotics: Effects on immunity. American J. of Clin. Nutrit. 73: 444-450.

Isolauri, E., P.V. Kirjavainen and S. Salminen. 2002. Probiotics – A role in the treatment of intestinal infection and inflammation. Gut. 50: 54-59.

Jayaraman, P., M.B. Holowko, J.W. Yeoh, S. Lim and C.L. Poh. 2017 . Repurposing a two-component system-based biosensor for the killing of *Vibrio cholerae*. ACS Syn. Bio. 6: 1403-1415.

Jimenez, P.N., G. Koch, J.A. Thompson, K.B. Xavier, R.H. Cool, W.J. Quax et al. 2012. The multiple signaling systems regulating virulence in *Pseudomonas aeruginosa*. Microbio. Mol. Bio. Rev. 76: 46-65.

Jin, L.Z., R.R. Marquardt and X. Zhao. 2000. A strain of *Enterococcus faecium* (18C23) inhibits adhesion of enterotoxigenic *Escherichia coli* K88 to porcine small intestine mucus. Appl. Env. Microbio. 66: 4200-4204.

Joeres-Nguyen-Xuan, T.H., S.K. Boehm, L. Joeres, J. Schulze, W. Kruis et al. 2010. Survival of the probiotic *Escherichia coli* Nissle 1917 (*EcN*) in the gastrointestinal tract given in combination with oral mesalamine to healthy volunteers. Inflamm. Bowel Diseas. 16: 256-262.

Jones, K. 2010. Probiotics: Preventing antibiotic-associated diarrhea. J. Special. Ped. Nurs. 15(2): 160-162.

Kahouli, I., M. Malhotra, M.A. Jamali and S. Prakash. 2015. *In-vitro* characterization of the anti-cancer activity of the probiotic bacterium *Lactobacillus fermentum* NCIMB 5221 and potential against colorectal cancer cells. J. Cancer Sci. and Ther. 7: 224-235.

Kaper, J.B. and V. Sperandio. 2005. Bacterial cell-to-cell signaling in the gastrointestinal tract. Infect. Immunol. 73: 3197-3209.

Karimi, G., M.R. Sabran, R. Jamaluddin, K. Parvaneh, N. Mohtarrudin, Z. Ahmad et al. 2015. The anti-obesity effects of *Lactobacillus casei* strain Shirota versus Orlistat on high fat diet-induced obese rats. Food and Nutrit. Res. 59: 1-8.

Kaur, B., N. Garg, A. Sachdev and B. Kumar. 2014. Effect of the oral intake of probiotic *Pediococcus acidilactici* BA28 on *Helicobacter pylori* causing peptic ulcer in C57BL/6 mice models. Appl. Biochem. and Biotech. 172: 973-983.

Kirjavainen, P.V., A.C. Ouwehand, E. Isolauri and S.J. Salminen. 1998. The ability of probiotic bacteria to bind to human intestinal mucus. FEMS Microbio. Lett. 167: 185-189.

Kotula, J.W., S.J. Kerns, L.A. Shaket, L. Siraj, J.J. Collins, J.C. Way et al. 2014. Programmable bacteria detect and record an environmental signal in the mammalian gut. Proceeds. Nation. Acad. of Sci. 111(13): 4838-4843.

Forceta, A., J.C. Paton, R. Morona, J. Cook and A.W. Paton. 2006. A recombinant probiotic for treatment and prevention of cholera. Gastroent. 130(6): 1688-1695.

Lan, R.Y., I.R. Mackay and M.E. Gershwin. 2007. Regulatory T cells in the prevention of mucosal inflammatory diseases: Patrolling the border. J. of Autoimm. 29: 272-280.

Larsen, N., L. Thorsen, E.N. Kpikpi, B. Stuer-Lauridsen, M.D. Cantor, B. Nielsen et al. 2014. Characterization of *Bacillus* spp. strains for use as probiotic additives in pig feed. Appl. Microbio. Biotech. 98(3): 1105-1118.

LeBlanc, J.G., G.S. de Giori, E.J. Smid, J. Hugenholtz and F. Sesma. 2007. Folate production by lactic acid bacteria and other food-grade microorganisms. Comm. Curr. Res. and Edu. Topics and Trends in Appl. Microbio. 1: 329-339.

Lee, J. and L. Zhang. 2015. The hierarchy quorum sensing network in *Pseudomonas aeruginosa*. Prot. Cell. 6: 26-41.

Lenoir-Wijnkoop, I., M.E. Sanders, M.D. Cabana, E. Caglar, G. Corthier, N. Rayes et al. 2007. Probiotic and prebiotic influence beyond the intestinal tract. Nutrit. Rev. 65: 469-489.

Li, D., G. Rosito and T. Slagle. 2013. Probiotics for the prevention of necrotizing enterocolitis in neonates: An 8-year retrospective cohort study. J. Clin. Pharm. and Therap. 38: 445-449.

Lilly, D.M. and R.H. Stillwell. 1965. Probiotics: Growth-promoting factors produced by microorganisms. Sci. 147: 747-748.

Lima-Filho, J.V.M., E.C. Vieira and J.R. Nicoli. 2000. *Saccharomyces boulardii* and *Escherichia coli* combinations against experimental infections with *Shigella flexneri* and *Salmonella enteritidis* subsp. Typhimurium. J. of Appl. Microbio. 88: 365-370.

Limdi, J.K., C. O'Neill and J. McLaughlin. 2006. Do probiotics have a therapeutic role in gastroenterology? World J. of Gastroent. 12: 5447-5457.

Lodinova-Zadnikova, R., B. Cukrowska and H. Tlaskalova-Hogenova. 2003. Oral administration of probiotic *Escherichia coli* after birth reduces frequency of allergies and repeated infections later in life (after 10 and 20 years). Int. Arch. Allerg. Imm. 131: 209-211.

Mangell, P., P. Nejdfors, M. Wang, S. Ahrne, B. Westrom, H. Thorlacius et al. 2002. *Lactobacillus plantarum* 299v inhibits *Escherichia coli*-induced intestinal permeability. Digest. Diseas. Sci. 47: 511-516.

Marteau, P., P. Pochart, B. Flourie, P. Pellier, L. Santos, J.F. Desjeux et al. 1990. Effect of chronic ingestion of a fermented dairy product containing *Lactobacillus acidophilus* and *Bifidobacterium bifidum* on metabolic activities of the colonic flora in humans. American J. of Clin. Nutrit. 52: 685-688.

Marteau, P., E. Cuillerier, S. Meance, M.F. Gerhardt, A. Myara, M. Bouvier et al. 2002. *Bifidobacterium animalis* strain DN-173 010 shortens the colonic transit time in healthy women: A double-blind, randomized, controlled study. Aliment Pharma. and Therap. 16: 587-593.

Martinez, R.C., S.A. Franceschini, M.C. Patta, S.M. Quintana, R.C. Candido, J.C. Ferreira et al. 2009. Improved treatment of vulvovaginal candidiasis with fluconazole plus probiotic *Lactobacillus rhamnosus* GR-1 and *Lactobacillus reuteri* RC-14. Lett. Appl. Microbio. 48: 269-274.

McCormick, S.P. 2013. Microbial detoxification of mycotoxins. J. Chem. Eco. 39: 907-918.

Medellin-Pena, M.J., H. Wang, R. Johnson, S. Anand and M.W. Griffiths. 2007. Probiotics affect virulence-related gene expression in *Escherichia coli* O157:H7. Appl. Env. Microbio. 73(13): 4259-4267.

Mego, M., J. Majek, R. Koncekova, L. Ebringer, S. Ciernikova, P. Rauko et al. 2005. Intramucosal bacteria in colon cancer and their elimination by probiotic strain *Enterococcus faecium* M-74 with organic selenium. Folia Microbio. (Praha). 50: 443-447.

Metchnikoff, E. 1908. Essaia optimists. pp. 161-183. *In*: P. C. Mitchell (ed.). Classics in Longevity and Aging. New York: Putmans Sons.

Metchnikoff, I.I. 2004. The Prolongation of Life: Optimistic Studies. New York: Springer Publishing Company.

Michail, S. and F. Abernathy. 2002. *Lactobacillus plantarum* reduces the *in vitro* secretory response of intestinal epithelial cells to enteropathogenic *Escherichia coli* infection. J. Ped. Gastroent. Nutrit. 35(3): 350-355.

Milton, D.L. 2006. Quorum sensing in vibrios: Complexity for diversification. Int. J. Med. Microbio. 296: 61-71.

Moro-García M.A., R. Alonso-Arias, M. Baltadjieva, C. Fernández Benítez, M.A. Fernández Barrial, E. Díaz Ruisánchez et al. 2013. Oral supplementation with *Lactobacillus delbrueckii* subsp. *bulgaricus* 8481 enhances systemic immunity in elderly subjects. Age (Dordr). 35: 1311-1326.

Neish, A.S. 2009. Microbes in gastrointestinal health and disease. Gastroent. 136(1): 65-80.

Ng, W.L. and B.L. Bassler. 2009. Bacterial quorum-sensing network architectures. Ann. Rev. Gen. 43: 197-222.

Niers, L., R. Martin, G. Rijkers, F. Sengers, H. Timmerman, N. van Uden et al. 2009. The effects of selected Probiotic strains on the development of eczema (the P and A study). Allergy 64: 1349-1358.

Oelschlaeger, T.A. 2010. Mechanisms of probiotic actions – A review. Int. J. of Med. Microbio. 300(1): 57-62.

Ortiz, L., F. Ruiz, L. Pascual and L. Barberis. 2014. Effect of two probiotic strains of *Lactobacillus* on *in vitro* adherence of *Listeria monocytogenes*, *Streptococcus agalactiae*, and *Staphylococcus aureus* to vaginal epithelial cells. Curr. Microbio. 68(6): 679-684.

Ouwehand, A.C. 2007. Antiallergic effects of probiotics. J. of Nutrit. 137: 794S-797S.

Palmer, J.D., E. Piattelli, B.A. McCormick, M.W. Silby, C.J. Brigham, V. Bucci et al. 2018. Engineered Probiotic for the inhibition of *Salmonella* via tetrathionate-induced production of microcin H47. ACS Infect. Diseas. 4(1): 39-45.

Parker, R.B. 1974. Probiotics, the other half of the antibiotic story. Animal Nutrit. Health 29: 4-8.

Parvez, S., K.A. Malik, S.A. Kang and H.Y. Kim. 2006. Probiotics and their fermented food products are beneficial for health. J. of Appl. Microbio. 100(6): 1171-1185.

Patil, M.B. and N. Reddy. 2006. Bacteriotherapy and probiotics in dentistry. KSDJ. 2: 98-102.

Paton, A.W., M.P. Jennings, R. Morona, H. Wang, A. Focareta, L.F. Roddam et al. 2005. Recombinant Probiotics for treatment and prevention of enterotoxigenic *E. coli* diarrhea. Gastroent. 128(5): 1219-1228.

Phavichitr, N., P. Puwdee and R. Tantibhaedhyangkul. 2013. Cost-benefit analysis of the probiotic treatment of children hospitalized for acute diarrhea in Bangkok, Thailand. The Southeast Asian J. of Trop. Med. and Pub. Health. 44: 1065-1071.

Pieniz, S., R. Andreazza, J.Q. Pereira, F.A. de Oliveira Camargo and A. Brandelli. 2013. Production of selenium-enriched biomass by *Enterococcus durans*. Bio. Trace Elem. Res. 155(3): 447-454.

Pinto, G.S., M.S. Cenci, M.S. Azevedo, M. Epifanio and M.H. Jones. 2014. Effect of yogurt containing *Bifidobacterium animalis* subsp. *lactis* DN-173010 probiotic on dental plaque and saliva in orthodontic patients. Caries Res. 48: 63-68.

Renault. P. 2002. Genetically modified lactic acid bacteria: Applications to food or health risk assessment. Biochimie. 84: 1073-1087.

Riglar, D.T., T.W. Giessen, M. Baym, S.J. Kerns, M.J. Niederhuber, R.T. Bronson et al. 2017. Engineered bacteria function in the mammalian gut as long-term live diagnostics of inflammation. Nat. Biotech. 35 (7): 653-658.

Ruan, Y., J. Sun, J. He, F. Chen, R. Chen, H. Chen et al. 2015. Effect of probiotics on glycemic control: A systematic review and meta-analysis of randomized, controlled trials. PLoS One. 10(7): e0132121.

Ruseler-van Embden, J.G., L.M. van Lieshou, M.J. Gosselink and P. Marteau. 1995. Inability of *Lactobacillus casei* strain GG, *L. acidophilus*, and *Bifidobacterium bifidum* to degrade intestinal mucus glycoproteins. Scandinav. J. of Gastroent. 30: 675-680.

Rutherford, S.T. and B.L. Bassler. 2012. Bacterial quorum sensing: Its role in virulence and possibilities for its control. Cold Spring Harbor Perspect. in Med. 2(11): a012427.

Saeidi, N., K.W. Choon, T.M. Lo, H.X. Nguyen, H. Ling, S.S.J. Leong et al. 2011. Engineering microbes to sense and eradicate *Pseudomonas aeruginosa*, a human pathogen. Mol. Sys. Bio. 7: 521.

Saikali, J., C. Picard, M. Freitas and P.R. Holt. 2004. Fermented milks, probiotic cultures, and colon cancer. Nutrit. and Cancer 49(1): 14-24.

Sanders, M.E. 1999. Probiotics. Food Tech. 53(11): 67-77.

Savard, P., B. Lamarche, M.E. Paradis, H. Thiboutot, É. Laurin, D. Roy et al. 2011. Impact of *Bifidobacterium animalis* ssp. *Lactis* BB-12 and, *Lactobacillus acidophilus* LA-5-containing yoghurt, on fecal bacterial counts of healthy adults. Int. J. of Food Microbio. 149: 50-57.

Schachtsiek, M., W.P. Hammes and C. Hertel. 2004. Characterization of *Lactobacillus coryniformis* DSM 20001T surface protein Cpf mediating coaggregation with and aggregation among pathogens. Appl. and Env. Microbio. 70(12): 7078-7085.

Schatzmayr, G., F. Zehner, M. Taubel, D. Schatzmayr, A. Klimitsch, A.P. Loibner et al. 2006. Microbiologicals for deactivating mycotoxins. Mol. Nutrit. and Food Res. 50: 543-551.

Soccol, C.R., M.R.M. Prado, L.M.B. Garcia, C. Rodrigues, A.B.P. Medeiros and V. T. Soccol. 2014. Current developments in Probiotics. J. Microb. Biochem. Technol. 7(1): 011-020.

Soh, S.E., M. Aw, I. Gerez, Y.S. Chong, M. Rauff, Y.P. Ng et al. 2009. Probiotic supplementation in the first 6 months of life at risk Asian infants-effects on eczema and atopic sensitization at the age of 1 year. Clin. and Exp. Allerg. 39: 571-578.

Song, S., S.J. Lee, D.J. Park, S. Oh and K.T. Lim. 2016. The anti-allergic activity of *Lactobacillus plantarum* L67 and its application to yogurt. J. of Dairy Res. 99: 9372-9382.

S pitz, J., R. Yuhan, A. Koutsouris, C. Blatt, J. Alverdy and G. Hecht. 1995. Enteropathogenic *Escherichia coli* adherence to intestinal epithelial monolayers diminishes barrier function. American J. of Physiol. 268(2 Pt 1): G374-G379.

Steidler, L., W. Hans, L. Schotte, S. Neirynck, F. Obermeier, W. Falk et al. 2000. Treatment of Murine colitis by *Lactococcus lactis* secreting IL-10. Sci. 289(5483): 1352-1355.

Stiles, M.E. and W.H. Holzapfel. 1997. Lactic acid bacteria of foods and their current taxonomy. Int. J. of Food Microbio. 36: 1-29.

Szajewska, H., M. Kotowska, J.Z. Mrukowicz, M. Armánska and W. Mikolajczyk. 2001. Efficacy of *Lactobacillus* GG in prevention of nosocomial diarrhea in infants. J. of Ped. 138: 361-365.

Teitelbaum, J.E. and W.A. Walker. 2002. Nutritional impact of pre- and probiotics as protective gastrointestinal organisms. Ann. Rev. of Nutrit. 22: 107-138.

TMR. 2018. Probiotics market: Global industry analysis and opportunity assessment 2018-2016, Transparency Market Res. 2018. www.transparencymarketresearch. com.

Trapecar, M., T. Leouffre, M. Faure, H.E. Jensen, P.E. Granum, A. Cencic et al. 2011. The use of a porcine intestinal cell model system for evaluating the food safety risk of *Bacillus cereus* probiotics and the implications for assessing enterotoxigenicity. APMIS. 119(12): 877-884.

Tripathi, M.K. and S.K. Giri. 2014. Probiotic functional foods: Survival of probiotics during processing and storage. J. of Funct. Foods 9: 225-241.

Turner, P.C., Q.K. Wu, S. Piekkola, S. Gratz, H. Mykkanen, H. El-Nezami et al. 2008. *Lactobacillus rhamnosus* strain GG restores alkaline phosphatase activity in differentiating Caco-2 cells dosed with the potent mycotoxin deoxynivalenol. Food and Chem. Toxicol. 46: 2118-2123.

Upadrasta, A. and R.S. Madempudi. 2016. Probiotics and blood pressure: Current insights. Integrat. Blood Press. Cont. 9: 33-42.

Viljanen, M., M. Kuitunen, T. Haahtela, K. Juntunen-Backman, R. Korpela, E. Savilahti et al. 2005. Probiotic effects on faecal inflammatory markers and on faecal IgA in food allergic atopic eczema/dermatitis syndrome infants. Ped. Allerg. and Immun. 16: 65-71.

Wadher, K.J., J.G. Mahore and M.J. Umekar. 2010. Probiotics: Living medicines in health maintenance and disease prevention. Int. J. of Pharma Bio Sci. 1(3): 1-9.

Wickens, K., T.V. Stanley, E.A. Mitchell, C. Barthow, P. Fitzharris, G. Purdie et al. 2013. Early supplementation with *Lactobacillus rhamnosus* HN001 reduces eczema prevalence to 6 years: Does it also reduce atopic sensitization? Clin. and Exp. Allerg. 43: 1048-1057.

Wu, Z.J., X. Du and J. Zheng. 2013. Role of *Lactobacillus* in the prevention of *Clostridium difficile*-associated diarrhea: A meta-analysis of randomized controlled trials. Chinese Med. J. (Engl.) 126: 4154-4161.

Yamada, T., S. Nagata, S. Kondo, L. Bian, C. Wang, T. Asahara et al. 2009. Effect of continuous probiotic fermented milk intake containing *Lactobacillus casei* strain Shirota on fever in mass infectious gastroenteritis rest home outbreak. Kansenshogaku Zasshi 83: 31-35.

Yan, F. and D.B. Polk. 2002. Probiotic bacterium prevents cytokine-induced apoptosis in intestinal epithelial cells. J. of Bio. Chem. 277: 50959-50965.

Yu, H., L. Liu, Z. Chang, S. Wang, B. Wen, P. Yin et al. 2013. Genome sequence of the bacterium *Bifidobacterium longum* strain CMCC P0001, a probiotic strain used for treating gastrointestinal disease. Genome Announc. 713-716.

Yun, B., S. Oh and M.W. Griffiths. 2014. *Lactobacillus acidophilus* modulates the virulence of *Clostridium difficile*. J. of Diary Sci. 97(8): 4745-4758.

Zhou, J.S., P.K. Gopal and H.S. Gill. 2001. Potential probiotic lactic acid bacteria *Lactobacillus rhamnosus* (HN001), *Lactobacillus acidophilus* (HN017) and *Bifidobacterium lactis* (HN019) do not degrade gastric mucin *in vitro*. Int. J. of Food Microbio. 63: 81-90.

Zokaeifar, H., N. Babaei, C.R. Saad, M.S. Kamarudin, K. Sijam, J.L. Balcazar et al. 2014. Administration of *Bacillus subtilis* strains in the rearing water enhances the water quality, growth performance, immune response, and resistance against *Vibrio harveyi* infection in juvenile white shrimp, *Litopenaeus vannamei*. Fish Shellfish Immun. 36(1): 68-74.

CHAPTER
8

Enzymatic Reaction for Oligosaccharides Production

Nor Hasmaliana Abdul Manas[1,2], Mohd Khairul Hakimi Abdul Wahab[2],
Nur Izyan Wan Azelee[1,2] and Rosli Md Illias[1,2]*

[1] Institute of Bioproduct Development (IBD), Universiti Teknologi Malaysia,
81310 UTM Skudai, Johor, Malaysia
[2] Department of Bioprocess and Polymer Engineering, School of Chemical and
Energy Engineering, Faculty of Engineering, Universiti Teknologi Malaysia,
81310 Skudai, Johor, Malaysia

1. Introduction

Oligosaccharides are a part of dietary fibres which are non-digestible carbohydrates. Among the non-digestible carbohydrates, oligosaccharides are gaining outstanding popularity owing to their physiological benefits to their consumers. They are carbohydrate polymers intermediate in nature between simple sugars and polysaccharides, consisting of a several number of monosaccharides normally from three to ten monomer units. Oligosaccharides are prepared from natural sources such as plant derivatives by techniques that take advantage of size, alkaline stability, or some combination of these and other properties of the molecule of interest. Besides, it can also be synthesized by physical, chemical or enzymatic methods. A part of oligosaccharides have been prepared by partially breaking down more complex carbohydrates (polysaccharides). Oligosaccharides' properties are determined by composition, degree of polymerization and glycosidic linkage(s) (Terrasan et al., 2019). The preparation of oligosaccharides usually results in diverse molecular sizes of the desired molecule. Oligosaccharides with a higher degree of polymerisation are named tri-, tetra-, penta-, etc. Structures of oligosaccharides may be predominately linear or branched.

There are several types of oligosaccharides that have been studied. For instance, lignocellulosic biomass has a high polysaccharides content which can be further extracted to different oligosaccharides depending on the chemical composition of the biomass. Xylooligosaccharide, galactooligosaccharide,

*Corresponding author: r-rosli@utm.my

arabinooligosaccharide and mannooligosaccharide are some of the examples of oligosaccharides that can be extracted from lignocellulosic biomass. Other oligosaccharides include fructooligosaccharide, gluco-oligosaccharide, isomaltooligosaccharide, soybean meal oligosaccharide and gentiooligosaccharide. Oligosaccharides have been used as prebiotics. Among established prebiotics that are supported by human trials data are fructans inulin, fructooligosaccharide, galactooligosaccharide and lactulose (Rastall, 2010).

1.1 Oligosaccharides as Prebiotics

Oligosaccharides are compounds that promote proliferation of probiotics (Qiang et al., 2009). Probiotics are live microorganisms which when administered in adequate amounts colonize as commensal in the host and conferring a health benefit. The biological functions of oligosaccharides appear to span the spectrum from the trivial, to those that are crucial for the development, growth, function or survival of an organism (Varki, 1993). Three general principles emerge as shown in Table 1. However, such oligosaccharide sequences are also more likely to be targets for recognition by pathogenic toxins and microorganisms.

Table 1: The principle of the biological functions of oligosaccharides

Principle	Description
First	The contribution to the organism is vague and the prediction of oligosaccharide functions on a given glycoconjugate is difficult.
Second	Similar oligosaccharides may act with different functions and at different locations even within the same organism.
Third	Specific biological functions of oligosaccharides are facilitated either by unique oligosaccharide sequences, unusual common terminal sequences, or by further modifications of the carbohydrate.

Most oligosaccharides are not fermented by human enzymes and oral microflora, therefore are non-calorific and non-cariogenic (Prapulla et al., 2000). However, oligosaccharides are fermentable by intestinal microflora. Therefore, oligosaccharides have been used as prebiotics to stimulate the growth and activity of gut microflora. According to the updated definition, a prebiotic can be defined as "a substrate that is selectively utilized by host microorganisms conferring a health benefit" (Gibson et al., 2017). Prebiotics have been reported to have increased nutrient absorption, improved immune response and prevented colorectal cancer (Morris and Morris, 2012).

1.2 Other Applications of Oligosaccharides

More recently, oligosaccharides have attracted considerable interest especially in the pharmaceutical and food sectors. The increase is due to the overwhelming demand for healthier food amongst consumers. Oligosaccharides are widely being used as food ingredients, in prebiotic supplements, cosmetics and drug delivery, agrochemicals and in animal feed. In the cosmetic industries, stabilizers and bulking agents commonly use oligosaccharides as one of the ingredients. Most oligosaccharides serve as antimicrobial agents and immunity providers, and provide other health benefits, as well. The non-digestible oligosaccharides (NDO) serve as dietary fibres for dietary supplements, weight controlling agents, sweeteners, and humectants in confectioneries, bakeries and breweries (Patel and Goyal, 2011). In 1991, Japan has categorised several oligosaccharides as "foods for specified health use" (FOSHU). Worldwide research findings have categorised NDOs under functional foods (Patel and Goyal, 2011). In food products, they are used as low-sweetness humectants and anticariogenic agents.

A synbiotic concept has been generated by recent progress on commercial prebiotic oligosaccharides and probiotic bacteria. Yogurt manufactures fortify their product with prebiotic oligosaccharides to benefit the human gut health. The mode of action of probiotics is suppressing pathogenic microorganisms by lowering the pH value of the large intestine, producing some vitamins, and stimulating the human immune system (Tuohy et al., 2005). The development of scale-up synthesis methods is vital for the large scale application of oligosaccharides in pharmaceutical, cosmetics, foods, animal feed and agrochemical industries (Patel and Goyal, 2011). Oligosaccharides have been widely introduced in animal feeds and are still generating more interest. Oligosaccharides which were originated from the *Saccharomyces cerevisiae* cell wall, which are mannooligosaccharides, are important feed ingredients for poultry and livestocks (Patel and Goyal, 2011). The interest in prebiotic oligosaccharides is driven by various industrial applications. With a galloping acceptance of oligosaccharides as functional foods and a conscious increase in the need for environmental friendliness, these oligosaccharides-fortified products seem to be having great promise.

2. Enzymatic Production of Oligosaccharides

Oligosaccharides can be extracted from natural sources using chemical extraction methods. For instance, raffinose oligosaccharides (Patel and Goyal, 2010) and soy oligosaccharides (Mussatto and Mancilha, 2007) are extracted from plant materials using water or aqueous methanol or ethanol solutions. However, to meet the increasing market demand, a more specific and efficient technique for large-scale production is used. Various approaches have been explored in the synthetic glycochemistry field for the synthesis of sugar-based compounds, specifically oligosaccharides. Since synthetic biotechnology progresses remarkably, enzymes have been recognized as alternative tools for oligosaccharides synthesis. Nowadays, the enzymatic method is opted for oligosaccharides synthesis as it offers stereochemical specificity and does not require tedious protection-deprotection steps as required in chemical synthesis (Wang and Huang, 2009). In addition, the use of enzymes is a more environment-friendly approach, operating without application of obnoxious chemicals and being aligned with the perspective of a biodegradable process (Wan Azelee et al., 2016). Oligosaccharides can be synthesized using glycosyl transferase and glycosyl hydrolase. Application of the former enzyme is limited by the poor availability of enzymes, the use of complex and expensive substrate and instability in solution (Prapulla et al., 2000). On the other hand, glycosyl hydrolase is widely distributed in nature, requires simple donor molecules, and is stable in aqueous solution (Hansson et al., 2001).

2.1 Glycosyl Hydrolases in Oligosaccharides Production

Oligosaccharides are synthesized by glycosyl hydrolases with hydrolysis and transglycosylation activity. Glycosyl hydrolases are carbohydrate-active enzymes that are known to catalyze stereospecific and regioselective reactions that enable a production of specific oligosaccharides without the need of protection-deprotection steps as required in conventional glycochemical reactions. As shown in Table 2, enzymes in glycosyl hydrolase family have been recognized to produce various oligosaccharides depending on the type of substrate and its stereochemical and regioselectivity. A wide chemical diversity of oligosaccharides are witnessed to have different types of glycosidic linkages, polymerization degrees and various structures such as linear, branched or cyclic. Glycosyl hydrolases are able to form specific structures of oligosaccharides owing to the regioselectivity and stereochemical control of the enzymes (Abdul Manas et al., 2018).

Table 2: Oligosaccharides and glycosyl hydrolase producers

Oligosaccharides	Glycosyl hydrolase producers	Reference
Chitooligosaccharides	Chitosanase	Song et al., 2014
	Endochitinase	Vaikuntapu et al. 2018
Fructooligosaccharides	β-fructofuranosidase	Lorenzoni et al., 2014
	Levansucrase	Li et al., 2015
	Inulosucrase	Díez-Municio et al., 2013
Galactooligosaccharides	α-galactosidase	Hinz et al., 2006
	β-galactosidase	Wu et al., 2013
Maltooligosaccharides	Maltogenic amylase	Abdul Manas et al., 2014
	Neopullulanase	Kuriki et al., 1996
	β-glucosidase	Hansson et al., 2001
	α-glucosidase	Fernández-Arrojo et al., 2007
Mannooligosaccharides	Mannanase	Arnling Bååth et al., 2018
Xylooligosaccharides	Xylanase	Chang et al., 2017
	α-L-arabinofuranosidase	Arab-Jaziri et al., 2015

Glycosyl hydrolases are widely distributed in microorganisms. For instance, Sabater and co-workers (2019) have isolated microbial β-galactosidase of dairy propionibacteria from Argentinean foods to synthesis galactooligosaccharides from lactose and its derived oligosaccharides (OsLu) from lactulose. Dairy propionibacteria are microorganisms that are traditionally used by industry for the manufacture of Swiss-type cheeses and the biological production of propionic acid. Additionally, several probiotic effects have been reported for dairy propionibacteria, which could be due to their ability to modulate gut physiology, microbiota composition and host immunity in a beneficial manner (Rabah et al., 2017). β-fructofuranosidase has been isolated from fungi, *Penicillium citreonigrum* and has been used for Fructooligosaccharides production (Nascimento et al., 2016).

2.2 Oligosaccharides Production via Polysaccharides Hydrolysis

Polysaccharides are abundant in nature and are good substrates for oligosaccharides production. Glycosyl hydrolases specifically hydrolyze polysaccharides into smaller saccharide units. An example of a polysaccharide is hemicellulose, which is a part of lignocellulosic biomass that can be hydrolyzed into fermentable sugars and oligosaccharides. Xylooligosaccharides are an example of prebiotic nutraceuticals that can be sourced from lignocellulosic biomass, such as agro-residues and have gained attention due to their multidimensional beneficial influences on human health and livestock (Adebola et al., 2014). Xylooligosaccharides are produced from xylan by hydrolysis with xylanase as the main enzyme (Prapulla et al. 2000). Xylooligosaccharides with DP between 2 and 5 have been produced from brewers' spent grain using the endo-1,4-β-xylanase M3 from *Trichoderma longibrachiatum* (Amorim et al., 2019).

While xylan is mostly found in hardwoods, mannan is a prominent type of hemicellulose found in softwoods. β-mannanases are enzymes that can hydrolyze manan into mannooligosaccharides. Production of mannooligosaccharides from mannan has been described using bacterial β-mannanase (Zang et al., 2015). Mannooligosaccharides derived from mannan have shown improved growth of probiotic bacteria (Singh et al., 2018).

Fructooligosaccharide can be synthesized by hydrolysis of inulin using inulinase (Singh et al., 2016). Inulin is a naturally occurring polyfructan produced by plants and is the second most abundant polysaccharide after starch. Many researchers have found that controlled enzymatic hydrolysis of inulin by inulinase produces a significant amount of fructooligosaccharide. Besides that, fructooligosaccharide has also been produced from maple syrup by the action of high levansucrase activity (Thompson et al., 1994). Li and co-worker (2015) have determined the ability of *Bacillus amyloliquefaciens* levansucrase to synthesize a variety of hetero-fructooligosaccharides in maple syrups enriched with various disaccharides, with lactose being the preferred fructosyl acceptor. Other than maple syrup, fructoligosaccarides could also be found in other plants such as banana, artichoke, onion, chicory, garlic, asparagus, yacon root and blue agave.

Chitooligosaccharides are degraded products of chitin and chitosan. Specifically, chitin is one of the abundant natural polysaccharide while chitosan is its most important derivative. Regardless of having various bioactivities, the water insolubilities of both chitin and chitosan limit their applications in many industries. The physical, chemical or enzymatic depolymerization of chitin and chitosan deliver chitooligosaccharides which are water-soluble and low molecular weight derivatives. Astonishing bioactivities of chitooligosaccharides established them as desired molecules for industries. Multisize chitooligosaccharides have been successfully produced in a continuous process by chitosanase coated silica gel (Song et al., 2014). It is one of the efficient methods for the bioconversion of high molecular weight chitosan to chitooligosaccharides.

2.3 Oligosaccharides Production via Glycosyl Transfer Reaction

Glycosyl hydrolase enzymes are a variety of enzymes that can catalyze two reactions, which are hydrolysis that cleaves the substrate to smaller products, or transglycosylation that joins two molecules to produce a larger or longer product. Apart from hydrolysis, oligosaccharides can be synthesized using a single enzyme through transglycosylation mechanism of glycosyl hydrolases that involves the glycosyl transfer reaction. The enzyme transfers a glycosidic residue from a glycoside donor to a carbohydrate acceptor instead of a water molecule in hydrolysis. In such a reaction, monosaccharides or disaccharides are normally used as acceptor molecules. For example, fructooligosaccharide is synthesized by the transfructosylation activity of β-fructofuranosidase using sucrose as a glycosyl molecule. Meanwhile, galactooligosaccharides can be prepared from lactose by an enzymatic reaction using β-galactosidase (Patel and Goyal, 2011). α-Galactosidase from *Bifidobacterium adolescentis* DSM 20083 produced oligosaccharides using melibiose as glycosyl substrate (Hinz et al. 2006). Maltooligosaccharides enriched in trisaccharides and tetrasaccharides were produced by the transglycosylation activity of α-glucosidase from *Xanthophyllomyces dendrorhous* using maltose as glycosyl donor (Fernández-Arrojo et al., 2007).

These enzymes with multi-functionality have been gaining a lot of interest among researchers. Both activities are kinetically controlled, thus, the enzyme can act as a biochemical switch whose response is regulated by changes in the surrounding conditions (Jeffery, 2004). The switching point can be explained with a structural basis and an understanding of this biological role will lead to a subsequent improvement of the existing biochemical catalyst. Extensive research has been carried out as it has the potential to be exploited in synthetic applications and the development of a novel synthesis pathway through reaction kinetic control and protein engineering. The transglycosylation activity of glycosyl hydrolases offers a great advantage in the industry for the synthesis of oligosaccharides. However, the bottleneck is that the synthesis reaction competes with hydrolysis resulting in a low yield of oligosaccharides produced. The products of transglycosylation are usually susceptible to hydrolysis by the enzyme, therefore resulting in a poor yield (Wang and Huang, 2009). The transglycosylation efficiency of glycosyl hydrolase highly dependant on the competition of the sugar acceptor with the water molecule in the active site (Wang and Huang, 2009). Therefore, many strategies have been extensively studied in order to favor transglycosylation.

3. Approaches to Oligosaccharides Production

The hydrolysis and transglycosylation reactions catalyzed by glycosyl hydrolases are highly influenced by water activity. In the presence of a suitable nucleophile other than water, the enzyme can transfer the glycosyl residues to form a glycosidic linkage (Hansson et al., 2001). The reaction equilibrium and kinetic control approach is carried out based on the fact that the hydrolysis activity is a reversible reaction. The use of an enzyme mixture or direct fermentation is advantageous when complex polysaccharides are used as a substrate. The optimization of the enzyme structure by protein engineering is also promising as massive data on structure-function data of glycosyl hydrolases is available. Finally, the use of a bioprocessing strategy and enzyme immobilization are also capable of providing a continuous and sustainable production of oligosaccharides.

3.1 Manipulation of Reaction Conditions and Kinetics

Formation of oligosaccharides via the enzymatic method is influenced by various reaction conditions such as substrate, substrate concentration, enzyme concentration, temperature, pH, and time. Formerly, lactose was the only substrate for GOS production (Cardelle-Cobas et al., 2008). However, many studies have later on shown that lactulose, a non-absorbable sugar and other carbohydrates are also good substrates for the enzymatic synthesis of prebiotic oligosaccharides using microbial β-galactosidase. Sabater and co-workers (2019) have achieved maximum GOS formation, which accounted for 23.6% (w/w) of total sugars by β-galactosidase of *P. acidipropionici* LET 120 after letting the reaction proceed for 24 hours. In contrast, a shorter reaction time of only 5 h was needed to reach a maximum GOS formation of 25.0% (w/w) of the total sugars using a commercial enzyme preparation, Lactozym. Lactulose was also hydrolyzed into galactose and fructose by β-galactosidase and subsequently transgalactosylated into OsLu (di- and tri-saccharides) (Sabater et al., 2019).

Empirical models have been proposed for easier quantification of reactants and during any reaction in particular (Kim et al., 1996). The development of a kinetic model was initiated by the enzymatic reaction mechanism which was highly subject to its source and purity. A five-step, ten-parameter kinetic model was proposed by Khandekar et al. (2014) based on the basic Michaelis–Menten concept to understand fructooligosaccharides (FOS) production from sucrose reaction. Various sucrose concentrations were attained ranging from 210-850 g/L. Invertase enzyme shows higher trans-fructosylation activity using greater substrate concentrations (Hidaka et al., 1988). Lower sucrose concentrations exhibit higher hydrolysis activity of invertase enzyme compared to the trans-fructosylation activity (Yun, 1996). A work by Khandekar and co-workers (2014) shows that the FOS yield reaches a maximum value after 2–3 h of reaction time and started to decline with a prolonged reaction due to the hydrolysis reaction, when some portion of FOS started to hydrolyze to form sucrose and fructose. The active sites of invertase for trans-fructosylation activity was reported to be inhibited by glucose and this situation is called product inhibition. By applying Lineweaver–Burk method, the kinetic parameters for FOS production were determined to be $K_m = 1.08$ M and $V_m = 0.2619 \times 10^{-3}$ M/s (Khandekar et al., 2014).

Kinetic analysis of galactooligosaccharides formation has been published by Mueller et al., 2018. The kinetics of the formation of prebiotic galactooligosaccharides (GOS) from lactose was investigated utilising galactosidase as the catalyst. The kinetic models were developed considering several stages which were the hydrolysis of lactose, formation of GOS and inhibiting effects of the side products, glucose and galactose. By implementing the *Christiansen* methodology and assuming pseudo-first order rate laws for all elementary steps, a general equilibrium limited rate approach without a rate determining step was obtained. The simple Michaelis–Menten kinetics are likely to be chosen by some previous authors for the ease of mathematical description and modelling of the GOS formation process. The influence of temperature (high and low temperatures) affect the kinetics by increasing the hydrolysis reaction rates more than the transgalactosylation reaction (Mueller et al., 2018). Chen et al. (2003) found that temperatures above 37 °C resultd in larger amounts of allolactose (galactooligosaccharides with a degree of polymerisation (DP) of 2) combined with a simultaneous inhibition of GOS formation with a higher DP. From his work, Chen and co-workers (2003) have successfully obtained the maximum GOS yield with a highest lactose conversion of 80%. Higher conversion rates result in a reduction of GOS yield. The optimum GOS yield with 49.05% was reached after six hours reaction time at pH 7.4, 40 °C, 14 U mL^{-1} enzyme concentration with an initial 150 g L^{-1} lactose concentration. This was explained by a low lactose to water ratio which lead to a higher hydrolysis activity.

Response surface methodology has also been used to study the correlation between parameters affecting oligosaccharides production. The influence of the main reaction parameters such as temperature, enzyme concentration, pH, initial lactose concentration and reaction time on the synthesis of GOS by β-galactosidase and their interaction to maximise the yield has been deciphered by mathematical models. The models have predicted an optimum yield of 12.18% GOS at 40 °C, 5 U/ml enzyme concentration, PH 7.0, 250 g/l initial lactose concentration and 3 hours of reaction time. Additionally from the analysis, it was understood that the PH made the most significant impact on the transgalactosylation/hydrolysis ratio. The generated models are useful in predicting galato-oligosaccharides production under specified reaction conditions including conditions required to maximise the yield of galato-oligosaccharides of a specific degree of polymerization (González-Delgado et al., 2016).

Most of the kinetic models developed for oligosaccharides production through enzymatic reactions showed a good correlation with the experimental data. This proved that the models are reliable and useful for optimization of oligosaccharides production.

3.2 Enzyme Cocktails and Direct Fermentation

Production of oligosaccharides from complex substrates such as biomass, normally requires multiple enzymes for efficient hydrolysis. Lignocellulosic biomass for example, is composed of a highly crystalline polymer of cellulose, hemicellulose and lignin. Expansin can strongly bind to cellulose and hemicellulose, and disrupts the lignocellulosic network to allow more efficient enzymatic hydrolysis of the polysaccharides (Kerff et al., 2008). An expansin-endoglucanase fusion enzyme was reported to increase cellulose degradation to produce sugar (Nakashima et al., 2014). Production of Xylooligosaccharides from corncob was also enhanced using expansin and endo-xylanase enzyme fusion (Chang et al., 2017). The approach offers a cost-effective method as the pre-treatment cost is significantly reduced.

A range of enzymes are required for complete hydrolysis of hemicellulose that has a complex structure. O-acetylgalactoglucomannan is a major hemicellulose component of softwood. Among enzymes that are needed to completely hydrolyze this highly decorated and acetylated polysaccharide into monosaccharides are endo-β-mannanases, β-mannosidases, β-glucosidases, α-galactosidases, and acetyl esterases. However, if mannooligosaccharides are desired, β-mannanase plays a vital role in mannan degradation (Arnling Bååth et al., 2018). A combination of *B. stearothermophilus* maltogenic amylase and α-glucanotransferase from *Thermotoga maritime* significantly improved the yield of isomaltooligosaccharide produced from liquefied corn syrup. A 10% increment in the yield was obtained when an enzyme cocktail was used compared to a single enzyme (Lee et al., 2002). Besides, the use of *Bacillus subtilis* saccharifying α-amylase simultaneously with neopullulanase also improved the yield of isomaltooligosaccharide (Kuriki et al., 1993).

Fermentation of biomass using a specific microorganism is another approach in producing oligosaccharides. Amorim and co-workers (2019) have developed a one-step process for producing prebiotic xylooligosaccharides from brewer's spent grain without the use of a previous conventional pre-treatment. Brewer's spent grain was directly used in fermentation of *Trichoderma viride* and *Trichoderma reesei*. This residue provided a large number of bands associated with xylooligosacchairdes production and a lower concentration of xylose in the fermentation medium. Under optimal conditions (3 days, pH 7.0, 30 °C and 20 g/L of brewer's spent grain), xylooligosaccharides production yield of 38.3 ± 1.8 mg/g (xylose equivalents/g of brewers' spent grain) was successfully achieved.

3.3 Protein Engineering

Protein engineering is considered one of the advanced techniques to improve enzyme activity for oligosaccharide production. An enzyme that is tailor-made to produce a specific oligosaccharide structure is possibly constructed by protein engineering. The availability of vast structural data of glycosyl hydrolase gives an advantage to the process. Figure 1 shows two main pathways to the construction of a mutant enzyme with a high transglycosylation activity, which is preferred for oligosaccharide synthesis. Random mutagenesis could be used when there is a lack of structural data available to predict the mutation site. Random mutagenesis might involve high throughput screening of mutant libraries to identify the improved mutants. This is followed by genetic and structural analysis to examine the mutation site that changes the enzyme activity. Random mutagenesis could also be used to provide a platform for a better understanding of the structure-function relationship before site-directed mutagenesis could be performed. In contrast, site-directed mutagenesis starts with structural analysis to identify potential mutation sites based on the understanding of the enzyme structure-function. This process is called rational design that might involve thorough structure analysis with homology modeling, docking and molecular dynamics study (Manas et al., 2016).

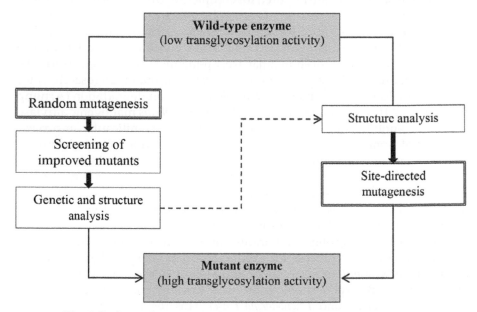

Fig. 1: Pathway for mutant construction using protein engineering

Random mutagenesis was successfully performed on α-L-arabinofuranosidase of *Thermobacillus xylanilyticus* involving high a throughput screening of 30,000 mutants' library using a colorimetric substrate. Two-step screening was employed; the first and second steps involved selecting hydrolytically-impaired mutants, followed by mutants with an improved transglycosylation activity using Xylooligosaccharides as acceptors. Mutants that exhibited double synthesis activity compared to the wild-type had amino acid substitutions in subsite-1 of the active site. This mutation caused modulation in the active site topology and subsequent substrate binding to impair hydrolysis enhances its ability to perform transarabinofuranosylation (Arab-Jaziri et al., 2015). In another study, a directed-evolution technique has been used to produce β-glycosidase variants with improved transglycosylation activity. Structural analysis revealed that the amino acid substitutions were located in the vicinity of subsite-1, which helped in better binding of the acceptor molecule (Feng et al., 2005). Random mutagenesis has also been performed on arabinofuranosyl hydrolase (Pennec et al., 2015) and β-glucosidase (Lundemo et al., 2016) to improve their transglycosylation activities.

Oligosaccharide structures are highly dependant on the active site topology of the enzyme. Random mutagenesis could provide a clue on the important amino acid residues that play an important role in hydrolysis and transglycosylation activities. Elucidation of crystal structure of an enzyme with a bound substrate also provides valuable knowledge on the molecular network inside the active site. Site-directed mutagenesis is a great tool to redesign the active site topology based on structural data to specifically produce oligosaccharides NiceN with desired characteristics. Based on site-directed mutagenesis studies which have been carried out on glycosyl hydrolases, there are generally four common approaches that have been employed; reduce steric interference, increase active site hydrophobicity, mutation at positive (+n) and negative (–n) subsites, respectively. The significance of each technique is summarized in Table 3.

Table 3: Site-directed mutagenesis strategy for glycosyl hydrolases and its significant

Mutation strategy	Significance
Reduce steric interference.	Improve degree of polymerization, improve acceptor binding.
Increase active site hydrophobicity.	Impair hydrolysis by expelling water molecules.
Mutation at positive (+n) subsite.	Enhance acceptor binding.
Mutation at negative (-n) subsite.	Decrease binding affinity for enhanced acceptor binding.

The size of the active site could determine the degree of polymerization of oligosaccharides produced. Many studies have shown that reducing steric interference in the active site could produce oligosaccharides with a higher degree of polymerization. High degree of polymerization is desired as it cannot be fermented by intestinal enzymes when used as a prebiotic. Engineered cyclodextrin-glucanotransferase by a single mutation at subsite −3 by substitution to a shorter side-chain amino acid (H43T) provided a bigger room for synthesis of γ-cyclodextrin compared to the parent enzyme that mainly produced (β-cyclodextrin) from tapioca starch (Goh et al., 2009). Similar findings were reported when mutation of Trp359 to Phe at subsite +1 reduced the steric interference effect and resulted in better accommodation of longer acceptor molecules for transglycosylation (Abdul Manas et al., 2015). In addition, reduction of steric interference also improved acceptor binding. Enlargement of α-galactosidase active site entrance by F328A mutation improved acceptor binding by having closer contact with the anomeric carbon atom C1 of the donor galactosy. This increased transglycosylation activity 16-fold (Bobrov et al., 2013).

A route to improving transglycosylation activity is, reducing the competing hydrolysis reaction by expelling water molecules from the active site. This can be achieved by increasing the active site hydrophobicity through mutation of hydrophilic amino acids that bind to the water molecule. These amino acids normally reside at subsite +1 (Durand et al., 2016). Among amino acids that strongly bind with water molecules in the active site are Tyr, Trp, and His. Mutation of His497 of α-galactosidase to Met improved the transglycosylation activity as Met did not form a hydrogen bond with water molecule; therefore favouring carbohydrate molecules as acceptors (Hinz et al., 2006). Mutations of Trp359 and Tyr377 of maltogenic amylase also demonstrated a similar effect (Abdul Manas et al., 2015). Water molecules can also be expelled by making the active site entrance of the water pathway more hydrophobic. Such mutations have been successfully performed on neopullulanase (Kuriki et al., 1996).

Active sites of glycosyl hydrolases generally consist of negative and positive subsites separated by catalytic residues. As shown in Figure 2, each

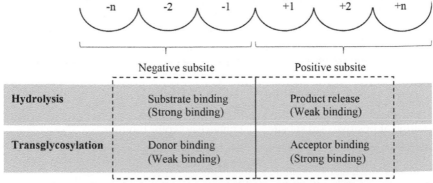

Fig. 2: General active site structure of glycosyl hydrolases with hydrolysis and transglycosylation activities and role of subsites for each reaction.

subsite plays a different role in hydrolysis and transglycosylation. During hydrolysis, negative subsites are regions where the substrate bind themselves and positive subsites are where hydrolysis products are released after catalysis. During transglycosylation, donor molecules bind to the negative subsites and acceptor molecules bind to the positive subsites (Abdul Manas et al., 2018).

This discrepancy in accommodation of molecules in hydrolysis and transglycosylation enables rational design to be performed by manipulating the strength of glycosyl molecule binding in each subsite. Hydrolysis requires strong binding of glycosyl molecule at the negative subsite and weak binding at the positive subsite to ease product release. This is in contrast to transglycosylation, which requires strong binding of glycosyl molecule as acceptor at the positive subsites.

Strong binding affinity of acceptors could be achieved by increasing number of positive subsites with negative free energies and addition of amino acids which bind glycosyl molecules such as aromatic residues. α-amylase exhibited transglycosylation activity when Ala residues were substituted with aromatic Phe and Tyr at positive subsites (Saab-Rincón et al., 1999). Introduction of a new hydrogen bond between mutated Tyr and Gln in β-galactosidase and acceptor molecules at the positive subsite also improved transglycosylation activities (Wu et al., 2013). The hydrolysis activity was reduced when the binding affinity at the negative subsites was weakened. Mutation of amino acid that strongly binds the glycosyl molecule weakened the binding of substrates. This has been demonstrated by mutation of aromatic amino acids at negative subsites of chitinases that improve its transglycosylation activities (Aronson et al., 2006, Taira et al., 2010).

3.4 Enzyme Immobilization

Enzyme immobilization is a popular and successful approach to overcome a low yield of oligosaccharides production. It enables enzymes to be used multiple times and in large-scale industrial processes. In comparison with free enzymes, immobilized enzymes as biocatalysts have several superior qualities such as efficient separation of the biocatalyst, easier recovery of the heterogeneous products and enhanced stability of the biocatalyst under extreme operating or storage conditions (elevated temperature or pressure, alkaline or acidic pH). Immobilized enzymes have high reusability due to their stability and easy recovery, which ensure lower production costs and enable continuous production. Few modes of enzyme immobilization have been developed, for instance, attachment/adsorption, physical entrapment/encapsulation and cross-linking.

Figure 3 shows the illustration of these immobilization modes. These modes are not specific for particular enzymes or products. They vary widely depending on the compositions of enzymes and chemical characteristics, properties of products and substrates and the various uses of the products obtained (Homaei et al., 2013).

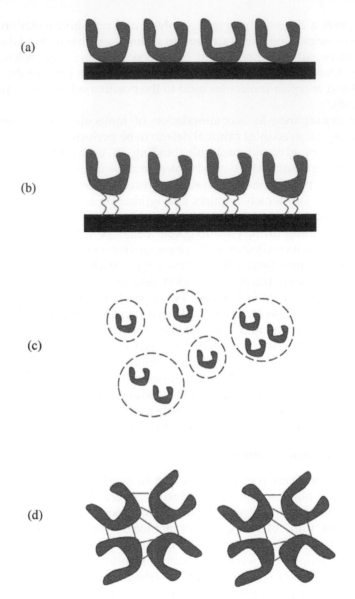

Fig. 3: Schematic diagram of four different modes of enzyme immobilization. (a) Adsorption. (b) Covalent immobilization. (c) Entrapment. (d) Cross-linking.

3.4.1 Enzyme Immobilization Using Support Binding (Adsorption, Covalent Attachment)

Immobilization of enzymes through support binding is the first ever enzyme immobilization method discovered in 1916 and it remains the most commonly used approach. Enzyme immobilization using support binding can be classified into two categories; adsorption and covalent attachment.

Adsorption is the simplest and basic method where the enzyme and suitable insoluble support are mixed together under appropriate condition of ionic strength and pH. The physical interaction between the enzyme and the insoluble support can be driven by hydrogen bonding, hydrophobic effect, electrostatic interaction, or a combination of these. Although this method is simple and relatively inexpensive, it has a major limitation as the adsorbed enzyme can be potentially and easily leached out into the aqueous solvents especially (Hanefeld et al., 2009; Brady and Jordaan, 2009). Covalent binding is similar to adsorption but is better and stronger compared to adsorption. Covalent binding between enzyme and support usually includes the involvement of lysine amino acid of which is frequently present on the surface of the enzyme (Brady and Jordaan, 2009). Enzyme leaching is minimized and low contamination of the products occurs using this immobilization method.

There are varieties of supports that have been used in the production of oligosaccharides depending on the enzyme used and properties of the product. Table 4 summarized various immobilization matrices which have been mentioned in literature. Endo-xylanase has been immobilized through ionic bonding on carboxymethyl-cellulose (CM-cellulose), and through covalent bonding on glyoxyl-agarose and cyanogen bromide-agarose (CNBr-agarose). In comparison of the three supports, glyoxyl-agarose showed the lowest immobilization yield but it is the most stable and able to retain more than 80% of relative activity after 5 cycles of use compared to the CM-cellulose and CNBr-agarose derivatives, which are only able to retain 29.5% and 31.3% of relative activity after the same number of cycles, respectively. By using multiple covalent immobilization on glyoxyl-agarose, the immobilized xylanase successfully produced Xylooligosaccharides with a degree of polymerization of 2 to 5 (Heinen et al., 2017). The use of glyoxyl-agarose has also been reported for immobilization of xylanase from *Aspergillus versicolor*. The immobilization is able to preserve 85% of its catalytic activity, approximately 700-fold stability at 60°C and reusability at least 10 times with full catalytic activity retention. Using this support, 18% of xylobiose to xylohexaose were successfully produced in batch hydrolysis whereby xylobiose was produced 2.5-folds higher than the soluble xylanase (Aragon et al., 2013). Covalent immobilization of xylanase A from *Bacillus subtillis* has also been reported in a separate study using agarose and chitosan activated with glutaraldehyde and glyoxal groups respectively. Similarly, agarose-glyoxal was found to be the best support to achieve 100% immobilization yield and 42.7% of recovered activities. The immobilized xylanase demonstrated high thermal stability, which allowed its use for more than 10 cycles to achieve 24% Xylooligosaccharide conversion with a degree of polymerization between 2 to 4 (Milessi et al., 2016). A study done by Urrutia et al., 2013 also showed that glyoxyl agarose exhibited the highest β-galactosidase immobilization yield and stability compared to amino-glyoxyl agarose, carboxy-glyoxyl agarose and chelate-glyoxyl agarose. After 10 repeated batches with galactooligosaccharide, yield and productivity remain unchanged, and the total product obtained was

Table 4: Immobilization support used for various oligosaccharides-producing enzymes

Enzyme	Supports	Best support	Oligosaccharide produced	Reference
Endo-xylanase from *Aspergillus tamarii*	CM-cellulose, CNBr-agarose, Glyoxyl-agarose.	Glyoxyl-agarose	Xylooligosaccharides with DP 2-5. Highest are X3 (39.7%) and X4 (29.4%)	(Heinen et al., 2017)
XynA from *Bacillus subtilis*	Agarose-glyoxal, chitosan-glyoxal and chitosan-glutaraldehyde.	Agarose-glyoxal	Xylooligosaccharides mainly X2, X3, X4	(Milessi et al., 2016)
Endo-inulase	Chitin, Chitosan, Activated alumina, Polystyrene anionic ion exchange resin (WA30), Polystyrene carrier material (UF93)	Polystyrene carrier material (UF93)	Inulo-oligosaccharide	(Yun et al., 2000)
Mannanase from *P. occitanis*	Chitin-glutaraldehyde	Chitin-glutaraldehyde	Mannooligosaccharide. Mainly mannotriose and mannotetraose	(Blibech et al., 2011)
β-galactosidase from *Kluyveromyces fragilis*	Magnetic poly (GMA-EDGMA-HEMA)* nanobeads	Magnetic poly (GMA-EDGMA-HEMA) nanobeads	Galactooligosaccharide (2240g/g immobilized β-galactosidase)	(Liu et al., 2012)
β-galactosidase from *Kluyveromyces fragilis*	Magnetic poly (GMA-EDGMA-HEMA)* nanospheres grafted with polyethyleneimine	Magnetic poly (glycidyl methacrylate-ethylene glycol dimethacrylate-hydroxyethyl methacrylate) nanospheres grafted with polyethyleneimine	Galactooligosaccharide (4500g/g immobilized β-galactosidase)	(Liu et al., 2012)

Xylanase from *Aspergillus versicolor*	Glyoxyl-agarose	Glyoxyl-agarose	Xylooligosaccharide. X2-X6 with 18% of hydrolysis rate.	(Aragon et al., 2013)
Xylanase from *Talaromyces thermophilus*	DEAE-Sephadex, Amberlite, Duolite, Florisil, Chitin, Gelatin, Polyacrylamide, Chitosan	Gelatin-glutaraldehyde	Xylooligosaccharide. X2 as main product followed by X3	(Maalej-achouri et al., 2009)

*(GMA-EDGMA-HEMA): (glycidyl methacrylate-ethylene glycol dimethacrylate-hydroxyethyl methacrylate)

56.6 g/g of immobilized enzyme. Glyoxyl-agarose is considered a well-known support that has been used numerous times either during support screening or as the most suitable support for certain enzyme (Terrasan et al., 2016; Rodriguez-colinas et al., 2016; Milessi et al., 2016).

Organic and natural polymers have also been used as immobilization matrices. A study by Gaur et al. (2006) used three different approaches of enzyme immobilization for the production of galactooligosaccharides which are in the form of cross-linked enzyme aggregates (CLEA), adsorped on celite and covalently coupled to chitosan. Among these three methods, the chitosan-immobilized β-galactosidase showed greater performance and can be consecutively used for four cycles. Using 20% lactose, chitosan-immobilized β-galactosidase can produce 17.3% trisaccharide within 2 hours of reaction time. The use of chitosan was also discovered for immobilization of *Lactobacillus reuteri* inulosucrase to produce inulin-type Fructooligosaccharides. Among hydrogel chitosan beads, core-shell chitosan beads and dried chitosan beads, immobilized inulosucrase on shell chitosan beads displayed higher capacity and stability compared to other types of chitosan (Charoenwongpaiboon et al., 2018). Cotton cloth was also used as a support to produce galactooligosaccharides from *Aspergillus oryzae* β-galactosidase. Using lactose as substrate, a 27% (w/w) maximum galactooligosaccharides production was achieved at 50% lactose conversion with 500 g/L of initial lactose. Tri-saccharides were the major oligosaccharides formed (Albayrak and Yang, 2002).

Immobilization supports activated with glutaraldehyde have been reported in oligosaccharides production. Glutaraldehyde provides a functional group on the matrix surface in order for the enzyme to form a covalent bond. Inulo-oligosaccharide was produced using immobilization of endo-inulase on polystyrene activated with glutaraldehyde as a support in a reactor. Polystyrene carrier material showed the best operational capacity and immobilization capacity compared to chitin, chitosan, activated alumina and polystyrene anionic ion exchange resin. Using the recommended operating conditions of a reactor, the inulo-oligosaccharide was produced with 82% yield with no significant loss of enzyme activity (Yun et al., 2000). In a production of mannooligosaccharides, mannanase was covalently immobilized on chitin activated with glutaraldehyde. The study reported successful immobilization with 94.81% yield and 72.17% activity recovery. Using locust bean gum as a substrate, the immobilized mannanase successfully produced mannotriose and mannotetraose as the main products (Blibech et al., 2011).

Nanomaterial is another emerging matrix for enzyme immobilization. A high production of galactooligosaccharide has been achieved using immobilized *Kluyveromyces fragilis* β-galactosidase on magnetic poly (GMA-EDGMA-HEMA) nanobeads. These novel nanobeads were prepared from a combination of glycidyl methacrylate (GMA), ethylene glycol dimethacrylate (EDGMA) and hydroxyethyl methacrylate (HEMA) using emulsifier free emulsion polymerization. With 145.6 mg/g of enzyme attached and 72.6%

of activity recovery, this support was improved by 2.6-fold compared to commercial Eupergit-C. Thus, it can produce a total of 2,240 g of galactooligosaccharide per gram of immobilized β-galactosidase after 10 batch reactions and is able to retain 81.5% of its original activity (Liu et al., 2012). The polyethyleneimine-grafted magnetic poly (GMA-EDGMA-HEMA) on the other hand, displayed 3.2-fold adsorption capacity of β-galactosidase in comparison with commercial ion exchange resin DEAE-Sepharose and could adsorb 86.7 mg/g of enzyme with 92.5% of recovered enzyme activity. By recycling the immobilized β-galactosidase for 15 consecutive batch reactions, 4.5 kg of galactooligosaccharide can be produced per gram of adsorbed enzyme using lactose as a substrate with 84.6% retention of its initial activity (Liu et al., 2012).

Immobilization on magnetic materials offers high and easy recovery for re-utilization of the enzyme. Researchers have explored this important material for enzyme immobilization in production of oligosaccharides. For example, epoxy-coated magnetic beads were used as a support for purification and immobilization of β-galactosidase from *Thermotoga maritima*. The enzyme was attached through multiple covalent linkages which involved different subunits. As determined by NMR, galactooligosaccharide was produced with β-3′-galactosyl lactose as the main product (Marín-Navarro et al., 2014). In another study, chitosan-coated magnetic Fe_3O_4 nanoparticles were utilized for the production of fructo-oligosacharides by immobilization with β-fructofuranosidase. The immobilized enzyme can be reused for up to 10 days in batchwise conditions, and the maximum fructo-oligosacharides yield is 59.5% (w/w) from 50% (w/v) sucrose solution where 1-kestose and nystose are the main oligosaccharides produced (Chen et al., 2013).

Immobilized enzymes are not only used for soluble substrates but can also be used with insoluble substrates. A study done by Maalej-achouri et al., 2009 showed that the hydrolysis of several agro-industrial residues (ground wheat, corn and barley, wheat kernel, rabbit food, rice straw and wheat bran) for the produced Xylooligosaccharides using immobilized xylanase. Xylanase was covalently immobilized on several supports. The most suitable and efficient support was achieved by gelatin and glutaraldehyde as the cross-linkers with 98.8% and 99.2% of immobilization yield and xylanase activity recovery, respectively. The hydrolysis of the agro-industrial residues showed that, xylobiose was the main product followed by xylotriose.

3.4.2 Enzyme Immobilization Using Entrapment/Encapsulation

Entrapment or encapsulation is an irreversible enzyme immobilization technique, in contrast to covalent immobilization. The enzyme of interest is restricted in a confined network or space, for instance in porous gels, beads and fibers (Sardar and Ahmad, 2015; Mohamad et al., 2015). Chitosan and alginate beads are examples of commonly used materials for this immobilization technique. Both materials have been compared for entrapment of fructosyltransferase from *Aspergillus flavus* for continuous and efficient production of fructooligosaccharide. The performances of chitosan and alginate beads were determined based on transfructosylating activity and production of fructooligosaccharide after multiple uses. Based on the results it was concluded that, immobilization of fructosyltransferase in alginate beads produced higher transfructosylating yield compared to chitosan beads. Up to 7 days, fructooligosaccharide formation from alginate beads was consistent with an average yield of 65.37% (w/w) without much loss in fructosyltransferase activity. In constrast, fructooligosaccharide formation from chitosan beads was observed below 20% with continuous loss until the 15th cycle (Ganaie et al., 2014)

Although calcium alginate is well-known as a good entrapment matrix for enzyme immobilization, the gel is brittle and easily ruptured during the enzymatic catalysis process. The gel hardness could be enhanced by incorporating other substances into the gel. Rajagopalan and Krishnan (2008) reported that maltooligosaccharides were produced by immobilized α-amylase from *Bacillus subtilis* KCC103 in calcium alginate beads. With 3% (w/v) of alginate and 4% (w/v) of $CaCl_2$, the immobilized α-amylase can be repeatedly used up to 10 times with only 25% loss of efficiency. Interestingly, the reusability of the immobilized α-amylase is doubled when 1% silica gel is added to alginate prior to polymerization. The addition of silica gel increases the gel hardness to make it more robust and relatively hard for the enzyme and product to be leached out. It was found that immobilization profoundly affected the degree of polymerization of maltooligosaccharide produced. Maltooligosaccharide with a degree of polymerization from 1 to 4 was mainly produced from soluble starch by the free α-amylase. However, when the immobilized α-amylase was used to hydrolyze insoluble starch, the product distribution became narrow. On the other hand production of glucose was high in comparison to the free α-amylase approach.

Calcium alginate beads strengthened with silica were used for production of Xylooligosaccharides. Endoxylanase from solventogenic *Clostridium* strain BOH3 was immobilized in 4% (w/v) of calcium alginate and 3.5% (w/v) of silica gel to increase the gel hardness. Before addition of silica, enzyme tended to leach out after few cycles. Similar to what Rajagopalan and Krishnan (2008) achieved, adding silica gel, enhanced the hardness of calcium alginate beads and reduced the leaching of enzyme. Formation of Xylooligosaccharides from hydrolysis of beechwood and teakwood xylan as substrates was analyzed. Interestingly, only xylobiose (51.79 ± 1.5%), xylotriose (46.12 ± 0.65%) and a very small amount of xylose (<1%) were detected after 24 hours of hydrolysis (Rajagopalan et al., 2016).

A combination of enzyme and cell immobilization was successfully employed in the production of galactooligosaccharide. β- Galactosidase and yeast *Kluyveromyces marrianus* were immobilized in Lentikats® lens-shaped polyvinylalcohol (PVA) capsules. The galactooligosaccharide was produced in two stages. First, the immobilized β-galactosidase produced a low-content mixture of galactooligosaccharide having a concentration of 71.7 g/L and a final purity of 22.7%. Subsequently, the low-content galactooligosaccharide was used in 20 repeated batch runs with immobilized yeast to increase the galactooligosaccharide content. Although immobilized β-galactosidase produces a low-content of galactooligosaccharide, it can be used repeatedly, thus reducing production costs (Tokošová and Rosenberg, 2016).

3.4.3 Enzyme Immobilization Using Cross-Linking

The adsorption technique of immobilization creates a strong binding of the enzyme by virtue of multiple covalent bonds, which reduce the conformational flexibility of the enzyme. Eventually the attached enzyme usually has low catalytic activity (Hanefeld et al., 2009). On the other hand, the entrapment technique is accompanied with enzyme leakage when it is used multiple times. The use of entrapped enzyme narrowed the product distribution was also documented (Rajagopalan and Krishnan, 2008). The presence of a support matrix in the vicinity of the enzyme active site might hinder the entrance of substrates, especially when insoluble ones are used. Furthermore, the increase of enzyme rigidity caused by the formation of covalent bonds between the enzyme and the support matrix might limit substrate and product distribution. So, it is not effective if oligosaccharides with a high degree of polymerization are desired or insoluble substrates are to be used.

Support-free is becoming a popular method of enzyme immobilization to overcome these limitations. The two most widely used are cross-linked enzyme crystals (CLECs) and cross-linked enzyme aggregates (CLEAs) (Sheldon, 2011). These two methods differ in their preparation methods. The CLECs are prepared by crystallization of pure enzymes while CLEAs involve enzyme aggregation prior to cross-linking (Garcia-Galan et al., 2011). An inherent drawback of CLECs is the need of crystallization, which is usually a tedious and a high-cost procedure requiring a high purity enzyme. This is the reason for, very limited research on CLEC technology (Cui and Jia, 2015). In contrast, CLEAs technology involves easy procedures compatible with most proteins since the target enzyme can be precipitated under appropriate conditions, specific to it. Other than that, CLECs technology could hardly be used to obtain co-immobilized enzymes due to the difficulty of co-crystallizing pure enzymes. However, CLEAs enable the co-aggregation of several enzymes to have combi CLEAs that might be beneficial to catalyse cascade processes (Stressler et al., 2015).

CLEAs have many environmental and economic benefits for the industrial processes. Differing from other immobilization methods, 'carrier-free' enzymes do not have any special needs for binding to supports, and this circumvents the costs of expensive carriers (Sheldon, 2007). CLEAs can be easily prepared from crude enzyme extracts, directly taken from the fermentation broth without much purification as the methodology essentially combines two unit processes, purification and immobilization, into a single operation (Sheldon, 2007). CLEAs are proved to be significantly more stable to denaturation by heat, organic solvents and proteolysis than the corresponding soluble enzyme (Cui and Jia, 2015, Yang et al., 2012, Sheldon, 2011). The CLEA methodology offers a low cost, robust

Table 5: Enzyme immobilization techniques used in oligosaccharides production

Technique	Mechanism	Advantage	Disadvantage
Adsorption	Hydrogen bonding, hydrophobic effect, electrostatic interaction, ionic interaction	Well-established method Simple Inexpensive	Enzyme leaching especially in aqueous solvent Changes of pH or ionic strength affects enzyme binding
Covalent attachment	Covalent binding (Lysine and surface-functionalized matrix)	Strong binding Minimize enzyme leaching Low contamination of enzyme in the products	Multiple covalent bonds reduce enzyme conformational flexibility Reduced enzyme activity Limit substrate and product distribution
Entrapment/encapsulation	Enzyme is restricted in a confined network or space	Well-documented method Inexpensive Environmental friendly (natural polymer can be used)	Brittle beads and easily rupture cause enzyme leaching Diffusion limitation Low product distribution Not suitable for insoluble substrate
Cross-linked enzyme aggregates (CLEAs)	Enzyme precipitation using cross-linker	Maintain enzyme conformational flexibility Suitable for production of oligosaccharides with high degree of polymerization are Suitable for insoluble substrate Compatible with most protein Applicable for co-immobilization of two or more enzymes Circumvent the costs of expensive carriers	A new technique Aggregation might cause enzyme to lose activity

and reusable biocatalyst that exhibits a highly concentrated enzyme in the catalyst, high activity retention, enhanced operational stability and better tolerance to non-aqueous solvents. Moreover, the technique is applicable for the co-immobilization of two or more enzymes (combi-CLEAs) which can be advantageously used in catalytic cascade processes (Talekar et al., 2013).

Oligosaccharide production processes using CLEAs have been documented in this literature. A novel β-*galactosidase* (Bgal1-3) isolated from marine metagenomic library was immobilized using the CLEAs approach. The immobilized Bgal1-3-CLEAs showed the same optimum pH and temperature, pH 7 and 50°C, respectively, with the free enzyme. However, the immobilized Bgal1-3-CLEAs showed higher pH stability and improvement in thermostability where it was able to retain and constantly maintained its activity at more than 90% at 40 °C for 120 hours, whereas the free enzyme was only able to retain approximately 60% of its activity under the same conditions. The ability of Bgal1-3-CLEAs to produce galacto-oligosacharides using lactose as substrate was assessed in an organic-aqueous biphasic system. By comparing the galacto-oligosacharide contents produced by free Bgal1-13 and Bgal1-3-CLEAs, similar patterns and compositions were analyzed with galacto-oligosacharide syrups synthesized by both free Bgal1-3 and Bgal1-3-CLEAs containing at least 7 different oligosaccharides with degrees of polymerization ranging between 3 and 9. From the analysis, galacto-oligosacharide yield of Bgal1-3-CLEAs (59.4 ± 1.5%) was slightly higher than that of free Bgal1-3 (57.1 ± 1.7%) (Li et al., 2015).

New advancements in CLEA technology have taken place with the use of magnetic particles. This novel approach was discovered by Talekdar et al., 2012 by incorporating a magnetic particle that contained ferrous sulphate and ferric chloride into the mixture. The separation of CLEAs with magnetic particles could from the reaction mixture is easily achievable. This separation approach is less tedious and easily reaches 100% separation compared to other techniques using filters or centrifuges (Purohit et al., 2017). Two years after its discovery, a study by Bhattacharya and Pletschke (2014) introduced a similar concept, but this time the magnetic CLEAs particle was used for lignocellulolytic enzyme, xylanase, used to hydrolyze sugarcane bagasse. The magnetic-xylanase-CLEAs particle produced was highly effective in hydrolyzing the sugarcane bagasse with 7.4- and 9.0-fold sugar release from pre-treated lime and sugar bagasse pre-treated with NH-OH, respectively. Another study of the implementation of magnetic CLEAs particles was done by Purohit et al. (2017), where the prepared magnetic-xylanase-CLEAs was able to produce Xylooligosaccharide from two different insoluble substrates; rice straw and corn cob. After 1 hour of hydrolysis, the magnetic-xylanase-CLEA was able to produce 143 mg/g of rice straw and 152 mg/g of corn cob with a predominant amount of xylopentose and xylohexose. In comparison, the free enzyme was only able to produce 85.8 mg of Xylooligosaccharide/g of rice straw after 1 hour of hydrolysis and 126.7 mg of Xylooligosaccharide/g of corn cob after 2 hours of hydrolysis.

CLEAs technique discards the use of support material for immobilization. This technique could maintain enzyme flexibility to allow access to a wide range of substrates. Hence, CLEAs can be used for both soluble and insoluble substrates. The absence of a support carrier also reduces substrate diffusion limitations, thus enhancing enzyme catalytic efficiency. Table 5 summarizes all enzyme immobilization techniques available for oligosaccharides production that have been discussed in this chapter. Each technique has its own advantages and disadvantages and choosing the best technique would depend on the process employed, characteristics of the enzyme, substrate and product desired.

4. Conclusion

Research on enzymatic production of oligosaccharides is continuously growing with many techniques being explored. This chapter has discussed approaches including manipulation of reaction conditions, the use of enzyme cocktails, direct fermentation, protein engineering, and immobilization technology. Each technique has its own advantages and disadvantages, thus finding the right technique totally depends on the characteristics of the enzyme, substrate, product and conditions of the process. A wide range of methods presented in this chapter give an overview on all available approaches, hence helping researchers to choose their production strategy.

References

Abdul Manas, N.H., S. Pachelles, N.M. Mahadi and R.M. Illias. 2014. The characterisation of an alkali-stable maltogenic amylase from *Bacillus lehensis* G1 and improved malto-oligosaccharide production by hydrolysis suppression. PLoS One 9(9): e106481.

Abdul Manas, N.H., M.A. Jonet, A.M. Abdul Murad, N.M. Mahadi and R.M. Illias. 2015. Modulation of transglycosylation and improved malto-oligosaccharide synthesis by protein engineering of maltogenic amylase from *Bacillus lehensis* G1. Process Biochem. 50(10): 1572-1580.

Abdul Manas, N.H., R.M. Illias and N.M. Mahadi. 2018. Strategy in manipulating transglycosylation activity of glycosyl hydrolase for oligosaccharide production. Crit. Rev. Biotechnol. 38(2): 272-293.

Adebola, O.O., O. Corcoran and W.A. Morgan. 2014. Synbiotics: The impact of potential prebiotics inulin, lactulose and lactobionic acid on the survival and growth of lactobacilli probiotics. J. Funct. Foods. 10: 75-84.

Albayrak, N. and S.-T. Yang. 2002. Production of galactooligosaccharides from lactose by *Aspergillus oryzae* β-galactosidase immobilized on cotton cloth. Biotechnol. Bioeng. 77: 8-19.

Amorim, C., S.C. Silvério and L.R. Rodrigues. 2019. One-step process for producing prebiotic arabino-xylooligosaccharides from brewer's spent grain employing *Trichoderma* species. Food Chem. 270: 86-94.

Arab-Jaziri, F., B. Bissaro, C. Tellier, M. Dion, R. Faure, M.J. O'Donohue et al. 2015. Enhancing the chemoenzymatic synthesis of arabinosylated Xylooligosaccharides by GH51 α-L-arabinofuranosidase. Carbohydr. Res. 401: 64-72.

Aragon, C.C., A.F. Santos, A.I. Ruiz-Matute, N. Corzo, J.M. Guisan, R. Monti et al. 2013. Continuous production of xylooligosaccharides in a packed bed reactor with immobilized – Stabilized biocatalysts of xylanase from *Aspergillus versicolor*. J. Mol. Catal. B Enzym. 98: 8-14.

Arnling Bååth, J., A. Martínez-Abad, J. Berglund, J. Larsbrink, F. Vilaplana, L. Olsson et al. 2018. Mannanase hydrolysis of spruce galactoglucomannan focusing on the influence of acetylation on enzymatic mannan degradation. Biotechnol. Biofuels. 11: 114.

Aronson, N.N. Jr., B.A. Halloran, M.F. Alexeyev, X.E. Zhou, Y. Wang, E.J. Meehan et al. 2006. Mutation of a conserved tryptophan in the chitin-binding cleft of *Serratia marcescens* chitinase a enhances transglycosylation. Biosci. Biotechnol. Biochem. 70: 243-251.

Bhattacharya, A. and B.I. Pletschke. 2014. Magnetic cross-linked enzyme aggregates (CLEAs): A novel concept towards carrier free immobilization of lignocellulolytic enzymes. Enz. Microb Technol. 61-62: 17-27.

Blibech, M., F. Chaari, F. Bhiri, I. Dammak, R.E. Ghorbel, S.E. Chaabouni et al. 2011. Production of mannooligosaccharides from locust bean gum using immobilized *Penicillium occitanis* mannanase. J. Mol. Catal. B Enzym. 73(1-4): 111-115.

Bobrov, K.S., A.S. Borisova, E.V. Eneyskaya, D.R. Inaven, K.A. Shabalin, A.A. Kulminskaya et al. 2013. Improvement of the efficiency of transglycosylation catalyzed by α-galactosidase from *Thermotoga maritima* by protein engineering. Biochem. (Moscow). 78(10): 1112-1123.

Brady, D. and J. Jordaan. 2009. Advances in enzyme immobilisation. Biotechnol. Lett. 31(11): 1639-1650.

Cardelle-Cobas, A., C. Martínez-Villaluenga, M. Villamiel, A. Olano and N. Corzo. 2008. Synthesis of oligosaccharides derived from lactulose and Pectinex Ultra SP-l. J. Agric. Food Chem. 56(9): 3328-3333.

Chang, S., J. Chu, Y. Guo, H. Li, B. Wu , B. He et al. 2017. An efficient production of high-pure xylooligosaccharides from corncob with affinity adsorption-enzymatic reaction integrated approach. Bioresour. Technol. 241: 1043-1049.

Charoenwongpaiboon, T., K. Wangpaiboon, R. Pichyangkura and M.H. Prousoontorn. 2018. Highly porous core-shell chitosan beads with superb immobilization efficiency for *Lactobacillus reuteri* 121 inulosucrase and production of inulin-type fructooligosaccharides. RSC Adv. 80(3): 17008-17016.

Chen, C.W., C.-C. Ou-Yang and C.-W. Yeh. 2003. Synthesis of galactooligosaccharides and transgalactosylation modeling in reverse micelles. Enz. Microb. Technol. 33(4): 497-507.

Chen, S.-C., D.-C. Sheu and K.-J. Duan. 2014. Production of fructooligosaccharides using β-fructofuranosidase immobilized onto chitosan-coated magnetic nanoparticles. J. Taiwan Inst. Chem. Eng. 45: 1105-1110.

Cui, J.D. and S.R. Jia. 2015. Optimization protocols and improved strategies of cross-linked enzyme aggregates technology: Current development and future challenges. Crit. Rev. Biotechnol. 8551(1): 15-28.

Díez-Municio, M., B. de las Rivas, M.L. Jimeno, R. Muñoz, F.J. Moreno, M. Herrero et al. 2013. Enzymatic synthesis and characterization of fructooligosaccharides and novel maltosylfructosides by inulosucrase from *Lactobacillus gasseri* DSM 20604. Appl. Environ. Microbiol. 79(13): 4129-4140.

Durand, J., X. Biarnés, L. Watterlot, C. Bonzom, V. Borsenberger, A. Planas et al. 2016. A single point mutation alters the transglycosylation/hydrolysis partition, significantly enhancing the synthetic capability of an endo-glycoceramidase. ACS Catal. 6(12): 8264-8275.

Feng, H.-Y., J. Drone, L. Hoffmann, V. Tran, C. Tellier, C. Rabiller et al. 2005. Converting a β-glycosidase into a β-transglycosidase by directed evolution. J. Biol. Chem. 280(44): 37088-37097.

Fernández-Arrojo, L., D. Marín, A. Gómez De Segura, D. Linde, M. Alcalde, P. Gutiérrez-Alonso et al. 2007. Transformation of maltose into prebiotic isomaltooligosaccharides by a novel α-glucosidase from *Xantophyllomyces dendrorhous*. Process Biochem. 42(11): 1530-1536.

Ganaie, M.A., H.K. Rawat, O.A. Wani, U.S. Gupta and N. Kango. 2014. Immobilization of fructosyltransferase by chitosan and alginate for efficient production of fructooligosaccharides. Process Biochem. 49: 840-844.

Garcia-Galan, C., Á. Berenguer-Murcia, R. Fernandez-Lafuente and R.C. Rodrigues. 2011. Potential of different enzyme immobilization strategies to improve enzyme performance. Adv Synth. Catal. 353(16): 2885-2904.

Gaur, R., H. Pant, R. Jain and S.K. Khare. 2006. Galactooligosaccharide synthesis by immobilized *Aspergillus oryzae* β-galactosidase. Food Chem. 97: 426-430.

Gibson, G.R., R. Hutkins, M.E. Sanders, S.L. Prescott, R.A. Reimer, S.J. Salminen et al. 2017. Expert consensus document: The International Scientific Association for Probiotics and Prebiotics (ISAPP) consensus statement on the definition and scope of prebiotics. Nat. Rev. Gastroenterol. Hepatol. 14(8): 491-502.

Goh, K.M., N.M. Mahadi, O. Hassan, R.N.Z. Raja Abdul Rahman, R.M. Illias et al. 2009. A predominant β-CGTase G1 engineered to elucidate the relationship between protein structure and product specificity. J. Mol. Catal. B Enz. 57(1): 270-277.

González-Delgado, I., M.-J. López-Muñoz, G. Morales and Y. Segura. 2016. Optimisation of the synthesis of high galactooligosaccharides (GOS) from lactose with β-galactosidase from *Kluyveromyces lactis*. Int. Dairy J. 61: 211-219.

Hanefeld, U., Gardossi, L. and Magner, E. 2009. Understanding enzyme immobilisation. Chem. Soc. Rev. 38(2): 453-468.

Hansson, T., T. Kaper, J. van der Oost, W.M. de Vos and P. Adlercreutz. 2001. Improved oligosaccharide synthesis by protein engineering of β-glucosidase CelB from hyperthermophilic *Pyrococcus furiosus*. Biotechnol. Bioeng. 73(3): 203-210.

Heinen, P.R., M.G. Pereira, C.G.V. Rechia, P.Z. Almeida, L.M.O. Monteiro, T.M. Pasin et al. 2017. Immobilized endo-xylanase of *Aspergillus tamarii* Kita: An interesting biological tool for production of xylooligosaccharides at high temperatures. Process Biochem. 53: 145-152.

Hidaka, H., M. Hirayama and N. Sumi. 1988. A fructooligosaccharide-producing enzyme from *Aspergillus niger* ATCC 20611. Agric. Biol. Chem. 52(5): 1181-1187.

Hinz, S.W.A., C.H.L. Doeswijk-Voragen, R. Schipperus, L.A.M. van den Broek, J.-P. Vincken, A.G.J. Voragen et al. 2006. Increasing the transglycosylation activity of α-galactosidase from *Bifidobacterium adolescentis* DSM 20083 by site-directed mutagenesis. Biotechnol. Bioeng. 93(1): 122-131.

Homaei, A.A., R. Sariri, F. Vianello and R. Stevanato. 2013. Enzyme immobilization: An update. J. Chem. Biol. 6(4): 185-205.

Jeffery, C.J. 2004. Molecular mechanisms for multitasking: Recent crystal structures of moonlighting proteins. Curr. Opin. Struc. Biol. 14(6): 663-668.

Kerff, F., A. Amoroso, R. Herman, E. Sauvage, S. Petrella, P. Filée et al. 2008. Crystal structure and activity of *Bacillus subtilis* YoaJ (EXLX1), a bacterial expansin that promotes root colonization. Proceedings of the National Academy of Sciences of the United States of America 105(44): 16876-16881.

Khandekar, D.C., A. Palai, A. Agarwal and P.K. Bhattacharya. 2014. Kinetics of sucrose conversion to Fructooligosaccharides using enzyme (invertase) under free condition. Bioproc. Biosyst. Eng. 37: 2529-2537.

Kim, M.-H., M.-J. In, H.J. Cha and Y.J. Yoo. 1996. An empirical rate equation for the fructooligosaccharide-producing reaction catalyzed by β-fructofuranosidase. J. Ferment. Bioeng. 82(5): 458-463.

Kuriki, T., M. Yanase, H. Takata, Y. Takesada, T. Imanaka S. Okada et al. 1993. A new way of producing isomaltooligosaccharide syrup by using the transglycosylation reaction of neopullulanase. Appl. Environ. Microbiol. 59(4): 953-959.

Kuriki, T., H. Kaneko, M. Yanase, H. Takata, J. Shimada, S. Handa et al. 1996. Controlling substrate preference and transglycosylation activity of neopullulanase by manipulating steric constraint and hydrophobicity in active center. J. Biol. Chem. 271(29): 17321-17329.

Lee, H.-S., J.-H. Auh, H.-G. Yoon, M.-J. Kim, J.-H. Park, S.-S. Hong et al. 2002. Cooperative action of α-glucanotransferase and maltogenic amylase for an improved process of isomaltooligosaccharide (IMO) production. J. Agric. Food Chem. 50(10): 2812-2817.

Li, M., S. Seo and S. Karboune. 2015. *Bacillus amyloliquefaciens* levansucrase-catalyzed the synthesis of fructooligosaccharides, oligolevan and levan in maple syrup-based reaction systems. Carbohydr. Polym. 133: 203-212.

Li, L., G. Li, L.-C. Cao, G.-H. Ren, W. Kong, S.-D. Wang et al. 2015. Characterization of the cross-linked enzyme aggregates of a novel β-galactosidase, a potential catalyst for the synthesis of galactooligosaccharides. J. Agric. Food. Chem. 63 (3): 894-901.

Liu, H., J. Liu and B. Tan. 2012. Covalent immobilization of *Kluyveromyces fragilis* β-galactosidase on magnetic nanosized epoxy support for synthesis of galactooligosaccharide. Bioproc. Biosyst. Eng. 35(8): 1287-1295.

Liu, J.-F., H. Liu, B. Tan, Y.-H. Chen, R.-J. Yang et al. 2012. Reversible immobilization of *K. fragilis* β-galactosidase onto magnetic polyethylenimine-grafted nanospheres for synthesis of galactooligosaccharide. J. Mol. Catal. B Enz. 82: 64-70.

Lorenzoni, A.S.G., L.F. Aydos, M.P. Klein, R.C. Rodrigues and P.F. Hertz. 2014. Fructooligosaccharides synthesis by highly stable immobilized β-fructofuranosidase from *Aspergillus aculeatus*. Carbohydr. Polym. 103: 193-197.

Lundemo, P., E.N. Karlsson and P. Adlercreutz. 2016. Eliminating hydrolytic activity without affecting the transglycosylation of a GH1 β-glucosidase. Appl. Microbiol. Biotechnol. 101(3): 1121-1131.

Maalej-achouri, I., M. Guerfali, A. Gargouri and H. Belghith. 2009. Production of Xylooligosaccharides from agro-industrial residues using immobilized *Talaromyces thermophilus* xylanase. J. Mol. Catal. B Enz. 59: 145-152.

Marín-Navarro, J., D. Talens-Perales, A. Oude-Vrielink, F.J. Cañada and J. Polaina. 2014. Immobilization of thermostable β-galactosidase on epoxy support and its use for lactose hydrolysis and galactooligosaccharides biosynthesis. World J. Microbiol. Biotechnol. 30(3): 989-998.

Manas, N.H.A., F.D.A. Bakar and R.M. Illias. 2016. Computational docking, molecular dynamics simulation and subsite structure analysis of a maltogenic amylase from *Bacillus lehensis* G1 provide insights into substrate and product specificity. J. Mol. Graph. Model. 67: 1-13.

Milessi, T.S.S., W. Kopp, M.J. Rojas, A. Manrich, A. Baptista-Neto, P.W. Tardiol et al. 2016. Immobilization and stabilization of an endoxylanase from *Bacillus subtilis* (XynA) for xylooligosaccharides (XOs) production. Catal. Today 259: 130-139.

Mohamad, N.R., N.H. Che Marzuki, N.A. Buang, F. Huyop and R. Abdul Wahab. 2015. An overview of technologies for immobilization of enzymes and surface analysis techniques for immobilized enzymes. Biotechnol. Biotechnol. Equip. 29(2): 205-220.

Morris, C. and G.A. Morris. 2012. The effect of inulin and fructooligosaccharide supplementation on the textural, rheological and sensory properties of bread and their role in weight management: A review. Food Chem. 133(2): 237-248.

Mueller, I., G. Kiedorf, E. Runne, A. Seidel-Morgenstern and C. Hamel. 2018. Synthesis, kinetic analysis and modelling of galactooligosaccharides formation. Chem. Eng. Res. Des. 130: 154-166.

Mussatto, S.I. and I.M. Mancilha. 2007. Non-digestible oligosaccharides: A review. Carbohydr. Polym. 68(3): 587-597.

Nakashima, K., K. Endo, N. Shibasaki-Kitakawaa and T. Yonemoto. 2014. A fusion enzyme consisting of bacterial expansin and endoglucanase for the degradation of highly crystalline cellulose. RSC Adv. 4(83): 43815-43820.

Nascimento, A.K.C., C. Nobre, M.T.H. Cavalcanti, J.A. Teixeira and A.L.F. Portoa. 2016. Screening of fungi from the genus *Penicillium* for production of β-fructofuranosidase and enzymatic synthesis of fructooligosaccharides. J. Mol. Catal. B Enz. 134: 70-78.

Patel, S. and A. Goyal. 2011. Functional oligosaccharides: Production, properties and applications. World J. Microbiol. Biotechnol. 27: 1119-1128.

Pennec, A., R. Daniellou, P. Loyer, C. Nugier-Chauvin and V. Ferrières. 2015. Araf51 with improved transglycosylation activities: One engineered biocatalyst for one specific acceptor. Carbohydr Res. 402: 50-55.

Prapulla, S.G., V. Subhaprada and N.G. Karanth. 2000. Microbial production of oligosaccharides: A review. Adv. Appl. Microbiol. 47: 299-343.

Purohit, A., S.K. Rai, M. Chownk, R.S. Sangwan and S.K. Yadav. 2017. Xylanase from *Acinetobacter pittii* MASK 25 and developed magnetic cross-linked xylanase aggregate produce predominantly xylopentose and xylohexose from agro biomass. Bioresour. Technol. 244: 793-799.

Qiang, X., Y.L. Chao and Q.B. Wan. 2009. Health benefit application of functional oligosaccharides. Carbohydr. Polym. 77: 35-44.

Rabah, H., F.L. Rosa do Carmo and G. Jan. 2017. Dairy Propionibacteria: Versatile Probiotics. Microorganisms. 5(2). pii: E24.

Rajagopalan, G. and C. Krishnan. 2008. Immobilization of malto-oligosaccharide forming α-amylase from *Bacillus subtilis* KCC103: Properties and application. J. Chem. Technol. Biotechnol. 83: 1511-1517.

Rajagopalan, G., K. Shanmugavelu and K.-L. Yang. 2016. Production of xylooligosaccharides from hardwood xylan by using immobilized endoxylanase of *Clostridium* strain BOH3. RSC Adv. 6(85): 81818-81825.

Rastall, R.A. 2010. Functional oligosaccharides: Application and manufacture. Ann. Rev. Food Sci. Technol. 1(1): 305-339.

Rodriguez-colinas, B., L. Fernandez-Arrojo, P. Santos-Moriano, A.O. Ballesteros and F.J. Plou. 2016. Continuous packed bed reactor with immobilized β-galactosidase for production of galactooligosaccharides (GOS). Catalysts. 6(12): 189.

Saab-Rincón, G., G. del-Río, R.I. Santamaría, A. López-Munguía and X. Soberón. 1999. Introducing transglycosylation activity in a liquefying α-amylase. FEBS Lett. 453(1-2): 100-106.

Sabater, C., A. Fara, J. Palacios, N. Corzo, T. Requena, A. Montilla et al. 2019. Synthesis of prebiotic galactooligosaccharides from lactose and lactulose by dairy propionibacteria. Food Microbiol. 77: 93-105.

Sardar, M. and R. Ahmad. 2015. Enzyme immobilization: An overview on nanoparticles as immobilization matrix. Biochem. Anal. Biochem. 4: 178.

Sheldon, R.A. 2007. Cross-linked enzyme aggregates (CLEA®s): Stable and recyclable biocatalysts. Biochem. Soc. Trans. 35(6): 1583-1587.

Sheldon, R.A. 2011. Characteristic features and biotechnological applications of cross-linked enzyme aggregates (CLEAs). Appl. Microbiol. Biotechnol. 92(3): 467-477.

Singh, R.S., R.P. Singh and J.F. Kennedy. 2016. Recent insights in enzymatic synthesis of fructooligosaccharides from inulin. Int. J. Biol. Macromol. 85: 565-572.

Singh, S., A. Ghosh and A. Goyal. 2018. Mannooligosaccharides as prebiotic-valued products from agro-waste. pp. 205-221. *In*: S.J. Varjani, B. Parameswaran, S. Kumar and S.K. Khare (eds.). Biosynthetic Technology and Environmental Challenges. Singapore: Springer.

Song, J.Y., M. Alnaeeli and J.K. Park. 2014. Efficient digestion of chitosan using chitosanase immobilized on silica-gel for the production of multisize chitooligosaccharides. Process Biochem. 49(12): 2107-2113.

Stressler, T., J. Ewert, T. Eisele and L. Fischer. 2015. Cross-linked enzyme aggregates (CLEAs) of PepX and PepN – Production, partial characterization and application of combi-CLEAs for milk protein hydrolysis. Biocatal. Agric. Biotechnol. 4(4): 752-760.

Taira, T., M. Fujiwara, N. Dennhart, H. Hayashi, S. Onaga, T. Ohnuma et al. 2010. Transglycosylation reaction catalyzed by a class V chitinase from cycad, *Cycas revoluta*: A study involving site-directed mutagenesis, HPLC, and real-time ESI-MS. Biochim. Biophys. Acta. 1804(4): 668-675.

Talekar, S., V. Ghodake, T. Ghotage, P. Rathod, P. Deshmukh, S. Nadar et al. 2012. Novel magnetic cross-linked enzyme aggregates (magnetic CLEAs) of alpha amylase. Bioresour. Technol. 123: 542-547.

Talekar, S., S. Desai, M. Pillai, N. Nagavekar, S. Ambarkar, S. Surnis et al. 2013. Carrier free co-immobilization of glucoamylase and pullulanase as combi-cross linked enzyme aggregates (combi-CLEAs). RSC Adv. 3(7): 2265-2271.

Terrasan, C.R.F., C.C. Aragon, D.C. Masui, B.C. Pessela, G. Fernandez-Lorente, E.C. Carmona et al. 2016. β-xylosidase from *Selenomonas ruminantium*: Immobilization, stabilization, and application for xylooligosaccharide hydrolysis. Biocatal. Biotransfor. 34(4): 161-171.

Terrasan, C.R.F., W.G. de Morais Junior and F. Contesini. 2019. Enzyme immobilization for oligosaccharide production. pp. 415-423. *In*: L. Melton, F. Shahidi and P. Varelis (eds.). Encyclopedia of Food Chemistry. Oxford: Academic Press.

Thompson, J.D., D.G. Higgins and T.J. Gibson. 1994. CLUSTAL W: Improving the sensitivity of progressive multiple sequence alignment through sequence weighting, position-specific gap penalties and weight matrix choice. Nucleic Acids Res. 22(22): 4673-4680.

Tokošová, S., H. Hronská and M. Rosenberg. 2016. Production of high-content galactooligosaccharides mixture using β-galactosidase and *Kluyveromyces marxianus* entrapped in polyvinylalcohol gel. Chem. Pap. 70: 1445.

Tuohy, K.M., G.C.M. Rouzaud, W.M. Brück and G.R. Gibson. 2005. Modulation of the human gut microflora towards improved health using prebiotics – Assessment of efficacy. Curr. Pharm. Des. 11(1): 75-90.

Urrutia, P., C. Mateo, J.M. Guisan, L. Wilson and A. Illanes. 2013. Immobilization of *Bacillus circulans* B-galactosidase and its application in the synthesis of galactooligosaccharides under repeated-batch operation. Biochem. Eng J. 77: 41-48.

Vaikuntapu, P.R., M.K. Mallakuntla, S.N. Das, B. Bhuvanachandra, B. Ramakrishna, S.R. Nadendla et al. 2018. Applicability of endochitinase of *Flavobacterium johnsoniae* with transglycosylation activity in generating long-chain chitooligosaccharides. Int. J. Biol. Macromol. 117: 62-71.

Varki, A. 1993. Biological roles of oligosaccharides: All of the theories are correct. Glycobiol. 3(2): 97-130.

Wan Azelee, N.I., J.M. Jahim, A.F. Ismail, S.F.Z.M. Fuzi, R.A. Rahman et al. 2016. High xylooligosaccharides (XOS) production from pretreated kenaf stem by enzyme mixture hydrolysis. Ind. Crop Prod. 81: 11-19.

Wang, L.-X. and W. Huang 2009. Enzymatic transglycosylation for glycoconjugate synthesis. Curr. Opin. Chem. Biol. 13(5-6): 592-600.

Wu, Y., S. Yuan, S. Chen, D. Wu, J. Chen, J. Wu et al. 2013. Enhancing the production of galactooligosaccharides by mutagenesis of *Sulfolobus solfataricus* β-galactosidase. Food Chem. 138(2-3): 1588-1595.

Yang, X., P. Zheng, Y. Ni and Z. Sun. 2012. Highly efficient biosynthesis of sucrose-6-acetate with cross-linked aggregates of Lipozyme TL 100 L. J. Biotechnol. 161(1): 27-33.

Yun, J.W. 1996. Fructooligosaccharides – Occurrence, preparation, and application. Enz. Microb. Technol. 19(2): 107-117.

Yun, J.W., J.P. Park, C.H. Song, C.Y. Lee, J.H. Kim, S.K. Song et al. 2000. Continuous production of inulo-oligosaccharides from chicory juice by immobilized endoinulinase. Bioproc. Eng. 22(3): 189-194.

Zang, H., S. Xie, H. Wu, W. Wang, X. Shao, L. Wu et al. 2015. A novel thermostable GH5_7 β-mannanase from *Bacillus pumilus* GBSW19 and its application in mannooligosaccharides (MOS) production. Enz. Microb. Technol. 78: 1-9.

9

Encapsulation Process for Probiotic Survival during Processing

Shahrulzaman Shaharuddin, Nurzyyati Mohd Noor, Ida Idayu Muhamad*, Nor Diana Hassan, Yanti Maslina Mohd Jusoh and Norhayati Páe

School of Chemical and Energy Engineering, Faculty of Engineering, Universiti Teknologi Malaysia, Skudai, Johor Bahru, Johor, Malaysia

1. Introduction

Initially, 'probiotic' was related to the word 'antibiotic' which was defined as 'for life' (Hamilton-Miller et al., 2003). According to the Food and Agriculture Organization (FAO) of the United Nations and the World Health Organization (WHO), probiotic means live microorganisms which are capable of providing health benefits to hosts when administrated in sufficient amounts (FAO/WHO 2002). Another common description of probiotic is microorganisms which maintain and improve the balance of human or animal intestinal microbial activities which give beneficial effects to the host (Goldin, 1998). Many advantageous modes of action were identified. Table 1 shows the modes of action in ruminants for probiotics in ruminant rumen and post-rumen GIT.

Various adverse environmental conditions which might affect probiotic viability and survivability could be faced by applying the concept of providing physical barriers to the probiotic (Kailasapathy, 2009). Many methods have been applied to ensure their survivability and viability which includes selection of resistant strains, stress adaptation, incorporation of micronutrients and microencapsulation (Rokka and Rantamaki, 2010). Methods such as immobilization are a common technique used in order to protect probiotics in adverse environments.

2. Probiotic

Probiotics are live microbial feed supplements that may contribute beneficial effects to the host by assisting in digestion of food, producing micronutrients

*Corresponding author: idaidayu@utm.my

Table 1: Modes of action for probiotics in ruminant rumen and post-rumen GIT
(modified from Seo et al. 2010)

Location	Modes of action	References
Rumen	**Lactic acid producing bacteria** 1. Provision of a constant lactic acid supply. 2. Adaptation of overall microflora to the lactic acid accumulation. 3. Stimulation of lactate utilizing bacteria. 4. Stabilization of ruminal pH.	Abe et al., 1995, Abu-Tarboush et al., 1996, Seo et al., 2010
	Lactic acid utilizing bacteria 1. Conversion of lactate to volatile fatty acid (VFA). 2. Production of propionic acid rather than lactic acid. 3. Increase of feed efficiency. 4. Decrease of methane production. 5. Increase of ruminal pH.	Weiss et al., 2008
	Other bacteria (*Bacillus* spp.) 1. Increase acetate and total VFA concentration. 2. Enhance humoral immunity. 3. Inhibit gastrointestinal infection by pathogens or producing antimicrobial.	Roos et al., 2010
	Fungal DFM 1. Reduction of oxygen in the rumen. 2. Prevention of excess lactic acid in the rumen. 3. Provision of growth factors such as organic acid and vitamin B. 4. Increase of rumen microbial activity and numbers. 5. Improvement of ruminal end products (e.g., VFA, rumen microbial protein). 6. Increase of ruminal digestibility.	Roger et al., 1990, Lynch and Martin, 2002
Post-rumen GIT	1. Production of antibacterial compounds (acids, bacteriocins, antibiotics). 2. Competition with pathogens for colonization of mucosa and/or for nutrients. 3. Production and/or stimulation of enzymes. 4. Stimulation of immune response by host. 5. Metabolism and detoxification of toxic compounds.	Matsuguchi et al., 2003, Cotter et al., 2005, Dicks and Botes, 2010

and modulating the immune system (Mayer et al., 2015). Therapeutic benefits have led to an increase in the incorporation of probiotic bacteria such as *lactibaciili* and *bifidobacteria* in dairy products especially yoghurts. The efficiency of added probiotic bacteria depends on dose level and their viability must be maintained throughout storage and shelf-life of products and they must survive the gut environment (Kailasapathy, 2006). They are also defined as living microorganisms which resist gastric juices, bile and pancreatic secretions, attach to epithelial cells and colonize the human intestine (Del Piano et al., 2006). There are foods that contain live bacteria which are beneficial to health (Zubillaga et al., 2001).

The production of ruminant populations for human consumption especially goats and cows could be increased by creating better animal feed by adding value adding additives to it. Other materials or substances which could improve the value of the feed are beneficial microorganisms or probiotics such as lactic acid bacteria, non-lactic acid bacteria and yeast (Burgain et al., 2011). Probiotics have been incorporated into animal feed for better growth and health. Many beneficial effects or actions of probiotics can be offered for intestinal microbial balance (Vasiljevic and Shah, 2008) and host health directly (Burgain et al., 2011). The most common probiotic that has been utilised for many years in animal feed processing is lactic acid bacteria (LAB) (Vasiljevic and Shah, 2008, Burgain et al., 2011). *Lactobacillus* species are categorized under the large group of LAB. This genus is a Gram-positive organism and generates lactic acid during fermentation (Giraffa et al., 2010) and their growth is significantly affected by temperature and pH (Kok et al., 2012).

3. Prebiotic

A prebiotic is defined as 'a non-digestible food ingredient that beneficially affects the host by selectively stimulating the growth and/or activity of one or a limited number of bacteria in the colon that can improve health (Gibson and Roberfroid, 1995, Stanton et al., 2005).

Hi-maize, derived from high amylose maize (corn) is also a prebiotic that improves the survival of some probiotic organisms through the digestive process. As a prebiotic, Hi-maize is also used as a fuel for native bacteria in the colon. Hi-maize in the presence of the probiotic organism *Clostridium butyricum*, was found to significantly decrease the number of crypt foci, an indicator for colon cancer, better than *C. butyricum* in the absence of the prebiotic (Nakanishi et al., 2003).

Prebiotics are classified as functional food ingredients under the categories of inulin-type fructans and synthetic fructooligosaccharides. Inulin and their synthetic derivatives, the fructooligosaccharides (FOS), are predominantly obtained from the chicory plant (*Cichorium intybus*). Inulin has been defined as a polydisperse carbohydrate material consisting mainly, if not exclusively, of b-(2-1) fructosyl-fructose links. Inulin producing plant species are found in several monocotyledonous and dicotyledonous families, including Liliaceae,

Amaryllidaceae, Gramineae, and Compositae. However, only one inulin-containing plant species (chicory, *Cichorium intybus*) is used to produce inulin industrially. In chicory inulin, both Gpy-Fn (a-D-glucopyranosyl-[b-D-fructofuranosyl]n21 b-D-fructofuranoside) compounds are considered to be included under the same nomenclature, and the number of fructose units varies from 2 to >70 (Roberfroid, 2000). Inulin is a natural functional food ingredient commonly found in varying percentages in dietary foods. Inulin is a term applied to a heterogeneous blend of fructose molecules polymer which has a range of chains from 2 to 60 units long and serves in nature as an energy-reserve carbohydrate stored in plants and may act as an osmoprotectant (Tungland, 2000). Several commercial grades of inulin have a neutral, clean flavor and are used to improve the mouthfeel, stability and acceptability of low fat foods.

Inulin is less soluble than oligofructose and has the ability to form microcrystals when sheared in water or milk. There are many types of plants containing inulin and oligofructose. Inulin is a prebiotic which supports the growth of friendly bacteria in the human gut thus improving the health of the host. Normally, inulin is used as an extra supplement in dairy products such as milk, yogurt and creamers in the dairy food industry (Mirdawati, 2006). Inulin is a complex carbohydrate and a highly effective prebiotic, which can stimulate the growth of beneficial probiotic bacteria in the gut. Inulin passes through the deodenum and stomach undigested and is readily available to the gut bacterial flora (Roberfroid, 2000). Inulin is a term applied to a prebiotic indigestible carbohydrate that reaches the colon unchanged and aids many of human body's key metabolic functions. It is a natural functional food ingredient. Inulin falls under the general class of carbohydrates called fructans which contain fructose polymers. It is produced in over 36,000 plants worldwide naturally (Tungland, 2000).

4. Immobilization

Probiotic immobilization has been frequently applied in the food industry as the demand for functional foods has increased dramatically (Blaiotta et al., 2013, Martins et al., 2013). Plenty of basic immobilization techniques have been used in the previous studies as illustrated in Figure 1 (Kourkoutas et al., 2004). Basically, there are four main techniques namely attachment, entrapment, aggregation and encapsulation.

Most of the functional foods involve fermentation and consist of probiotics (Blaiotta et al., 2013). In the past, fermented dairy products were most commonly used as carriers for probiotics (Martins et al., 2013). In addition, other food matrices were also used in delivering the probiotic such as infant formulas, confectionaries, sweetened milk and fruit drinks (Patel et al., 2004). For probiotic carriers, having suitable characteristics such as physicochemical aspects, buffering capacity and pH were important in order to ensure the survival of the probiotic (Blaiotta et al., 2013).

Other alternative carriers were also needed to be examined in order to solve the problem of dairy products such as lactose intolerance, dyslipidemia, and vegetarianism such as vegetables and fruits (Peres et al., 2012). Various matrices of probiotic immobilization were reported in previous studies such as minimally processed fruit, fermented vegetables, fruit juices, fruit smoothies, snack products, cereal beverages and olives (Martins et al., 2013).

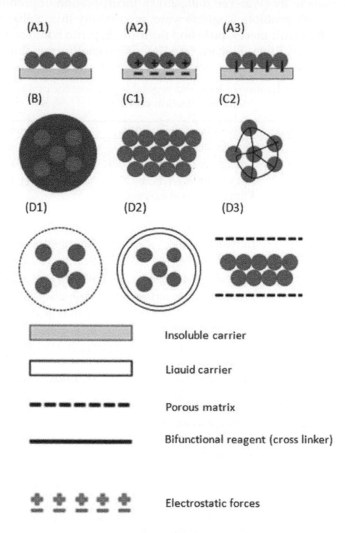

Fig. 1: Basic immobilization techniques: A) Immobilization on the surface of a solid carrier; 1) Adsorption on a surface. 2) Electrostatic binding on a surface. 3) Covalent binding on a surface. B) Entrapment within a porous matrix. C) Cell flocculation (aggregation): 1) Natural flocculation. 2) Artificial flocculation (cross-linking). D) Mechanical containment behind a barrier; 1) Microencapsulation. 2) Interfacial microencapsulation. 3) Containment between microporous membranes. (Adapted from Kourkoutas et al., 2004).

The most applied strains of probiotics in using immobilization for vegetables are *Lactobacillus acidophilus, Lactobacillus casei, Lactobacillus plantarum, Lactobacillus rhamnosus* and *Bifidobacterium lactis* (Martins et al., 2013). This technique has been widely used in enhancing and maintaining viability of probiotics during product processing and storage. Larger surface area to volume ratio may benefit in terms of heat protection by reducing the thermal conductivity (Wai-Yee *et al.*, 2011). Incorporation of prebiotic fibers as nutrients with probiotic yoghurt were successfully immobilized on fruit matrices such as fruit pieces, pulp and flour (do Espirito Santo et al., 2011).

As shown in Table 2, many studies of probiotic immobilization on solid carriers have been previously performed using various biofibers

Table 2: Utilization of fruit and vegetable matrices for probiotic immobilization carriers

Immobilization matrix	Probiotic microorganism	Reference
Apple	*Lactobacillus casei*	Kourkoutas et al., 2005
	Lactobacillus plantarum	Corbo et al., 2013
	Saccharomyces cerevisiae, Kluyveromyces marxianus, Lactobacillus casei	Kourkoutas et al., 2006
	Lactobacillus rhamnosus	Roble et al., 2010
	Lactobacillus casei	Kourkoutas et al., 2006
Quince	*Lactobacillus casei*	Kourkoutas et al., 2005
Wheat	*Lactobacillus casei*	Bosnea et al., 2009
Cereal	*Lactobacillus plantarum*	Michida et al., 2006
	Lactobacillus rhamnosus	Saarela et al., 2006
	Lactobacillus acidophilus Bifidobacterium animalis Lactobacillus rhamnosus	Helland et al., 2004
Bacterial cellulose	*Lactobacillus bulgaricus, Lactobacillus plantarum, Lactobacillus delbrueckii, Lactobacillus acidophilus* and *Lactobacillus casei*	Jagannath et al., 2010
Watermelon rind	*Saccharomyces cerevisiae*	Reddy et al., 2008
Pear	*Lactobacillus casei*	Kourkoutas et al., 2006
Cashew	*Lactobacillus rhamnosus, Lactobacillus casei, Lactobacillus paracasei, Streptococcus thermophilus, Streptococcus macedonicus*	Blaiotta et al., 2013
Cocoa	*Lactobacillus helveticus* and *Bifidobacterium longum*	Possemiers et al., 2010
Soybean	*Lactobacillus acidophilus*	Chen and Mustapha, 2012
Yam	*Lactobacillus acidophilus*	Lee et al., 2011
Oat bran	*Lactobacillus casei*	Guergoletto et al., 2010

(Kourkoutas et al., 2005, Kourkoutas et al., 2006, Michida et al., 2006, Reddy et al., 2008, Bosnea et al., 2009, Jagannath et al., 2010, Corbo et al., 2013). Most of the biofibers used were harvested from plants. They have a multi-phase system which consists of intricate internal cells, intercellular spaces, pores and capillaries. Plant cell walls are an excellent immobilization medium for microorganisms due to their small pore sizes (0.1 to 5.0 µm) which allows bacterial access when contact occurs (Alzamora et al., 2005). Entrapment and attachment of microorganisms at plant trichomes may occur due to the presence of lenticels and tissue lesions (Martins et al., 2013). Several studies on the immobilization of *Lactobacillus casei* have been previously performed using apple pieces (Kourkoutas et al., 2003). It corroborated with a study conducted by Kourkoutas et al. (2005) that demonstrated the usefulness of supporting the survivability of immobilized cells. Most of the carriers contain non-digestible components that may become prebiotic for probiotic growth on the substrate. Nutrients may be released through the holes formed in plant tissues and supply the needed nutrients for microbial growth. Thus, the areas of a plant which consist of pores, holes and irregular surfaces are the preferred intact areas for microorganisms (Martins et al., 2013). For instance, cereal have been considered as good substrates for probiotic immobilization due to their many apertures (Michida et al., 2006).

Residues from commodity crops that consist of fiber are cheap, abundant and are a type of renewable resource (Costa et al., 2013). Tonnes of sugarcane bagasse was produced annually, and this subsequently caused an increase in residue production (Mulinari et al., 2009). Reuse of residues promotes an environmentally friendly ecosystem. These residues can be used as probiotic carriers through the immobilization process. In addition, processed vegetables and fruits can be considered as ideal matrices that may provide substrates such as minerals, sugars, vitamins and other nutrients for probiotics (Soccol et al., 2010).

4.1 Heat Protection via Immobilization

The immobilization technique promotes the concept of a physical barrier that offers better protection of probiotics and safe delivery to targeted areas (Burgain et al., 2011). A number of delivery pathways may cause the probiotics to be exposed to high heat before reaching the targeted area. According to Nagar et al. (2012), about 8% of the activities of immobilized xylanase were preserved after exposure at 75 °C for 2 hours. However, they also found that almost all the activities of free xylanase (non-immobilized) were lost. The resistance of xylanase towards high pH and temperature is due to its mechanical attributes after the immobilization process. Kourkoutas et al. (2005) elaborated on the effects of various temperatures (30 °C, 37 °C and 45 °C) on immobilized biocatalysts using *Lactobacillus casei* where the biocatalysts' activities were maintained and no loss of activities was reported.

According to Quan et al. (2014), a glycopolymer may be used as an immobilization medium for an enzyme (catalase) in conditions of heat exposure. After 5 minutes of heat treatment (60 °C), the covalently immobilized catalase retained approximately 60% of its catalase activity. However, free catalase was found to be inactive after the same treatment. Kannoujia et al. (2009) performed enzyme immobilization using an activated polystyrene microtiter plate via a pressure technique. Enhancements of the thermal and storage stabilities were demonstrated in immobilized enzymes activity compared to the free enzymes. The enzymes activity remained at 64% and 33% after 30 minutes of heat treatment at 60 °C and 90 °C, respectively.

5. Encapsulation

Microencapsulation is a suitable method which is substantially applicable in the packaging and sealing of solid, liquid or gaseous materials, and their contents can be released under specific conditions (Anal and Stevens, 2005, Kailasapathy and Masondole, 2005, Anal et al. 2006). Microencapsulation technique promote the concept of a physical barrier that offers better protection to the probiotic and its safe delivery to the target (Burgain et al., 2011), thus effectively protecting the bacterial cell (Borgogna et al., 2010).

Various products have been integrated with probiotics including food (Tamine et al., 2005, Anal and Singh, 2007) and feed (Gaggia et al., 2010). Nevertheless, there is still the main concern of low viability of free probiotics in products which consequently reduces the health benefits they provide (Burgain et al., 2011). Microencapsulation techniques promote the concept of a physical barrier that offers better protection to the probiotic and safely delivers them to the target area (Burgain et al., 2011). Characteristics of the microcapsule obtained can vary from a thickness of a few microns to 1 mm, in a spherical, thin and strong walled shape (Anal and Singh, 2007). The performance of an encapsulated cell in a porous encapsulant is dependant on the porosity, pore size distribution, mechanical properties and encapsulant stability (Wong and Abdul-aziz, 2008).

Generally, the main objective of the microencapsulation of probiotics is to enhance their survivability and maintain viability to the maximum log CFU corresponding to at least the minimum concentration needed for delivering their beneficial effects to the host (Anal and Singh, 2007, Rokka and Rantamaki, 2010). Probiotics can resist adverse and stressful conditions such as acidity, oxygen and gastric environments using encapsulated walls (Kailasapathy, 2009, Borgogna et al., 2010, Rokka and Rantamaki, 2010, Burgain et al., 2011). In addition, it may also create a control for the release of active ingredients in the wall as it is dependent upon certain conditions thus enabling it to reach the targeted delivery area (Anal and Singh, 2007, Ding and Shah 2007).

5.1 Methodology of Encapsulation

According to Burgain et al. (2011), there are three phases involved in the encapsulation of probiotics. Based on their study, Burgain et al. (2011)

demonstrated that the process starts with the incorporation of bioactive components (core) in a matrix; solid or liquid. The incorporation of a solid core was conducted through agglomeration or adsorption while a liquid core form was incorporated through dissolution or dispersion. The procedure was continued to the second phase which involved the dispersion of liquid matrix while the solid matrix was pulverized. In the final phase, the microcapsules were stabilized (Poncelet and Dreffier, 2007). The flowchart of the whole process is shown in Figure 2.

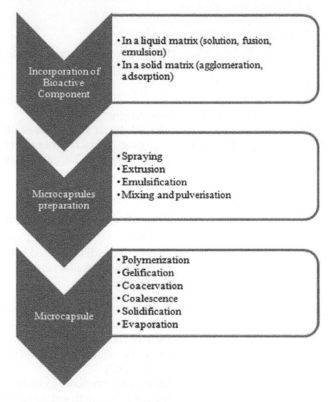

Fig. 2: Flow-chart of bioactive microencapsulation (Adapted from Burgain et al., 2011)

5.2 Encapsulants, Stabilizer and Protectors

There are various encapsulants for probiotic microencapsulation. Examples of common agents include *kappa*-carrageenan, alginate, starch and its derivatives, gum Arabic, gellan gum, xanthan gum, and milk as well as gelatin which is obtained from seaweed, other plants, bacteria and animal proteins, respectively (Rokka and Rantamaki, 2010). The types, contents, advantages and disadvantages of several examples of encapsulants for probiotic microencapsulation are demonstrated in Table 3.

Table 3: Type, composition and advantage of encapsulation agent for probiotic microencapsulation

Type	Composition	Advantage
Alginate	β-D-mannuronic and α-L-guluronic acids.	Simple, non-toxicity, biocompatibility and low cost (Krasaekoopt et al., 2003).
Gellan gum and xanthan gum	Repeating unit of four monomers that are glucose, glucuronic acid, glucose and rhamnose (Chen and Chen, 2007).	High resistance towards acid conditions (Sun and Griffiths, 2000).
Kappa-Carrageenan	Repeating D-galactose-4-sulphate units and 3,6-anhydro-D-galactose.	Keeps bacteria in a viable state (Dinakar and Mistry, 1994).
Cellulose acetate phthalate	Half of the hydroxyls are esterified with acetyls, a quarter are esterified with one or two carboxyls of a phthalic acid (Olaru and Olaru, 2005).	Safe nature Not soluble at acidic (pH less than 5) but it is soluble at pH higher than 6. Good protection for microorganisms in simulated gastro-intestinal conditions (Fávaro-Trindade and Grosso, 2002).
Chitosan	Glucosamine units.	Preferably use as a coat (Mortazavian et al. 2008).
Starch	Amylase and amylopectin (Sajilata et al., 2006).	Good enteric delivery and better release characteristic (Sajilata et al., 2006, Anal and Singh, 2007). Resistant starch can be use by probiotic bacteria in the large intestine (Mortazavian et al., 2008). Ideal surface for the adherence of the probiotic cells to the starch (Anal and Singh, 2007). Maintaining probiotic viability and metabolically active state (Crittenden et al., 2001).
Gelatin	Mixture of peptides and proteins.	Amphoteric nature. Strong interaction with the negatively charged such as gellan gum if pH below the isoelectric point (Krasaekoopt et al., 2003, Anal and Singh, 2007). High viscosity in water.
Milk proteins	Consists of casein and whey.	Natural vehicles as well as delivery system and biocompatibility (Livney, 2010).
Gum Arabic	Simple sugars and 2% protein.	Ability to act as an emulsifier (Vega and Roos, 2006).

6. Application of Microencapsulated Probiotics

6.1 Application in High Temperature Processing

In farming and agroindustry, feeding pellets for ruminants have become a demand of the current age. In the process of manufacturing probiotic feed, formulation, processing and subsequently production steps are involved. Each step may cause stress to the probiotic cell and the standard method in evaluating the feed involves assessing the number of viable cells that remain after production. In practice, pelleted animal feed has been used in feeding which involves a high degree of thermal treatment. This is to ensure that the pellets are robust enough to not only withstand the pelleting process but also its storage conditions (Seo et al., 2010). However, this condition has resulted in a low viability of probiotic after the pelleting process due to its heat-sensitive characteristics. Seo et al. (2010) explained in their study, the importance of viability of maintenance of probiotics in order to deliver the full benefits to the host's health and growth. Thus, thermotolerance of a probiotic is crucial for the pelleting process. This challenge created an opportunity for the development of a thermal protection technique which may reduce the effect of exposure to high temperatures. Therefore, a high quality encapsulated probiotic with beneficial additives for pelleting is necessary to enhance ruminant pellet value. In addition, other active materials need the thermotolerance characteristic for their pertinent processes. For instance, enzymes or biocatalysts were vulnerable to heat exposure (Kourkoutas et al., 2005, Kannoujia et al., 2009, Nagar et al., 2012, Quan et al., 2014). This exposure could have reduced the activity of active material. Many studies have been performed to protect the cells from heat using the microencapsulation technique (Table 4). Most of the protected cells were from the group of *Bifidobacterium* and *Lactobacillus*.

The effects of gelified bacteria and their density was studied on sub-lethal injuries through mild heating of *Escherichia coli, Salmonella typhimurium* and *Listeria innocua* (Noriega et al., 2013). Their findings stated that gelified bacteria suffered less sub-lethal injuries of viable colonies compared to non-gelified bacteria, though it was also stated that effects of density were bacteria dependent (Noriega et al., 2013). Karakus and Pekyardimci (2009) reported that heat protection on pectinesterase was enhanced through encapsulation using porous glass beads. This study indicated that the heat stability pattern showed increase tendency of enzymes to stabilize. In addition, 50% of encapsulated pectinesterase activity was retained after 30 days and it demonstrated longer storage stability as compared to control.

An optimized combination of coating materials was generated using 2% sodium alginate mixed with 1% gellan gum after evaluation using the response surface method and the sequential quadratic programming technique (Chen et al., 2007). This combination produced the best protectant for *Bifidobacterium bifidus* towards heat treatment (75 °C, 1 min). Encapsulation

Table 4: Study on microencapsulation in heat protectant of probiotic

Material	Encapsulant	References
Escherichia coli, Salmonella typhimurium and *Listeria innocua*	Alginate.	Noriega et al., 2013
Pectinesterase	Porous glass beads for enzyme immobilization, membranes cut off 12,000.	Karkus and Peyardimci, 2009
Bifidobacterium bifidum	Gellan gum and sodium alginate.	Chen et al., 2007
Lactobacillus reuteri	Aluminum carboxymethyl cellulose-rice bran (AlCMC–RB) composite.	Chitprasert et al., 2012
Lactobacillus rhamnosus, Bifidobacterium longum, L. salivarius, L. plantarum, L. acidophilus, L. paracasei, L. lactis	Sodium alginate.	Ding and Shah, 2007
Lactobacillus casei	Sodium alginate.	Mandal et al., 2006
Lactobacillus acidophilus	Sodium alginate and starch.	Sabikhi et al., 2010
Bifidobacterium breve, Bifidobacterium longum, Lactobacillus acidophilus	Freeze-dried bacteria.	Picot and Lacroix, 2003

using two walled materials (sodium alginate and gellan gum) enhanced the heat resistance of *Bifidobacterium bifidum* (Chen et al., 2007).

Chitprasert et al. (2012) studied the effects of aluminum carboxymethyl cellulose-rice bran (AlCMC–RB) composites at weight ratios of 1:0, 1:1 and 1:1.5. They found that both free cells and microencapsulated cells were almost fully destroyed after heat treatment. However, the amount of microencapsulated cells that survived was significantly higher as compared to the free cells. Thus, composites of AlCMC–RB showed high potential as wall materials in protection against heat (Chitprasert et al., 2012). Alginate matrics were discovered to be a suitable medium for enhancement of bacteria viability against acid, bile and heat exposures (Ding and Shah, 2007). In the study, they conducted the heat resistance test at 65 °C for 1 hour. They found that the viability for both free and microencapsulated probiotics was reduced to a large extent due to heat exposure. However, the viability of the free probiotics was much lesser compared to that of microencapsulated ones (Ding and Shah, 2007).

Another study concerning the effects of different conditions on encapsulated *Lactobacillus casei* NCDC-298 was performed using different alginate concentrations (2%, 3% or 4%), a pH as low as 1.5, high bile salt concentration (1% or 2%) and heat processing at temperatures of 55 °C, 60 °C or 65 °C for a duration of 20 minutes (Mandal et al., 2006). Encapsulated probiotics demonstrated higher survivability than those of free cells after being subjected to low pH conditions, high bile salt concentration and heat treatment. The increase of the alginate concentrations caused the enhancement of probiotic viability and similar release in colonic solution was obtained in all alginate concentrations used.

Sabhiki et al. (2010) reported the tolerance of probiotic *Lactobacillus acidophilus* LA1 against selected processing and simulated gastrointestinal conditions in their study. They used sodium alginate and starch as microencapsulation wall materials which enhanced the probiotic survivability after undergoing heat treatments at 72 °C, 85 °C and 90 °C (and exposure to high salt concentrations of 1%, 1.5% and 2%). Both free and microencapsulated probiotics suffered loss of viability, however, it was found that the microencapsulated probiotics may have reduced effects of heat exposure as some of them survived while none of the free probiotics survived at 90 °C. After high salt treatment of 2% (w/v), the probiotic viability was reduced to 5.67 and 2.30 for free and microencapsulated probiotics, respectively. Sodium alginate showed tremendous potential as an encapsulant for probiotics to give better survivability in extreme processing conditions and under simulated gastrointestinal environments.

A study using a spiral jet mill as a grinding system in producing multiphase low-diameter microcapsules of probiotics by emulsification and spray drying techniques was conducted by Picot and Lacroix (2003). They used a probiotic model of three strains of probiotic freeze-dried bacteria which were *Bifidobacterium breve* R070 (BB R070), *Bifidobacterium longum* R023 (BL R023) and *Lactobacillus acidophilus* R335 (LA R335). This study managed to reduce the microcapsules size which then affected the viability of the probiotic after the microencapsulation process. The best condition for preparing the powders of probiotic freeze-dried cultures was obtained using grind air pressure of 4 bar with a product feed rate of 300 g h/1. A thermo-tolerance test was performed using three different temperatures (45 °C, 65 °C and 80 °C) and only two micronized bifidobacteria strains were used. Protection of *Bifidobacterium breve* R070 from the thermal exposure significantly increased with particle-size reduction. Meanwhile, *Bifidobacterium longum* R023 showed intrinsic low heat resistance. They concluded that the micronization technique could be considered as a better method in producing small particle sizes of probiotic cultures and subsequently providing sufficient protection during heat exposure.

6.1.1 Heat Transfer and Microencapsulation

Heat transfer was defined as the transfer of thermal energy that was absorbed or liberated in a sensible heating method (Nolan, 2012). Heat can be transferred from one body to another via three mechanisms which are conduction, convection and radiation. Conduction was referred as the heat that was transferred after the initial molecular action without involving the motion of the medium. This mechanism occurs in all physical states solid, liquid or gas. Most of the conduction occurs in the solid state. Alternatively, the heat transferred by the actual motion of the medium is known as convection. Convection usually occurs in liquid or gas based media. Last off all, radiation is defined as the transfer of heat induced by electromagnetic waves.

The main factor in determining the heat transfer rate is the characteristics of the boundary layer which allow the heat transfer (Hetsroni et al., 2001). Another factor that may affect the heat and mass transfer of particles is the neighbouring surface surrounding them (Hetsroni et al., 2001). Presence of thermal insulators may interrupt the flow of heat transfer. The importance of thermal insulation is to protect a material from heat flow and providing a resistance path to it with an insulating material (Gertrude, 2011). The insulating material was the crucial point in decreasing the rate of heat transfer and several other factors to be considered were selection of insulator material and its thickness and the exposure of insulated surfaces to ambient conditions.

A study of the microwave effect toward metal-alginate beads demonstrated encapsulation potential in protecting minerals during heat treatment (Campanone et al., 2014). They evaluated the efficiency of four different wall materials; zinc sulphate heptahydrate, ferrous sulphate heptahydrate, calcium chloride and sodium alginate. After microwave treatment, calcium alginate beads showed lower temperature compared to other beads. This was due to the decrease in heating rate caused by the microstructure of the beads containing calcium.

A study using sodium alginate (NaA), poly-acrylic acid (AA) and glutaraldehyde (GA) in producing a network film was performed to evaluate its heat transfer behaviour (Bekin et al., 2014). The conductivity of the SA-AA-GA film was reduced with the incorporation of AA in SA and GA as crosslinker. Increased AA concentration may decrease the potential in an electric field. The concentration of 15 M NaCl solution was the best solution in obtaining the maximum total bending of SA-AA-GA film.

6.1.2 *Microencapsulation of Probiotics using Sugarcane Bagasse-Aliginate Composite as Encapsulant Agent*

Fibrous materials from agricultural waste such as sugarcane bagasse are a potential candidates for various applications due to their valuable characteristics especially in the immobilization and microencapsulation processes (Shaharuddin et al., 2014). They also show promising potential in heat resistance system applications. The advantages of probiotics as additives in animal feed were well documented in previous reports and studies. Inclusion of some probiotic species in the pelleting process has led to a negative effect as it was found to be non-viable in the final pellet. This issue could be solved through the application of heat protectant via immobilization and microencapsulation. Based on their thermotolerance, immobilization and microencapsulation techniques have been frequently performed in order to preserve the probiotic from detrimental heat effects. Sugarcane bagasse was chosen as the probiotic carrier in the immobilization step. This step was to be performed by the adsorption method according to the plan. The selected methodology and wall material for the microencapsulation process were extrusion and alginate, respectively. Both methodologies were selected based on their simple and low-cost procedures. Also, the series model of effective thermal conductivity was selected for model fitting evaluation (Bujard et al., 1994). An assumption of this model is unidirectional heat transfer. The effective thermal conductivity values are limited in the range of 0-5 W/m °C. The selection of this model is based on several reasons; a) both studies use heterogenous or composite materials as samples, b) data from this study is compatible with this particular model compared to other models, c) both studies have similar conditions and assumptions of the direction of heat flow. These reviews could be beneficial as references and could lead to the development of a novel thermotolerance microcapsule in this study.

As reported by Shaharuddin (2015), sugarcane bagasse (SB) is an excellent biomaterial for immobilizing *Lactobacillus rhamnosus* NRRL 442 and very supporting for a high level of cell viability retention. It also acts as a good filler in alginate microcapsules and has the potential to improve the survivability of *L. rhamnosus* during the microencapsulation process. The presence of immobilized probiotics changes the structure of sugarcane bagasse. Immobilization using sugarcane bagasse successfully preserves high cell viability after immobilization. Hence, this process help to retain probiotics during manufacturing and storage of the ruminant feed. The immobilization of model probiotic, *L. rhamnosus* on SB successfully enhances the survivability of the probiotic during heat exposure by a combination of immobilization and microencapsulation, with NaA as the encapsulation agent. Overall, increasing the ratio of NaA:SB significantly increases the cell survivability after microencapsulation and heat exposure (Shaharuddin and Muhamad, 2015). The immobilization step using adsorption before the microencapsulation process significantly increases the cell survivability after microencapsulation and heat exposure. This beneficial step contributes to the aim of achieving high efficiency even with the usage of NaA concentration as low as 1%. As a conclusion, the immobilized probiotic is well microencapsulated in an alginate-bagasse composite microcapsule.

Double heat protection via microencapsulation using sugarcane bagasse-alginate of *L. rhamnosus* significantly increases the cell survivability after heat exposure. The maximum survivability was demonstrated by microcapsules prepared by using a higher alginate concentration and a higher filler (Sugarcane Bagasse) ratio. It proved that the combination of immobilization via adsorption and microencapsulation produced greater protection of the probiotic from the heat. During heat exposure, heat loss occurred due to heat transferred to the multilayers of alginate and Sugarcane Bagasse. Heat was transferred from outside the microcapsule towards the centre of the medium by conduction. The heat loss occurred due to the heat that was transferred to the layer of alginate which was supported with the low thermal conductivity of alginate. Subsequently, it conductively transferred within the area of Sugarcane Bagasse which also has low thermal conductivity. Heat movement during the heat exposure was illustrated and the mechanism of heat resistance towards the microencapsulation using sugarcane bagasse-alginate of *L. rhamnosus* was demonstrated. These conclusions demonstrated great potential in the synthesis of heat resistant alginate-bagasse microcapsules for probiotics and for inclusion as probiotic additives. This study could be useful in the production of pelleted feed or other products that necessitate heat treatment thermo tolerance.

6.2 Application in Low Temperature Processing: Survival of Free and Encapsulated Bacteria in Ice Cream

Ice-cream is an ideal vehicle for delivery of these organisms in the human diet (Akin et al., 2007, Kailasapathy and Sultana, 2003). *Lactobacillus* and *Bifidobacterium* are the most common species of lactic acid bacteria used as probiotics for fermented dairy products (Akin et al., 2007). In frozen ice milk, 40% more lactobacilli survived when they were entrapped in calcium alginate beads (Sheu et al., 1993). Since ice cream is a whipped product, oxygen is incorporated in large amounts suiting the *L bulgaricus* which are aerobic facultative. Noor (2013) found that the survival of bacteria against unfavorable conditions such as oxygen toxicity or freezing and storage at lower temperatures in ice cream, is species dependent, that is in agreement with Haynes and Playne (2002) and Kailasapathy and Sultana (2003). In addition, it was demonstrated that, encapsulated cells required a longer time to decrease one log cycle in viable counts. These results indicate that alginate matrices not only protect entrapped cells from damage during frozen storage but also protect them from mechanical damage in the ice cream freezer (Sheu et al., 1993).

Probiotics as dairy food additives must be able to survive at low storage temperatures of refrigeration and freezing. In order to screen probiotic bacteria strains for addition to long shelf life foods, the effect of low temperatures (–20 °C) and survival rate of bacteria in low temperatures they must be studied (Homayouni et al., 2008). The lactic acid content of the samples was found to increase with prolonged fermentation upto 18 hours (Witthuhn et al., 2005, Noor, 2013) obtained the three best formulations for coated probiotic ice creams in order to get the most optimum formulation of encapsulation according to the highest log CFU/ml count of the bacteria *L. bulgaricus* growth after 2 days as shown in Table 5.

Lian et al. (2003) who encapsulated *B. longum* and *B. infantis* in gelatine reported that the bacteria remained relatively stable during the entire exposure period. Gelatine in the mixtures may protect the bacteria cell from the acidic conditions (Annan et al., 2008). In addition, ice-cream is an ideal matrix for delivery of probiotic organisms to the human body as compared to fermented dairy products (Akin et al., 2007, Haynes and Playne, 2002, Kailasapathy and Sultana, 2003). Other than that, coconut oil is also one of

Table 5: Three best formulations chosen from three different
percentage of stabilizer

Formulation	Stabilizer %	Skim milk %	Starch %	Sugar %	Inulin %	Fresh milk %	Cream %	Viability (coated) after two days at –20°C (log CFU/ml)
A	0.1	5	1.35	14.35	0.3	58.7	20	9.6021
B	0.3	5	1	17.3	0.3	55.9	20	9.7782
C	0.5	6	1	17.3	0.15	54.85	20	6.8451

the most digestive-friendly compounds because it is rich with medium-chain triglycerides (MCT). Coconut oil is remarkably stable and remains unaltered in acidic conditions and may protect the bacteria cell during transit in the gastrointestinal tract. Figure 3 shows the morphology of of gram stained *L. bulgaricus* strains used in the experiment. Beneficial *L. bulgaricus* colonies form a hostile environment for pathogenic (disease-causing) germs and play a major detoxification role in removing potentially harmful germs that travel through gastrointestinal tract.The results indicate that 9.60 log CFU/ml is the viability for formulation A by using 0.1% gelatine, 9.78 log CFU/ml for formulation B by using 0.3% gelatine, and 6.85 log CFU/ml for formulation C by using 0.5% gelatine. Based on the results, formulation B obtained the highest viability by using 0.3% gelatin which gives the result 9.78 log CFU/ml. In their work, halal gelatin from Halagel (M) Sdn. Bhd was used as a stabilizer in the ice-cream. It was found that encapsulated bacteria survived well compared to non-encapsulated free bacterial cells. The viability during 90 days of frozen storage of coated bacteria in formulation A was 7.98 log CFU/ml, in formulation B was 8.34 log CFU/ml and in formulation C was 4.88 log CFU/ml.

The number of viable probiotic bacteria in coated ice cream were between 10^8 and 10^9 cfu/g at the end of three months of storage which is its normal shelf life. This viable cell number is higher than that recommended by the International Dairy Federation (10^7cfu/g), indicating that the high initial number of probiotics can provide the recommended number in the final product. It was observed that the addition of gelatin had a significant effect on viscosity values between uncoated and coated ice-cream samples for formulation A (FA), formulation B (FB) and formulation C (FC). It showed that the viscosity values increased slightly with increasing gelatin levels. Various investigations have shown that probiotic cultures can better maintain their stability at appropriate levels in frozen food products compared to probiotic fermented milks. Highest protein content of 0.5% gelatin in formulation C has low viability due to higher viscosity and gives low values of desired characteristics like stability as compared to formulations A and B, due to higher acidity and lower pH. The highest fat content (%) in formulation B gives the highest viability compared to formulations A and C. The highest

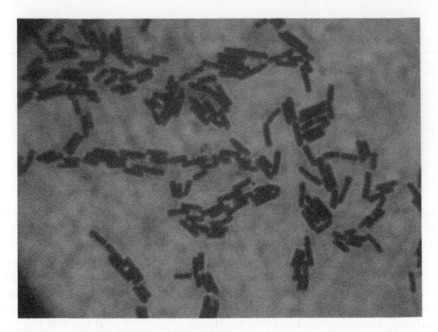

Fig. 3: Morphology of gram-stained *Lactobacillus bulgaricus* used in the experiment (×1000) showing bacilli or rod like bacteria.

viscosity and melting rate in formulation C gives low viability compared to formulations A and B due to high viscosity which did not improve the efficiency of bacteria. It can be concluded that ice-cream can be stored for a long time without changes in its attributes and it is a very popular product worldwide. Ice-cream is an ideal matrix for delivery of probiotic organisms to the human body compared to fermented dairy products.

7. Conclusions

Microencapsulation of probiotic cells has been shown to preserve them from detrimental environmental factors and can be utilized in low and high processing. Other advantages of microencapsulation are, improvement and stabilization of sensory properties and immobilization of the cells, which facilitate homogeneous distribution throughout the product.

　　The cell survivability after heat exposure could be enhanced using double heat protection with microencapsulation and immobilization of sugarcane bagasse-alginate. This discovery could be useful in the production of pelleted feed or other products that necessitate heat-treatment and thermo tolerance.

　　The design and development of functional probiotic ice-creams is an expensive and multistage process that takes into account many factors, such as sensory acceptance, physicochemical and microbial stability, price and other intrinsic functional properties in order to be successful in the marketplace (Fogliano and Vitaglione, 2005). This study indicates that encapsulation can significantly improve the survival of probiotic bacteria in ice cream. Milk products such as ice cream and frozen desserts may serve as carriers for delivering the probiotic bacteria into the human gut. The high solids level in ice cream including fat and milk solids can provide protection for the probiotic bacteria. Further studies are needed to evaluate the protection effect of microencapsulationon on probiotic survival in the gastrointestinal tract. Ice-cream is an ideal matrix for delivery of probiotic organisms to the human body compared to fermented dairy products. The incorporation of probiotic bacteria into different types of ice-cream is highly advantageous: in addition to making a food rich in health benefits, the ice-cream itself contains dairy raw material, vitamins and minerals and it is consumed by the general population. Thus, ice-cream can serve as excellent medium to deliver probiotics into the human intestine. Production and frozen storage of ice-cream have relatively little effect on probiotic survival compared to fermented milk products, and studies have shown that bacterial cultures remain at levels sufficient to offer the suggested therapeutic effects. Ice-cream can be supplemented with prebiotics to improve probiotic stability as well as the sensory and physiochemical characteristics of synbiotic ice-cream.

References

Abe, F., N. Ishibashi and S. Shimamura. 1995. Effect of administration of Bifidobacteria and lactic acid bacteria to newborn calves and piglets. J. Dairy Sci. 78: 2838-2846.

Abu-Tarboush, H.M., M.Y. Al-Saiady and A.H. Keir El-Din. 1996. Evaluation of diet Containing *Lactobacilli* on performance, fecal coliform, and *Lactobacilli* of young dairy calves. Anim. Feed Sci. Technol. 57: 39-49.

Akın, M.B., M.S. Akın and Z. Kırmacı. 2007. Effects of inulin and sugar levels on the viability of yogurt and probiotic bacteria and the physical and sensory characteristics in probiotic ice-cream. Food Chem. 104: 93-99.

Alzamora, S.M., D. Salvatori, M.S. Tapia, A. López-Malo, J. Welti-Chanes, P. Fito et al. 2005. Novel functional foods from vegetable matrices impregnated with biologically active compounds. J. Food Eng. 67(1-2): 205-214.

Anal, A.K. and W.F. Stevens. 2005. Chitosan-alginate multilayer beads for controlled release of ampicillin. Int. J. Pharmaceut. 290: 45-54.

Anal, A.K., W.F. Stevens and C. Remunan-Lopez. 2006. Ionotropic cross-linked chitosan microspheres for controlled release of Ampicillin. Int. J. Pharmeceut. 312: 166-173.

Anal, A.K. and H. Singh. 2007. Recent advances in microencapsulation of probiotics for industrial applications and targeted delivery. Trends Food Sci. Technol. 18: 240-251.

Annan, N.T., A.D. Borza and L.T. Hansen. 2008. Encapsulation in alginate-coated gelatin microspheres improves survival of the probiotic *Bifidobacterium adolescentis* 15703T during exposure to simulated gastro-intestinal conditions. Food Res. Int. 41: 184-193.

Bekin, S., S. Sarmad, K. Gürkan, G. Keçeli and G. Gürdağ. 2014. Synthesis, characterization and bending behavior of electro-responsive sodium alginate/ poly(acrylic acid) interpenetrating network films under an electric field stimulus. Sensors and Acuators B. Chemicals 202: 878-892.

Blaiotta, G., B. La Gatta, M. Di Capua, A. Di Luccia, R. Coppola, M. Aponte et al. 2013. Effect of chestnut extract and chestnut fiber on viability of potential probiotic *Lactobacillus* strains under gastrointestinal tract conditions. Food Microbiol. 36: 161-169.

Bosnea, L.A., Y. Kourkoutas, N. Albantaki, C. Tzia, A.A. Koutinas, M. Kanellaki et al. 2009. Functionality of freeze-dried *L. casei* cells immobilized on wheat grains. LWT- Food Sci. Technol. 42(10): 1696-1702.

Bujard, P., G. Kuhnlein, S. Ino and T. Shiobara. 1994. Thermal conductivity of molding compounds for plastic packaging. IEEE Transact. Component, Pack. and Manufact. Technol. 17(4): 527-532.

Borgogna, M., B. Bellich, L. Zorzin, R. Lapasin and A. Cesaro. 2010. Food microencapsulation of bioactive compounds: Rheological and thermal characterization of non-conventional gelling system. Food Chem. 122(2): 416-423.

Burgain, J., C. Gaiani, M. Linder and J. Scher. 2011. Encapsulation of probiotic living cells: From laboratory scale to industrial applications. J. Food Eng. 104(4): 467-483.

Campanone, L., E. Bruno and M. Martino. 2014. Effect of microwave treatment on metal-alginate beads. J. Food Eng. 135: 26-30.

Chen, M.J. and K.N. Chen. 2007. Applications of probiotic encapsulation in dairy products. pp. 83-107. In: Lakkis, Jamileh M. (ed.). Encapsulation and Controlled Release Technologies in Food Systems. Wiley-Blackwell, USA.

Chen, M., K. Chen and Y. Kuo. 2007. Optimal thermotolerance of *Bifidobacterium bifidum* in gellan-alginate microparticles. Biotechnol. Bioeng. 98(2): 411-419.

Chen, M. and A. Mustapha. 2012. Survival of freeze-dried microcapsules of α-galactosidase producing probiotics in a soy bar matrix. Food Microbiol. 30(1): 68-73.

Chitprasert, P., P. Sudsai and A. Rodklongtan. 2012. Aluminum carboxymethyl cellulose-rice bran microcapsules: Enhancing survival of *Lactobacillus reuteri* kub-ac5. Carbohydr. Polym. 90(1): 78-86.

Corbo, M.R., A. Bevilacqua, M. Gallo, B. Speranza and M. Sinigaglia. 2013. Immobilization and microencapsulation of *Lactobacillus plantarum*: Performances and *in vivo* applications. Innov. Food Sci. Emerg. Technol. 18: 196-201.

Costa, S.M., P.G. Mazzola, J.C.A.R. Silva, R. Pahl, A. Pessoa, S.A. Costa et al. 2013. Use of sugar cane straw as a source of cellulose for textile fiber production. Ind. Crops and Products 42: 189-194.

Crittenden, R., A. Laitila, P. Forsell, J. Matto, M. Saarela, T. Mattila-Sandholm et al. 2001. Adhesion of Bifidobacteria to granular starch and its implications in probiotic technologies. Appl. Environ. Microbiol. 67(8): 3469-3475.

Dinakar, P. and V.V. Mistry. 1994. Growth and viability of *Bifidobacterium bifidum* in Cheddar cheese. J. Dairy Sci. 77(10): 2854-2864.

Ding, W.K. and N.P. Shah. 2007. Acid, bile, and heat tolerance of free and microencapsulated probiotic bacteria. J. Food Sci. 72(9): 446-450.

Do Espirito Santo, A.P., P. Perego, A. Converti and M.N. Oliveira. 2011. Influence of food matrices on probiotic viability – A review focusing on the fruity bases. Trends in Food Sci. and Technol. 22(7): 377-385.

FAO/WHO. 2001. Evaluation of health and nutritional properties of powder milk and live lactic acid bacteria. Food and Agriculture Organization of the United

Nations and World Health Organization Expert Consultation Report. http://www.fao.org/es/ESN/Probio/Probio.htm

Fogliano V. and P. Vitaglione. 2005. Functional foods: Planning and development. Mol. Nutr. Food Res. 49: 256-262.

Gaggia, F., P. Mattarelli and B. Biavati. 2010. Probiotics and prebiotics in animal feeding for safe food production. Int. J. Food Microbiol. 141: 15-28.

Gertrude, A. 2011. Thermal Properties of Selected Materials for Thermal Insulation Available in Uganda. Master thesis. Makerere University.

Giraffa, G., N. Chanishvili and Y. Widyastuti. 2010. Importance of *Lactobacilli* in food and feed biotechnology. Res. Microbiol. 161: 480-487.

Goldin, B.R. 1998. Health benefits of probiotics. Br. J. Nutr. 5(1): 203-207.

Guergoletto, K.B., M. Magnani, J.S. Martin, C.G.T.D.J. Andrade and S. Garcia. 2010. Survival of *Lactobacillus casei* (Lc-1) Adhered to prebiotic vegetal fibers. Innov. Food Sci. Emerg. Technol. 11(2): 415-421.

Hamilton-Miller, J.M.T., G.R. Gibson and W. Bruck. 2003. Some insight into the derivation and early uses of the word 'Probiotic'. Brit. J. Nutr. 90: 845.

Haynes, I.N. and M.J. Playne. 2002. Survival of probiotic cultures in low fat ice-cream. Aust. J. Dairy Technol. 57: 10-14.

Hetsroni, G., C.F. Li, A. Mosyak and I. Tiselj. 2001. Heat transfer and thermal pattern around a sphere in a turbulent boundary layer. Int. J. Multiph. Flow. 27(7): 1127-1150.

Homayouni, A., A. Azizi, M.R. Ehsani, M.S. Yarmand and S.H. Razavi. 2008. Effect of microencapsulation and resistant starch on the probiotic survival and sensory properties of synbiotic ice cream. Food Chem. 111: 50-55.

Jagannath, A., P.S. Raju and A.S. Bawa. 2010. Comparative evaluation of bacterial cellulose (Nata) as a cryoprotectant and carrier support during the freeze drying process of probiotic lactic acid bacteria. LWT- Food Sci. Technol. 43: 1197-1203.

Kailasapathy, K. and K. Sultana. 2003. Survival and b-D-galactosidase activity of encapsulated and free *Lactobacillus acidophilus* and *Bifidobacterium lactis* in ice-cream. Aust. J. Dairy Technol. 58(3): 223-227.

Kailasapathy, K. and L. Masondole. 2005. Survival of free and microencapsulated *Lactobacillus acidophilus* and *Bifidobacterium lactis* and their effect on texture of feta cheese. Aust. J. Dairy Technol. 60: 252-258.

Kailasapathy, K. 2009. Encapsulation technologies for functional foods and nutraceutical product development. CAB reviews: Perspectives in agriculture. Vet. Sci. Nutr. Nat. Resour. 4(33): 1-19.

Kannoujia, D.K., S. Ali and P. Nahar. 2009. Pressure-induced covalent immobilization of enzymes onto solid surface. Biochem. Eng. J. 48(1): 136-140.

Karakus, E. and S. Pekyardımcı. 2009. Immobilization of apricot Pectinesterase (*Prunus armeniaca* l.) on porous glass beads and its characterization. J. Mol. Cat. B: Enz. 56(1): 13-19.

Kok, F.S., I.I. Muhamad, C.T. Lee, F. Razali, N. Pa'e, S. Shaharuddin et al. 2012. Effects of pH and temperature on the growth and β-Glucosidase activity of *Lactobacillus rhamnosus* NRRL 442 in anaerobic fermentation. Int. Rev. Chem. Eng. 4(3): 293-299.

Kourkoutas, Y., M. Kanellaki, A.A. Koutinas, I.M. Banat and R. Marchant. 2003. Storage of immobilized yeast cells for use in wine-making at ambient temperature. J. Agric. Food Chem. 51(3): 654-658.

Kourkoutas, Y., A. Bekatorou, I. Banat, R. Marchant and A. Koutinas. 2004. Immobilization technologies and support materials suitable in alcohol beverages production: A review. Food Microbiol. 21(4): 377-397.

Kourkoutas, Y., V. Xolias, M. Kallis, E. Bezirtzoglou and M. Kanellaki. 2005. *Lactobacillus casei* cell immobilization on fruit pieces for probiotic additive, fermented milk and lactic acid production. Process Biochem. 40: 411-416.

Kourkoutas, Y., M. Kanellaki and A.A. Koutinas. 2006a. Apple pieces as immobilization support of various microorganisms. LWT- Food Sci. Technol. 39(9): 980-986.

Kourkoutas, Y., L. Bosnea, S. Taboukos, C. Baras, D. Lambrou, M. Kanellaki et al. 2006b. Probiotic cheese production using *Lactobacillus casei* cells immobilized on fruit pieces. J. Dairy Sci. 89(5): 1439-1451.

Krasaekoopt, W., B. Bhandari and H. Deeth. 2003. Evaluation of encapsulation techniques of probiotics for yoghurt. Int. Dairy J. 13(1): 3-13.

Lian, W.C., H.C. Hsiao and C.C. Chou. 2003. Viability of microencapsulated bifidobacterium simulated gastric juice and bile solution. Int. J. Food Microbiol. 86: 293-301.

Livney, Y.D. 2010. Milk proteins as vehicles for bioactives. Curr. Opin. Coll. Interf. Sci. 15(1-2): 73-83.

Lynch, H.A. and S.A. Martin. 2002. Effects of *Saccharomyces cerevisiae* culture and *Saccharomyces cerevisiae* live cells on *in vitro* mixed ruminal microorganism fermentation. J. Dairy Sci. 85: 2603-2608.

Mayer, E.A., K. Tillisch and A. Gupta. 2015. Gut/brain axis and the microbiota. J. Clin. Invest. 125: 926-938.

Mandal, A. and D. Chakrabarty. 2011. Isolation of nanocellulose from waste sugarcane bagasse (SCB) and its characterization. Carbohyd. Polym. 86(3): 1291-1299.

Martins, E.M.F., A.M. Ramos, E.S.L. Vanzela, P.C. Stringheta, C.L. de Oliveira Pinto, J.M. Martins et al. 2013. Products of vegetable origin: A new alternative for the consumption of probiotic bacteria. Food Res. Int. 51(2): 764-770.

Michida, H., S. Tamalampudi, S.S. Pandiella, C. Webb, H. Fukuda, A. Kondo et al. 2006. Effect of cereal extracts and cereal fiber on viability of *Lactobacillus plantarum* under gastrointestinal tract conditions. Biochem. Eng. J. 28(1): 73-78.

Mortazavian, A.M., A. Azizi, M.R. Ehsani, S.H. Razavi, S.M. Mousavi, S. Sohrabvandi et al. 2008. Survival of encapsulated probiotic bacteria in Iranian yogurt drink (Doogh) after the product exposure to simulated gastrointestinal conditions. Milchwissenschaft 63(4): 427-429.

Nagar, S., A. Mittal, D. Kumar, L. Kumar and V.K. Gupta. 2012. Immobilization of xylanase on glutaraldehyde activated aluminum oxide pellets for increasing digestibility of poultry feed. Process Biochem. 47(9): 1402-1410.

Noor, N.M. 2013. Probiotic Viability and Sensory Properties of Ice-cream Supplemented with Encapsulated *Lactobacillus bulgaricus*. Master thesis. Universiti Teknologi Malaysia.

Noriega, E., E. Velliou, E. Van Derlinden, L. Mertens and J.F.M. Van Impe. 2013. Effect of cell immobilization on heat-induced sublethal injury of *Escherichia coli, Salmonella typhimurium* and *Listeria innocua*. Food Microbiol. 36(2): 355-364.

Olaru, N. and L. Olaru. 2005. Phthaloylation of cellulose acetate in acetic acid and acetone media. Iran. Polym. J. 14(12): 1058-1065.

Picot, A. and C. Lacroix. 2003. Effects of micronization on viability and thermotolerance of probiotic freeze-dried cultures. Int. Dairy J. 13(6): 455-462.

Poncelet, D. and C. Dreffier. 2007. Les Méthodes De Microencapsulation. De A À Z (Ou Presque). pp. 23-33. *In*: T. Vandamme, D. Poncelet and P. Subra-Paternault (eds.). Microencapsulation: Des Sciences Aux Technologies. Edition Tec and doc. Paris.

Possemiers, S., M. Marzorati, W. Verstraete and T. Van de Wiele. 2010. Bacteria and chocolate: A successful combination for probiotic delivery. Int. J. Food Microbiol. 141(1-2): 97-103.

Quan, J., Z. Liu, C. Branford-White, H. Nie and L. Zhu. 2014. Fabrication of glycopolymer/MWCNTs composite nanofibers and its enzyme immobilization applications. Collods and Surfaces B: Biointerf. 121: 417-424.

Reddy, L.V., Y.H.K. Reddy, L.P.A. Reddy and O.V.S. Reddy. 2008. Wine production by novel yeast biocatalyst prepared by immobilization on watermelon (*Citrullus vulgaris*) rind pieces and characterization of volatile compounds. Process Biochem. 43(7): 748-752.

Roßle, C., M.A.E. Auty, N. Brunton, R.T. Gormley and F. Butler. 2010. Evaluation of fresh-cut apple slices enriched with probiotic cacteria. Innov. Food Sci. Emerg. Technol. 11(1): 203-209.

Roger, V., G. Fonty, S. Komisarczuk-Bony and P. Gouet. 1990. Effects of physicochemical factors on the adhesion to cellulose avicel of the ruminal bacteria *Ruminococcus flavefaciens* and *Fibrobacter succinogenes* subsp. *succinogenes*. Appl. Environ. Microbiol. 56: 3081-3087.

Rokka, S. and P. Rantamaki. 2010. Protecting probiotic bacteria by microencapsulation: Challenges for industrial applications. Eur. Food Res. Technol. 23: 1-12.

Roos, T.B., V.C. Tabeleão, L.A. Dummer, E. Schwegler, M.A. Goulart, S.V. Moura et al. 2010. Effect of *Bacillus cereus* var. *toyoi* and *Saccharomyces boulardii* on the immune response of sheep to vaccines. Food Agric. Immunol. 21: 113-118.

Saarela, M., I. Virkajarvi, L. Nohynek, A. Vaari and J. Matto. 2006. Fibers as carriers for *Lactobacillus rhamnosus* during freeze-drying and storage in apple juice and chocolate-coated breakfast cereals. Int. J. Food Microbiol. 112: 171-178.

Sabikhi, L., R. Babu, D.K. Thompkinson and S. Kapila. 2010. Resistance of microencapsulated *Lactobacillus acidophilus* LA1 to processing treatments and simulated gut conditions. Food Bioproc. Technol. 3(4): 586-593.

Sajilata, M.G., R.S. Singhal and P.R. Kulkarni. 2006. Resistant starch – A review. Rev. Food Sci. and Food Safety. 5(1): 1-17.

Seo, J.K., S. Kim, M.H. Kim, S.D. Upadhaya, D.K. Kam, J.K. Ha et al. 2010. Direct-fed microbials for ruminant animals. Asian-Aust. J. Animal Sci. 23(12): 1657-1667.

Shaharuddin, S., I.I. Muhamad, F.S. Kok, K.A. Zahan and N. Khairuddin. 2014. Potential use of biofibers for functional immobilization of *Lactobacillus rhamnosus* NRRL 442. Key Eng. Mater. 594: 231-235.

Shaharuddin, S. 2015. Thermotolerance *Lactobacillus rhamnosus* NRRL 442 Encapsulated in Bagasse-alginate Microcapsule. PhD thesis. Universiti Teknologi Malaysia.

Shaharuddin, S. and I.I. Muhamad. 2015. Microencapsulation of alginate-immobilized bagasse with *Lactobacillus rhamnosus* NRRL 442: Enhancement of survivability and thermotolerance. Carbohydr. Polym. 119: 173-181.

Soccol, C.R., L. Porto, D.S. Vandenberghe, M.R. Spier, A. Bianchi, P. Medeiros et al. 2010. The potential of probiotics: A review. Food Technol. Biotechnol. 48(4): 413-434.

Sun, W. and M.W. Griffiths. 2000. Survival of Bifidobacteria in yogurt and simulated gastric juice following immobilization in gellan-xanthan beads. Int. J. Food Microbiol. 61(1): 17-25.

Vasiljevic T. and N.P. Shah. 2008. Probiotics – From Metchnikoff to bioactives. Int. Dairy J. 18: 714-728.

Vega, C. and Y.H. Roos. 2006. Spray-dried dairy and dairy-like emulsions-compositional considerations. J. Dairy Sci. 89: 383-401.

Wai-Yee, F., Y. Kay-Hay and L. Min-Tze. 2011. Agrowaste-based nanofibers as a probiotic encapsulant: Fabrication and characterization. J. Agric. Food Chem. 59: 8140-8147.

Weiss, W.P., D.J. Wyatt and T.R. McKelvey. 2008. Effect of feeding Propionibacteria on milk production by early lactation dairy cows. J. Dairy Sci. 91: 646-652.

Witthuhn, R.C., T. Schoeman, A. Cilliers and T.J. Britz. 2005. Impact of preservation and different packaging conditions on the microbial community and activity of Kefir grains. Food Microbiol. 22: 337-344.

Probiotics Applications in Agriculture

Ahmed Kenawy¹*, **Gaber Abo-Zaid²**, **Hamada El-Gendi²**, **Ghada Hegazy³**,
Ting Ho⁴ and Hesham El Enshasy²,⁵,⁶

¹ Nucleic Acids Research Department, Genetic Engineering and Biotechnology
  Research Institute, City of Scientific Research and Technological Applications
  (SRTA city), Universities and Research Institutes Zone, New Borg El-Arab,
  Alexandria 21934, Egypt
² Bioprocess Development Department, Genetic Engineering and Biotechnology
  Research Institute, City of Scientific Research and Technology Applications
  (SRTA city), Universities and Research Institutes Zone, New Borg El-Arab,
  Alexandria 21934, Egypt
³ Microbiology Department, Marine Environment Branch, National Institute
  of Oceanography and Fisheries (NIOF), Alexandria, Egypt
⁴ Global Bio-Innovation Limited, Tsuen Wan, Hong Kong SAR
⁵ Institute of Bioproduct Development (IBD), Universiti Teknologi Malaysia
  (UTM), Skudai, Johor Bahru, Malaysia
⁶ School of Chemical and Energy Engineering, Faculty of Engineering,
  Universiti Teknologi Malaysia (UTM), Skudai, Johor Bahru, Malaysia

1. Introduction

Plant probiotics are a group of microorganisms that can promote soil
health, stimulate plant growth, control plant pathogens, and enhance plant
tolerance against various abiotic and biotic stresses. This group combines
diverse natural-occurring living microbes including bacteria, archaea, fungi,
etc. However, an exploration of their probiotic potentials and pathogenicity
is needed before using them on a commercial scale. Therefore, isolation and
identification of these organisms are required at morphological, physiological,
biochemical, and molecular levels. In addition, their interaction with the
surrounding biotic and abiotic components of the hosting environment
should be clearly understood to ensure a successful functionality, when they
get in competition with the native microbiota in the rhizosphere. Recently,
an unambiguous amount of research efforts have been spent to explore the
diversity of soil microbes using different microbiology and molecular biology

*Corresponding author: aatta75@yahoo.com

techniques in order to unfold the hidden values of this group of microbes and discover the hidden secrets of soil microbiomes. Plant growth-promoting bacteria (PGPBs) and soil beneficial fungi are good examples for plant probiotics, since their activities increase plant growth, immunity, and yield. The microbial activities of this group of microorganisms vary according to the type of microbes and the adopted mechanisms (direct or indirect). These mechanisms include; biological nitrogen fixation, plant growth regulators production, mineral bio-solubilization, biological control of plant pathogens, and finally plant metabolic efficiency enhancement under stress conditions.

Despite the fact that probiotics are abundant and they have been in the focus of hundreds of researchers in the last few decades, many challenges are associated with the expansion of their use at larger scales, including technical, social awareness, as well as legal issues. Therefore, in this chapter, we will introduce plant probiotics and present their types, natural interactions, and the mechanisms employed to improve plant health. In addition, we will try to cover the most important applications of these bacteria in the agriculture field. Finally, we will present the challenges that encounter the extended use of this promising group of microorganisms for sustainable agriculture practices.

2. Biology of Plant Probiotics

The internationally recognized WHO definition of probiotics is "live microorganisms which, when administered in adequate amounts, confer a health benefit on the host". However, other definitions appeared that have a restrictive meaning according to specific mechanisms, site of action, delivery method, or the type of host. Therefore, plant probiotics are microorganisms which when added, separately or in consortia, to the plant/soil in adequate amounts will improve soil quality, and benefit plant health.

Over the last few decades, several microorganisms were studied extensively after they demonstrated stimulating effects on plant growth, development, yield, immunity, and performance under stress conditions among others. The stimulating effects of these microorganisms were found to be the result of complicated relationships, such as plant-microbe interactions, microbe-microbe interactions, and their relationship with the surrounding environment (Figure 1). For example, bacteria has the ability of fixing the atmospheric nitrogen into ammonia by a process called biological nitrogen fixation. In this process, a symbiotic bacteria such as different rhizobial genera or free living nitrogen fixers establish a close association with plant root in which microbes benefit from plant resources and in exchange fix the available form of nitrogen for plant metabolism (Borriss, 2015). In this process, both plants and microbes use a plethora of metabolites for preparing themselves and their microenvironment to establish beneficial associations. Some complicated mechanisms are also included, such as cellular signaling, cellular metabolic modifications, and structural modifications. For example, for a successful root nodulation and symbiotic nitrogen fixation, plants produce root exudates to stimulate specific rhizobial strains to nodulate their roots. Once these chemical messages reach the bacteria, a cascade of chemical signals between the microbe and the host will start using a battery of extracellular polysaccharides, flavonoids, and plant hormones. As a result, root nodules, bacteroids, as well as leghemoglobin are ready to fix N for plant. When the microbe is a free-living organism such as *Azospirillium* and *Azotobacter*, the microbe will use polysaccharides to adhere itself to the root and form a capsule as environmental stress protection and decreases the oxygen inhibitory effect on the nitrogenase enzyme-complex.

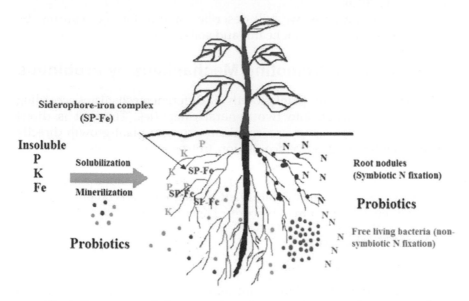

Fig. 1: Some types of plant-microbe interactions between plant roots and soil probiotics, including free and symbiotic nitrogen fixation, and minerals solubilization relationships

Some bacteria belonging to Rhizobiacaea were found inside the root tissues of rice and wheat as endophytes, which is different from symbiosis in which bacteria lives inside the plant and not into specific structures, such as root nodules, and promotes their growth (Yanni et al., 2016). In addition, some species belonging to the genera *Gluconacetobacter*, *Azospirillum* and *Herbaspirillum* are frequent sugarcane endophytes that act as nitrogen fixers, and hence contribute to plant nutrition (Alves et al., 2015). Probiotics were also found thriving in plants rhizosphere to ensure adequate nutrient supplies of phosphorus, potassium, and iron to the plants (Shaikh and Sayyed, 2015, Mhlongo et al., 2018). Probiotic microorganisms have developed several mechanisms to solubilize insoluble minerals using organic acids, siderophores, and specific enzymes to increase the availability of different elements as illustrated in Figure 1 (Velázquez et al., 2016, Etesami et al., 2017). On the other hand, it was found that these microbes could play better roles in enhancing plant health, when they exist in consortia rather than as a sole players. Pathogen controlling is another example of the extraordinary beneficial effects that probiotics can benefit plants with, as they help in destroying pathogenic organisms and preventing them from causing any damage to the plants (Kumari et al., 2018). In addition, probiotics can synthesise plant growth regulators which promote plant growth under stress (Aslantas et al., 2007, Dimkpa et al., 2009), and contribute to reducing the harmful level of some indigenous hormones (e.g. ethylene using ACC deaminase enzyme) to mitigate the harmful effects of stress response (Zerrouk et al., 2016).

In the following parts, we will describe some of the mechanisms, by which probiotics enhance plant health and soil quality.

3. Plant Growth Promoting Mechanisms by Probiotics

Soil probiotics can promote plant growth via different mechanisms according to their mode of action into two general categories. The first is direct mechanisms, in which soil microorganisms promote plant growth directly either by facilitating resource acquisition (nitrogen, phosphate, potassium and essential minerals), or modulating plant hormones. Meanwhile, the second is indirect mechanisms, where microorganisms promote plant growth indirectly by decreasing the inhibitory effects of pathogens on plant growth and development in the form of biocontrol agents (Alejandro et al., 2017)

3.1 Direct Mechanisms

3.1.1 Biological Nitrogen Fixation

Rich soil contains macro and micronutrients that are required for enhanced plant growth and improved plant yield. Among all known macronutrients, nitrogen (N) has an essential role for all living organisms, as it contributes to the formation of micro and macromolecules in the biological system, including proteins, amino acids, and nucleic acids and secondary metabolites among others. Despite its presence in a high percent in the air (78%) as N_2 molecules, most living organisms are unable to use N in this form. Hence, N_2 is considered a non-reactive, metabolically useless molecule to the majority of living organisms. Therefore, converting it to the mineral or organic forms is a vital demand for life on earth. In nature, a few microorganisms have developed the capability to covert N_2 into usable forms in processes generally called biological nitrogen fixation.

Biological nitrogen fixation is a process, by which the atmospheric nitrogen is converted into ammonia (NH_3) or other organic forms of nitrogenous compounds. These compounds are further utilized by other living organisms in their biological reactions. However, biological nitrogen fixation is an energy demanding process. In order to fix one molecule of N by the microorganism, 16 adenosine triphosphate ATP molecules are required (Hubbell and Kidder, 2009). The enzyme complex responsible for the completion of this process is called nitrogenases. This enzyme complex contains iron, and often a second metal, sometimes vanadium, but usually molybdenum (Wanger, 2011). This enzyme is very sensitive to oxygen (O_2). Therefore, if the organism got access to nitrogen fertilizers or the enzyme got exposed to O_2, the microorganism will stop the N fixation to conserve its energy resources (Atta, 2005).

Two types of biological nitrogen fixation occurs in natural ecosystems via soil probiotics, namely, symbiotic, and non-symbiotic nitrogen fixation. Members of the following genera, *Azotobacter, Bacillus, Beijerinckia, Azospirillum, Paenibacillus, Gluconoacetobacter, Herbaspirillum,* and *Burkholderia* were reported as free nitrogen-fixing microorganisms. Meanwhile, members belonging to Rhizobiacaea are able to invade the roots of legumes and fix the nitrogen symbiotically inside specific root structures called root nodules. This family contains an increasing number of genera such as *Rhizobium, Bradyrhizobium, Sinorhizobium, Allorhizobium, Neorhizobium, Pararhizobium, and Azorhizobium*. In addition, the genus *Azospirillum* is generally associated with cereals in temperate zones, increasing crop yields in most of them, as well as in some legumes and sugarcane (Sahoo et al., 2014, Yadegari et al., 2010). Members of the genus *Azotobacter* are capable to fix nitrogen in rice crops (Habibi et al., 2014). Some species belonging to the genera *Gluconoacetobacter, Azospirillum,* and *Herbaspirillum* are frequent sugarcane endophytes and act as nitrogen fixers, and hence contribute to plant nutrition (Alves et al., 2015). The genus *Herbaspirillum* has also been identified as a nitrogen fixing endophyte for several legumes (Puri et al., 2015, Peix et al., 2015). Moreover, some of these free-living bacteria may enter roots of some crops, such as species of *Azoarcus, Azospirillum,* and *Burkholderia* in rice roots, which increase the nitrogen concentration in these crops (Santi et al., 2013). Likewise, the nitrogen-fixing *Azorhizobium* strains in wheat plants (Bhattacharyya and Jha, 2012) increase nitrogen concentration in them. Interestingly, some strains of the genera *Rhizobium* and *Bradyrhizobium* were found in association with rice and wheat roots, increasing nutrient concentration in these plants and improved their yields (Yanni et al., 2016). On the other hand, certain diazotrophic bacteria are able to establish symbiotic relationships within plant tissues, mainly through the formation of root nodules. For example, symbiosis occurs between rhizobia and legumes, *rhizobia* and *Parasponia, Frankia* and actinorhizal plants and cyanobacteria and cycads. Nitrogen fixation occurs between some plants and fungi (Atta, 2005, El-Akhal et al., 2013).

3.1.2 Phosphate Solubilizing Bacteria (PSB)

Phosphorous (P) is an essential nutrient for plant growth and development. Unlike nitrogen, this element is insoluble in soil, and occurs in the form of inorganic rocks and/or organic phosphates. Accordingly, mineralization of soil organic P plays an important role in phosphorus cycling. Organic P may represent 4-90% of the total soil P (Sharma et al., 2013). Almost half of the known microorganisms in the rhizosphere possess phosphatases enzymes that are used for P mineralization. Alkaline and acid phosphatases convert organic phosphate as a substrate into inorganic forms (Sharma et al., 2013). The principal mechanism for P mineralization is the production of acids by microorganisms, by which P is liberated and converted to the soluble form. Meanwhile, the principal mechanism for organic P mineralization is the production of acid phosphatases. Phosphatase enzymes are present in almost all organisms. Nevertheless, some bacteria, fungi, and few algae are able to produce extracellular enzymes that participate in the mineralization of organic phosphate compounds in the rhizosphere (Yadav and Tarafdar, 2011).

Therefore, the use of probiotics, especially P-solubilizing microbes (PSM), as inocula might represent a green substitute for these environment-damaging chemical P fertilizers. Soil bacteria such as the genera *Bacillus*, *Azotobacter*, *Pseudomonas*, *Micrococcus*, *Paenibacillus*, *Klebsiella*, *Pantoea*, *Deftia*, and *Flavobacterium*, among others, have been reported to be efficient phosphate solubilizers (Dastager et al., 2010, Pindi and Satyanarayana, 2012). Moreover, several examples exist in literature that documents the isolation of phosphate-solubilizing rhizobial strains, which promote the growth of several crops, such as *Daucuscarota*, *Lactuca sativa* and legumes (Flores-Felix et al., 2013; Brígido et al., 2013, Flores-Félix et al., 2015a). In addition, a *Phyllobacterium* strain able to solubilize phosphates improves the quality of strawberry growth and yield (Flores-Félix et al., 2015b). Garcia-Fraile et al. (2012) reported two *Rhizobium leguminosarum* strains as phosphate solubilizing probiotics and both were recommended as proper PGPR for pepper and tomato plants. The genus *Mesorhizobium* also has strains which are good P solubilizers and have the ability to promote the growth of chickpea and barley (Brígido et al., 2013).

3.1.3 *Potassium Solubilizing Bacteria (KSB)*

After nitrogen and phosphorous, potassium (K) is the third macronutrient which is essential for plant growth and yield improvement. Potassium is found in the soil in huge quantities but is in an insoluble form, in clay as mineral crystals and rocks (e.g., biotite, feldspar, illite, muscovite, orthoclase, and mica). Plants cannot use the potassium in these crystals until these structures are broken down by soil probiotics and become available to plants (Jaiswal et al., 2016). There is quite a diversity of K-solubilizing bacterial genera (KSB), such as *Acidothiobacillus ferrooxidans*, *Paenibacillus* spp., *Bacillus mucilaginosus*, *B.edaphicus*, and *B. circulans*, which have the capacity to solubilize K (Velázquez et al., 2016, Etesami et al., 2017). KSB can dissolve the insoluble K through the production of organic and inorganic acids, complexolysis, polysaccharides, exchange reactions, acidolysis, and chelation. Among the KSB, *Bacillus* and *Paenibacillus* are two of the best-reported KSB. As an example for probiotic bacterium, *Bacillus edaphicus*, has been reported to increase potassium uptake in wheat and *Paenibacillus glucanolyticus* was found to increase the dry weight of black pepper (Sangeeth et al., 2012). Likewise, wheat and maize plants inoculated with *Bacillus mucilaginosus* under laboratory-controlled conditions showed increased plant biomass (Singh et al., 2010). Moreover, *Bacillus mucilaginosus* was used with the phosphate-solubilizing *Bacillus megaterium* as probiotic mixed inoculum that promoted the growth and yield of different plants, including groundnut, eggplant, pepper, and cucumber (Velázquez et al., 2016). Furthermore, the genus *Pseudomonas* was described as KSB that improved the growth and development of tea plants (*Camellia sinensis*) (Bagyalakshmi et al., 2012) and tobacco (Zhang and Kong, 2014). Similarly, a strain of the genus *Frauteria* was described as KSB that enhanced the growth of tobacco (Subhashini, 2015).

3.1.4 Siderophores Producing Probiotics

In the soil, iron occurs mainly as Fe^{3+} in insoluble hydroxides and oxyhydroxides forms; therefore, iron is not available to both probiotics and plants (Rajkumar et al., 2010). Plant probiotics such as PGPR produce iron-chelating low molecular weight (200–2000 Da) compounds that are identified as siderophores (Shaikh and Sayyed, 2015, Mhlongo et al., 2018). More than 500 types of siderophores are identified, of which 270 have been structurally characterized (Boukhalfa et al., 2003). The common bacterial siderophores are catecholates (i.e. enterobactin), some are carboxylates (i.e. rhizobactin), and hydroxamates (i.e. ferrioxamine B) (Matzanke, 1991). However, some bacterial siderophores contain a mix of the above-mentioned main functional groups (i.e. pyoverdine) (Cornelis, 2010). Plant can obtain iron from probiotics by siderophores by two mechanisms; (i) Chelating Fe(III) in the soil occurs followed by transferring the microbial Fe(III) – siderophores complex into the apoplast of the plant root, where siderophore reduction occurs (Mengel, 1995). As a result, Fe(II) is trapped in the apoplast, leading to a high iron concentration in the root tissues (Kosegarten et al., 1999); (ii) Microbial siderophores can bind iron from soils and then exchange ligands with phytosiderophores (Masalha et al., 2000).

Furthermore, siderophores have the ability to chelate other heavy metals that are found in the soil, such as Al, Cd, Cu, Ga, In, Pb, and Zn. Therefore, siderophores-producing probiotics can act as soil bio remediators (Neubauer et al., 2000, Kiss and Farkas, 1998). Many examples exist in the literature about the successful use of probiotic bacteria in siderophores applications as biological control agents. For instant, a large numbers of probiotics produce siderophores with a high efficiency and adequate quantity, such as *Bacillus* spp., which were isolated from the maize rhizosphere by Bjelić et al. (2018). Moreover, Kumar et al. (2013) had used a large number of siderophores-producing probiotics to control various plant pathogens biologically. For example, siderophore-producing *Bacillus* spp. elicited the biocontrol of black scurf and stem canker caused by *R. solani*. In addition, some other rhizobacterial species, such as *Acinetobacter baumannii*, *Aeromonashydrophila*, *Acinetobacter* sp., *P. alcaliphila*, *Klebsiella pneumoniae*, and *P. brassicacearum* which produce siderophores also have significant antagonistic effects against the phytopathogen *Fusarium oxysporum*, and nevertheless also had plant growth promotion effects (Singh and Varma, 2015).

3.1.5 Probiotics and Phytohormones Biosynthesis

Many PGPR bacteria are able to synthetize phytohormones, which are organic molecules involved in several processes of plant growth and development, and stress response. The biosynthesis of these phytohormones by certain microorganisms might be involved in plant pathogenesis. However, a wide spectrum of beneficial bacteria are also able to produce these hormones as plant stimulators (Spaepen, 2015). Microbes can produce a variety of plant growth regulators, such as auxins, cytokinins, gibberellins, and ethylene, as illustrated in Figure 2). Each one of these phytohormones regulates specific plant developmental stages (Martínez-Hidalgo et al., 2015). For example, *Acetobacter diazotrophicus* and *Herbaspirillum seropedicae* are indole-3-acetic acid (IAA) producers (Bastian et al., 1998). Gibberellic acid (GA3) is produced by *Azospirillum lipoferum* strain op33 (Bottini et al., 1989), meanwhile, the cytokinins (zeatin), as well as ethylene are produced by *Azospirillum* sp. and *Bradyrhizobium* sp. (Strzelczyk et al., 1994). Meanwhile, abscisic acid (ABA) was found to be produced by *Azospirillum brasilense* strains Cd and Az39 (Perrig et al., 2005). Some effects of plant growth regulators will be presented in the following parts.

Auxins are a group of plant growth regulators produced by several bacteria. However, these compounds play essential roles as signaling molecules between microbes to coordinate their activities. Indole-3-acetic acid (IAA) is the most popular auxin in plants, and auxin-producing *Bacillus* spp. have been reported to exert a positive effect in the development of several crops, such as *Solanum tuberosum* (potato) or *Oryza sativa* (rice) (Ahmed and Hasnain, 2010, Sokolova et al., 2011). The relationship of auxin production by *Bradyrhizobium elkanii* and root nodulation was studied. Mutant strains deficient in the biosynthesis of IAA induced few nodules on soybean roots, but nodulation was promoted by exogenous IAA application (Fukuhara et al., 1994).

Cytokinins are widely distributed growth regulators in higher plants, algae, and bacteria. In planta, it promotes cytokinesis, vascular cambium sensitivity, and vascular differentiation. In addition, cytokinins stimulate cell division, cell enlargement, and tissue expansion, indicating a beneficial effect on plant growth and yield (Hanano et al., 2006, Weyens et al., 2009). The ability of rhizobacteria to produce cytokinin was believed to be a main factor that affects plant growth and improving plant resistance to abiotic stresses such as drought stress (Liu et al., 2013). However, less information is available on cytokinins, which are produced by plant-associated bacteria. In addition, members of the genus *Bacillus* were reported as cytokinin producing probiotic bacteria (Sokolova et al., 2011, Liu et al., 2013). Moreover, *Bacillus megaterium* and *Azotobacter chroococcum* strains were found to produce cytokinins and promote cucumber growth. Enriching plants with cytokinins, which are produced by PGPR, showed enhanced

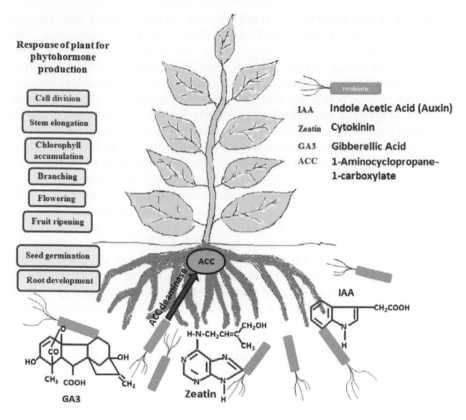

Response of plant for phytohormone production

Cell division

Stem elongation

Chlorophyll accumulation

Branching

Flowering

Fruit ripening

Seed germination

Root development

IAA   **Indole Acetic Acid (Auxin)**

Zeatin   **Cytokinin**

GA3   **Gibberellic Acid**

ACC   **1-Aminocyclopropane-1-carboxylate**

Fig. 2: Production of plant growth regulators by plant probiotics and their roles in plant growth and development

drought resistance in container seedlings. As the oriental thuja seedlings were inoculated with cytokinin-producing *Bacillus subtilis* strains, seedlings conferred better resistance to drought stress (Liu et al., 2013). This enhancing effect of cytokinins might be attributed to their potential effects on stomatal opening, reducing leaf damage, and improving photosynthesis, altogether resulting in accumulating more plant biomass (Liu et al., 2013).

The genus *Sphingomonas* was also reported to produce the phytohormone gibberellin. Tomato plants inoculated with the gibberellin-producing *Sphingomonas* sp. LK11 strain showed improved plant growth phenotypes (Kong et al., 2015). Moreover, Asaf et al. (2017) reported the positive effect of using a collection of phytohormones producing strains of *Sphingomonas* and *Serratia*, and enterobacteria on soybean plant growth and development.

Ethylene is an important plant hormone responsible for several physiological processes in planta, including fruit ripening, and flower blooming, promotes seed germination, secondary root formation, and root-hair elongation. Ethylene can be produced by probiotic microorganisms, such as PGPR bacteria (Gamalero and Glick, 2015). On the other hand, PGPRs were found to have an enhancing effect on plant growth under stresses. A large number of rhizobial and related species are able to produce 1-aminocyclopropane-1-carboxylate (ACC) deaminase, an enzyme responsible for the cleavage of the ethylene precursor, ACC, into α-ketobutyrate and ammonia (Honma and Shimomura, 1978). The genus *Pseudomonas* is the first producer of this particular enzyme (Glick, 2014). Indeed, ACC deaminase was first purified from a *Pseudomonas* strain (Nascimento et al., 2016). Magnucka and Pietr (2015) reported that various strains of ACC-producing *Pseudomonas* benefit the growth of wheat seedlings. Zerrouk et al. (2016) showed that a *Pseudomonas* strain isolated from date palm rhizosphere was able to improve the development of maize plants under two different stresses, as revealed by the increase in all parameters measured under both salt and aluminum stress. The mode of action of this enzyme is to reduce ACC levels in plants by the PGPR organisms, hence decreasing the high ethylene levels in plants (Glick et al., 1998, 2007), which inhibits plant growth in high concentrations and may cause plant death (Glick, 2014). As an example for the successful use of different probiotics in plant growth stimulation, two *Rhizobium* and *Pseudomonas* ACC-deaminase-producing strains resulted in significant improvements in growth parameters, physiology, and quality of mung bean plants under saline stress conditions. Similar results were found when *Pisum sativum* (pea) plants were cultivated on alluvial soils (Ahmad et al., 2013). *Serratia* and *Pseudomonas* ACC-deaminase-producing strains were able to improve the yield of wheat plants growing under saline conditions (Magnucka and Pietr, 2015).

3.2 Indirect Mechanisms

3.2.1 Mycoparasitism

Mycoparasitism can be defined as the interaction between a parasitic fungus and another host fungus, which is eventually utilized as a food by the first one. A large number of fungi have the ability of mycoparasitism that plays substantial roles in the biocontrol of several biotic plant diseases. Four steps were explained by Chet et al. (1981) for the development of mycoparasitism: chemotropism (chemotaxes) and recognition, attachment and coiling, cell wall penetration and host cell digestion. The production of lytic enzymes by fungi is an essential strategy for mycoparasitism. As an example for the mycoparasitism relationship, chitinase produced by *Trichoderma harzianum* T32 was an important component for bio-controlling *Ganoderma boninense* upm001, when *T. harzianum* T32 mycoparasitism activity was tested against *G. boninense* upm001 (Naher et al., 2018). Meanwhile, Mokhtari et al. (2018) investigated the *in vitro* mycoparsitism activities of *T. afroharzianum* (T8A4), *T. reesei* (T9i12) and *T. guizouhense* (T4) against *Phytophthora capsici,* and *R. solani.* They showed a high density of coils of Trichodermal hyphae surrounding the mycelia of plant pathogenic fungi that was used in their study. In addition, they detected a lytic degradation of *Rhizoctonia solani* mycelium in the presence of T8A4 strain, indicating a successful mycoparasitism in the studied fungi.

3.2.2 Competition

Philippot et al. (2013) defined competition as the indirect interaction between pathogens or microbes that exist in the rhizosphere, in which they compete with each other for food and/or essential nutrients and physical site occupation. In other words, we can describe microbial competition as the ability of microbes to fulfill their need of essential nutrients, water, stimulants, and space faster than other microbes using their bioweapons. Since microbes grow in environments where resources are scarce, but have developed a plethora of competition strategies, including rapid growth to benefit from the limited available resources, direct aggression to eliminate competitors, and/or adopting other metabolism plans which benefit from the presence of competitors (Ghoul and Metri, 2013).

Two limiting factors determine the survival of a microbe in a specific environment and are the motivation behind microbial competition; space and nutrients, including carbon, phosphorus, nitrogen, sulfur, calcium, hydrogen, iron, and other metals. One of the important metals that has a distinctive role in microbial competition is iron. Iron as a metal is a very important nutrient for all microorganisms in the rhizosphere and it represents a main target for competition between microorganisms (Sivasakthi et al., 2014). The concentration of iron that microbes need is 10^{-6} molar (M), yet, the available quantity of iron in the rhizosphere is $\sim 10^{-8}$ M. So, microbes have developed the ability to produce low molecular weight compounds, named siderophores that have a high affinity to bind iron from the surrounding environment (Scavino and Pedraza, 2013). As a means of competition, probiotics inhibit the growth of plant pathogens by reducing iron availability in the rhizosphere. Siderophores produced by *P. fluorescens* and *Serratia* spp. contribute to the suppression of plant fungi by consuming the available iron in the rhizosphere (Khilyas et al., 2016, Trapet et al., 2016). Furthermore, competition between microbes may occur due to the production of a specific stimulant (elicitor) by the germinating seeds or growing roots, such as: fatty acids, polysaccharides, volatile compounds, and flavonoids that induce specific strains of beneficial bacteria (Gorecki et al., 1985, Paulitz, 1991).

3.2.3 Antibiosis

The term antibiosis was created in 1889 by Vuillemin to symbolize the antagonistic relationship between living organisms (Bentley and Bennett, 2003). This relationship can be defined as the capability of microorganisms to produce low molecular weight compounds or antibiotics that have a direct effect on other microorganisms. There are several antibiotics produced by plant probiotics that play important roles in controlling plant pathogens. For example, pseudomonds produce several antifungal compounds [phenazines, phenazine-1-carboxylic acid (PCA), phenazine-1-carboxamide (PCN), pyrrolnitrin, pyoluteorin, 2,4diacetylphloroglucinol, rhamnolipids, oomycin A, cepaciamide A, ecomycins, viscosinamide, butyrolactones, N-butylbenzene sulfonamide, pyocyanin], bacterial antibiotics (pseudomonic acid and azomycin), antitumor antibiotics (FR901463 and cepafungins) and antiviral antibiotics (Karalicine) (Ramadan et al., 2016). In addition, *Bacillus* spp. produce a great number of antifungal and antibacterial compounds that are synthesized from both non-ribosomal, and ribosomal sources. Ribosomal antibiotics include: bacteriocins, subtilosin A, subtilin, sublancin, and ericin, while non-ribosomal antibiotics include chlorotetain, bacilysin, mycobacillin, rhizocticins, difficidin, and bacillaene. Moreover, several lipopeptides are produced by *Bacillus* spp. such as surfactin, iturins, bacillomycin among others (Wang et al., 2015). The antifungal activities of both PCA and PCN produced by *Pseudomonas* spp. were observed against a number of plant pathogenic fungi such as *Botrytis cinerea* (Zhang et al., 2015), and *R. solani* (Olorunleke et al., 2015, Niu et al., 2016), *F. graminearum* (Hu et al., 2014). A novel strain of *Bacillus*, which was identified as *B. methylotrophicus* 39b, was efficient in controlling crown gall disease caused by *Agrobacterium tumefaciens* strains C58 and B6. The antibiotic compound produced by *B. methylotrophicus* 39b was identified as lipopeptide surfactin (Gargouri et al., 2017).

3.2.4 Production of Lytic Enzymes

Plant probiotics protect plants from pathogen attacks through different antagonisms. Pathogen antagonism is a complex process that could be achieved via different strategies, one of them being the production of lytic enzymes (Goswami et al., 2016). Lytic enzymes primarily target plant pathogenic fungi cell walls, hence they are known as cell wall degrading enzymes (Prapagdee et al., 2008). Two main groups of cell wall degrading enzymes were recognized to participate in this process, namely chitinase and gluconase that degrade both Chitin and β 1,3-glucans, respectively. These polysaccharides are dominant constituents of many higher class fungal (Ascomycetes, Basidiomycetes, Deuteromycetes) cell walls. Meanwhile, lower fungi, such as myxomycetes, and phycomycetes, have cellulose as the predominant constituent (Khare and Yadav, 2017). Chitinases attack the glycosidic bonds of the chitin polymer (Kim et al., 2003); hence, it was used to control many plant pathogenic fungi. The distribution of chitinase producers among microorganisms is very wide, including bacteria: *Bacillus pumilus* (Rishad et al., 2016), *Bacillus thuringiensis,* and *Bacillus licheniformis* (Gomaa, 2012), Actinomyces: *Thermobifidafusca* (Gaber et al., 2016), *Streptomyces pratensis* (Shivalee et al., 2018), and fungi: *Trichoderma* (Harman et al., 2004), *Humicolagrisea* (Kumar et al., 2017), and *Aspergillus flavus* (Rawway et al., 2018). Many research articles revealed the importance of these enzymes as efficient plant-disease controllers and their *in-vitro* roles in inhibiting various plant pathogenic fungi were described (Kim et al., 2008, Ramos et al., 2010, Malviya et al., 2018). Due to the absence of chitin in plant tissues, chitinase holds greater potential than glucanase as a biocontrol agent for plant pathogens (Chilukoti et al., 2010). Biocontrol agents based upon enzymatic destruction of the pathogenic fungal cell wall were extensively reported for different fungal species, including *Trichoderma* (Harman et al.,

2004), Actinomycets (Ashokvardhan et al., 2014), as well as members of both species *Bacillus* and *pseudomonas* (Kumar et al., 2010), *Bacillus* sp. or *Serratia* sp. (Manjula et al., 2004, Kishore et al., 2005). It is noteworthy to mention that over-expression of chitinase and β-gluconase genes by plants under fungal invasion conditions was reported as a protective response from plants against fungal infection (Kumari and Vengadaramana, 2017), which reflects the importance of these two groups of enzymes as biocontrol agents.

Proteases also depict another important group of lytic enzymes that contribute to plant protection against pathogens (Hasan et al., 2014). Proteases are assets in plant protection against pathogens, which directly target the pathogen's cell wall protein matrix, exposing the main fungal cell wall components, chitin and β-glucan, to cell wall degrading enzymes (Hasan et al., 2014), or indirectly by inactivating the extracellular enzymes of plant pathogens (Howell, 2003). In addition, some other hydrolytic enzymes might be involved in plant defense against pathogens, including cellulases, proteases, and lipases (Glick, 2012). Away from plant protection, hydrolytic enzymes from plant probiotics may promote plant growth through decomposition of non-living organic matters and cellulosic residues, which enrich the soil with different nourishing nutrients (Tariq et al., 2017).

3.2.5 *Production of Probiotic Exopolysaccharides*

Many plant probiotic bacteria produce extracellular polysaccharides (EPS). These biopolymers are biodegradable and of great physiological importance to the bacterial cells, as they provide protection and work as signaling cues between organisms. Two main types of EPS were characterized in soil bacteria, including capsules-forming polysaccharides (PS), and the second PS group which is secreted into the surrounding environment (Glick, 1997, Ngoufack et al., 2004, Sanlibaba and Cakmak, 2016). The structural composition and production levels of these biopolymers vary according to the bacteria type and the environmental conditions (Costa et al., 2018). Plants in the ecosystem are exposed to biotic and abiotic stresses that impact plant growth and development. Rhizobacterial EPS production was found to play an important role in promoting plant growth during abiotic stresses such as drought and salinity. As most of these EPSs are of hygroscopic nature, they play remarkable roles in drought tolerance, especially under low water potential conditions (Roberson and Firestone, 1992), maintaining soil moisture content, which sustains normal plant growth (Glick et al., 1997, Khan et al., 2017).

The high capability of rhizobacteria to promote plant growth and root nodulation under conditions of extremely low moisture content was reported in desert soil (Jenkins et al., 1987, Zhao and Running, 2010), and this capability was attributed to the production of EPSs as a response to drought stress (Sandhya et al., 2009, Niu et al., 2018). The production of EPSs in response to stress was studied by Sandhya and Ali (2015), who found that *Pseudomonas putida* EPS production increased when the bacteria were exposed to different abiotic stresses. Therefore, introducing EPS producing rhizobacteria to the soil can ultimately improve soil physical properties by enhancing its particles aggregation, and water holding capacity. Consequently, water infiltration and root penetration capabilities are enhanced (Bashan et al., 2004, Sandhya et al., 2009, Sandhya and Ali, 2014). Salinity tolerance is another mechanism, by which EPS producing bacteria are able to promote plant growth (Mohammed, 2018, Yadav et al., 2018). Many plant probiotics showed an increase in exopolysaccharide production and biofilm formation as a response to the increase in Na^+ ion concentration (Qurashi and Sabri, 2012, Deng et al., 2015, Kasim et al., 2016). The role of exopolysaccharide in salinity stress endurance could be attributed to chelating sodium ions and quenching their toxicity to plant tissues (Arora et al., 2013, Kasotia et al., 2016).

Some polysaccharide-producing rhizobacteria form biofilm around plant roots, and hence, increase the probability of successful plant–microbe interaction and salinity stress reduction (Geddie and Sutherland, 1993, Arora et al., 2010, Upadhyay et al., 2011, Mohammed, 2018). The formation of PS biofilm around plant roots can also promote plant growth and health by acting as a physical barrier against root pathogens and ensures efficient surface nutrients uptake (Batool and Hasnain, 2005, Pawar et al., 2016). In the case of root nodules formation by rhizobia, EPS works as an adhesive material that sticks the bacterial cells to the root hair and facilitates root invasion (Atta, 2005).

3.2.6 Induced Systemic Resistance (ISR)

Induced systemic resistance (ISR) was defined by Peer et al. (1991) as the mode of action of disease suppression by a non-pathogenic rhizobacteria (PGPR) that represents a group of plant probiotics, where jasmonic acid (JA) and/or ethylene are produced after the application of some nonpathogenic rhizobacteria (Figure 3). Two reactions were confirmed by Patel and Minocheherhomji (2018) for ISR in plants: one of them is the ability of a plant to defy a pathogen; and the second is the susceptibility of a plant to have interactions with a rhizobacterium. Rhizobacteria in ISR do not produce antifungals, antibiotics, and siderophores, but protect the plant by modifying the host defense system (Shaikh and Sayyed, 2015). Cell wall thickening or rapid death of diseased cells might occur as fast responses after defense elicitation occurs through ISR to stop the extension of pathogen invasion (Lugtenberg et al., 2002). Several fundamental elements are required to establish ISR, which have been reported by Shafi et al. (2017). These basic elements include the existence of phenolic compounds, genetic and structural modifications, plant resistance activators, and activation of enzymatic missiles (Shafi et al., 2017).

It was found that rhizobacteria-induced ISR depends on the bacterial strain and plant species. However, the level of ISR was found to be a correlated to stress conditions. Therefore, stress environmental conditions are

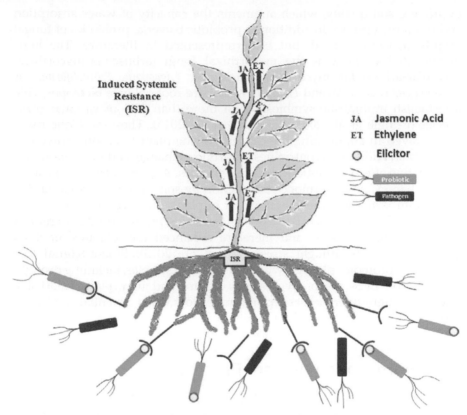

Fig. 3: Induction of the induced systemic resistance (ISR) in plant

detrimental for the development of ISR status (Shafi et al., 2017). Probiotics-induced ISR can be triggered by a diverse number of microorganisms, which have been used as biocontrol agents for controlling soil-borne and aerial plant diseases (Bakker et al., 2003). Whipps (2001) described the importance of non-pathogenic *Bacillus* spp. and *Pseudomonas* spp. in the biocontrol of plant diseases, by which, these probiotic bacteria are responsible for triggering ISR mechanism. In addition, some non-pathogenic fungi play a fundamental role in inducing systemic resistance in plants, such as *T. asperellum*, which produces phenylalanine, hydroperoxidelyase, and accumulates phytoalexin to induce plant resistance against *P. syringae* pv. *Lachrymans* (Yedidia et al., 2003). Moreover, induced changes on the systemic and physiological levels in tomato plants were observed, when *T. hamatum* colonized plant roots, where disease damage was prevented (Alfano et al., 2007).

3.2.7 Abiotic Stress Tolerance

Probiotics microorganisms were found capable of mitigating the effect of plant abiotic stresses through different mechanisms. One of them is improving soil quality, which augments the capacity of water absorption and nutrients up-take. In addition to probiotic bacteria, probiotics of fungal origins are widely used, but underrepresented in literature. The best-documented example is the mycorrhizal fungi (arbuscular-mycorrhiza, Vesicular-arbuscular mycorrhiza) including *Gigaspora, Funneliformis* or *Rhizophagus* (Glomus), and *Laccaria*, which are root obligate biotrophs, able to establish mutualistic symbiosis with more than 80% of vascular plant species (Pringle et al., 2009, Rouphael et al., 2015). These probiotic fungi are involved in augmenting the capacity of the plant to absorb water and nutrients; in addition, it contributes to carbon exchange, and thus modulates the negative effects of biotic and abiotic stresses. As it was presented earlier in this chapter, phytohormones production by some probiotic bacteria was of great importance in the enhancement of plant tolerance to drought. This effect was observed when *Bacillus subtilis* strains produced cytokinins and affected stomata closure, and therefore enhanced the tolerance of thuja seedlings towards drought stress (Liu et al., 2013). Rhizobacterial EPS production promotes plant growth during abiotic stresses, including drought and salinity. EPSs are of hygroscopic nature and they play remarkable roles in drought tolerance, especially under low water potential conditions (Roberson and Firestone, 1992), maintaining soil moisture content, which sustain normal plant growth (Glick et al., 1997, Khan et al., 2017)

4. Bioformulation of Plant Probiotics

Plant probiotics are microbes that benefit plant growth and could be in the form of biofertilizers, phytostimulators, rhizoremediators and biocontrol agents (Somers et al., 2004, Menendez, 2017). These beneficial microorganisms have been in the focus of scientific research for decades, and recently, the use of these microbes in enhancing plant growth is going to the commercial level. However, when research results are transferred from the lab scale to the open field, plant probiotics should be formulated with inert (inactive) carriers as bio-formulations for the ease of usage, storage, and commercialization. From bio-formulation perspectives, Burges and Jones,1998 indicated that bio-formulation includes assisting agents that protect organisms to deliver them to their targets in order to improve product efficiency. Meanwhile, the term of bio-formulation was defined by Arora et al. (2010), as a preparation of a microbe that may be a partial or complete alternative for chemical fertilizers/ pesticides. Many plant probiotic formulations are available in the market, but to maximize their efficiency in plant disease protection, plant growth promotion, and increasing crop production, improvement in product shelf-life, broad-spectrum of action by them, and consistent performance under field conditions must be achieved (Nakkeeran et al., 2004).

A good bio-formulation of plant probiotics must include three elements; an active ingredient, a carrier material and the addition of additives. The active ingredient is generally a viable organism in the form of living microbe/s or spores that can survive long time storage (Auld et al., 2003, Hynes and

Boyetchko, 2006). While, a suitable carrier material works as a support for the active ingredient and assures that the microbial cells are easily established in/or around the plant to provide a better chance for promoting its growth or controlling plant pests and pathogens. In addition, carrier materials may play important roles in increasing bio-formulation shelf life by protecting the microorganism from drought and subsequent death (Burges and Jones, 1998, Trivedi et al., 2005). The carrier material could be organic or inorganic, but it should be non-reactive, economical, eco-friendly, and easily available in bulk amounts for commercial utilization. Carrier materials such as clay, peat, vermiculite, talc, lignite, kaolinite, pyrophyllite, zeolite, montmorillonite, alginate, polyacrylamide beads, diatomaceous earth, press mud, sawdust among others are currently used for formulation purposes (Digat, 1989, El-Fattah et al., 2013, Zayed, 2016). Gums, silica gel, methyl cellulose, and starch are well-known additives that were used in the preparation of different bio-formulations that develop physical, chemical, and nutritional properties in them (Schisler et al., 2004, Hynes and Boyetchko, 2006). There are three available types of bio-formulations all over the world: liquid, solid, and encapsulated respectively.

4.1 Liquid Formulations

Plant probiotics are accessible in the market in various types of liquid formulations. Liquid inoculants or bio-formulations are dependant on aqueous broth cultures, mineral or organic oils and oil-in-water, and polymer-based suspensions (Schisler et al., 2004, Bharti et al., 2017). These types of liquid formulations are supported with some substances that develop stickiness, stabilization and dispersal abilities (Singleton et al., 2002). A typical liquid formulation contains 10-40% microorganisms, 1-3% suspender ingredient, 1-5% dispersant, 3-8% surfactant and 35-65% carrier liquid (oil or water) (Brar et al., 2006). Liquid formulations have some advantages over solid formulations. For example, ease of application is the major advantage of liquid formulations. In addition, liquid formulations provide cell cultures with sufficient nutrients and other materials that protect them; therefore, their shelf life is longer than formulations based on solid organic and inorganic carriers. In another example, a soybean oil based formulation of *Trichoderma asperellum* was developed by Mbarga et al. (2014) and had an enormous potential in the management of cacao black pod disease with increased half-life of the conidia compared with aqueous suspensions. In addition, oil-based formulations have been proving to be better ingredients in foliar sprays (Feng et al., 2004).

Different additives were used to prepare liquid formulations, including; sucrose, that improved survival of rhizobia; glycerol, which was utilized for preservation of the viability of *Pseudomonas fluorescens* cells in liquid formulations (Taurian et al., 2010). Carboxymethyl cellulose (CMC) is another additive that was used for the preparation of inoculants of *Brevibacillus brevis*, *Bacillus licheniformis*, *Micrococcus* sp., and *Acinetobacter calcoaceticus* for the growth improvement of *Jatrophacurcas* (Jha and Saraf,

2012). Gum Arabic, which is an adhesive, was used for rhizobial cells to protect the microbe from drought, and improving its survivability under stress conditions (Deaker et al., 2011). Polyvinylpyrrolidone (PVP) was also used to increase the survival rate of *Bradyrhizobium japonicum* and resembled gum Arabic in its functions (Singleton et al., 2002). Liquid formulations of *Azospirillum* spp. were developed using various cell protectants, such as trehalose, polyvinylpyrrolidone, and glycerol which were added to the formulations and increased their shelf life (Vendan and Thangaraju, 2006). A high percentage of PGPR *Azospirillum* spp. formulations have been commercialized in South America as liquid formulas and the shelf life of their liquid formulations exceed 6 months (Cassan and Diaz-Zorita, 2016).

4.2 Solid Formulations

Different forms of solid formulations were developed for producing products with a long shelf life and ease of use and transfer. Granules, wettable powders, and dust forms are amongst these solid formulations, and the advantages of each type are described as under.

4.2.1 Granules Formulations

Granules are defined as dry particles that consist of active ingredients, binders and suitable carriers (Mishra and Arora, 2016). Brar et al. (2006) reported that the concentration of the active ingredients in granule formulations is 5 to 20%. Granules are classified as coarse particles and microgranules based on particle size ranges of 100 to 1000 μm and 100 to 600 μm respectively. Granular formulations have more to do with storage and increased shelf life (Callaghan and Gerard, 2005). Several materials were used in formulating different probiotics products, such as: cottonseed flour (Ridgway et al., 1996); corn meal baits; granules formed with gelatinized cornstarch or flour (Tamez et al., 1996), gluten (Behle et al., 1997), semolina (durum), wheat flour (Andersch et al., 1998), wheat meal granules (Navon, 2000), gelatin or acacia gum (Maldonado et al., 2002), and earth diatoms (Batta, 2008). Clayton et al. (2004) revealed that granular formulations are more effective in increasing seed production compared to peat powder or liquid formulations. Semolina–kaolin based granular formulations of *P. trivialis* X33d show a higher suppression activity as a bio-herbicide towards weed (great brome) in wheat fields in comparison to kaolin-talc-based granular formulations (Mejri et al., 2013). BioShield™, a granule containing *Serratia entomophila*, was sold in New Zealand for controlling grass grub larvae in established pastures (Young et al. 2010). The main disadvantage of granular formulations is inactivation of the active ingredient in ultraviolet (UV) light which reduces their applicability, even though they are very effective. There are some UV protectants that can be added to the final products, such as Tinopal, folic, 2-hydroxy-4-methoxy-benzophenone among others, which reduce the UV inactivation impact on the active ingredients (Warrior et al., 2002, Cohen and Joseph, 2009). Currently, some of the entomopathogenic fungal strains, reported by Fernandes et al. (2015), have shown tolerance to UV radiation, and therefore, they are used safely in such formulations.

4.2.2 Wettable Powder Formulations

Wettable powder (WP) formulations are one of the oldest and most important categories of formulations. In order to accomplish preferred effective formulations, Brar et al. (2006) illustrated that such formulations should consist of 50-80% technical powder, 15-45% filler, 1-10% dispersant, and 3-5% surfactant. A WP formulation can have a shelf life exceeding 18 months with adequate control of moisture content. Khan et al. (2007) reported that many agricultural materials and industrial waste by-products could be used as a carrier in WP formulations such as wheat bran–sand mixture, sawdust–sand–molasses mixture, corn cob–sand–molasses mixture, bagasse–sand–molasses mixture, organic cakes, cow dung–sand mixture, compost/farm manure, inert charcoal, diatomaceous earth, and fly ash. In addition, other carriers can be used for the preparation of WP formulations, such as clay, peat, vermiculite, talc, lignite, kaolinite, among others. (Digat, 1989, El-Fattah et al., 2013, Zayed, 2016). As an example for a WP formulation, which was effective in controlling post-harvest disease with better efficiency than chemical treatment, Chen et al. (2015) prepared a formulation containing: 60% *B. cereus* freeze dried powder, 28.9% diatomite (carrier), 4% sodium lignin sulfonate (dispersal agent), 6% alkyl naphthalene sulfonate (wetting agent), 1% K_2HPO_4 (stabilizer), and 0.1% β-cyclodextrin (UV protectant).

Woo et al. (2014) reported that wettable powder formulations of *Trichoderma* with a percentage active content of 55.3% were commercialized. Moreover, Basyony and Abo-Zaid (2018) prepared a wettable powder formulation of PGPR *B. subtilis* B10 as a bionematicide for controlling *Meloidogyne incognita*. The results obtained revealed that WP formulations were effective in reducing the number of galls and egg-masses with reduction percentages of 69.7 and 71.2%, respectively. Moreover, Abdel-Gayed et al. (2019) utilized the same WP formulation with an active ingredient of the PGPR *B. subtilis* B4 as a biofungicide for reducing peanut diseases caused by *Rhizoctonia solani* and *Sclerotium rolfsii*.

4.2.3 Dust Formulations

Dust is a very finely ground combination of the active ingredient (usually 10%) with particles size varying from 50 to 100 μm (Jitendra and Arora, 2016). Microorganisms as active ingredients, can be formulated as dust formulations. These have been used for a long time, and have proven more effective,compared to other types in some cases. Handling problems associated with dust has restricted the use of such formulations (Harris and Dent, 2000, Ifoulis and Savopoulou-Soultani, 2004). However, the nonpathogenic fungus *Fusarium oxysporum* was formulated as a dust formulation under a commercial name of Biofox C and used in basil, cyclamen, tomato, and carnation (Kaur et al., 2010).

4.3 Encapsulated Formulations

Encapsulated formulations were made by coating or entrapping microbial cells within a polymeric material to create beads, which are permeable to nutrients, gases, and metabolites. The aim of this process is to improve cell viability within the beads (John et al., 2011). Encapsulated formulations were divided into two types according to the size of the polymeric bead produced. The first type is macro-encapsulation, in which size of beads varies from few millimeters to centimeters. The second type is a micro-encapsulation, in which the size of the beads varies from 1 to 1000µm, with a diameter of less than 200µm generally (Nordstierna et al., 2010). Encapsulation plays a vital role in the protection of active ingredients from harmful effects of inadequate environmental factors such as drought, freezing, and heat shock effects. There are several polymers, which could be used for the encapsulation of active ingredients. These polymers could be either:

(1) Natural polymers such as: alginate, gelatin, carrageenan, agarose, agar, starch, cellulose, among others or
(2) Synthetic polymers such as: polyacrylamides, polystyrene, polyurethane among others. (Amiet-Charpentier et al., 1998, Park and Chan, 2000, Cheze-Lange et al., 2002, Schoebit et al., 2013, Schoebitz and Belch, 2016).

The action of the encapsulated formulations is based on the gradual degradation of the polymer by microorganisms naturally existing in the soil, followed by the release of entrapped plant probiotics or active ingredients from the beads to the surrounding environment. Alginate is one of the important polymers that has been used for plant probiotics encapsulation, and it is naturally produced from brown algae (Draget et al., 2002, Yabur et al., 2007) and some bacteria such as *Azotobacter* and *Pseudomonas* (Sabra et al., 2001, Trujillo-Roldán et al., 2003). Alginate formulations have some advantages including their nontoxic nature, biodegradability, availability at reasonable costs, and slow release of the entrapped microorganisms or active ingredients into the soil that is managed by the polymeric structure (Bashan et al., 2002, Zohar-Perez et al., 2002). For example, alginate beads containing *B. subtilis* and *P. corrugate* were able to protect the bacteria for 3 years at a storage temperature of 4°C and their ability to induce and increase plant growth was stable (Trivedi and Pandey, 2008). Also, debris recovered from wastewater treatment stations which use alginate beads containing *Chlorella vulgaris*, *C. sorokiniana*, and *A. brasilense* was utilized to improve the growth of sorghum plants and enhance desert soil fertility (Trejo et al., 2012, Lopez et al., 2013). These examples indicate the importance of probiotics and their formulation techniques for the improvement of both plant and soil health.

5. Applications

5.1 Probiotics as Biofertilizers

Bio-fertilizers are formulations that contain living cells of efficient strains of nitrogen fixing, phosphate, potassium, and other elements solubilizing strains, plant growth stimulating, and pathogens-preventing microorganisms. The aim of using bio-fertilization is to increase the availability of nutrients to the plants, and to increase the number of useful microorganisms in the rhizosphere (Mahdi et al., 2010). Bio-fertilizers are eco-friendly, because they have the following advantages over traditional fertilizers. They are more cost-effective than chemical fertilizers; able to increase nitrogen content of soil and plants biologically; increase solubility of the insoluble phosphate and potassium from organic and inorganic sources. In addition, they secrete certain plant growth promoting substances to enhance it under normal and stressed conditions, and exhibit anti-pathogenic activities towards different plant pathogens protecting the plants from them (Mazid and Khan, 2014).

Examples of some types of biofertilizers used in agriculture:

- Nitrogen fixing bio-fertilizers (NFB) examples include *Rhizobium* spp. and *Cyanobacteria.*
- Phosphate solubilizing bio-fertilizer (PSB) examples include *Bacillus* spp., *Pseudomonas* spp., and *Aspergillus* spp.
- Phosphate mobilizing bio-fertilizers (PMB) examples include *Mycorrhiza.*
- Plant growth promoting bio-fertilizer (PGPB) examples include *Pseudomonas* spp.
- Potassium solubilizing bio-fertilizer (KSB) examples include *Bacillus* spp. and *Aspergillus niger*
- Potassium mobilizing bio-fertilizer (KMB) examples include *Bacillus* spp.
- Sulfur oxidizing bio-fertilizer (SOB) examples include *Thiobacillus* spp. (Itelima et al., 2018).

5.2 Plant Probiotics as Biocontrol Agents

There are several methods that are used for controlling plant diseases, such as crop rotation, use of resistant plant varieties, and chemical pesticides, etc.; however, the use of agrochemical pesticides is still a common approach to manage different plant pathogens. Several environmental and health hazards are associated with the use of agrochemical pesticides. Therefore, biocontrol agents are considered an eco-friendly approach that must be employed as an alternative management method of plant diseases. Plant probiotics play an undeniable role in controlling plant pathogens, including fungi, bacteria, nematodes, and viruses by different mechanisms, such as mycoparasitism, competition, antibiosis, production of lytic enzymes, siderophores production, and induced systemic resistance. For these reasons, plant probiotics act as environment friendly biocontrol agents (Jha and Saraf, 2015, Mishra et al., 2015, Myresiotis et al., 2015). *Bacillus, pseudomonads* and *Trichoderma* spp. are major groups of microorganisms among the most studied biocontrol agents because their broad spectrum of activities against different plant pathogens (Siddiqi, 2006, Woo et al., 2014, Shafi et al., 2017, Bjelićet et al., 2018, Przemieniecki et al., 2018).

Unfortunately, the number of applications, which use probiotics as biocontrol agents, still represent a low percentage of the agricultural approaches used to manage plant diseases. Both successful commercialization of plant probiotics as biocontrol agents and improvement of interdisciplinary research in this field are needed. On the other hand, scaling-up production levels using fermentation technology, development of formulation methods, and improvement of marketing strategies to increase end-user knowledge is also required.

5.3 Impact of Plant Probiotics on Crop Yield and Quality

The increasing demands for human food and animal feed, as well as the escalating deterioration in the area and quality of agricultural land, altogether have paved the way for the extensive use of chemical fertilizers and pesticides in the agriculture sector. The extensive use of these agro-chemicals not only causes a severe damage to human and animal health, but also deteriorates the quality of the environment (García-Fraile et al., 2015). Yet, adopting condensed farming practices allows farmers to increase land productivity, and has also intensified the use of chemical fertilizers and pesticides (Timsina, 2018). Eliminating these agro-chemicals from the agricultural process will have adverse impacts on crop yield and quality. Therefore, searching for alternative substitutions for mineral fertilizers and pesticides remains a challenge. Recently, Organic farming and the use of plant probiotics was introduced as a safe, environmentally friendly substitution of these troublesome chemicals (Ahemad and Kibret, 2014). The role of plant probiotics in increasing soil nitrogen and phosphorus availability to plants naturally, as biofertilizers, may be one of the factors that enhanced plant health and crop quality (Yildirim et al., 2011, Bhardwaj et al., 2014, Agbodjato et al., 2016, Kuan et al., 2016). In addition, the usage of several bacterial species and fungi as biocontrol agents, and phytostimulators is now a stable alternative in the biological control field and organic farming practices. This has reduced the amount of agro-chemical usage and its adverse effects on the health of living organisms.

The positive effect of probiotics on crop quality enhancement and how we benefit from it are presented in several research papers (Ochoa-Velasco et al., 2016, Ordookhani et al., 2010). For example, plant probiotics, belonging to the *Phyllobacterium* genus, was reported to enhance vitamin C content of strawberry fruits; this increase was associated with an increase in the content of nitrogen, phosphorous, potassium, and iron compared to non-treated plants (Flores-Félix et al., 2015). When the same crop was treated with two different strains of probiotics, *Paraburkholderia fungorum* BRRh-4 and *Bacillus amylolequefaciens* BChi1, Rahman et al. (2018) reported a significant increase in plant yield and fruit quality, including the following fruit traits: higher titre of phenolics, flavonoids, carotenoids, and anthocyanins with a higher total antioxidant activity than the untreated control sample. Cordero et al. (2018) used probiotics as an alternative to chemical fertilizers on tomatoes. They found enhancement in N, P, and K content under unstressed conditions.

In addition, when broccoli plants received probiotics, their chlorophyll content as well as N, K, Ca, S, P, Mg, Fe, Mn, Zn, and Cu contents increased (Yildirim et al., 2011). For example, banana fruits showed increased levels of N, P, K, Ca, and Mg uptake (Vayssières et al., 2009). Moreover, pepper with improved antioxidant activity was obtained upon treatment with probiotics (Silva et al., 2014). Likewise, flavonoid content has also enhanced in blackberry plants through application of plant probiotics (García-Seco et al., 2013, Ramos-Solano et al., 2014). Altogether, all this indicates the importance of probiotics as an alternative agricultural practice that improves plant health and crop value.

5.4 Role of Soil Probiotics in Soil Bioremediation

Bioremediation is a sequential process that uses microorganisms, plants, or their enzymes to detoxify contaminants from the environment. By the end of this process, all the intermediate compounds are converted into nontoxic compounds, as shown in Figure 4. Probiotic microorganisms were found to have an excellent efficacy to remediate soil contaminants, enhance native microbiota performance, reduce stress on plant, and improve bio-fertilizers efficiency. In addition, PGPRs help in maintaining soil structure, detoxify chemicals, achieve heavy metals sequestration, nutrients recycling, and control pests and diseases (Figure 4). The activities of these microorganisms led to the improvement of plant health and increased plant tolerance to stress. In turn, plants provide microbes with important sources of nutrients in the form of root exudates, such as free amino acids, proteins, carbohydrates, alcohols, vitamins, and hormones, and therefore, facilitate plant growth (Han et al., 2005, Babalola, 2010, Tak et al., 2013). Many factors affect the rate of this process, including soil moisture, soil pH, availability of oxygen, availability of nutrients, the level of contaminants and presence of suitable bio-remediating microorganisms (Cai et al., 2013). As an example for heavy metal bioremediation, the beneficial rhizobacteria *Alcaligenes feacalis* RZS2 and *Pseudomonas aeruginosa* RZS3 were introduced into contaminated soil. The results indicated that these probiotics have the ability to bio-remediate heavy metal from the soil by producing siderophores and promoted plant growth in the contaminated soil (Patel et al., 2016). In another example, bio-augmentation of heavy metals using *Pseudomonas aeruginosa* has led to the reduction of metal concentration and translocation. In addition, the removal of petroleum hydrocarbons was achieved using the same *Pseudomonas aeruginosa* strain. However, better results were achieved when a bio-augmentation-plant assisted technique was employed (Agnello et al., 2016).

6. Challenges and Obstacles in Plant Probiotics Applications

With the massive increase in human population, the demand for more resources is natural. The agricultural sector represents an essential economic

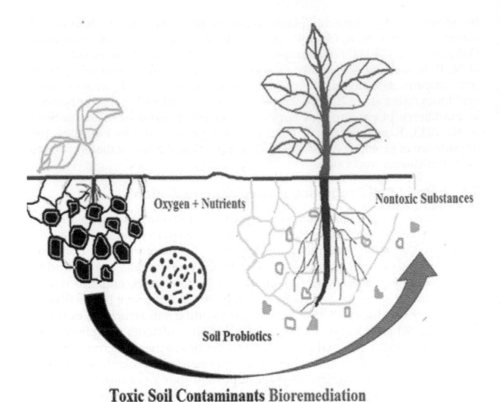

Toxic Soil Contaminants Bioremediation

Fig. 4: Bioremediation of soil contamination using probiotics

growth base in both developed and developing countries (Bhattacharyya and Jha, 2012, Malusá et al., 2012). Meanwhile, with growing environmental issues, the need for novel farming practices to avoid the excessive use of hazardous chemicals is imperative. Recently, more attention has been given to plant probiotics as a promising tool for directing agricultural sustainability, as indicated by its growing global demand (García-Fraile et al., 2015). Though, many plant probiotics formulations are already on the shelf (Vessey, 2003, Woo et al., 2014, García-Fraile et al., 2015), many challenges and limitations are being faced in their widespread application and marketing. One of the most important limiting factors is the selective nature of plant probiotics for hosts contrary to chemical fertilizers, which have a broad-spectrum of applicability to diverse plant species and have proved effective against different plant diseases (Timmusk et al., 2017). Hence, both host selectivity and bio-controlling spectra of diseases utilizing plant probiotic formulations have to be addressed with a wider perspective. Consequently, many research articles emphasize the importance of co-inoculation using more than one plant probiotic strain (consortia) within the same formulation for field applications (Abd-Alla et al., 2001, Atta, 2005, Suneja et al., 2007,

Malusa et al., 2012, Kong et al., 2018, Wang et al., 2018). Bradácová et al. (2019) reporting an enhancement in tomato plant production inoculated with a microbial consortium as an example compared with single inoculants under different stress conditions. Recent studies reported the impact of introducing microbial consortia on the rhizobial population diversity, and how such changes in the rhizosphere microbiota could affect the targeted plant (Wallenstein, 2017, Bonanomi,2018, Kong et al., 2018).

As many factors of either biotic or abiotic nature can affect the diversity of the rhizobacterial microbiota, regular microbiology techniques are limiting our knowledge about plant probiotic inoculants for survival or interaction within the community. Therefore, an introduction of new modern molecular approaches opens the doors to intensified the research in this field, which could enhance discoveries of plant probiotic formulations and sustainable agricultural approaches. In this regard, many attempts were made to explore the microbial diversity of soil probiotics (Soni et al., 2017, Ventorino et al., 2018). Field applications of research results represent another major challenge in plant probiotics implementation. This is due to the enormous environmental factors involved during filed application, such as soil properties, inoculation site, nature of plant microbial community, and the competition between the native microbiota and the introduced strains. One of the main obstacles for plant probiotics utilization is ensuring the viability of the end-product formulations, under field application conditions, by amending the inoculants with a suitable protectants and stabilizing agents. This indicates the importance of the non-living components of the formulation, including fillers and carriers (Herrmann and Lesueur, 2013).

As it was indicated earlier in this chapter, different types of plant probiotic formulations include peat, liquids, powders and granules can be formulated (Arora et al., 2011). De Gregorio et al. (2017) reported a higher enhancement of stability and viability in plant rhizobacteria *Pantoea agglomerans* and *Burkholderia caribensis* upon immobilization in nanofibers. The quality of plant probiotic products is also of great importance, including cell viability in the final formulation with effective cell numbers, product purity, and the safety of humans and the environment (Nakkeeran et al., 2005, Chauhan et al., 2015, Lesueur et al., 2016, Bharti et al., 2017). Additionally, the contamination of probiotic products is another problem that affects product quality. Herrmann et al. (2015) reported the adulteration of many plant probiotic commercial products with *Bacillus* sp. and human pathogens.

Another challenge for plant probiotics marketing is the difficulties encountered during new product registration under strict legislation. These difficulties could be attributed to the unconsolidated registration requirements among countries, complicity of regulations, long time periods required for registration, and the high expenses involved in the process. Altogether, it is important to highlight the importance of generating unified, faster, and easier registration procedures for new plant probiotic products registration (Herrmann et al., 2015, Timmusk et al., 2017).

References

Abd-Alla, M.H., S.A. Omar and S.A. Omar. 2001. Survival of rhizobia/bradyrhizobia and a rock-phosphate-solubilizing fungus *Aspergillus niger* on various carriers from some agro-industrial wastes and their effects on nodulation and growth of faba bean and soybean. J. Plant Nut. 24: 261-272.

Abdel-Gayed, M.A., G.A. Abo-Zaid, S.M. Matar and E.E. Hafez. 2019. Fermentation, formulation and evaluation of PGPR *Bacillus subtilis* isolate as a bioagent for reducing occurrence of peanut soil-borne diseases. J. Integrat. Agric. 18: 2-4.

Agbodjato, N.A., P.A. Noumavo, A. Adjanohoun, L. Agbessi and L. Baba-Moussa. 2016. Synergistic effects of plant growth promoting rhizobacteria and chitosan on *in vitro* seeds germination, greenhouse growth, and nutrient uptake of Maize (*Zea mays* L.). Biotechnol. Res. Intern. 2016, Article ID7830182.

Agnello, A.C., M. Bagard, E.D. van Hullebusch, G. Esposito and D. Huguenot. 2016. Comparative bioremediation of heavy metals and petroleum hydrocarbons co-contaminated soil by natural attenuation, phytoremediation, bioaugmentation and bioaugmentation-assisted phytoremediation. Sci. Total Environ. 563-564: 693-703.

Ahemad, M. and M. Kibret. 2014. Mechanisms and applications of plant growth promoting rhizobacteria: Current perspective. J. King Saud Univ. Sci. 26: 1-20.

Ahmad, M., Z.A. Zahir, M. Khalid, F. Nazli and M. Arshad. 2013. Efficacy of *Rhizobium* and *Pseudomonas* strains to improve physiology, ionic balance and quality of mung bean under salt-affected conditions on farmer's fields. Plant Physiol. Biochem. 63: 170-176.

Ahmed, A. and S. Hasnain. 2010. Auxin producing *Bacillus* sp.: Auxin quantification and effect on the growth *Solanum tuberosum*. Pure Appl. Chem. 82: 313-319.

Alejandro, J.G., C.L. Lorena, F.B. María and R. Raúl. 2017. Plant probiotic bacteria enhance the quality of fruit and horticultural crops. AIMS Microbiol. 3: 483-501.

Alfano, G., M.L. Ivey, C. Cakir, J.I. Bos, S.A. Miller, L.V. Madden et al. 2007. Systemic modulation of gene expression in tomato by *Trichoderma hamatum* 382. Phytopathol. 97: 429-437.

Alves, G.C., S.S. Videira, S. Urquiaga and V.M. Reis. 2015. Differential plant growth promotion and nitrogen fixation in two genotypes of maize by several *Herbaspirillum* inoculants. Plant and Soil 387: 307-321.

Amiet-Charpentier, C., P. Gadille, B. Digat and J.P. Benoit. 1998. Microencapsulation of rhizobacteria by spray-drying: Formulation and survival studies. J. Microencapsul. 15: 639-659.

Andersch, W., R. Hain and M. Kilian. 1998. Granulates containing microorganisms. US Patent 5,804,208 filed 15 Nov 1995: 8.

Arora, M., A. Kaushik, N. Rani and C.P. Kaushik. 2010. Effect of cyanobacterial exopolysaccharides on salt stress alleviation and seed germination. J. Environ. Biol. 31: 701-704.

Arora, N.K., E. Khare and D.K. Maheshwari. 2011. Plant growth promoting rhizobacteria: Constraints in bioformulation, commercialization, and future strategies. pp. 97-116. *In*: D.K. Maheshwari (eds.). Plant Growth and Health Promoting Bacteria. Microbiol. Monographs. Verlag Berlin Heidelberg: Springer.

Arora, N.K., S. Tewari and R. Singh. 2013. Multifaceted plant-associated microbes and their mechanisms diminish the concept of direct and indirect PGPRs. pp. 411-449. *In*: N.K. Arora (eds.). Plant Microbe Symbiosis: Fundamentals and Advances. Springer.

Asaf, S., M.A. Khan, A.L. Khan, M. Waqas, R. Shahzad, A-Y. Kim et al. 2017. Bacterial endophytes from arid land plants regulate endogenous hormone content and promote growth in crop plants: An example of *Sphingomonas* sp. and *Serratia marcescens*. J. Plant Interactions. 12: 31-38.

Ashokvardhan, T., A.B. Rajithasri, P. Prathyusha and K. Satyaprasad. 2014. Actinomycetes from *Capsicum annuum* L. rhizosphere soil have the biocontrol potential against pathogenic fungi. Inter. J. Curr. Microbiol. Appl. Sci. 3: 894-903.

Aslantas, R., R. Cakmakci and F. Sahin. 2007. Effect of plant growth promoting rhizobacteria on young apple tree growth and fruit yield under orchard conditions. Scientia Horticult. 111: 371-377.

Atta, A.M. 2005. Molecular biodiversity and symbiosis of French bean rhizobia in some Egyptian soils. M.Sc. Thesis. Inst. of Grad. Stud. and Res., Alex. Univ, Egypt.

Auld, B.A., S.D. Hetherington and H.E. Smith. 2013. Advances in bioherbicide formulation. Weed Biol. and Mang. 3: 61-67.

Babalola, O.O. 2010. Beneficial bacteria of agricultural importance. Biotechnol. Letters 32: 1559-1570.

Bagyalakshmi, B., P. Ponmurugan and S. Marimuthu. 2012. Influence of potassium solubilizing bacteria on crop productivity and quality of tea (*Camellia sinensis*). Afric. J. Agric. Res. 7: 4250-4259.

Bakker, P.A.H.M., L.X. Ran, C.M.J. Pieterse, L.C. van Loon et al. 2003. Understanding the involvement of rhizobacteria-mediated induction of systemic resistance in biocontrol of plant diseases. Can. J. Plant Pathol. 25: 5-9.

Bashan, Y., G. Holguin and L.E. de-Bashan. 2004. Azospirillum-plant relationships: Physiological, molecular, agricultural, and environmental advances. Can. J. Microbiol. 50: 521-577.

Bashan, Y., J.P. Hernandez, L.A. Leyva and M. Bacilio. 2002. Alginate microbeads as inoculant carrier for plant growth-promoting bacteria. Biol. Fertil. Soils 35: 359-368.

Bastian, F., A. Cohen, P. Piccoli. V. Luna, R. Bottini, R. Baraldi et al. 1998. Production of indole-3-acetic acid and gibberellins A_1 and A_3 by *Acetobacter diazotrophicus* and *Herbaspirillum seropedicae* in chemically-defined culture media. Plant Growth Reg. 24: 7-11.

Basyony, A.G. and G.A. Abo-Zaid. 2018. Biocontrol of the root-knot nematode, *Meloidogyne incognita*, using an eco-friendly formulation from *Bacillus subtilis*, lab. and greenhouse studies. Egyptian J. Biol. Pest Cont. 28: 1-13.

Batool, R. and S. Hasnain. 2005. Growth stimulatory effects of Enterobacter and Serratia located from biofilms on plant growth and soil aggregation. Biotechnol. 4: 347-353.

Batta, Y.A. 2008. Control of main stored-grain insects with new formulations of entomopathogenic fungi in diatomaceous earth dusts. Intern. J. Food Eng. 4: 9.

Behle, R.W., M.R. McGuire, R.L. Gillespie and B.S. Shasha. 1997. Effects of alkaline gluten on the insecticidal activity of *Bacillus thuringiensis*. J. Econom. Entomol. 90: 354-360.

Bentley, R. and J.W. Bennett. 2003. What is an antibiotic? revisited. Adv. Appl. Microbiol. 52: 303-331.

Bhardwaj, D., M.W. Ansari, R.K. Sahoo and N. Tuteja. 2014. Biofertilizers function as key player in sustainable agriculture by improving soil fertility, plant tolerance and crop productivity. Microb. Cell Fact. 13: 66.

Bharti, N., S.K. Sharma, S. Saini, A. Verma, Y. Nimonkar, O. Prakash et al. 2017. Microbial plant probiotics: Problems in application and formulation. pp. 317-335.

In: V. Kumar, M. Kumar, S. Sharma, R. Prasad (eds.). Probiotics and Plant Health. Singapore: Springer.

Bhattacharyya, P.N. and D.K. Jha. 2012. Plant growth-promoting rhizobacteria (PGPR): Emergence in agriculture. World J. Microbiol. Biotechnol. 28: 1327-1350.

Bjelić, D., J. Marinković, B. Tintor and N. Mrkovački. 2018. Antifungal and plant growth promoting activities of indigenous rhizobacteria isolated from maize (*Zea mays* L.) rhizosphere. Commun. Soil Sci. and Plant Anal. 49: 88-98.

Bonanomi, G., M. Lorito, F. Vinale and S.L. Woo. 2018. Organic amendments, beneficial microbes, and soil microbiota: Toward a unified framework for disease suppression. Ann. Rev. Phytopathol. 56: 1-20.

Borriss, R. 2015: Bacillus, a plant-beneficial bacterium. pp. 379-391. *In*: B. Lugtenberg (ed.). Principles of Plant-Microbe Interactions. International Publishing: Springer.

Bottini, R., M. Fulchieri, D. Pearce and R.P. Pharis. 1989. Identification of gibberellins A_1, A_3 and iso-A_3, in cultures of *Azospirillum lipoferum*. Plant Physiol. 90: 45-47.

Boukhalfa, H., J. Lack, S.D. Reilly, L. Hersman and M.P. Neu. 2003. Siderophore production and facilitated uptake of iron and plutonium in *P. putida*. AIP Conf. Proceedings 673: 343-344.

Bradácová, K., A.S. Florea, A. Bar-Tal, D. Minz, U. Yermiyahu, R. Shawahnae et al. 2019. Microbial consortia versus single-strain inoculants: An advantage in PGPM-assisted tomato production. Agron. 9: 105.

Brar, S.K., M. Verma, R.D. Tyagi and J.R. Valéro. 2006. Recent advances in downstream processing and formulations of *Bacillus thuringiensis* based biopesticides. Process Biochem. 41: 323-342.

Brígido, C., F.X. Nascimento, J. Duan, B.R. Glick and S. Oliveira. 2013. Expression of an exogenous 1aminocyclopropane-1-carboxylate deaminase gene in *Mesorhizobium* spp. reduces the negative effects of salt stress in chickpea. FEMS Microbiol. Lett. 349: 46-53.

Burges, H.D. and K.A. Jones (eds). 1998. Formulation of Microbial Biopesticides: Beneficial Microorganisms, Nematodes and Seed Treatments. Kluwer Academic Publishers, Dordrecht, pp. 411.

Cai, M., J. Yao, H. Yang, R. and K. Masakorala. 2013. Aerobic biodegradation process of petroleum and pathway of main compounds in water flooding well of Dagang oil field. J. Bioresour. Technol. 144: 100-106.

Callaghan, M.O. and F.M. Gerard. 2005. Establishment of Serratia entomophila in soil from a granular formulation. New Zealand Plant Protec. 58: 122-125.

Cassan, F. and M. Diaz-Zorita. 2016. *Azospirillum* sp. in current agriculture: From the laboratory to the field. Soil Biol. Biochem. 103: 117-130.

Chauhan, H., D.J. Bagyaraj, G. Selvakumar and S.P. Sundaram. 2015. Novel plant growth promoting rhizobacteria—Prospects and potential. Appl. Soil Ecol. 95: 38-53.

Cheng, H., L. Li, J. Hua, H. Yuan and S. Cheng. 2015. A preliminary preparation of endophytic bacteria CE3 wettable powder for biological control of postharvest diseases. Notulae botanicae Horti Agrobotanici Cluj-Napoca 43: 159-164.

Chet, I. 1987. Trichoderma application, mode of action, and potential as a biocontrol agent of soil-borne plant pathogenic fungi. pp. 137-160. *In*: I. Chet (ed.). Innovative Approaches to Plant Disease Control. New York: John Wiley and Sons.

Cheze-Lange, H., D. Beunard, P. Dhulster, D. Guillochon, A.M. Caze, M. Morcellet et al. 2002. Production of microbial alginate in a membrane bioreactor. Enz. Microb. Technol. 30: 656-661.

Chilukoti, N., A. Kondreddy, P. Pallinti, S. Katta, P.V.S.R.N. Sarma et al. 2010. Biotechnological approaches to develop bacterial chitinases as a bioshield against fungal diseases of plants. Crit. Rev. Biotechnol. 30: 231-241.

Clayton, G.W., W.A. Rice, N.Z. Lupwayi, A.M. Johnston, G.R. Lafond, C.A. Grant et al. 2014. Inoculant formulation and fertilizer nitrogen effects on field pea: Nodulation, N_2 fixation and nitrogen partitioning. Can. J. Plant Sci. 84: 79-88.

Cohen, E. and T. Joseph. 2009. Photostabilization of *Beauveria bassiana* conidia using anionic dyes. Appl. Clay Sci. 42: 569-574.

Cordero, I., L. Balaguer, A. Rincon and J.J. Pueyo. 2018. Inoculation of tomato plants with selected PGPR represents a feasible alternative to chemical fertilization under salt stress. J. Plant Nutr. Soil Sci. 181: 694-703.

Cornelis, P. 2010. Iron uptake and metabolism in pseudomonads. Appl. Microbiol. Biotechnol. 86: 1637-1645.

Costa, O.Y.A., J.M. Raaijmakers and E.E. Kuramae. 2018. Microbial extracellular polymeric substances: Ecological function and impact on soil aggregation. Front. Microbiol. 9: 1636.

Dastager, S.G., C.K. Deepa and A. Pandey. 2010. Isolation and characterization of novel plant growth promoting *Micrococcus* sp. NII-0909 and its interaction with cowpea. Plant Physiol. and Biochem. 48: 987-992.

De Gregorio, P.R., G. Michavila, L.R. Muller, C.D.S. Borges, M.F. Pomares, E.L. Saccol de Sá et al. 2017. Beneficial rhizobacteria immobilized in nanofibers for potential application as soybean seed bioinoculants. PLoS One 12: e0176930.

Deaker R., M.L. Kecskés, M.T. Rose, K. Amprayn, K. Ganisan, T.K.C. Tran et al. 2011. Practical methods for the quality control of inoculants biofertilisers. ACIAR Monograph Series No. 147, Canberra, pp. 101.

Deng, J., E.P. Orner, J.F. Chau, E.M. Anderson, A.L. Kadilak, R.L. Rubinstein et al. 2015. Synergistic effects of soil microstructure and bacterial EPS on drying rate in emulated soil micromodels. Soil Biol. Biochem. 83: 116-124.

Digat, B. 1989. Strategies for seed bacterization. Acta Horticulturae 253: 121-130.

Dimkpa, C.O., D. Merten, A. Svatos, G. Büchel and E. Kothe. 2009. Siderophores mediate reduced and increased uptake of cadmium by Streptomyces tendae F4 and sunflower (*Helianthus annuus*), respectively. J. Appl. Microbiol. 107: 1687-1696.

Draget, K.I., O. Smidsrød and G. Skjåk-Bræk. 2002. Alginates from algae. pp. 215-240. *In*: A. Steinbüchel, S. De Daets and E.J. Vandame (eds.). Biopolymers. Weinheim: Wiley-VCH.

El-Akhal, M.R., A. Rincon, T. Coba de la Pena, M.M. Lucas, N. El Mourabit, S. Barrijal et al. 2013. Effects of salt stress and rhizobial inoculation on growth and nitrogen fixation of three peanut cultivars. Plant Biol. (Stuttgart) 15: 415-421.

El-Fattah, D.A.A., W.E. Eweda, M.S. Zayed and M.K. Hassanein. 2013. Effect of carrier materials, sterilization method, and storage temperature on survival and biological activities of *Azotobacter chroococcum* inoculant. Ann. Agric. Sci. 58: 111-118.

Etesami, H., S. Emami and H.A. Alikhani. 2017. Potassium solubilizing bacteria (KSB): Mechanisms, promotion of plant growth, and future prospects – A review. J. Soil Sci. and Plant Nut. 17: 897-911.

Feng, M.G., X.Y. Pu, S.H. Ying and Y.G. Wang. 2004. Field trials of an oil-based emulsifiable formulation of *Beauveria bassiana* conidia and low application rates of imidacloprid for control of false-eye leafhopper *Empoasca vitis* on tea in southern China. Crop Prot. 23: 489-496.

Fernandes, E.K., D.N. Rangel, G.L. Braga and D. Roberts. 2015. Tolerance of entomopathogenic fungi to ultraviolet radiation: A review on screening of strains and their formulation. Curr. Genet. 61: 427-440.

Flores-Felix, J.D., E. Menendez, L.P. Rivera, M. Marcos-García, P. Martínez-Hidalgo, P.F. Mateos et al. 2013. Use of *Rhizobium leguminosarum* as a potential biofertilizer for *Lactuca sativa* and Daucuscarota crops. J. Plant Nut. Soil Sci. 176: 876-882.

Flores-Félix, J.D., M. Marcos-García, L.R. Silva, E. Menéndez, E. Martínez-Molina, P.F. Mateos et al. 2015a. Rhizobium as plant probiotic for strawberry production under microcosm conditions. Symbiosis. 67: 25-32.

Flores-Félix, J.D., L.R. Silva, L.P. Rivera, M. Marcos-García, P. García-Fraile, E. Martínez-Molina et al. 2015b. Plants probiotics as a tool to produce highly functional fruits: The case of Phyllobacterium and vitamin C in strawberries. PLoS One 10: e0122281.

Fukuhara, H., Y. Minakawa, S. Akao and K. Minamisawa. 1994. The involvement of indole-3-acetic acid produced by *Bradyrhizobium elkanii* in nodule formation. Plant and Cell Physiol. 35: 1261-1265.

Gaber, Y., S. Mekasha, G. Vaaje-Kolstad, V.G. Eijsink and M.W. Fraaije. 2016. Characterization of a chitinase from the cellulolytic actinomycete *Thermobifida fusca*. Biochim. et Biophys. Acta (BBA) Proteins and Proteomics 1864: 1253-1259.

Gamalero, E. and B.R. Glick. 2015. Bacterial modulation of plant ethylene levels. Plant Physiol. 169: 13-22.

García-Fraile, P., E. Menéndez and R. Rivas. 2015. Role of bacterial biofertilizers in agriculture and forestry. AIMS Bioeng. 2: 183-205.

Garcia-Fraile, P., L. Carro, M. Robledo, M.H. Ramírez-Bahena, J.D. Flores-Félix, M.T. Fernández et al. 2012. Rhizobium promotes non-legumes growth and quality in several production steps: towards a biofertilization of edible raw vegetables healthy for humans. PloS One 7(5): e38122.

García-Seco, D., A. Bonilla, E. Algar, A. García-Villaraco, J.G. Mañero, B. Ramos-Solano et al. 2013. Enhanced blackberry production using *Pseudomonas fluorescens* as elicitor. Agronomy Sust. Develop. 33: 385-392.

Gargouri, O.F., D.B. Abdallah, I. Ghorbel, I. Charfeddine, L. Jlaiel, M.A. Trik et al. 2017. Lipopeptides from a novel *Bacillus methylotrophicus* 39b strain suppress Agrobacterium crown gall tumours on tomato plants. Pest Mang. Sci. 73: 568-574.

Geddie, J.L. and I.W. Sutherland. 1993. Uptake of metals by bacterial polysaccharides. J. Appl. Bacteriol. 74: 467-472.

Ghoul, M. and S. Mitri. 2016. The ecology and evolution of microbial competition. Trends in Microbiol. 24: 833-845.

Glick, B.R., C. Liu, S. Ghosh and E.B. Dumbroff. 1997. Early development of canola seedlings in the presence of the plant growth-promoting rhizobacterium *Pseudomonas putida* GR12-2. Soil Biol. and Biochem. 29: 1233-1239.

Glick, B.R., D.M. Penrose and J. Li. 1998. A model for the lowering of plant ethylene concentrations by plant-growth promoting bacteria. J. Theor. Biol. 190: 63-68.

Glick, B.R., B. Todorovic, J. Czarny, Z. Cheng, J. Duan, B. McConkey et al. 2007. Promotion of plant growth by bacterial ACC Deaminase. Critical Reviews in Plant Sciences 26: 227-242.

Glick, B.R. 2012. Plant growth-promoting bacteria: Mechanisms and applications. Scientifica 2012, Article ID 963401.

Glick, B.R. 2014. Bacteria with ACC deaminase can promote plant growth and help to feed the world. Microbiol. Res. 169: 30-39.

Gomaa, E.Z. 2012. Chitinase production by *Bacillus thuringiensis* and *Bacillus licheniformis*: Their potential in antifungal biocontrol. J. Microbiol. 50: 103-111.

Gorecki, R.J., G.E. Harman and L.R. Mattick. 1985. The volatile exudates from germinating pea seeds of different viability and vigor. Can. J. Bot. 63: 1035-1039.

Goswami, D., J.N. Thakker and P.C. Dhandhukia. 2016. Portraying mechanics of plant growth promoting rhizobacteria (PGPR): A review. Cogent Food & Agric. 2: 1-19.

Habibi, S., S. Djedidi, K. Prongjunthuek, M.F. Mortuza, N. Ohkama-Ohtsu, H. Sekimoto et al. 2014. Physiological and genetic characterization of rice nitrogen fixer PGPR isolated from rhizosphere soils of different crops. Plant and Soil 379: 51-66.

Han, J., L. Sun, X. Dong, Z. Cai, X. Sun, H. Yang et al. 2005. Characterization of a novel plant growth-promoting bacteria strain *Deftia tsuruhatensis* HR4 both as adiazotroph and a potential biocontrol agent against various plant pathogens. Syst. Appl. Microbiol. 28: 66-76.

Hanano, S., M.A. Domagalska, F. Nagy and S.J. Davis. 2006. Multiple phytohormones influence distinct parameters of the plant circadian clock. Genes Cells. 11: 1381-1392.

Harman, G.E., C.R. Howell, A. Viterbo, I. Chet and M Lorito. 2004. Trichoderma species-opportunistic, avirulent plant symbionts. Nature Rev. Microbiol. 2: 43-56.

Harris, J. and D. Dent. 2000. Priorities in Biopesticide Research and Development in Developing Countries. Centre for Agriculture and Bioscience International Publishing. Wallingford UK, pp. 69. ISBN 085 1994792.

Hasan, S., G. Gupta, S. Anand and H. Kaur. 2014. Lytic enzymes of Trichoderma: Their role in plant defense. International J. Appl. Res. Studies. 3: 1-5.

Herrmann, L. and D. Lesueur. 2013. Challenges of formulation and quality of biofertilizers for successful inoculation. Appl. Microbiol. Biotechnol. 97: 8859-8873.

Herrmann, L., M. Atieno, L. Brau and D. Lesueur. 2015. Microbial quality of commercial inoculants to increase BNF and nutrient use efficiency. pp. 1031-1040. *In*: D.B.J Frans (ed.). Biological Nitrogen Fixation. Hoboken: John Wiley & Sons.

Honma, M. and T. Shimomura. 1978. Metabolism of 1-amincyclopropane-1-carboxylic acid. Agric. and Biol. Chem. 42: 1825-1831.

Howell, C.R. 2003. Mechanisms employed by Trichoderma species in the biological control of plant diseases: The history and evolution of current concepts. Plant Dis. 87: 4-10.

Hu, W., Q. Gao, M.S. Hamada, D.H. Dawood, J. Zheng, Y. Chen et al. 2014. Potential of *Pseudomonas chlororaphis* subsp. *aurantiaca* Strain Pcho10 as a biocontrol agent against *Fusarium graminearum*. Phytopathol. 104: 1289-1297.

Hubbell, D.H. and G. Kidder. 2009. Biological Nitrogen Fixation. University of Florida IFAS Extension Publication SL 16. 10: 1-4.

Hynes, R.K. and S.M. Boyetchko. 2006. Research initiatives in the art and science of biopesticide formulations. Soil Biol. Biochem. 38: 845-849.

Ifoulis, A.A. and M. Savopoulou-Soultani. 2004. Biological control of *Lobesia botrana* (Lepidoptera: Tortricidae) larvae by using different formulations of *Bacillus thuringiensis* in 11 vine cultivars under field conditions. J. Econ. Entomol. 97: 340-343.

Itelima, J.U., W.J. Bang, M.D. Sila, I.A. Dryimba and O.J. Egbere. 2018. Bio-fertilizers as key player in enhancing soil fertility and crop productivity: A review. Dir. Res. J. Agric. Food Sci. 2: 22-28.

Jaiswal, D.K., J.P. Verma, S. Prakash, V.S. Meena and R.S. Meena. 2016. Potassium as an important plant nutrient in sustainable agriculture: A state of the art. pp. 21-29. *In*: V.S. Meena, B.R. Maurya, J.P. Verma, R.S. Meena (eds.). Potassium Solubilizing Microorganisms for Sustainable Agriculture. Springer.

Jenkins, M.B., R.A. Virginia and W.M. Jarrell. 1987. Rhizobial ecology of the woody legume mesquite (*Prosopis glandulosa*) in the Sonoran desert. Appl. Environ. Microbiol. 53: 36-40.

Jha, C.K. and M. Saraf. 2012. Evaluation of multispecies plant-growth promoting consortia for the growth promotion of *Jatropha curcas* L. J. Plant Growth Regul. 31: 588-598.

Jha, C.K. and M. Saraf. 2015. Plant growth promoting rhizobacteria (PGPR): A review. E3 J. Agric. Res. and Dev. 5: 108-119.

John, R.P., R.D. Tyagi, S.K. Brar, R.Y. Surampalli and D. Prévost. 2011. Bio-encapsulation of microbial cells for targeted agricultural delivery. Crit. Rev. Biotechnol. 31: 211-226.

Kasim, W.A., R.M. Gaafar, R.M. Abou-Ali, M.N. Omar and H.M. Hewait. 2016. Effect of biofilm forming plant growth promoting rhizobacteria on salinity tolerance in barley. Ann. Agric. Sci. 61: 217-227.

Kasotia, A., A. Varma, N. Tuteja and D.K. Choudhary. 2016. Amelioration of soybean plant from saline-induced condition by exopolysaccharide producing pseudomonas-mediated expression of high affinity K^+ transporter (HKT1) gene. Curr. Sci. 111: 25.

Kaur, R., J. Kaur and R.S. Singh. 2010. Nonpathogenic fusarium as a biological control agent. J. Plant Pathol. 9: 79-91.

Khan, M.S., A. Zaidi and P.A. Wani. 2007. Role of phosphate-solubilizing microorganisms in sustainable agriculture – A review. Agron. for Sust. Develop. 27: 29-43.

Khan, N., A. Bano and M.A. Babar. 2017. The root growth of wheat plants, the water conservation and fertility status of sandy soils influenced by plant growth promoting rhizobacteria. Symbiosis. 72: 195-205.

Khare, E. and A. Yadav. 2017. The role of microbial enzyme systems in plant growth promotion. Clim. Change and Environ. Sust. 5: 122-145.

Khilyas, I.V., T.V. Shirshikova, L.E. Matrosova, A.V. Sorokina, M.R. Sharipova, L.M. Bogomolnaya et al. 2016. Production of siderophores by *Serratia marcescens* and the role of MacAB efflux pump in siderophores secretion. BioNano Sci. 6: 480-482.

Kim, K.J., Y.J. Yang and J.G. Kim. 2003. Purification and characterization of chitinase from Streptomyces sp. M-20. J. Biochem. Mol. Biol. 36: 185-189.

Kim, Y.C., H. Jung, K.Y. Kim and S.K. Park. 2008. An effective biocontrol bioformulation against Phytophthora blight of pepper using growth mixtures of combined chitinolytic bacteria under different field conditions. Eur. J. Plant Pathol. 120: 373-382.

Kishore, G.K., S. Pande and A.R. Podile. 2005. Biological control of late leaf spot of peanut (Arachis hypogaea) with chitinolytic bacteria. Phytopathol. 95: 1157-1165.

Kiss, T. and E. Farkas. 1998. Metal-binding ability of desferrioxamine B. J. Incl. Pheno. and Mol. Recog. in Chem. 32: 385-403.

Kong, W., D.R. Meldgin, J.J. Collins and T. Lu. 2018. Designing microbial consortia with defined social interactions. Nature Chem. Biol. 14: 821-829.

Kong, Z., B.R. Glick, J. Duan, S. Ding, J. Tian, B.J. McConkey et al. 2015. Effects of 1-aminocyclopropane-1-carboxylate (ACC) deaminase-overproducing *Sinorhizobium meliloti* on plant growth and copper tolerance of Medicago lupulina. Plant Soil 391: 383-398.

Kosegarten, H., F. Grolig, A. Esch, K.H. Glüsenkamp and K. Mengel. 1999. Effects of NH_4^+, NO_3^- and HCO_3^- on apoplast pH in the outer cortex of root zones of maize, as measured by the fluorescence ratio of fluorescein boronic acid. Planta. 209: 444-452.

Kuan, K., B.R. Othman, K. Abdul Rahim and H. Shamsuddin. 2016. Plant growth-promoting rhizobacteria inoculation to enhance vegetative growth, nitrogen fixation and nitrogen remobilisation of maize under greenhouse conditions. PloS One 11: e0152478.

Kumar, H., V.K. Bajpai, R.C. Dubey, D.K. Maheshwaria and S.C. Kang. 2010. Wilt disease management and enhancement of growth and yield of *Cajanus cajan* (L.) var. Manak by bacterial combinations amended with chemical fertilizer. Crop Prot. 29: 591-598.

Kumar, M., A. Brar, V. Vivekanand and N. Pareek. 2017. Production of chitinase from thermophilic *Humicola grisea* and its application in production of bioactive chitooligosaccharides. Int. J. Biol. Macromolec. 104: 1641-1647.

Kumar, S.S., K.R. Ram, D.R. Kumar, S. Panwar and C.S. Prasad. 2013. Biocontrol by plant growth promoting rhizobacteria against black scurf and stem canker disease of potato caused by *Rhizoctonia Solani*. Arch. Phytopathol. Plant Prot. 46: 487-502.

Kumari, Y.S.M.A.I. and A. Vengadaramana. 2017. Stimulation of defense enzymes in tomato (*Solanum lycopersicum* L.) and chilli (*Capsicum annuum* L.) in response to exogenous application of different chemical elicitors. Universal J. Plant Sci. 5: 10-15.

Lesueur, D., R. Deaker, L. Herrmann, L. Bräu and J. Jansa. 2016. The production and potential of biofertilizers to improve crop yields. pp. 71-92. *In*: N.K. Arora, M. Samina and B. Raffaella (eds.). Bioformulations: For Sustainable Agriculture. India: Springer.

Liu, F., S. Xing, H. Ma, Z. Du and B. Ma. 2013. Cytokinin-producing, plant growth-promoting rhizobacteria that confer resistance to drought stress in *Platycladus orientalis* container seedlings. Appl. Microbiol. Biotechnol. 97: 9155-9164.

Lopez, B.R., Y. Bashan, A. Trejo and L.E. de-Bashan. 2013. Amendment of degraded desert soil with waste water debris containing immobilized *Chlorella sorokiniana* and *Azospirillum brasilense* significantly modifies soil bacterial community structure, diversity, and richness. Biol. and Fertil. Soils. 49: 1053-1063.

Lugtenberg, B.J., T.F. Chin-A-Woeng and G.V. Bloemberg. 2002. Microbe-plant interactions: Principles and mechanisms. Antonie van Leeuwenhoek, Int. J. Gen. Mol. Microbiol. 81: 373-383.

Magnucka, E.G. and S.J. Pietr. 2015. Various effects of fluorescent bacteria of the genus *Pseudomonas* containing ACC deaminase on wheat seedling growth. Microbiol. Res. 181: 112-119.

Mahdi, S.S., G.I. Hassan, S.A. Sammoon, H.A. Rather, A.D. Showkat, B. Zehra et al. 2010. Bio-fertilizers in organic agriculture. J. Phytol. 2: 42-54.

Maldonado Blanco, M.G., L.J. Galan Wong, C. Rodriguez Padilla and H. Quiroz Martinez. 2002. Evaluation of polymer-based granular formulations of *Bacillus thuringiensis* israelensis against larval *Aedes aegypti* in the laboratory. J. American Mosquito Cont. Assoc. 18: 352-358.

Malusá, E., L. Sas-Paszt and J. Ciesielska. 2012. Technologies for beneficial microorganisms inocula used as biofertilizers. The Sci. World J. 2012, Article ID 491206.

Malviya, K.M., P. Trivedi and A. Pandey. 2018. Chitinase and glucanase activities of antagonistic *Streptomyces* spp. isolated from fired plots under shifting cultivation in northeast India. J. Adv. Res. Biotechnol. 3: 1-7.

Manjula, K., G.K. Kishore and A.R. Podile. 2004. Whole cells of Bacillus subtilis AF1 proved more effective than cell-free and chitinase based formulations in biological control of citrus fruit rot and groundnut rust. Can. J. Microbiol. 50: 737-744.

Martínez-Hidalgo, P., J. García and M.J. Pozo. 2015. Induced systematic resistance against *Botrytis cinereal* by Micromonospora strains isolated from root nodules. Front. Microbiol. 6: 922.

Masalha, J., H. Kosegarten, Ö. Elmaci and K. Mengel. 2000. The central role of microbial activity for iron acquisition in maize and sunflower. Biol. and Fertil. of Soils 30: 433-439.

Matzanke, B.F. 1991. Structures, coordination chemistry and functions of microbial iron chelates. pp. 15-64. *In*: G. Winkelmann (ed.). CRC Handbook of Microbial Iron Chelates. F.L. Boca Raton, USA: CRC Press.

Mazid, M. and T.A. Khan. 2014. Future of bio-fertilizers in Indian agriculture: An overview. International J. Agric. and Food Res. 3: 10-23.

Mbarga, J.B., B.A.D. Begoude, Z. Ambang, M. Meboma, J. Kuate, B. Schiffers et al. 2014. A new oil-based formulation of *Trichoderma asperellum* for the biological control of cacao black pod disease caused by *Phytophthora megakarya*. Biol. Cont. 77: 15-22.

Mejri, D., E. Gamalero and T. Souissi. 2013. Formulation development of the deleterious rhizobacterium *Pseudomonas trivialis* X33d for biocontrol of brome (*Bromus diandrus*) in durum wheat. J. Appl. Microbiol. 114: 219-228.

Menendez, E. and P.G. Fraile. 2017. Plant probiotic bacteria: Solutions to feed the world. AIMS Microbiol. 3: 502-524.

Mengel, K. 1995. Iron availability in plant tissues – Iron chlorosis on calcareous soils. pp. 389-397. *In*: J. Abadia (ed.). Iron Nutrition in Soils and Plants. Dordrecht, The Netherlands: Kluwer.

Mhlongo, M.I., L.A. Piater, N.E. Madala, N. Labuschagne and I.A. Dubery. 2018. The chemistry of plant-microbe interactions in the rhizosphere and the potential for metabolomics to reveal signaling related to defense priming and induced systemic resistance. Front. Plant Sci. 9: 112.

Mishra, J. and N.K. Arora. 2016. Bioformulations for plant growth promotion and combating phytopathogens: A sustainable approach. pp. 3-33. *In*: N. Arora, S. Mehnaz and R. Balestrini (eds.). Bioformulations: For Sustainable Agriculture. Springer, New Delhi.

Mishra, S., A. Singh, C. Keswani, A. Saxena, B.K. Sarma and H.B. Singh.. 2015. Harnessing plant-microbe interactions for enhanced protection against phytopathogens. pp. 111-125. *In*: N. Arora (ed.). Plant Microbes Symbiosis: Applied Facets. Springer, New Delhi.

Mohammed, A.F. 2018. Effectiveness of exopolysaccharides and biofilm forming plant growth promoting rhizobacteria on salinity tolerance of faba bean (*Vicia faba* L.). Afr. J. of Microbial. Res. 12: 399-404.

Mokhtari, W., M. Achouri, B. Hassan and R. Abdellah. 2018. Mycoparasitism of *Trichoderma* spp. against *Phytophthora capsici* and *Rhizoctonia solani*. Int. J. Pure Appl. Biosci. 6: 14-19.

Myresiotis, C.K., Z. Vryzas and E. Papadopoulou-Mourkidou. 2015. Effect of specific plant growth-promoting rhizobacteria (PGPR) on growth and uptake of neonicotinoid insecticide thiamethoxam in corn (*Zea mays* L.) seedlings. Pest Mang. Sci. 71: 1258-1266.

Naher, L., K.U. Yusuf, S.H. Habib, H. Ky and S. Siddiquee. 2018. Mycoparasitism activity of *Trichoderma harzianum* associated with chitinase expression against *Ganoderma boninense*. Pakistan J. Botany. 50: 1241-1245.

Nakkeeran, S., K. Kavitha, S. Mathiyazhagan, W.G.D. Fernando, G. Chandrasekar, P. Renukadevi et al. 2004. Induced systemic resistance and plant growth promotion by *Pseudomonas chlororaphisstrain* PA-23 and *Bacillus subtilis* strain CBE4 against rhizome rot of turmeric (*Curcuma longa* L.). Can. J. Plant Pathol. 26: 417-418.

Nakkeeran, S., W.D. Fernando and Z.A. Siddiqui. 2005. Plant growth promoting rhizobacteria formulations and its scope in commercialization for the management of pests and diseases. pp. 257-296. *In*: Z.A. Siddiqui (ed.). PGPR: Biocontrol and Biofertilization. Dordrecht: Springer.

Nascimento, F.X., C. Brígido, B.R. Glick and M.J. Rossi. 2016. The role of rhizobial ACC deaminase in the nodulation process of leguminous plants. Int. J. Agron. 2016, Article ID 1369472.

Navon, A. 2000. *Bacillus thuringiensis* application in agriculture. pp. 355-369. *In*: J.F. Charles, A. Delécluse, C.L. Roux (eds.). Entomopathogenic bacteria: From laboratory to field application. Dordrecht: Springer.

Neubauer, U., B. Nowak, G. Furrer and R. Schulin. 2000. Heavy metal sorption on clay minerals affected by the siderophore desferroixamine B. Environ. Sci. Technol. 34: 2749-2755.

Ngoufack, F.Z., A.N. El-Noda, F.M. Tchouanguep and M. El-Soda. 2004. Effect of ropy and capsular exopolysaccharides producing strain of *Lactobacillus plantarum* 162RM on characteristics and functionality of fermented milk and soft Kareish type cheese. Afr. J. Biotechnol. 3: 512-518.

Niu, J., J. Chen, Z. Xu, X. Zhu, Q. Wu, J. Li et al. 2016. Synthesis and bioactivities of amino acid ester conjugates of phenazine-1-carboxylic acid. Bioorg. Medic. Chem. Lett. 26: 5384-5386.

Niu, X., L. Song, Y. Xiao and W. Ge. 2018. Drought-tolerant plant growth-promoting rhizobacteria associated with Foxtail Millet in a semi-arid agro-ecosystem and their potential in alleviating drought stress. Front. Microbiol. 8: 2580.

Nordstierna, L., A.A. Abdalla, M. Nordin and M. Nyden. 2010. Comparison of release behavior from microcapsules and microspheres. Prog. Org. Coat. 69: 49-51.

Ochoa-Velasco, C.E., R. Valadez-Blanco, R. Salas-Coronado, F. Sustaita-Rivera, B. Hernández-Carlos, S. García-Ortega et al. 2016. Effect of nitrogen fertilization and *Bacillus licheniformis* biofertilizer addition on the antioxidants compounds and antioxidant activity of greenhouse cultivated tomato fruits (*Solanum lycopersicum* L. var. Sheva). Scientia Horticul. 201: 338-345.

Olorunleke, F.E., G.K.H. Hua, N.P. Kieu, Z. Ma and M. Höfte. 2015. Interplay between orfamides, sessilins and phenazines in the control of Rhizoctonia diseases by *Pseudomonas* sp. CRM12a. Env. Microbiol. Rep. 7: 774-781.

Ordookhani, K., K. Khavazi, A. Moezzi and F. Rejali. 2010. Influence of PGPR and AMF on antioxidant activity, lycopene and potassium contents in tomato. Afr. J. Agric. Res. 5: 1108-1116.

Park, J.K. and H.N. Chang. 2000. Microencapsulation of microbial cells. Biotechnol. Advances 18: 303-319.

Patel, P.R., S.S. Shaikh and R.Z. Sayyed. 2016. Dynamism of PGPR in bioremediation and plant growth promotion in heavy metal contaminated soil. Indian J. of Exp. Biol. 54: 286-290.

Patel, S.T. and F.P. Minocheherhomji. 2018. Review: Plant growth promoting rhizobacteria – Blessing to agriculture. Int. J. Pure and Appl. Biosci. 6: 481-492.

Paulitz, T.C. 1991. Effect of *Pseudomonas putida* on the stimulation of *Pythium ultimum* by seed volatiles of pea and soybean. Phytopathol. 81: 1282-1287.

Pawar, S.T., A.A. Bhosale, T.B. Gawade and T.R. Nale. 2016. Isolation, screening and optimization of exo-polysaccharide producing bacterium from saline soil. J. Microbiol. Biotechnol. Res. 3: 24-31.

Peer, R.V., G.J. Niemann and B. Schippers. 1991. Induced resistance and phytoalexin accumulation in biological control of Fusarium wilt of carnation by *Pseudomonas* sp. Strain WCS417r. Phytopathol. 81: 728-734.

Peix, A., M.H. Ramírez-Bahena, E. Velázquez and E.J. Bedmar. 2015. Bacterial associations with legumes. J. Crit. Rev. Plant Sci. 34: 17-42.

Perrig, D., M.L.O. Boiero, Masciarelli, C. Penna, O.A. Ruiz, F.D. Cassán et al. 2007. Plant-growth-promoting compounds produced by two agronomically important strains of *Azospirillum brasilense* and implications for inoculant formulation. Appl. Microbiol. Biotechnol. 75: 1143-1150.

Philippot, L., J.M. Raaijmakers, P. Lemanceau and W.H. Van der Putten. 2013. Going back to the roots: The microbial ecology of the rhizosphere. Nature Rev. Microbiol. 11: 789-799.

Pindi, P.K. and S.D.V. Satyanarayana. 2012. Liquid microbial consortium – A potential tool for sustainable soil health. J. Biofert. Biopesticides 3: 1-9.

Prapagdee, B., C. Kuekulvong and S. Mongkolsuk. 2008. Antifungal potential of extracellular metabolites produced by *Streptomyces hygroscopicus* against phytopathogenic fungi. Int. J. Biol. Sci. 4: 330-337.

Pringle, A., J.D. Bever, M. Gardes, J.L. Parrent, M.C. Rillig and J.N. Klironomos. 2009. Mycorrhizal symbioses and plant invasions. Ann. Rev. of Ecol. Evol. & Syst. 40: 699-715.

Przemieniecki, S.W., T.P. Kurowski, M. Damszel and K. Krawczyk. 2018. Effectiveness of the Bacillus sp. SP-A9 strain as a biological control agent for spring wheat (*Triticum aestivum* L.). J. Agric. Sci. and Technol. 20: 609-619.

Puri, A., K.P. Padda and C.P. Chanway. 2016. Evidence of nitrogen fixation and growth promotion in canola (*Brassica napus* L.) by an endophytic diazotroph *Paenibacillus polymyxa* P2b-2R. Biol. & Fert. of Soils 52: 119-125.

Qurashi, A.W. and A.N. Sabri. 2012. Bacterial exopolysaccharide and biofilm formation stimulate chickpea growth and soil aggregation under salt stress. Braz. J. Microbiol. 43: 1183-1191.

Rahman, M., A.A. Sabir, J.A. Mukta, Md. M.A. Khan, M. Mohi-Ud-Din, Md. G. Miah et al. 2018. Plant probiotic bacteria *Bacillus* and *Paraburkholderia* improve growth, yield and content of antioxidants in strawberry fruit. Sci. Reports 8: 2504.

Rajkumar, M., N. Ae, M.N.V. Prasad and H. Freitas. 2010. Potential of siderophore-producing bacteria for improving heavy metal phytoextraction. Trends Biotechnol. 28: 142-149.

Ramadan, E.M., A.A. Abdel-Hafez, E.A. Hassan and F.M. Saber. 2016. Plant growth promoting rhizobacteria and their potential for biocontrol of phytopathogens. Afr. J. Microbiol. 10: 486-504.

Ramos-Solano, B., A. Garcia-Villaraco, F.J. Gutierrez-Mañero, J.A. Lucas, A. Bonilla, D. Garcia-Seco et al. 2014. Annual changes in bioactive contents and production in field-grown blackberry after inoculation with *Pseudomonas fluorescens*. Plant Physiol. and Biochem. 74: 1-8.

Ramos-Solano, B., J.A.L. García, A.G. Villaraco, E. Algar, J.G. Cristobal, F.J.G. Mañero et al. 2010. Siderophore and chitinase producing isolates from the rhizosphere of *Nicotiana glauca* Graham growth and induce systemic resistance in *Solanum lycopersicum* L. Plant Soil 334: 189-197.

Rawway, M., E.A. Beltagy, U.M. Abdul-Raouf, M.A. Elshenawy and M.S. Kelany. 2018. Optimization of process parameters for chitinase production by a marine *Aspergillus flavus* MK20. J. Ecol. of Health & Env. 6: 1-8.

Ridgway, R.L., V.L. Illum, R.R. Farrar, D.D. Calvin, S.J. Fleischer, M.N. Inscoe et al. 1996. Granular matrix formulation of *Bacillus thuringiensis* for control of the European corn borer (Lepidoptera: Pyralidae). J. Econ. Entomol. 89: 1088-1094.

Rishad, K., S. Rebello, V.K. Nathan, S. Shabanamol and M. Jisha. 2016. Optimised production of chitinase from a novel mangrove isolate, *Bacillus pumilus* MCB-7 using response surface methodology. Biocatal. & Agric. Biotechnol. 5: 143-149.

Roberson, E.B. and M.K. Firestone. 1992. Relationship between desiccation and exopolysaccharide production in a soil *Pseudomonas* sp. Appl. Env. Microbiol. 58: 1284-1291.

Rouphael, Y., P. Franken, C. Schneider, D. Schwarz, M. Giovannetti, M. Agnolucci et al. 2015. Arbuscular mycorrhizal fungi act as biostimulants in horticultural crops. Scientia Horticul. 196: 91-108.

Sabra, W., A.P. Zeng and W.D. Deckwer. 2001. Bacterial alginate: Physiology, product quality and process aspects. Appl. Microbiol. Biotechnol. 56: 315-325.

Sahoo, R.K., M.W. Ansari, M. Pradhan, T.K. Dangar, S. Mohanty, N. Tuteja et al. 2014. Phenotypic and molecular characterization of native *Azospirillum* strains from rice fields to improve crop productivity. Protoplasma. 251: 943-953.

Sandhya, V. and S.Z. Ali. 2015. The production of exopolysaccharide by *Pseudomonas putida* GAPP45 under various abiotic stress conditions and its role in soil aggregation. Microbiol. 84: 512-519.

Sandhya, V.Z., M. Grover, G. Reddy and B. Venkateswarlu. 2009. Alleviation of drought stress effects in sunflower seedlings by the exopolysaccharides producing *Pseudomonas putida* strain GAP-P45. Biol. and Fert. of Soils 46: 17-26.

Sangeeth, K.P., R.S. Bhai and V. Srinivasan. 2012. *Paenibacillus glucanolyticus*, a promising potassium solubilizing bacterium isolated from black pepper (*Piper nigrum* L.) rhizosphere. J. Spices and Arom. Crops 21: 118-124.

Sanlibaba, P. and G.A. Cakmak. 2016. Exo-polysaccharides production by lactic acid bacteria. Appl. Microbiol. 2: 1-5.

Santi, C., D. Bogusz and C. Franche. 2013. Biological nitrogen fixation in non-legume plants. Ann. Bot. 111: 743-767.

Scavino, A.F. and R.O. Pedraza. 2013. The role of siderophores in plant growth-promoting bacteria. pp. 265-285. *In*: D.K. Maheshwari, M. Saraf and A. Aeron (eds.). Bacteria in Agrobiology: Crop Productivity. Verlag Berlin Heidelberg: Springer.

Schisler, D.A., P.J. Slininger, R.W. Behle and M.A. Jackson. 2004. Formulation of *Bacillus* spp. for biological control of plant diseases. Phytopathol. 94: 1267-1271.

Schoebitz, M. and M.D. Belchı. 2016. Encapsulation techniques for plant growth-promoting rhizobacteria. pp. 251-265. *In*: N.K. Arora, M. Samina and B. Raffaella (eds.). Bioformulations: For Sustainable Agriculture. Springer.

Schoebitz, M., M.D. López and A. Roldán. 2013. Bioencapsulation of microbial inoculants for better soil-plant fertilization: A review. Agron. Sust. Dev. 33: 751-765.

Shafi, J., H. Tian and M. Ji. 2017. Bacillus species as versatile weapons for plant pathogens: A Review. Biotechnol. Biotechnol. Equip. 31: 446-459.

Shaikh, S. and R.Z. Sayyed. 2015. Role of plant growth-promoting rhizobacteria and their formulation in biocontrol of plant diseases. pp. 37-51. *In*: N. Arora (ed.). Plant Microbes Symbiosis: Applied Facets. Springer.

Sharma, S.B., R.Z. Sayyed, M.H. Trivedi and T.A. Gobi. 2013. Phosphate solubilizing microbes: Sustainable approach for managing phosphorus deficiency in agricultural soils. Springer Plus 2: 587-127.

Shivalee, A., K. Lingappa and D. Mahesh. 2018. Influence of bioprocess variables on the production of extracellular chitinase under submerged fermentation by Streptomyces pratensis strain KLSL55. J. Genet. Eng. Biotechnol. 16: 421-426.

Siddiqi, Z. 2006. Prospective biocontrol agents of plant pathogens. 111-142. *In*: Z. Siddiqi (ed.). PGPR: Biocontrol and Biofertilization. Netherlands: Springer.

Silva, L.R., J. Azevedo, M.J. Pereira, L. Carro, E. Velazquez, A. Peix et al. 2014. Inoculation of the non-legume *Capsicum annuum* (L.) with Rhizobium strains. 1. Effect on bioactive compounds, antioxidant activity, and fruit ripeness. J. Agric. & Food Chem. 62: 557-564.

Singh, G., D.R. Biswas and T.S. Marwaha. 2010. Mobilization of potassium from waste mica by plant growth promoting rhizobacteria and its assimilation by maize (*Zea mays*) and wheat (*Triticum aestivum* L.): A hydroponics study under phytotron growth chamber. J. Plant Nutr. 33: 1236-1251.

Singh, N. and A. Varma. 2015. Antagonistic activity of siderophore producing rhizobacteria isolated from the semi-arid regions of southern India. Int. J. Curr. Microbiol. Appl. Sci. 4: 501-510.

Singleton, P., H. Keyser and E. Sande. 2002. Development and evaluation of liquid inoculants. pp. 52-66. *In*: D. Herridge (ed.). Inoculants and nitrogen fixation of legumes in Vietnam. ACIAR Proceedings 109e: Canberra.

Sivasakthi, S., G. Usharani and P. Saranraj. 2014. Biocontrol potentiality of plant growth promoting bacteria (PGPR) – *Pseudomonas fluorescens* and *Bacillus subtilis*: A review. Afr. J. Agric. Res. 9: 1265-1277.

Sokolova, M.G., G.P. Akimova and O.B. Vaishlya. 2011. Effect of phytohormones synthesized by rhizosphere bacteria on plants. J. Appl. Biochem. Microbiol. 47: 274-278.

Somers, E., J. Vanderleyden and M. Srinivasan. 2004. Rhizosphere bacterial signalling: A love parade beneath our feet. Crit. Rev. Microbiol. 30: 205-240.

Soni, R., V. Kumar, D.C. Suyal, L. Jain and R. Goel. 2017. Metagenomics of plant rhizosphere microbiome. pp. 193-205. *In*: R. Singh, R. Kothari, P. Koringa and S. Singh (eds.). Understanding Host-Microbiome Interactions – An Omics Approach. Singapore: Springer.

Spaepen, S. 2015. Plant hormones produced by microbes. pp. 247-256. *In*: B. Lugtenberg (ed.). Principles of Plant-Microbe Interactions. Cham: Springer.

Strzelczyk, E., M. Kampert and R. Pachlewski. 1994. The influence of pH and temperature on ethylene production by mycorrhizal fungi of pine. Mycorrhiza 4: 193-196.

Subhashini, D.V. 2015. Growth promotion and increased potassium uptake of tobacco by potassium-mobilizing bacterium *Frateuria aurantia* grown at different potassium levels in vertisols. Com. Soil Sci. and Plant Anal. 46: 210-220.

Suneja, P., S.S. Dudeja and N. Narula. 2007. Development of multiple coinoculants of different biofertilizers and their interaction with plants. Arch. Agron. Soil Sci. 53: 221-230.

Tak, H.I., F. Ahmad and O.O. Babalola. 2013. Advances in the application of plant growth-promoting rhizobacteria in phytoremediation of heavy metals. pp. 33-52. *In*: D. Whitacre (ed.). Reviews of Environmental Contamination and Toxicology. New York: Springer.

Tamez-Guerra, P., M.R. McGuire, H. Medrano-Roldan, L.J. Galan-Wong, B.S. Shasha, F.E. Vega et al. 1996. Sprayable granule formulations for *Bacillus thuringiensis*. J. Econ. Entomol. 89: 1424-1430.

Tariq, M., M. Noman, T. Ahmed, A. Hameed, N. Manzoor, M. Zafar et al. 2017. Antagonistic features displayed by plant growth promoting rhizobacteria (PGPR): A review. J. Plant Sci. and Phytopathol. 1: 038-043.

Taurian, T., M.S. Anzuay, J.G. Angelini, M.L. Tonelli, L. Ludueña, D. Pena et al. 2010. Phosphate-solubilizing peanut associated bacteria: Screening for plant growth promoting activities. Plant Soil 329: 421-431.

Timmusk, S., L. Behers, J. Muthoni, A. Muraya and A.-C. Aronsson. 2017. Perspectives and challenges of microbial application for crop improvement. Front. Plant Sci. 8: 49.

Timsina, J. 2018. Can organic sources of nutrients increase crop yields to meet global food demand? Agronomy. 8: 214.

Trapet, P., L. Avoscan, A. Klinguer, S. Pateyron, S. Citerne, C. Chervin et al. 2016. The *Pseudomonas fluorescens* siderophore pyoverdine weakens *Arabidopsis thaliana* defense in favour of growth in iron-deficient conditions. Plant Physiol. 171: 675-693.

Trejo, A., L.E. De-Bashan, A. Hartmann, J.P. Hernandez, M. Rothballer, M. Schmid et al. 2012. Recycling waste debris of immobilized microalgae and plant growth-promoting bacteria from wastewater treatment as a resource to improve fertility of eroded desert soil. Environ. Exp. Botany 75: 65-73.

Trivedi, P., A. Pandey and L.M.S. Palni. 2005. Carrier-based preparations of plant growth-promoting bacterial inoculants suitable for use in cooler regions. World J. Microbiol. and Biotechnol. 21: 941-945.

Trivedi, P. and A. Pandey. 2008. Recovery of plant growth-promoting rhizobacteria from sodium alginate beads after 3 years following storage at 4 degrees. J. Ind. Microbiol. Biotechnol. 35: 205-209.

Trujillo-Roldán, M.A., S. Moreno, D. Segura, E. Galindo and G. Espín. 2003. Alginate production by an Azotobacter vinelandii mutant unable to produce alginate lyase. Appl. Microbiol. Biotechnol. 60: 733-737.

Upadhyay, S.K., J.S. Singh and D.P. Singh. 2011. Exopolysaccharide-producing plant growth-promoting rhizobacteria under salinity condition. Pedosphere 21: 214-222.

Vayssières, J.F., G. Goergen, O. Lokossou, P. Dossa and C. Akponon. 2009. A new bactrocera species in benin among mango fruit fly (*Diptera tephritidae*) species. Acta Horticul. 820: 581-588.

Velázquez, E., L.R. Silva, M.H. Ramírez-Bahena and A. Peix. 2016. Diversity of potassium-solubilizing microorganisms and their interactions with plants. pp. 99-110. *In*: V.S. Meena, B.R. Maurya, J. Prakash Verma and R.S. Meena (eds.). Potassium Solubilizing Microorganisms for Sustainable Agriculture. India Journal: Springer.

Vendan, R. and M. Thangaraju. 2006. Development and standardization of liquid formulation for *Azospirillum bioinoculant*. Indian J. Microbiol. 46: 379-387.

Ventorino, V., A. Pascale, P. Adamo, C. Rocco, N. Fiorentino, M. Mori et al. 2018. Comparative assessment of autochthonous bacterial and fungal communities and microbial biomarkers of polluted agricultural soils of the Terra dei Fuochi. Sci. Reports 8: 14281.

Vessey, J.K. 2003. Plant growth promoting rhizobacteria as biofertilizers. Plant Soil 255: 571-586.

Wagner, S.C. 2011. Biological Nitrogen Fixation. Nature Edu. Knowledge 3: 15.

Wallenstein, M.D. 2017. Managing and manipulating the rhizosphere microbiome for plant health: A systems approach. Rhizosphere 3: 230-232.

Wang, J., Q. Li, S. Xu, W. Zhao, Y. Lei, C. Song et al. 2018. Traits-based integration of multi-species inoculants facilitates shifts of indigenous soil bacterial community. Front. Microbiol. 9: 1692.

Wang, X., D.V. Mavrodi, L. Ke, O.V. Mavrodi, M. Yang, L.S. Thomashow et al. 2015. Biocontrol and plant growth-promoting activity of rhizobacteria from Chinese fields with contaminated soils. Microb. Biotechnol. 8: 404-418.

Warrior, P., K. Konduru and P. Vasudevan. 2002. Formulation of biological control agents for pest and disease management. *In*: S.S. Gnanamanickam (ed.). Biological Control of Crop Diseases. New York: Marcel Dekker Inc.

Weyens, N., D.V.D. Leile, S. Taghavi and J. Vangronsveld. 2009. Phytoremediation: Plant-endophyte partnerships take the challenge. Curr. Opin. Biotechnol. 20: 248-254.

Weyens, N., D.V.D. Lelie, S. Taghavi, L. Newman and J. Vangronsveld. 2009. Exploiting plant-microbe partnerships to improve biomass production and remediation. Trends Biotechnol. 27: 591-598.

Whipps, J.M. 2001. Microbial interactions and biocontrol in the rhizosphere. J. Exp. Botany 52: 487-511.

Woo, S.L., M. Ruocco, F. Vinale, M. Nigro, R. Marra, N. Lombardi et al. 2014. Trichoderma-based products and their widespread use in agriculture. The Open Mycol. J. 8: 71-126.

Yabur, R., Y. Bashan and G. Hernández-Carmona. 2007. Alginate from the macroalgae *Sargassum sinicola* as a novel source for microbial immobilization material in wastewater treatment and plant growth promotion. J. Appl. Phycol. 19: 43-53.

Yadav, B.K. and J.C. Tarafdar. 2011. *Penicillium purpurogenum*, unique P mobilizers in arid agro-ecosystems. Arid Land Res. Mang. 25: 87-99.

Yadav, S.N., A.K. Singh, J.K. Peter, H. Masih, J.C. Benjamin, D.K. Singh et al. 2018. Study of exopolysaccharide containing pgprs on growth of okra plant under water stress conditions. Int. J. Cur. Microbiol. Appl. Sci. 7: 3337-3374.

Yadegari, M., H.A. Rahmani, G. Noormohammadi and A. Ayneband. 2010. Plant growth promoting rhizobacteria increase growth, yield and nitrogen fixation in *Phaseolus vulgaris*. J. Plant Nut. 33: 1733-1743.

Yanni, Y.G., F.B. Dazzo, A. Squartini, M. Zanardo, M.I. Zidan, A.E.Y. Elsadany et al. 2016. Assessment of the natural endophytic association between Rhizobium and wheat and its ability to increase wheat production in the Nile delta. Plant and Soil 407: 367-383.

Yedidia, I., M. Shoresh, K. Kerem, N. Benhamou, Y. Kapulnik, I. Chet et al. 2003. Concomitant induction of systemic resistance to *Pseudomonas syringae* pv. lachrymans in cucumber by *Trichoderma asperellum* (T-203) and the accumulation of phytoalexins. Appl. Environ. Microbiol. 69: 7343-7353.

Yildirim, E., H. Karlidag, M. Turan, A. Dursun and F. Goktepe. 2011. Growth, nutrient uptake, and yield promotion of broccoli by plant growth promoting rhizobacteria with manure. Hort. Sci. 46: 932-936.

Young, S.D., R.J. Townsed, J. Swaminathan and M.O. Callaghan. 2010. Serratia entomophila–coated seed to improve ryegrass establishment in the presence of grass grubs. New Zealand Plant Protec. 63: 229-234.

Zayed, M.S. 2016. Advances in formulation development technologies. pp. 219-223. *In*: D. Singh, H. Singh and R. Prabha (eds.). Microbial inoculants in sustainable agricultural productivity. India: Springer.

Zerrouk, I.Z., M. Benchabane, L. Khelifi, K. Yokawa, J. Ludwig-Müller, F. Baluska et al. 2016. A pseudomonas strain isolated from date palm rhizospheres improves root growth and promotes root formation in maize exposed to salt and aluminum stress. J. Plant Physiol. 191: 111-119.

Zhang, C. and F. Kong. 2014. Isolation and identification of potassium-solubilizing bacteria from tobacco rhizospheric soil and their effect on tobacco plants. Appl. Soil Ecol. 82: 18-25.

Zhang, J., W. Liu, X. Yang, A. Gao, X. Li, W. Xiaoyang et al. 2011. Isolation and characterization of two putative cytokinin oxidase genes related to grain number per spike phenotype in wheat. Mol. Biol. Reports 38: 2337-2347.

Zhang, Y., C. Wang, P. Su and X. Liao. 2015. Control effects and possible mechanism of the natural compound Phenazine-1-Carboxiamide against *Botrytis cinerea*. PLoS One 10: 1-17.

Zhao, M. and S.W. Running. 2010. Drought-induced reduction in global terrestrial net primary production from 2000 through 2009. Science 329: 940-943.

Zohar-Perez, C., E. Ritte, L. Chernin, I. Chet and A. Nussinovitch. 2002. Preservation of chitinolytic *Pantoae agglomerans* in a viable form by cellular dried alginate-based carriers. Biotechnol. Progr. 18: 1133-1140.

Probiotics in Aquaculture Applications

Fagr Kh. Abdel-Gawad[1]*, Samah M. Bassem[1] and Hesham El Enshasy[2,3,4]

[1] Water Pollution Research Dept., Environmental Research Division, Centre of Excellence for Advanced Science, National Research Centre, Dokki, Cairo, Egypt

[2] Institute of Bioproduct Development (IBD), Universiti Teknologi Malaysia (UTM), Skudai, Johor Bahru, Malaysia

[3] School of Chemical and Energy Engineering, Faculty of Engineering, Universiti Teknologi Malaysia (UTM), Skudai, Johor Bahru, Malaysia

[4] Bioprocess Development Department, Genetic Engineering and Biotechnology Research Institute, City of Scientific Research and Technology Applications (SRTA city), Universities and Research Institutes zone, New Borg El-Arab, Alexandria 21934, Egypt

1. Introduction

Recently, aquaculture has become a rapidly growing industry in both developed and developing countries. Such growth is supposed to increase in the future at a faster rate which will be encouraged by a depletion of fisheries and the market forces that globalize food supply sources. However, such rapid industrial growth leads to some severe impacts on animal and human health with the relase of high amounts of chemicals into the aquatic environment (Boxall et al., 2004, Sampantamit et al., 2020).

Many researchers considered different strategies taken for modulation of gut microbiota composition to improve immunity, obtain better growth and digestion in addition to the enhancement of disease resistance of hosts in various livestocks and also in human beings (Burr et al., 2007, Kechagia et al., 2013, Sornplang and Piyadeatsoontorn, 2016). Using gut microbiota in different food supplements of beneficial microbes is considered a modern approach not only from a nutritional viewpoint but also as an alternative therapeutic form to avoid the adverse impacts of drugs. Such useful microorganisms are usually known as probiotics that can colonize, multiply in host gut after administration which leads to many beneficial impacts by modulating different biological systems in the host (Pérez-Sánchez et al., 2013, Lazado et al., 2015, Sanders et al., 2018).

---

*Corresponding author: fagrabdlgawad@gmail.com

Probiotics may be defined as cultured products or live microbial feed supplements that affect hosts beneficially by improving microbial balance in addition to their potential to be of commercialized (Fuller, 1987). A more refined definition of probiotics is given by the World Health Organization (WHO) and also the Food and Agriculture Organization (FAO); live microorganisms that whenever managed in appropriate amounts grant a health benefit on different hosts (FAO/WHO Report 2001). Only a few publications consider the verification of probiotics viability by adding to feed, affecting health benefits significantly. The use of probiotics in the field of aquaculture was considered in many published reviews (Farzanfar, 2006, Gram and Ringø, 2005, Qi et al., 2009, Yousefian and Amiri, 2009, Hai, 2015, Shefat, 2018, Mamun et al., 2019). It may be thought that the origin of probiotic applications in aquaculture is uncertain, while evidence proposes comprehensive use of fish in addition to other invertebrate cultures (Austin and Austin, 2012. Preetha et al., 2007) particularly in South America and also in Asian countries India and China especially. Most of the publications dealing with probiotic applications in aquaculture follow the definition of FAO/WHO that involves a wide range of microorganisms; bacteriophage, Gram-negative and Gram-positive bacteria, yeast and microalgae (Kumar et al., 2006, Silva et al., 2013, Ringø et al., 2020). Probiotics show different mechanisms of action in delivering their benefits to hosts as shown in (Figure 1) which shows that research in the field of probiotics in aquatic animals is a dynamic field (Zorriehzahra et al., 2016).

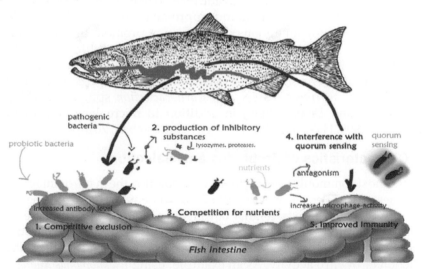

Fig. 1: Overall mechanism of action of probiotics in the host. (1) Competitive exclusion: Ability of probiotic agents to adhere to host's gut so prohibiting colonization of the pathogenic bacteria, (2) Production of inhibitory components: production of some inhibitory components blocking the pathogenic organisms, (3) Competition for nutrients: Probiotics agents use nutrients leading to nutrient deficiency for pathogens, (4) Interference with quorum sense: Probiotics produce components related to quorum sense action, (5) Improved Immunity: Raise the macrophage efficiency in addition to the antibody percent.

2. Probiotics in Aquaculture

2.1 Probiotics and Sustainable Aquaculture

The aquaculture sector is known to be one of the the most growing and promising since it provides food security for people all over the world (De et al., 2014). Fish and other aquatic animals are subjected to stressful conditions during intense practices of aquaculture (Dawood et al., 2016a,b, Hossain et al., 2016) affecting the aquaculture industry significantly through disease outbreaks. Antibiotics and other chemicals are used broadly for infectious disease control in the field of aquaculture (Adel et al., 2016, Dawood et al., 2017). However, extra use of antibiotics negatively affects hosts in addition to their neighbouring environment. Therefore, to develop and sustain the aquaculture industry, it is essential to have more effective and safe substitutes for antibiotics (Abdelkhalek et al., 2017). For achieving a safe and sustainable aquaculture industry, probiotics are considered one the most promising alternatives to antibiotics that enhance growth, nutritional efficiency, immunity and full health status in many fish farms (Dawood et al., 2016 a,b, Van Doan et al., 2017). Different strategies were investigated in animals for modulation of gut microbiota which are responsible for food digestion status, to improve the general health of hosts (Burr *et al.*, 2007). Probiotics were considered as valuable growth supplements for the functional promotion of host diets in many aquaculture farms (Ouwehand et al., 2002, Dawood and Koshio, 2016). Climate change in addition to density and stress intensity management may represent immunosuppressive impacts, while the feed additives qualified for immunomodulation may promote the efficiency of any organism's response to different stresses (Magnadottir, 2010). Many forms of probiotics were used in aquaculture such as (*Lactococcus* sp., *Bacillus* sp., *Micrococcus* sp., *Weissella* sp., *Vibrio* sp., *Enterococcus* sp., *Lactobacillus* sp., *Paenibacillus* sp., *Streptococcus* sp. and *Carnobacterium* sp.), yeasts (*Phaffia*, *Saccharomyces* and *Debaryomyces*) in addition to microalgae (Tetraselmis) (Gatesoupe, 1999, He et al., 2011, Jahangiri and Esteban, 2018).

2.2 Characteristics of Probiotics and Selection:

From the most commonly used methods of probiotics, the in vitro-antagonism techniques were selected, where pathogens were subjected to the selected probiotic or even their extracellular outputs in a solution or a solid medium (Sotomayor and Balcázar, 2003). Moreover, it was proposed that in vitro efficiency in well-diffusion tests and also broth cultures couldn't be used for prediction of in vitro activity. As an example, using *Pseudomonas fluorescens* (strain AH2) against *Aeromonas salmonicida* in case of in vitro antagonism didn't confer protection to furunculosis in Atlantic salmon, but was an efficient probiotic in case of rainbow trout, producing protection against vibriosis (Gram et al., 2001). Therefore, it's essential to define the origin (it's better to utilize the isolated strains from hosts), safety (non-pathogenic strains) and also the strain's ability to move through the host's gastrointestinal tract. The capability of microorganisms to aggregate is referred to in the main selection parameters for effective probiotics, as the effective adherence to the intestinal epithelial cells to minimize or even prohibit pathogen colonization (Vine et al., 2004; Pérez-Sánchez et al., 2013). Moreover, effective probiotic strains must present their beneficial acts in hosts. So, probiotics must be active under ordinary storage cases and must be suitable for industrial processes. Finally, selection of probiotics for application in aquaculture should include the following: knowledge of the background information, gaining effective probiotics, estimation of the capability of effective probiotics to compete with the pathogenic strains, evaluation of pathogenicity of the effective probiotics, estimation of different effects of potential probiotics in hosts, economic value / benefit analyses (Gomez-Gil et al., 2000, El Baz et al., 2018, Chauhan and Singh, 2019). Probiotics may be applied to hosts or offered to aquatic environments through various ways; provided with live feed (Gomez-Gil et al., 1998), added to culture water (Spanggaard et al., 2001), bathing (Austin et al., 1995), provided with artificial feed (Rengpipat et al., 2000). As an example; the daily inoculations of larval white shrimp (*Litopenaeus vannamei*) containers by probiotic bacteria upto a density of 10^5 cfu ml $^{-1}$ prohibited colonization by pathogenic bacteria especially in larval aquacultures (Peeters and Rodríguez, 1999).

2.3   Probiotics Application and Benefits in Aquaculture

Using probiotics in the field of aquaculture is believed to be the most important sustainable alternative to the conventional use of other synthetic antimicrobials which lead to deleterious impacts on the environment. Probiotics don't act only as means for disease control like vaccines and immunostimulants but have many other benefits especially for aquatic animals (Verschuere et al., 2000, Wang et al., 2008). The most common and general concept of probiotics is that "one or more species of microorganisms with advantageous impacts for the host, are able to stick in the digestive tract due to its modulation to acid and bile salts" (Irianto and Austin, 2002). However application of probiotics in aquaculture has been discovered recently, great interest in their use has been generated due to their impact on disease control (Wang et al., 2008). There are many benefits probiotics are known for (summarized in Table 1)

3.   Probiotics Commercial Preparations

Application of probiotics as an alternative environmentally friendly tool has recently gained interest globally. Several preparations of probiotics were found in the market which contain some live microorganisms, which were used for the improvement of aquaculture cultivation. Also, probiotics may be used in food additives by directly adding them to culture tanks or even mixed with food. From the initial evaluations of commercial products, focus was on preparing bacteria known as Biostar derived from some *Bacillus*

Table 1: Benefits of probiotics application in aquaculture

Benefits	Probiotic strain	Aquatic species	Reference
Growth promoter	Alteromonas CA2	Crassostrea gigas	Douillet and Langdon, 1994
	Bacillus sp. S11	Penaeus monodon	Rengpipat et al., 1998
	Bacillus sp.	Ictalucus punctatus	Queiroz and Boyd, 1998
	Bacillus coagulans	Cyprinus carpio koi	Lin et al., 2012
	Bacillus NL 110, Vibrio NE 17	Macrobrachium rosenbergii	Rahiman et al., 2010
	Carnobacterium divergens	Gadus morhua	Gildberg et al., 1997
	Lactobacillus helveticus	Scophthalmus maximus	Gatesoupe, 1999
	Lactobacillus lactis AR21	Brachionus plicatilis	Harzeveli et al., 1998
	Lactobacillus casei	Poeciliopsis gracilis	Hernandez et al., 2010
	Streptomyces	Xiphophorus helleri	Dharmaraj and Dhevendaran, 2010
	Streptococcus thermophilus	Scophthalmus maximus	Gatesoupe, 1999
Pathogen inhibition	Bacillus sp.	Penaeids	Moriarty, 1998
	Bacillus spp., Enterococcussp.	Farfantepenaeus brasiliensis	Moreira et al., 2011
	Carnobacterium sp. Hg4-03	Hepialus gonggaensis larvae	Youping et al., 2011
	Enterococcus faecium SF 68	Anguilla anguilla	Chang and Liu, 2002
	Lactobacillus rhamnosus ATCC 53103	Oncorhynchus mykiss	Nikoskelainen et al., 2001
	Lactobacillus acidophilus	Clarias gariepinus	Abdullah et al., 2011
	Lactococcus lactis	Epinephelus coioides	Zhang et al., 2012
	Micrococcus luteus A1-6	Oncorhynchus mykiss	Irianto and Austin, 2002
	Pseudomonas fluorescens	Oncorhynchus mykiss	Gram et al. 1999
	Pseudomonas fluorescens AH2	Oncorhynchus mykiss	Gram et al., 2001

	Pseudomonas sp.	*Oncorhynchus mykiss*	Spanggaard *et al.*, 2001
	Roseobacter sp. BS. 107	*Scallop larvae*	Ruiz-Ponte et al., 1999
	Saccharomyces cerevisiae, S. exiguous, Phaffia rhodozyma	*Litopenaeus vannamei*	Scholz et al., 1999
	Tetraselmis suecica	*Salmo salar*	Austin et al., 1992
	Vibrio alginolyticus	*Salmonids*	Austin *et al.*, 1995
	Vibrio fluvialis	*Oncorhynchus mykiss*	Irianto and Austin, 2002
Nutrient digestibility	*Bacillus NL 110, Vibrio NE 17*	*Macrobrachium rosenbergii*	Rahiman *et al.*, 2010
	Carnobacterium sp. Hg4-03	*Hepialus gonggaensis larvae*	Youping et al., 2011
	Lactobacillus acidophilus	*Clarias gariepinus*	Al-Dohail et al., 2009
	Lactobacillus helveticus	*Scophthalmus maximus*	Gatesoupe, 1999
	Shewanella putrefaciens Pdp11	*Solea senegalensis*	Tapia et al., 2012
Water quality	*Bacillus sp. 48*	*Penaeus monodon*	Wang et al., 2008
	Bacillus coagulans SC8168	*Pennaeus vannamei*	Zhou et al., 2009
	Bacillus NL 110, Vibrio sp. NE 17	*Macrobrachium rosenbergii*	Rahiman et al., 2010
	Bacillus sp., Saccharomyces sp.	*Penaeus monodon*	Shishehchian et al., 2001
	Lactobacillus acidophilus	*Clarias gariepinus*	Al-Dohail et al., 2009
Stress tolerance	*Alteromonas sp.*	*Sparus auratus*	Varela et al., 2010
	Bacillus subtilis, Lactobacillus acidophilus, Saccharomyces cerevisiae	*Paralichthys olivaceus*	Taoka et al., 2006a
	Lactobacillus delbrueckii	*Dicentrarchus labrax*	Carnevali et al., 2006
	Lactobacillus casei	*Poecilopsis gracilis*	Hernandez et al., 2010
	Pediococcus acidilactici	*Litopenaeus stylirostris*	Castex et al., 2009
	Shewanella putrefaciens Pdp11	*Makimaki*	Tapia et al., 2012

(Contd.)

Table 1: (*Contd.*)

Benefits	Probiotic strain	Aquatic species	Reference
Reproduction improvement	Bacillus subtilis	Poecilia reticulata, Xiphophorus maculatus	Ghosh et al., 2007
	Lactobacillus rhamnosus	Danio rerio	Gioacchini et al., 2010
	Lactobacillus acidophilus, Lactobacillus casei, Enterococcus faecium, Bifidobacterium thermophilum	Xiphophorus helleri	Abasali, and Mohamad, 2010

isolates. Such preparations were used with catfish aquaculture to examine inoculum concentration impacts (Queiroz and Boyd, 1998). Application of commercial probiotic strains derived from *Bacillus* spp. affected viability and quality of shrimp aquaculture (Moriarty, 1998.). The impact of *Bacillus toyoi* and *Enterococcus faecium* SF68 isolates found in Toyocerin and Cernivet LBC respectively were evaluated for decreasing mortality of European eel due to edwards ielosis for detection of high efficiency with *E. faecium* SF68 (Chang and Liu, 2002). It was noted that *E. faecium* was known to be used as a probiotic for humans, while *B. toyoi* was used for terrestrial animals. A feed supplement was also used with *Oreochromis niloticus* containing a *B. subtilis* strain joined by hydrolytic enzymes that produce Biogen (El-Haroun et al., 2006).

Lactic acid producing bacteria have gained attention for use as probiotics. The human probiotic, *Lactobacillus rhamnosus* ATCC 53103 was applied to rainbow trout experimentally for 51 days for mortality reduction by *Aeromonas salmonicida*, responsible for the fish disease "furunculosis" (important fish disease in many fish farms all over the world). The mortality decreased from 52.6 to 18.9 % when 10^9 cells g^{-1} were provided with food, while if probiotic dose increased reaching 10^{12} cells g^{-1} of feed the mortality reached 46.3% (Nikoskelainen et al., 2001). Application of high doses of mixed cultures of probiotics (*L. acidophilus, L. casei, E. faecium,* and *B. thermophilum* (Primalac)) elevated the gonadosomatic index and also fingerlings production of reproductive age (Abasali and Mohamad, 2010). Also using mixed cultures of probiotics in *Penaeus vannamei* farms increase feed convention, survival and also production of shrimp in aquaculture farms (Wang et al., 2005). Another probiotic preparation, mixed cultures of bacterial strains (*Bacillus subtilis, Lactobacillus acidophilus,* and *Clostridium butyricum*) and yeast (*Saccharomyces cerevisiae*)), presented in different forms of Alchem Poseidon and Alchem Korea CO. and Wonju Korea CO. were used with tilapia *Oreochromis niloticus*. Results indicated enhancement of nonspecific immune characters such as migration of neutrophils, and lysozyme activity in addition to plasma bactericidal activity, which resulted in improved resistance to infection with *Edwardsiella tarda* (Taoka et al., 2006a). Previous research studies concerning the effect of probiotics on humans and animals revealed their capability to enhance the efficiency of beneficial bacteria in the colon (Martínez et al., 2008, Tuohy et al., 2003). There are some products of commercial aquaculture products known to include prebiotics in their formula such as yucca extract, glucans and mannans which enhance product benefits (Smart Microbials Inc, 2012). Some commercial products containing Bacillus spores, for example, Biostart a biocontrol product are presented in (Table 2). As Baciluus species are used in biocontrol products, they often contain mixtures of species (Liqualife and Biostart as seen in Table 2) (Gatesoupe, 1999). Other probiotic products were developed involving single species; Paciflor and Toyocerin (Verschuere et al., 2000). An efficient strategy used in some developing countries is *Bacillus* species isolation from shrimp ponds and then using them in producing commercial products, for example, PF was present in shrimp food with a *Bacillus* sp labelled S11 (Rengpipat et al., 2003). However, in the recent years, more studies have been conducted to develop a combined administration of probiotics with prebiotics (synbiotics) to enhance the viability and sustainability of living probiotics in aquculture systems. The widely used oligosaccahrides prebiotics such as inulin and fructrooligosacharide showed significant positive results when added to different types of probiotics (Azimirad et al., 2016). In addition, to their role for supporting probiotic growth, prebiotics can also act as immunostimulants for host animals (Carbone and Faggio, 2016).

This became an essential part of the probiotic formula to support better cell growth in the applied prorbiotic strains and thus increasing their effectiveness when applied aquaculture systems.

4. Mode of Action of Different Aquaculture Probiotics

4.1 Contention for Space

Pathogenic bacteria always needs to be attached to the mucosal layer of the gastrointestinal tract of the host for disease initiation (Adams, 2010). Probiotic bacteria have a major action mechanism related to competition for adhesion site exclusion. There is a very important criterion which should be considered when selecting probiotics; the capability of bacteria to colonize the gut and then to adhere to the epithelial surface to intervene with pathogen adhesion (Lazado et al., 2011). For example, the non-pathogenic intestinal microbes such as Lactobacilli compete with the pathogens for adhesion points on intestinal surfaces, especially intestinal villi and enterocytes (Brown, 2011). However, in many cases probiotic attachment could be nonspecific, referring to different physicochemical factors, or specific factors due to probiotics adhesion on the adherent bacterial surface and also various receptors on epithelial cells (Lazado et al., 2015).

Table 2: Examples of Aquaculture commercial probiotic products

Probiotics species / Strain	Commercial product	Country	Target animal	Reference
Bacillus. subtilis Wu-T and Wu-S at 10^8 CFU g^{-1} also, more strains were included; Lactobacillus and Saccharomyces spp	BaoZyme-Aqua	Sino-Aqua Corp., Kaohsiung, Taiwan	shrimps	Hong et al., 2005
A mixture of many strains; B. licheniformis, B. megaterium, Paenibacillus polymyxa and also two strains of B. subtilis	Biostart	Advanced Microbial Systems, Shakopee, MN, USA and Microbial Solutions, Johannesburg, South Africa	Aquaculture (Atlantic salmon)	Verschuere et al., 2000
Many Bacillus sp.	Liqualife	Cargill, Animal Nutrition Division, Minneapolis, MN United States	Aquaculture (Salmon, Tilapia, Shrimp catfish, crab and others)	Gatesoupe, 1999 www.cargill.com
A mixture of 4 strains of B. subtilis	Premarin	Sino-Aqua, Kaohsiung, Taiwan	Aquaculture (shrimps)	Urdaci and Pinchuk, 2004
B. cereus var toyoi mixed with calcium carbonate and maize flour	Toyocerin	Asahi Vet. S.A., Tokyo, Japan	European eel	Scan, 2001
A mixture of Bacillus species	Sanoguard	Inve Aquaculture Belgium	Shrimp	http://www.hatcheryfeed.com/

4.2 Production of Inhibitory Compounds

Different probiotic bacteria are known for their ability to produce various substances with bacteriostatic or bactericidal impacts on other pathogens (Servin, 2004). These compounds include: hydrogen peroxide, bacteriocins, lysozymes and many other types of bioinhibitory molecules (Tinh et al., 2007). Thes play a critical role in animal health and have a natural protection mechanism against pathogens in aquaculture system.

4.2.1 Probiotics Antibacterial Activity

Many probiotics, especially in aquaculture, are recorded as having antibacterial activity against many pathogens. Probiotic *L. lactis* RQ516, is an example, which is widely used in tilapia (*Oreochromis niloticus*) farms showing inhibitory effects against A. hydrophila (Zhou et al., 2010). Also, *Leuconostoc mesenteroides* inhibit the growth of various pathogens in tilapia fish (Zapata and Lara-Flores, 2013). Moosavi-Nasab et al. (2014) studied the effectiveness of many lactic acid bacterial species (*Sterptococcus salivarius, Lactobacillus buchneri, Lactobacillus fermentum and Lactobacillus acidophilus*)which were isolated from Spanish mackerel (*Scomberomorus commerson*) and their ability for growth inhibition of pathogenic *Listeria innocua* was recorded. The isolated Lactobacilli from the intestine of different fish species; Gender fish (*Punitus carnaticus*), catfish (*Clarias Orientalis*), Jillabe fish (*Oreochromis* sp.) Rohu fish (*Labeo rohita*) and Hari fish (*Anguilla* sp.) demonstrated noticeable antibacterial activity against both *Vibrio and Aeromonas* sp. (Dhanasekaran et al., 2008).

4.2.2 Probiotics Antiviral Activity

Recently, the antiviral activity of probiotics gained more attention. Some bacteria having probiotic properties such as *Aeromonas, Vibrio* and *Pseudomonas* sp. showed antiviral impacts against the hematopoietic necrosis virus (IHNV) (Kamei et al., 1988). Also, it was reported that using the *Bacillus megaterium* probiotic strain raised the resistance of the shrimp *Litopenaeus vannamei* to the white spot syndrome virus (WSSV) (Li et al., 2009). Moreover, some probiotics such as *Vibrio* and *Bacillus* sp positively affected the shrimp *L. vannamei* against WSSV. Disease resistance, especially against lymphocystis viral disease, could be enhanced in olive flounder (*Paralichthys olivaceus*) by using *Lactobacillus* probiotics either as a single strain or mixed with Sporolac (Harikrishnan et al., 2010). It was also reported that use of probiotics in aquaculture systems can reduce many viral diseases in shrimps such as: white spot syndrome, white tail disease, yellow head disease, Taura syndrome, and myconecrosis which are caused by different types of DNA and RNA viruses such as WSSV, TSV, YHV, MrNV and other types (Lakshmi et al., 2013).

4.2.3 Probiotics Antifungal Activity

Comapred to the large number of published work of the antibacterial and antifungal properties of aquaculture probiotics, little information is available for antifungal properties of probiotics in aquaculture systems. However, it have been reported that *Aeromonas* media (strain A 199) was isolated from eel (*Anguilla australis*) showed antifungal properties against *Saprolegnia* sp. In addition, the application of *Janthinobacterium* sp. M169, *Pseudomonas* sp. M162 and *Pseudomonas* sp. M174 increased the immunity of rainbow trout against saprolegniasis (Lategan et al., 2004). It was also reported that using *L. plantarum* FNCC 226 as a probiotic in aquaculture systems can control the fungal pathogen *Saprolegnia parasitica* A3 in catfish (Atira et al., 2012). It was also reported in patented research, that a newly isolated probiotic bacterial strain belonging to *Bacillus* spp. Also showed a high capacity for controlling the growth of *Saprolengia* sp. (Kang et al., 2012). Recent research also showed that a marine bacteria identified as *Bukholderia* sp. HD05 exhibited antifungal activity against *Saprolegnia* sp. and can decrease the fungal infection rate of grass carp (*Ctenopharyngodon idella*) by 53.3% when used in aquaculture systems (Zhang et al., 2019). The antifungal activity of this bacteria was mediated through the production of the glutamic acid derivative (2-pyrrolidone-5-carboxylic acid).

4.3 Water Quality Improvement

Use of some gram-positive probiotic bacteria has also been reported to contribute to the improvement of water quality. For example: *Bacillus* spp, exhibs high capacity in the conversion of organic matter to carbon, while, most of the gram-negative bacteria contribute to reducing organic matter pollution by converting it to bacterial biomass (slime) (Mohapatra et al., 2012). Some probiotic bacteria have a considerable algicidal effect particularly on many species of microalgae (Fukami et al., 1997). Moreover, addition of probiotics in aquculture have been proven as an efficient approach in the manipulation of microbial diversity in aquaculture pond systems and shift towards safer and more efficient microbial consortiums (Mohapatra et al., 2012, Padmavathi et al., 2012). It has also been reported that, dissolved oxygen, hydrogen sulfide, ammonia, and pH of water were found to be of better quality by the addition of probiotics. This helps in maintaining a healthy environment for the growth of healthy shrimps (Aguirre-Guzman et al., 2012). Addition of bacterial strains with some probiotic properties such as *Nitrobacter* and *Nitrosomonas* to fish ponds can increase the biochemical properties and reduce the growth of pathogenic strains of *Pseudomonas* (Sunitha and Krishna, 2016). Other studies proved clearly that the addition of a probiotic consortium of *Bacillus lichineformis*, *B. subtilis*, and *Pseudomonas* s. can play a significant role in bioremediation of toxic compounds and improves different water quality parameters in aquaculture systems (Kshatri et al., 2018).

4.4 Nutrients and Enzymatic Supply

It was found that some microorganisms have a positive impact during the digestive process of aquatic animals (Balcazar et al., 2006). Some bacteria participate in digestion through the production of extracellular enzymes, such as lipase, protease and also some growth promoting agents (Wang et al., 2000). Previous studies revealed that some probiotics, particularly from *Clostridium* and *Bacteroides* sp., have the capacity to produce vitamins, essential amino acids and fatty acids for the host (Balcazar et al., 2006, Tinh et al., 2007). Application of auxinic algae (*Isochrysis galbana*) with CA2 bacterial strain to Gnotobiotic oyster larvae (*Crassostrea gigas*) not only supports higher growth rates but also involves efficient utilization of nutrients as well (Douillet and Langdon, 1994).

4.5 Immunomodulation Mechanisms of Probiotics:

4.5.1 Fish

The probiotics stimulation of immune systems have been repored in many studies. This effect is considered as one of the most positive effects when probiotics are applied to feed or food chains. Probiotics stimulate the host immune system, involving pro-inflammatory cytokines stimulation on the activity of immune cells and raising the phagocytic activity of leucocytes (Pirarat et al., 2006). Moreover, probiotics can raise many immune systems biomolecules such as acid phosphatase, lysozymes (Lara-Flores and Aguirre-Guzman, 2009), gamma interferon (IFN-g), IL-10, transforming growth factor b), cytokines (interleukin-1 (IL-1), IL-6, IL-12 (Nayak, 2010) in addition to complementing antimicrobial peptides (Mohapatra et al., 2012). In addition, using probiotics leads to inhibition of pathogens impeding colonization in fish digestive tracts, improving microbial balance in the intestine, releasing inhibitory compounds such as; hydrogen peroxides, bacteriocins, proteases, lysozymes and siderophores (Saurabh et al., 2005) and increasing the activity of the digestive enzymes and the production of vitamins and fatty acids (Ringø et al., 1995) and major amino acids which are important to lactic acid bacteria (Ringø and Gatesoupe, 1998) and can improve growth performance, and also elevate resistance of some pathogens in shrimp and fish (Lakshmi et al., 2013). Using probiotics in tilapia farms (*O. niloticus*) increased lysozyme activity, bacterial activity, neutrophil migration and as a consequence improved fish resistance to *Edwardsiella tarda* infection (Taoka et al., 2006b).

4.5.2 Shrimp

Probiotics affected shrimps significantly and improved their natural or non-specific immunity. Using probiotics leads to the production of different cellular compounds like oxidase enzyme, agglutinins, anticoagulant proteins (Lakshmi et al., 2013), hydrogen peroxide, tyrotricidin, antimicrobial peptides (chemokines and defensins), proteases, monostatins, siderophores, polymyxin organic acids increasing competitive exclusion, encapsulation, phagocytosis, formation of humoral components and nodules (Balcazar et al., 2007). Also, probiotics have a critical role in the enhancement of shrimp resistance to some known diseases; *A. hydrophila* infection, white spot disease and vibriosis (Liu et al., 2010; Zokaeifar et al., 2014).

4.5.3 Mechanism of Immunomodulation in the Gut Immune System

The gut immune system is linked with the gut-associated lymphoid tissues with some variations towards Peyer's patches, antigen-transporting M cells and secretory IgA in the intestine of different mammals gut immune systems (Nayak, 2010). Although macrophages, lymphoid cells, mucus IgM and granulocytes were detected in the fish intestine, the impact of probiotics on intestinal immune cells is little known (Bakke-McKellep et al., 2007, Nayak, 2010). There are limited studies covering probiotic applications and their ability for the enhancement of the piscine gut immune system (Nayak, 2010, Lazado and Caipang 2014a). The most prevalent research is related to human and also terrestrial vertebrates (Lazado and Caipang, 2014b). Moreover, research studies have revealed the ability of probiotics to enhance the piscine gut immune system by raising the proportion of acidophilic granulocytes (AGs) and IgC-cells (Picchietti et al., 2009, Salinas et al., 2008). It was found that adding LAB (*Lactobacillus rhamnosus* GG, human origin) in the feed of tilapia, *O. niloticus* modulates the intestinal immune cells population. Furthermore, adding AGs and intra-epithelial lymphocytes significantly stimulates the immune system in the probiotic-fed group (Pirarat et al., 2011). Artemia and rotifers were used in the supplementation of *L. delbrueckii* ssp. *delbrueckii* (AS13B) to the larvae of sea bass, *Dicentrarchus labrax* as live vectors (Picchietti et al., 2009). Results revealed that populations of AGs and T-cells in the intestinal mucosa increased significantly in fish fed with probiotics.

5. Probiotics Safety Considerations

Probiotics used in the food industry are supposed to be safe, traditionally, and no human hazards have been detected, which is the best evidence for their safety (Nomoto, 2005). Theoretically speaking, probiotics may lead to different types of side effects especially to susceptible individuals: redundant immune stimulation, gene transfer, systemic infections and harmful metabolic actions. Moreover, no definite evidence proving it has been found. Practically, limited reports have considered bacteremia in humans, while probiotic isolation from different infections have resulted from opportunistic contagion brought by skin lesions, chronic illnesses or drug-induced aberrations. Such circumstances have lead to reduced intestinal barriers which foster bacterial passage through mucosal epithelium. Therefore, such microorganisms are transferred to mesenteric lymph nodes in addition to other organs causing bacteremia which lead to septicemia (Ishibashi and Yamazaki, 2001). Many previous studies detected bacteremia in patients suffering from chronic illnesses or a weakened immune system (Lahtinen et al., 2009, Soccol et al., 2010). In some lactic acid bacteria which are used as probiotics, antibiotic resistance could be related to different genes in chromosomes, transposons or plasmids. Moreover, there is no adequate information about the conditions in which such genetic elements can be mobilized (FAO/WHO 2006). Furthermore, it was noted that some enterococci might cause virulence properties and are also able to transmit antibiotic resistance elements, so these organisms are not recommended to be used as probiotics especially in humans unless otherwise, their producer declares that such a strain cannot transmit antibiotic resistance or promote infection (Lahtinen et al., 2009).

Concerning the safety measures of different aquaculture products, many studies in Asia, especially in Latin America aquaculture farms (Penaeus monodon), found that bacterial white spot syndrome proceeded due to frequent application of probiotics (*Bacillus subtilis*). Such spots resulted from white spot viral syndrome, which is very dangerous, and propagates rapidly as a deadly disease and leads to great losses in shrimp aquaculture (Chou et al., 1995, Wang et al., 1999). Moreover, as most people consume aquaculture products either raw or half cooked, many concerns arise if probiotic residues lead to human infection. The effect of probiotic application (*Shewanella algae*) in some shrimp farms was studied to detect its potential risk to humans. Mice were provided by about 10^{36} cfu of *S. algae* to reach LD_{50} and results indicated that it is safe to be used in mice and as a consequence in shrimp farms (Shakibazadeh et al., 2011). FAO and WHO stressed the need to state general guidelines for rating probiotics in food through a systematic approach to justify their health cases. Based on scientific evidence, an expert research working group studied how to state methodology and general criteria for probiotics evaluation (Pineiro and Stanton, 2007). Research results were declared in the "Guide for Evaluation of Probiotics in Food" summarizing important guidelines dealing with evaluation of probiotic impacts on nutrition and general health in food. Results declared that there are no detected virulent or pathogenic effects in lactococci, bifidobacteria or lactobacilli; however it is recommended that certain bacteremia cases are referred to some lactobacilli strains. However, their proportion does not increase with high concentrations of lactobacillus in probiotic formulations. Also, sometimes enterococci produce virulence properties; so it is not recommended for human use (FAO/WHO 2006). There is no specific published data concerning probiotic pathogenicity of probiotic microorganisms with different animals; rats, mice and fish in other aquatic biota, proposing their safety (Lahtinen et al., 2009). Thus, continuous research in such scientific fields is a must nowadays utilizing the following recommended approaches: (i) intrinsic property analyses of probiotic strains, ii) pharmacokinetics studies (activity, survival in the intestine, dose-response, and also recovery from mucosa), and (iii) Studying different interactions between probiotic microorganisms and their hosts (Soccol et al., 2010).

6. Future Perspectives

With the increased demand for healthy and safe food, probiotics become essential components of marine animal feed to enhance both animal growth and animal immunity to resist many pathogenic diseases. This also helps in a great measure to reduce the extensive use of hormones, chemicals and antibiotics which proved to have a negative impact on animals and human health once they entered the feed/food supply chain. Further research is now needed to study in depth the marine animal microbiota in terms of diversity, functional properties, and microbial sustainability in aquaculture systems. However, the big challenge so far is the availability of probiotic based products on a large scale with acceptable prices to replace the traditionally used chemicals and antibiotics in many regions of the world. Therefore, further research is needed to tackle many basic research and industrial issues to increase market acceptability of probiotic based products in aquaculture. However, with the increased awareness of the dangerous effect of the extensive use of antibiotics in aquaculture feed chains which leads to the increase of microbial resistance and develops a new generation of highly resistant strains for the currently used medical antibiotics, its expected that the business of biostimulants such as probiotics will grow further to support safe and healthy food production.

References

Abasali, H. and S. Mohamad. 2010. Effect of dietary supplementation with probiotic on reproductive performance of female livebearing ornamental fish. Res. J. Anim. Sci. 4 (4): 103-107.

Abdelkhalek, N.K., I.A. Eissa, E. Ahmed, O.E. Kilany, M. El-Adl, M.A.O. Dawood et al. 2017. Protective role of dietary *Spirulina platensis* against diazinon-induced oxidative damage in Nile tilapia; *Oreochromis niloticus*. Environ. Toxicol. Pharmacol. 54: 99-104.

Abdullah, A.M., R. Hashim and P.M. Aliyu. 2011. Evaluating the use of *Lactobacillus acidophilus* as a biocontrol agent against common pathogenic bacteria and the effects on the haematology parameters and histopathology in African catfish *Clarias gariepinus* juveniles. Aquacult. Res. 42(2): 196-209.

Adams, C.A. 2010. The probiotic paradox: Live and dead cells are biological response modifiers. Nutr. Res Rev. 23: 37-46.

Adel, M., S. Yeganeh, M. Dadar, M. Sakai and M.A.O. Dawood. 2016. Effects of dietary *Spirulina platensis* on growth performance, humoral and mucosal immune responses and disease resistance in juvenile great sturgeon (*Huso huso* Linnaeus, 1754). Fish & Shellfish Immunol. 56: 436-444.

Aguirre-Guzman, G., M. Lara-Flores, J.G. Sanchez-Martinez, A.I. Campa-Cordova and A. Luna-Gonzalez. 2012. The use of probiotics in aquatic organisms: A review. Afr. J. Microbiol. Res. 6(23): 4845-4857.

A Hatcheryfeed.com publication. 2016. Aquafeed.com LLC. http://www. hatcheryfeed.com/

Al-Dohail, M.A., R. Hashim and M. Aliyu-Paiko. 2009. Effects of the probiotic, *Lactobacillus acidophilus*, on the growth performance, haematology parameters and immunoglobulin concentration in African catfish (*Clarias gariepinus*, Burchell 1822) fingerling. Aquacult. Res. 40: 1642-1652.

Atira, N.J., I.N.P. Aryantha and I.D.G. Kadek. 2012. The curative action of *Lactobacillus plantarum* FNCC 226 to *Saprolegnia parasitica* A3 on catfish (*Pangasius hypophthalamus*, Sauvage). Int. Food Res. J. 19(4): 1723-1727.

Austin, B., E. Baudet and M. Stobie. 1992. Inhibition of bacterial fish pathogens by *Tetraselmis suecica*. J. Fish Dis. 90: 389-392.

Austin, B., L.F. Stuckey, P.A.W. Robertson, I. Effendi and D.R.W. Griffith. 1995. A probiotic strain of *Vibrio alginolyticus* effective in reducing diseases caused by *Aeromonas salmonicida*, *Vibrio anguillarum* and *Vibrio ordalii*, J. Fish Dis. 18(1): 93-96.

Austin, B. and D.A. Austin. 2012. Bacterial Fish Pathogens: Disease of Farmed and Wild Animals, 5th edn. Springer, Dordrecht, Netherlands.

Azimirad, M., S. Meshkini, N. Ahmadifard and S.H. Hoseinifar. 2016. The effects of feeding with symbiotic (*Pediococcus acidilactici* and fructooligosaccharide) enriched adult *Artemia* on skin mucus immune responses, stress resistance, intestinal microbiota and performance of angelfish (*Pterophyllum scalare*). Fish & Shellfish Immunol. 54: 516-522.

Bakke-McKellep, A.M., M.K. Froystad, E. Lilleeng, F. Dapra, S. Refstie, A. Krogdahl et al. 2007. Response to soy: T-cell-like reactivity in the intestine of Atlantic salmon, *Salmo salar* L. J. Fish Dis. 30: 13-25.

Balcazar, J.L., I. de Blas, I. Ruiz-Zarzuela, D. Cunningham, D. Vendrell, J.L. Múzquiz et al. 2006. The role of probiotics in aquaculture. Vet. Microbiol. 114: 173-186.

Balcazar, J.L., I. de Blas, I. Ruiz-Zarzuela, D. Vendrell, A.C. Calvo, I. Marquez et al. 2007. Changes in intestinal microbiota and humoral immune response following probiotic administration in brown trout (*Salmo trutta*). Br. J. Nutr. 97: 522-527.

Boxall, A.B., L.A. Fogg, P.A. Blackwell, P. Kay, E.J. Pemberton, A. Croxford et al. 2004. Veterinary medicines in the environment. Rev. Environ. Contam. Toxicol. 180: 1-91.

Brown, M. 2011. Modes of action of probiotics: Resent developments. J. Anim. Vet. Adv. 10(14): 1895-1900.

Burr, G., D. Gatlin and S. Ricke. 2007. Microbial ecology of the gastrointestinal tract of fish and the potential application of prebiotics and probiotics in fin fish aquaculture. J. World Aquacult. Soc. 36: 425-436.

Carbone, D. and C. Faggio. 2016. Importance of prebiotics in aquaculture as immunostimulants. Effects on immune system of *Sparus aurata* and *Dicentrarchus labrax*. Fish & Shellfish Immunol. 54: 172-178.

Carnevali, O., L. de Vivo, R. Sulpizio, G. Gioacchini, I. Olivotto, S. Silvi et al. 2006. Growth improvement by probiotic in European sea bass juveniles (*Dicentrarchus labrax*, L.), with particular attention to IGF-1, myostatin and cortisol gene expression. Aquaculture 258(1-4): 430-438.

Castex, M., P. Lemaire, N. Wabete and L. Chim. 2009. Effect of dietary probiotic *Pediococcus acidilactici* on antioxidant defences and oxidative stress status of shrimp *Litopenaeus stylirostris*. Aquaculture 294(3-4): 306-313.

Chang, C.I. and W.Y. Liu. 2002. An evaluation of two probiotic bacterial strains, *Enterococcus faecium* SF68 and *Bacillus toyoi*, for reducing edwardsiellosis in cultured European eel, *Anguilla anguilla* L. J. Fish Dis. 25(5): 311-315.

Chauhan, A. and R. Singh. 2019. Probiotics in aquaculture: A promising emerging alternative approach. Symbiosis 77: 99-113.

Chou, H.Y., X.Y. Huang, C.H. Wang, H.C. Chiang and C.F. Lo. 1995. Pathogenicity of a baculovirus infection causing white spot syndrome in cultured penaeid shrimp in Taiwan. Dis. Aquat. Org. 23(3): 165-173.

Dawood, M.A.O., S. Koshio, M. Ishikawa, S. Yokoyama, M.F. El Basuini, M.S. Hossain et al. 2016a. Effects of dietary supplementation of *Lactobacillus rhamnosus* or/and *Lactococcus lactis* on the growth, gut microbiota and immune responses of red sea bream, *Pagrus major*. Fish & Shellfish Immunol. 49: 275-285.

Dawood, M.A.O., S. Koshio, M. Ishikawa, M. El-Sabagh, M.A. Esteban, A.I. Zaineldin et al. 2016b. Probiotics as an environment friendly approach to enhance red sea bream, *Pagrus major* growth, immune response and oxidative status. Fish & Shellfish Immunol. 57: 170-178.

Dawood, M.A.O. and S. Koshio. 2016. Recent advances in the role of probiotics and prebiotics in carp aquaculture: A review. Aquaculture 454: 243-251.

Dawood, M.A.O., S. Koshio, M. Ishikawa, M. El-Sabagh, S. Yokoyama, W.L. Wang et al. 2017. Physiological response, blood chemistry profile and mucus secretion of red sea bream (*Pagrus major*) fed diets supplemented with *Lactobacillus rhamnosus* under low salinity stress. Fish Physiol. Biochem. 43(1): 179-192.

De, B.C., D.K. Meena, B.K. Behera, P. Das, P.D. Mohapatra and A.P. Sharma. 2014. Probiotics in fish and shellfish culture: Immunomodulatory and ecophysiological responses. Fish Physiol. Biochem. 40(3): 921-971.

Dhanasekaran, D., S. Saha, N. Thajuddin and A. Panneerselvam. 2008. Probiotic effect of *Lactobacillus* isolates against bacterial pathogens in *Clarias orientalis*. Med. Biol. 15(3): 97-102.

Dharmaraj, S. and K. Dhevendaran. 2010. Evaluation of Streptomyces as a probiotic feed for the growth of ornamental fish *Xiphophorus helleri*. Food Technol. Biotechnol. 48(4): 497-504.

Douillet, P.A. and C.J. Langdon. 1994. Use of a probiotic for the culture of larvae of the Pacific oyster (*Crassostrea gigas* Thunberg), Aquaculture 119(1): 25-40.

El Baz, A., H. El Enshasy, Y.M. Shetia, H. Mahrous, N.Z. Othman and A. El Yousef et al. 2018. Probiotic assessment and semi-industrial scale production of a new yeast, *Cryptococcus* sp. YMHS, isolated from the red sea. Probiotics Antimicrob. Proteins 10: 77-88.

El-Haroun, E.R., A.M.A.S. Goda and M.A.K. Chowdhury. 2006. Effect of dietary probiotic Biogen (R) supplementation as a growth promoter on growth performance and feed utilization of Nile tilapia *Oreochromis niloticus* (L.). Aquacul. Res. 37: 1473-1480.

FAO/WHO. 2001. Health and nutritional properties of probiotics in food including powder milk with liver lactic acid bacteria. Food and Agriculture Organization and World Health Organization Joint report (34 pp.).

FAO/WHO. 2006. Guidelines for the Evaluation of Probiotics in Food. https://www. who.int/foodsafety/fs_management/en/probiotic_guidelines.pdf

Farzanfar, A. 2006. The use of probiotics in shrimp aquaculture. FEMS Immunol. Med. Microbiol. 48: 149-158.

Fuller, R. 1987. A review, probiotics in man and animals. J. Appl. Bacteriol. 66: 365-378.

Fukami, K., T. Nishijima and Y. Ishida. 1997. Stimulative and inhibitory effects of bacteria on the growth of microalgae. Hydrobiol. 358: 185-191.

Gatesoupe, F.J. 1999. The use of probiotic in aquaculture. Aquacult. 180: 147-165.

Gildberg, A., H. Mikkelsen, E. Sandaker and E. Ringø. 1997. Probiotic effect of lactic acid bacteria in the feed on growth and survival of fry of Atlantic cod (*Gadus morhua*). Hydrobiol. 352(1-3): 279-285.

Ghosh, S., A. Sinha and C. Sahu. 2007. Effect of probiotic on reproductive performance in female livebearing ornamental fish. Aquacult. Res. 38(5): 518-526.

Gioacchini, G., F. Maradonna, F. Lombardo, D. Bizzaro, I. Olivotto, O. Carnevali et al. 2010. Increase of fecundity by probiotic administration in zebra fish (*Danio rerio*). Reproduction 140(6): 953-959.

Gomez-Gil, B., A. Roque and J.F. Turnbull. 2000. The use and selection of probiotic bacteria for use in the culture of larval aquatic organisms. Aquaculture 191: 259-270.

Gomez-Gil, B., M. Herrera-Vega, F. Abreu-Grobois and A. Roque. 1998. Bioencapsulation of two different Vibrio species in nauplii of the brine shrimp (*Artemia franciscana*). Appl. Environ. Microbiol. 64: 2318-2322.

Gram, L. and E. Ringø. 2005. Prospects of fish probiotics. pp. 379-417. *In*: W. Holzapfel, P. Naughton (eds.). Microbial Ecology in Growing Animals. Elsevier, Edinburgh, UK.

Gram, L., T. Løvold, J. Nielsen, J. Melchiorsen and B. Spanggaard. 2001. *In vitro* antagonism of the probiont *Pseudomonas fluorescens* strain AH2 against *Aeromonas salmonicida* does not confer protection of salmon against furunculosis. Aquaculture 199(1-2): 1-11.

Gram, L., J. Melchiorsen, B. Spanggaard, I. Huber and T.F. Nielsen. 1999. Inhibition of *Vibrio anguillarum* by *Pseudomonas fluorescens* AH2, a possible probiotic treatment of fish. Appl. Environ. Microbiol. 65(3): 969-973.

Hai, N.V. 2015. The use of probiotics in aquaculture. J. Appl. Microbiol. 119(4): 917-935.

Harikrishnan, R., C. Balasundaramb and M.S. Heo. 2010. Effect of probiotics enriched diet on *Paralichthys olivaceus* infected with lymphocystis disease virus (LCDV). Fish Shellfish Immunol. 29: 868-874.

Harzeveli, A.R.S., H. Van Duffel, P. Dhert, J. Swing and P. Sorgeloos. 1998. Use of a potential probiotic *Lactococcus lactis* AR21 strain for the enhancement of growth in the rotifer *Brachionus plicatilis* (Muller). Aquacult. Res. 29(6): 411-417.

He, S., W. Liu, Z. Zhou, W. Mao and P. Ren. 2011. Evaluation of probiotic strain *Bacillus subtilis* C-3102 as a feed supplement for koi carp (*Cyprinus carpio*). J. Aquacult. Res. Dev. 3(Suppl): S1-S5.

Hernandez, L.H.H., T.C. Barrera, J.C. Mejia, L. Héctor, T.C. Barrera, J.C. Mejía et al. 2010. Effects of the commercial probiotic *Lactobacillus casei* on the growth, protein content of skin mucus and stress resistance of juveniles of the Porthole livebearer *Poecilopsis gracilis* (Poecilidae). Aquacult. Nutr. 16(4): 407-411.

Hong, H.A., L.H. Duc and S.M. Cutting. 2005. The use of bacterial spore formers as probiotics. FEMS Microbiol. Rev. 29: 813-835.

Hossain, M.S., S. Koshio, M. Ishikawa, S. Yokoyama, N.M. Sony, M.A. Dawood et al. 2016. Efficacy of nucleotide related products on growth, blood chemistry, oxidative stress and growth factor gene expression of juvenile red sea bream, *Pagrus major*. Aquaculture 464: 8-16.

Irianto, A. and B. Austin. 2002. Probiotics in aquaculture. J. Fish Dis. 25: 633-642.

Ishibashi, N. and S. Yamazaki. 2001. Probiotics and safety. Am. J. Clin. Nutr. 73(2): 465S-470S.

Jahangiri, L. and M.Á. Esteban. 2018. Administration of probiotics in the water in finfish aquaculture systems: A review. Fishes 3: 33.

Kamei, Y., M. Yoshimizu, Y. Ezura and T. Kimura. 1988. Screening of bacteria with antiviral activity from fresh water salmonid hatcheries. Microbiol. Immunol. 32: 67-73.

Kang, H., H.S. Seo, S.H. Woo and S.Y. Yang. 2019. Probiotics for biological control against *Saprolegnis* sp. WO 2012/105804 A2.

Kechagia, M., D. Basoulis, S. Konstantopoulou, D. Dimitriadi, K. Gyftopoulou, N. Skarmoutsou et al. 2013. Health benefits of probiotics: A review. ISRN Nutr. 2013: Article ID, 481651.

Kshatri, J., C.V. Rao and V.S. Settaluri. 2018. Neutralization of toxins in aquaculture using probiotics. Int. J. Pharmaceut. Sci. Res. 9(6): 2484-2489.

Kumar, R., S.C. Mukherjee, K.P. Prasad and A.K. Pal. 2006. Evaluation of *Bacillus subtilis* as a probiotic to Indian major carp *Labeo rohita* (Ham.). Aquacult. Res. 37: 1215-1221.

Lahtinen, S.J., R.J. Boyle, A. Margolles, R. Frías and M. Gueimonde. 2009. Safety assessment of probiotics. Prebiotics and Probiotics Science and Technology, pp. 1193-1235.

Lakshmi, B., B. Viswanath and D.V.R. Sai Gopal. 2013. Probiotics as antiviral agents in shrimp aquaculture. J. Pathol. Article ID 424123.

Lara-Flores, M. and G. Aguirre-Guzman. 2009. The use of probiotic in fish and shrimp aquaculture: A review. pp. 4-16. *In*: N. Perez-Guerra and L. Pastrana-Castro (eds.). Probiotics: Production, Evaluation and Uses in Animal Feed. Chapter 4. Kerala: Research Signpost.

Lategan, M.J., F.R. Torpy and L.F. Gibson. 2004. Control of saprolegniosis in the eel (*Anguilla australis*, Richardson) by *Aeromonas media* strain A199. Aquaculture 240: 19-27.

Lazado, C.C., C.M.A. Caipang, M.F., Brinchmann and V. Kiron. 2011. *In vitro* adherence of two candidate probiotics from Atlantic cod and their interference with the adhesion of two pathogenic bacteria. Vet. Microbiol. 148: 252-259.

Lazado, C.C. and C.M.A. Caipang. 2014a. Mucosal immunity and probiotics in fish. Fish Shellfish Immunol. 39: 78-89.

Lazado, C.C. and C.M. Caipang. 2014b. Bacterial viability differentially influences the immunomodulatory capabilities of potential host-derived probiotics in the intestinal epithelial cells of Atlantic cod *Gadus morhua*. J. Appl. Microbiol. 116(4): 990-998.

Lazado, C.C., C.M. Caipang and E.G. Estante. 2015. Prospects of hostassociated microorganisms in fish and penaeids as probiotics with immunomodulatory functions. Fish Shellfish Immunol. 45(1): 2-12.

Li, J., B. Tan and K. Mai. 2009. Dietary probiotic *Bacillus* OJ and isomalto oligosaccharides influence the intestine microbial populations, immune responses and resistance to white spot syndrome virus in shrimp (*Litopenaeus vannamei*). Aquaculture 291: 35-40.

Lin, S.H., Y. Guan, L. Luo and Y. Pan. 2012. Effects of dietary chitosan oligosaccharides and *Bacillus coagulans* on growth, innate immunity and resistance of koi (*Cyprinus carpio koi*). Aquaculture 342(343): 36-41.

Liu, K.F., C.H. Chiu, Y.L. Shiu, W. Cheng and C.H. Liu. 2010. Effects of the probiotic, *Bacillus subtilis* E20, on the survival, development, stress tolerance, and immune status of white shrimp, *Litopenaeus vannamei* larvae. Fish Shellfish Immunol. 28(5-6): 837-844.

Magnadottir, B. 2010. Immunological control of fish diseases. J. Marine Biotechnol. 12: 36-379.

Mamun, A.A., S. Nasren, S.S. Rathore, M.J. Sidiq, P. Dharmakar, K.V. Anjusha et al. 2019. Assessment of probiotic in aquaculture: Functional changes impact on fish gut. Microbiol. Res. J. Int. 29(1): 1-10.

Martínez, C.P., L. Mayorga, T. Ponce, A. Roldán, E. Barranco, R. González, et al. 2008. Efecto de los fructooligosaćaridos en la poblaci´on bacteriana fecal de un neonato, crecida en cultivo por lote. Revista Mexicana De Ciencias Farmaćeuticas 39(1): 32-37.

Mohapatra, S., T. Chakraborty, V. Kumar, G. De Boeck and K.N. Mohanta. 2012. Aquaculture and stress management: A review of probiotic intervention. J. Anim. Physiol. Anim. Nut. 14: 1-26.

Moriarty, D.J.W. 1998. Control of luminous *Vibrio* species in penaeid aquaculture ponds. Aquaculture 164(1-4): 351-358.

Moreira, S.D., S.S. Medeiros, L.L. Pereira, L.A. Romano, W. Wasielesky et al. 2011. The use of probiotics during the nursery rearing or the pink shrimp *Farfantepenaeus brasiliensis* (Latreille, 1817) in a zero exchange system. Aquacult. Res. 43: 1828-1837.

Moosavi-Nasab, M., E. Abedi, S. Moosavi-Nasab, M.H. Eskandari. 2014. Inhibitory effect of isolated lactic acid bacteria from *Scomberomorus commerson* intestines and their bacteriocin on *Listeria innocua*. Iran Agr. Res. 33(1): 43-52.

Nayak, S.K. 2010. Probiotics and immunity: A fish perspective. Fish Shellfish Immunol. 29: 2-14.

Nikoskelainen, S., A. Ouwehand, S. Salminen and G. Bylund. 2001. Protection of rainbow trout (*Oncorhynchus mykiss*) from furunculosis by *Lactobacillus rhamnosus*. Aquaculture 198(3-4): 229-236.

Nomoto, K. 2005. Prevention of infections by probiotics. J. Biosci. Bioeng. 100: 583-592.

Ouwehand, A.C., S. Salminen and E. Isolauri. 2002. Probiotics: An overview of beneficial effects. Antoine Van Leewenhoek. 82: 279-289.

Padmavathi, P., K. Sunitha and K. Veeraiah. 2012. Efficacy of probiotics in improving water quality and bacterial flora in fish ponds. Afr. J. Microbiol. Res. 6(49): 7471-7478.

Peeters, M. and J. Rodríguez. 1999. Problemas bacterianos en la industria camaronera ecuatoriana, practicas de manejo y alternativas de control. El Mundo Acuicola. 5: 13-18.

Pérez-Sánchez, T., I. Ruiz-Zarzuela, I. de Bías and J.L. Balcázar. 2013. Probiotics in aquaculture: A current assessment. Rev. Aquaculture 5: 1-14.

Picchietti, S., A.M. Fausto, E. Randelli, O. Carnevali, A.R. Taddei, F. Buonocore et al. 2009. Early treatment with *Lactobacillus delbrueckii* strain induces an increase in intestinal T-cells and granulocytes and modulates immune-related genes of larval *Dicentrarchus labrax* (L.). Fish Shellfish Immunol. 26: 368-376.

Pirarat, N., T. Kobayashi, T. Katagiri, M. Maita and M. Endo. 2006. Protective effects and mechanisms of a probiotic bacterium *Lactobacillus rhamnosus* against experimental *Edwardsiella tarda* infection in tilapia (*Oreochromis niloticus*). Vet. Immunol. Immunopathol. 113: 339-347.

Pirarat, N., K. Pinpimai, M. Endo, T. Katagiri, A. Ponpornpisit and N. Chansue. 2011. Modulation of intestinal morphology and immunity in Nile tilapia (*Oreochromis niloticus*) by *Lactobacillus rhamnosus* GG. Ress Vet Sci. 91: 9-97.

Pineiro, M. and C. Stanton. 2007. Probiotic bacteria: Legislative framework-requirements to evidence basis. J. Nutr. 137: 850S–853S.

Preetha, R., N.S. Jayaprakash and I.S.B. Singh. 2007. Synechocystis MCCB 114 and 115 as putative probionts for *Penaeus monodon* post-larvae. Dis. Aquat. Org. 74: 243-247.

Qi, Z., X. Zhang, N. Boon and P. Bossier. 2009. Probiotics in aquaculture of China – Current state, problems and prospect. Aquaculture 290: 15-21.

Queiroz, J.F. and C.E. Boyd. 1998. Effects of a bacterial inoculum in channel catfish ponds. J. World Aquacult. Soc. 29(1): 67-73.

Rahiman, K.M.M., Y. Jesmi, A.P. Thomas and A.A.M. Hatha. 2010. Probiotic effect of *Bacillus* NL110 and *Vibrio* NE17 on the survival, growth performance and immune response of *Macrobrachium rosenbergii* (de Man). Aquacult. Res. 41(9): 120-134.

Rengpipat, S., W. Phianphak, S. Piyatiratitivorakul and P. Menasveta. 1998. Effects of a probiotic bacterium on black tiger shrimp *Penaeus monodon* survival and growth. Aquaculture 167(3-4): 301-313.

Rengpipat, S., A. Tunyanun, A.W. Fast, S. Piyatiratitivorakul and P. Menasveta. 2003. Enhanced growth and resistance to Vibrio challenge in pond-reared black tiger shrimp *Penaeus monodon* fed a *Bacillus* probiotic. Dis. Aquat. Organ. 55: 169-173.

Ringø, E., E. Strom and J. Tabacheck. 1995. Intestinal microflora of salmonids: A review. Aquacult. Res. 26: 773-789.

Ringø, E. and F.J. Gatesoupe. 1998. Lactic acid bacteria in fish: A review. Aquaculture 160: 177-203.

Ringø, E., H. Van Doan, S.H. Lee, M. Soltani, S.H. Hoseinifar, R. Harikrishnan et al. 2020. Probiotics, lactic acid bacteria and bacilli: Interesting supplementation for aquaculture. J. Appl. Microbiol. 129: 116-136.

Ruiz-Ponte, C., J.F. Samain, J.L. Śanchez and J.L. Nicolas. 1999. The benefit of a Roseobacter species on the survival of scallop larvae. Marine Biotechnol. 1(1): 52-59.

Salinas, I., R. Myklebust, M.A. Esteban, R.E. Olsen, J. Meseguer, E. Ringø et al. 2008. *In vitro* studies of *Lactobacillus delbrueckii* subsp. *lactis* in Atlantic salmon (*Salmo salar* L.) foregut: Tissue responses and evidence of protection against *Aeromonas salmonicida* subsp. *salmonicida* epithelial damage. Vet. Microbiol. 128: 167-177.

Sampantamit, T., L. Ho, C. Lachat, N. Sutummanwong, P. Sorgeloos, P. Goethals et al. 2020. Aquaculture production and its environmental sustainability in Thailand: Challenges and potential solutions. Sustainability 12: 2010.

Sanders, M.E., D. Merenstein, C.A. Merrifield and R. Hutkins. 2018. Probiotics for human use. Nutr. Bulletin. 43(3): 212-225.

Saurabh, S., A.K. Choudhary and G.S. Sushma. 2005. Concept of probiotics in aquaculture. Fish Chimes 25(4): 19-22.

SCAN. 2001. Report of the Scientific Committee on Animal Nutrition on product Toyocerin for use as feed additive. European Commission, Health and Consumer Protection Directorate- General. (SCAN) Scientific Committee on Animal Nutrition.

Scholz, U., D.G. Garcia, D. Riccque, S.L.E. Cruz, A.F. Vargas, J. Latchford et al. 1999. Enhancement of vibriosis resistance in juvenile *Penaeus vannamei* by supplementation of diets with different yeast products. Aquaculture 176(3-4): 271-283.

Servin, A. 2004. Antagonistic activities of lactobacilli and bifidobacteria against microbial pathogens. Microbiol. Rev. 28: 405-440.

Shakibazadeh, S., C.R. Saad, A. Christianus, M.S. Kamarudin, K. Sijam, P. Sinaian et al. 2011. Assessment of possible human risk of probiotic application in shrimp farming. Int. Food Res. J. 18: 433-437.

Shefat, S.H.T. 2018. Probiotic strains used in aquaculture. Int. Res. J. Microbiol. 7: 43-55.

Shishehchian, F., F.M. Yusoff and M. Shariff. 2001. The effects of commercial bacterial products on macrobenthos community in shrimp culture pond. Aquacult. Int. 9(5): 429-436.

Silva, E.F., M.A. Soares, N.F. Calazans, J.L. Vogeley, B.C. do Valle, R. Soares et al.

2013. Effect of probiotic (*Bacillus* spp.) addition during larvae and postlarvae culture of the white shrimp *Litopenaeus vannamei*. Aquacult. Res. 44: 13-21.

Smart Microbials Inc. 2012. Un universo de microbios inteligentes a su servicio. Technical details on the use of commercial probiotic product Aqua BOOSTER. http://smartmicrobialsinc.com/aqua.php.

Soccol, C., L. Porto, M. Rigon, A.B. Medeiros, C. Yamaguishi, J. Lindner et al. 2010. The potential of probiotics: A review. Food Technol. Biotechnol. 48(4): 413-434.

Sornplang P. and S. Piyadeatsoontorn. 2016. Probiotic isolates from unconventional sources: A review. J. Anim. Sci. Technol. 58: 26.

Sotomayor, M.A. and J.L. Balcázar. 2003. Inhibition of shrimp pathogenic vibrios by mixture of probiotic strain. Rev. Aquat. 19: 9-15.

Sunitha, K. and P.V. Krishna. 2016. Efficacy of probiotics in water quality and bacterial biochemical characterization of fish ponds. Int. J. Curr. Micrbiol. App. Sci. 5(9): 30-37.

Spanggaard, B., I. Huber, J. Nielsen, E.B. Sick, C.B. Pipper et al. 2001. The probiotic potential against vibriosis of the indigenous microflora of rainbow trout. Environ. Microbiol. 3(12): 755-765.

Tapia-Paniagua, S.T., P. Díaz-Rosales, R.J.M. Léon-Rubio, I. Garcia de la Banda, C. Lobo, F.J. Alarcón et al. 2012. Use of the probiotic *Shewanella putrefaciens* Pdp11 on the culture of Senegalenses sole (*Solea senegalensis*, Kaup 1858) and gilthead seabream (*Sparaus aurata* L.), Aquacult. Int. 21: 1-15.

Taoka, Y., H. Maeda, J.-Y. Jo, M.-J. Jeon, S.C. Bai, W.-J. Lee. 2006a. Growth, stress tolerance and non-specific immune response of Japanese flounder *Paralichthys olivaceus* to probiotics in a closed recirculating system. Fisheries Sci. 72(2): 310-321.

Taoka, Y., H. Maeda and J.Y. Jo. 2006b. Use of live and dead probiotic cells in tilapia (*Oreochromis niloticus*). Fisheries Sci. 72: 755-766.

Tinh, N.T.N., K. Dierckens, P. Sorgeloos and P. Bossier. 2007. A review of the functionality of probiotics in the larviculture food chain. Marine Biotechnol. 10: 1-12.

Tuohy, K.M., H.M. Probert, C.W. Smejkal and G.R. Gibson. 2003. Using probiotics and prebiotics to improve gut health. Drug Discov. Today 8(15): 692-700.

Urdaci, M.C. and I. Pinchuk. 2004. Antimicrobial activity of *Bacillus probiotics*. pp. 171-182. *In*: E. Ricca, A.O. Henriques and S.M. Cutting (eds.). Bacterial Spore Formers: Probiotics and Emerging Applications. Horizon Bioscience.

Van Doan, H., S.H. Hoseinifar, M.A. Dawood, C. Chitmanat and K. Tayyamath. 2017. Effects of *Cordyceps militaris* spent mushroom substrate and *Lactobacillus plantarum* on mucosal, serum immunology and growth performance of Nile tilapia (*Oreochromis niloticus*). Fish & Shellfish Immunol. 70: 87-94.

Varela, J.L., I. Ruíz-Jarabo, L. Vargas-Chacoff, S. Arijo, J.M. León-Rubio, I. García-Millán et al. 2010. Dietary administration of probiotic Pdp11 promotes growth and improves stress tolerance to high stocking density in gilthead seabream *Sparus auratus*. Aquaculture 309(1-4): 265-271.

Verschuere, L., G. Rombaut, P. Sorgeloos and W. Verstraete. 2000. Probiotic bacteria as biological control agents in aquaculture. Microbiol. Mol. Biol. Rev. 64: 655-671.

Vine, N.G., W.D. Leukes, H. Kaiser, S. Daya, J. Baxter, T. Hecht et al. 2004. Competition for attachment of aquaculture candidate probiotic and pathogenic bacteria on fish intestinal mucus. J. Fish Dis. 27: 319-326.

Wang, Y.G., M.D. Hassan, M. Shariff, S.M. Zamri and X. Chen. 1999. Histopathology and cytopathology of white spot syndrome virus (WSSV) in cultured *Penaeus monodon* from peninsular Malaysia with emphasis on pathogenesis and the mechanism of white spot formation. Dis. Aquat. Org. 39(1): 1-11.

Wang, X., H. Li, X. Zhang, Y. Li, W. Ji and H. Xu. 2000. Microbial flora in the digestive tract of adult penaeid shrimp (*Penaeus chinensis*). J Ocean Univ Qingdao. 30: 493-498.

Wang, Y.B., Z.R. Xu and M.S. Xia. 2005. The effectiveness of commercial probiotics in northern white shrimp *Penaeus vannamei* ponds. Fisheries Sci. 71(5): 1036-1041.

Wang, Y.B., J.R. Li and J. Lin. 2008. Probiotics in aquaculture: Challenges and outlook. Aquaculture 281: 1-4.

Youping, Y., M. Dongdong, L. Shijiang and W. Zhongkang. 2011. Effects on growth and digestive enzyme activities of the *Hepialus gonggaensis* larvae caused by introducing probiotics. World J. Microbiol. Biotechnol. 27(3): 529-533.

Yousefian, M. and M.S. Amiri. 2009. A review of the use of prebiotic in aquaculture for fish and shrimp. Afr. J. Biotechnol. 8: 7313-7318.

Zapata, A.A. and M. Lara-Flores. 2013. Antimicrobial activities of lactic acid bacteria strains isolated from Nile tilapia (*Oreochromis niloticus*) intestine. J. Biol. Life Sci. 4(1): 123-129.

Zhang, S., Y. Sing, M. Long and Z. Wei. 2012. Does dietary administration of *Lactococcus lactis* modulate the gut microbiota of grouper, *Epinephelus coioides*. J. World Aquacult. Soc. 43(2): 198-207.

Zhang, L., D. Xu, F. Wang and Q. Zhang. 2019. Antifungal activity of *Burkholderia* sp. HD05 against *Saprolegnia* sp. by 2-pyrrolidone-5-carboxylic acid. Aquaculture 511: 634198.

Zhou, X.X., Y.B. Wang and W.F. Li. 2009. Effect of probiotic on larvae shrimp (*Penaeus vannamei*) based on water quality, survival rate and digestive enzyme activities. Aquaculture 287(3-4): 349-353.

Zhou, X., Y. Wang, J. Yao and W. Li. 2010. Inhibition ability of lacticacid bacteria *Lactococcus lactis*, against *A. hydrophila* and study of its immunostimulatory effect in tilapia (*Oreochromis niloticus*). Int. J. Eng. Sci. Technol. 2(7): 73-80.

Zokaeifar, H., N. Babaei, C.R. Saad, M.S. Kamarudin, K. Sijam and J.L. Balcazar. 2014. Administration of *Bacillus subtilis* strains in the rearing water enhances the water quality, growth performance, immune response, and resistance against *Vibrio harveyi* infection in juvenile white shrimp, *Litopenaeus vannamei*. Fish Shellfish Immunol. 36: 68-74.

Zorriehzahra, M.J., T.D. Somayeh, A. Milad, T.K. Ruchi, Karthik, D. Kuldeep et al. 2016. Probiotics as beneficial microbes in aquaculture: An update on their multiple modes of action: A review. Vet. Q. 36(4): 228-241.

CHAPTER
12

Probiotics and Functional Feed

Guzin Iplıkcıoglu Cıl[1], Gaye Bulut[2], Duygu Budak[3], Guzin Camkerten[4]
and Ilker Camkerten[5]*

[1] Food Hygiene and Technology Department, Faculty of Veterinary Medicine,
Ankara University, Ankara 06110, Turkey
[2] Obstetric and Gynecology Department, Faculty of Veterinary Medicine,
Aksaray University, Aksaray 68100, Turkey
[3] Animal Nutrition and Nutritional Diseases Department, Faculty of Veterinary
Medicine, Aksaray University, Aksaray 68100, Turkey
[4] Technical Sciences Vocational School, Aksaray University, Aksaray 68100, Turkey
[5] Internal Medicine Department, Faculty of Veterinary Medicine, Aksaray
University, Aksaray 68100, Turkey

1. Introduction

Increased demand for livestock and aquaculture products is putting pressure
on the sector to produce more with limited resources. New methods in
animal breeding are being developed, aiming to increase animal yield,
production quality and food safety taking into account both the welfare
of the animal and respect for the environment. In the past, antibiotics and
other medicinal products had been broadly used as antimicrobial growth
promoters (AGPs), mainly in order to improve the microbiota and to increase
productivity and growth. On the other hand sensitive issues and bans on
antibiotics made probiotics and prebiotics alternatives to them having
features improving adaptation of animals to poor conditions, and increasing
their resistance against diseases and genetical potentials. In recent years this
has had remarkable effects on the decrease of metabolic disease risks (Vyas
et al., 2014). High hopes are evoked and the a clear consensus has started to
develop on the use of probiotics as natural and biotechnological functional
feed (Moharrery and Asadi, 2009, Markowiak and Śliżewska, 2018). The
current probiotic market has reached 33.19 billion dollars and expected to
reach 46.55 billion dollars by 2020. Besides human consumption, probiotics
are also increasingly used in the animal industry, with an annual market
growth rate of 7.7% (Park et al., 2016).

*Corresponding author: ilkercamkerten@hotmail.com

Gastrointestinal microbiota is not only involved in digestion and fermentation which are of particular importance in animals but also has considerable beneficial effects. The gut microbiota is responsible for the synthesis of vitamins; bioconversion of toxic compounds to nontoxic residues; stimulation of the immune system; maintenance of gut peristalsis and intestinal mucosal integrity, and plays the role of a barrier against colonization by pathogens. Probiotics are a significant component of microbiota and their beneficial effects have led to a better understanding of the major contribution of the gut microbiota to animal nutrition and health (Chaucheyras-Durand and Durand, 2010).

Probiotics are defined as live microorganisms, which when administered in adequate amounts, confer a health benefit on the host. For animals, the probiotic concept was started with the use of living cultures in the feed for various animals (FAO/WHO, 2009). The beneficial effects of probiotics added to functional feeds in animal production have been related to different modes of action. Probiotics are used widely in farm animals and in aquaculture, as they are mainly associated with reducing some clinical diseases and increasing growth and improving productivity including growth rate, feed intake, feed efficiency ratio, milk composition, egg production and reproduction performances (Collins et al., 2009).

This chapter provides information about the importance of gut microbiota and probiotics in animal nutrition, the effect of functional feed in different categories of livestock and aquaculture, additional safety and risk associated with the use of probiotics.

2. Gut Microbiota, Balance and Imbalance

The gut is a highly complex ecosystem where microbiota, nutrients, and host cells interact extensively. It has an important role in converting, storing and expending energy from the diet. For a long time, gut was considered only from this point of view and no attention was paid to gut bacteria. Now, more research is pointing out that, microbiota is an active element of a healthy physiology and changes in it is a cause or a consequence of the diseases (Butel, 2014).

The diverse union of bacteria, archaea, fungi, protozoa, viruses, and their collective genome found on and within the host comprises the microbiota. The microbiota has a dynamic structure and is subject to important changes during the life of the host in response to a variety of factors including diet, environment, medical interventions, and disease states (Barko et al., 2018).

The microbiota in the gastrointestinal tract of mammals is now accepted as a metabolically active organ with a large biodiversity in species that can reach a high number of cells of the order of 10^{14}. One of the key issues about microbiota is identification and the knowledge of the species present, in different animals. The microbiota of each individual has a specific "bacteria fingerprint", a profile of its own species which is different from other individuals (Gaggia et al., 2010). When we classify mammals as monogastric and polygastric, their microbiota cluster into groups that correspond to these categories. In monogastric animals such as pig, chicken, horse, and man, the major microbial groups are *Bacteroides, Clostridium, Bifidobacterium, Eubacterium, Lactobacillus, Enterobacteriaceae, Streptococcus, Fusobacterium, Peptostreptococcus* and *Propionibacterium*. In polygastric animals, such as cow, sheep, and lamb, the most important microbial ecosystem is rumen and major microbial groups are belonging to *Fibrobacter, Ruminococcus, Butyrivibrio,* and *Bacteroides* together with *Prevotella, Selenomonas, Streptococcus, Lactobacillus* and *Megasphaera* (Ley et al., 2009).

Gastrointestinal microbiota is not only involved in digestion and fermentation which is of particular importance in animals but it also has considerable beneficial effects. The gut microbiota is responsible for the synthesis of vitamins; bioconversion of toxic compounds to nontoxic residues; stimulation of the immune system; maintenance of gut peristalsis, intestinal mucosal integrity, and plays the role of a barrier against colonization by pathogens. These beneficial effects have led to a better understanding of the major contribution of the gut microbiota to animal nutrition and health. The modulation of the gut microbiota with probiotics as feed additives is quite an important issue in animal nutrition and breeding for providing host-protection to reduce the risk of intestinal diseases and removing specific microbial disorders (Chaucheyras-Durand and Durand, 2010, Gaggia et al., 2010).

Present studies support that, early stages of life are significant for the composition of microbiota. The balanced and diverse microorganism composition in early life leads to an optimal, healthy and stable microbiota that strengthens itself and the host. However, when disturbances occur in early life, for example, due to antibiotic use, host's inability to interact properly with the microbiota, a mismatch between the environment of parent and offspring, non-optimized feeding conditions or infections, organisms might be more susceptible to disease early and later in life (Brugman et al., 2018).

Under healthy, homeostatic conditions the microbiota is composed of various organisms, off which some are known to be beneficial and some harmful to the host, but they are in balance. It is commonly known that the homeostasis of microbiota influences many elements of host physiology, including development, metabolism, and immunity (Chan et al., 2013). Dysbiosis refers to an undesirable alteration of the microbiota resulting in an imbalance between protective and harmful bacteria or a disturbed interaction between bacteria and host. Loss of beneficial microorganisms, increase of pathogens, and loss of diversity are events that encompass dysbiosis (Lee and Hase, 2014, Sommer et al., 2017). Dysbiosis has been associated with many disease conditions in humans, including local gastrointestinal diseases and systemic diseases. In domestic animals, dysbiosis was practically unknown until the ban on antimicrobial growth promoters in animal feed in Europe on 1 January 2006. Since then, it is perceived as one of the most challenging problems in the field, especially in broilers and in newly weaned pigs (Ducatelle et al., 2015).

Numerous environmental factors are also able to affect the composition and functions of gut microbiota in livestock animals. Among them, feeding practices, the composition of animal diets, farm management, and productivity constraints are parameters which can cause the imbalance of microbiota in the gastrointestinal tract and consequently affect the health and welfare of the animal. In these situations, there are some therapeutic manipulations for the construction of homeostasis in the intestinal microbiome. Balance is generally achieved through modification of the diet, administration of probiotics, prebiotics, or antibiotics, and in some cases by fecal microbiome transplantation (Chaucheyras-Durand and Durand, 2010, Barko, 2018).

3. Probiotics Today in Functional Feed

Probiotics are defined as live microorganisms, which when administered in adequate amounts, confer a health benefit on the host (FAO/WHO, 2009). Although the term probiotic is a relatively new word, belief in beneficial bacteria is very ancient in human civilisation. For animals, the probiotic concept was started with the use of living cultures in feeds for various animals. Eckles and Williams et al. published in their report that they used yeast as a supplementary feed for lactating cows, in 1924. In the 1940s, the use of *Streptomyces aureofaciens* probiotics resulted in weight gain in animals. In 1947, Møllgaard indicated improvements in health and skeletal formation in pigs with impaired mineral absorption supplemented with lactic acid bacillus. The milestone in the development of a probiotic approach to animal health was competitive exclusion (CE), also known as the 'Nurmi concept' originating from the finding that newly hatched chicks could be protected against *Salmonella* colonization of the gut by dosing them with a suspension of gut content prepared from healthy adult chickens (Eckles and Williams, 1924, Møllgaard, 1947, Nurmi and Rantala, 1973, Musa et al., 2009).

The definition of probiotics was also affected simultaneously with the use of microorganisms in feeds. Parker defined probiotics as 'organisms and substances which contribute to intestinal microbial balance' in 1974, but this definition was thought to be connoting antibiotics as well. After that Fuller in 1989 redefined probiotics as 'a live microbial feed supplement which benefits the host animal by improving its intestinal microbial balance'. This revised definition, added the need for a probiotic to be viable. In 2008, it was proposed that the same growth-promoting effects in animals may have been seen in human by the use of probiotics (Fuller, 1989, Angelakis, 2017).

Animal health and production should not be considered separately from nutrition and human health. Year after year, new methods in animal breeding are being developed, aiming at quality and safety of food while taking into account the welfare of the animal and respect for the environment (Markowiak and Slizewka, 2018). One of the important components of animal breeding is feed. Studies have shown that feed and feed additives have a great effect on the host gut microbiota, so it makes them a significant part of healthy livestock production (Kogut and Arsenault, 2016).

In the past, antibiotics and other medicinal products had been broadly used as antimicrobial growth promoters (AGPs), mainly in order to improve the microbiota and to increase productivity and growth. By some possible mechanisms such as reduction in total bacterial load, suppression of pathogens, thinning of the mucosal layer, and direct modulation of the immune system antibiotics improve the performance of animals, however, the whole mechanism is not clear (Hardy, 2002). Long term use of AGPs has led to developing some safety concerns about them. Risk of antibiotic resistance, especially the spread of resistance genes to human and animal pathogens, the release of antibiotics into the environment and persistence of chemical residues in animal products are the main concerns. Regarding these negative effects, European nations have implemented bans on the use of AGPs, and the USA has increased the regulatory and political scrutiny and the search for natural alternatives ensuring similar benefits has started (Allen et al., 2013, Uyeno et al., 2015).

The addition of growth promoters into animal feed has made significant contributions to the improvement of productivity and the decrease in food production costs. Additionally, growth promoters are also being enhanced due to the shortage of valuable resources, like animals, feed, water, and land. It became obvious that feed additives, including antibiotics, probiotics, and prebiotics can manipulate the animal gut microbiota through diet. On the other hand sensitive issues regarding antibiotics and bans on them have made made probiotics and prebiotics the best alternative to them. High hopes are evoked and a clear consensus has started to develop on the use of probiotics as functional feeds (Newbold and Hillman, 2004, Angelakis, 2017).

4. Required Properties of Microorganisms in Feed

Before a probiotic is approved as a feed additive its effectiveness has to be proven and its safety must be demonstrated. Survival in the gastrointestinal tract after ingestion, adhesion to the related parts of the body, antimicrobial activity, and antibiotic susceptibility are among the main probiotic properties that should be analyzed to assess functionality and safety (Simon et al., 2001, Gaggia et al., 2010).

The first requirement for a good probiotic agent is reaching the site of action, usually, the gut, and its ability to survive physiological stresses such as stomach acidity, low gut pH and biliary salts in the gastrointestinal tract. Probiotics should be able to compete with normal microflora. Furthermore, probiotics should have the ability to adhere to gut epithelial tissue and colonize, persist and withstand short periods in the gastrointestinal tract (Musa et al., 2009, Angelakis, 2017). Probiotics should be genetically stable and should exhibit antimicrobial activity or antagonistic behavior towards pathogenic bacteria. They need to modify the intestinal microbiota and at least have one proven health benefit (Ezema, 2013).

In the definition of probiotics, their viability and beneficial effects are highlighted, which are contingent on their presence in adequate numbers. Therefore, the probiotic strain must reach the intestine in a viable form and in sufficient numbers. This requires, maintainance of the probiotic characteristics and stability during the feed processing, including pelleting by heat in many applications. Probiotics must be conserved in the matrix in which they were incorporated. Moreover, they need to sustain their stability in the feed during the storage spanning weeks (Simon, 2005, Butel, 2014).

For safety, probiotics should be nonpathogenic and nontoxic. Their ingestion must not present any risk to the host. Probiotics are subject to legal regulations thus they should be safe for human and animal health. In the USA they are regulated by the FDA and the microorganisms used for consumption should have the Generally Regarded As Safe (GRAS) status. Qualified Presumption of Safety (QPS) term is used for safety assessment in Europe, which is introduced by EFSA. The QPS concept is similar to GRAS status but it involves some additional criteria such as, including the history of safe usage and absence of the risk of acquired resistance to antibiotics (Anadon et al., 2006, Markowiak and Slizewka, 2018).

The advanced molecular methods are implemented to define the phenotypic and genetic properties of probiotics that are useful for industrial production. For instance, for identification of probiotics strains molecular tools based on 16S ribosomal DNA sequencing and PCR techniques have been developed or microarrays are allowing the detection of multiple characteristics of probiotics (Collins et al., 2009).

5. Application of Probiotics in Livestock and Aquaculture

Probiotics are used widely in farm animals and in aquaculture, as they are mainly associated with reducing some clinical diseases and increasing growth and improving productivity (Collins et al., 2009). The beneficial effects of probiotics in animal production have been related to different modes of action. The first one is related to the imbalance of the intestinal ecosystem and could be a risk factor for pathogen infections. Intensive farming systems due to confinement of large numbers of animals in small areas, stress, and competition for feed, reduced maternal contact, unatural feeding, and transport are the factors can cause this imbalance. The other action in animals is related to productivity. These include growth rate, feed intake, feed efficiency, milk composition, egg production and reproduction in ruminants, poultry, pigs and horses. Each species of livestock and aquaculture has its critical point in the production chain (Iplikcioglu Cil, 2018).

5.1 Ruminants

Application of probiotics in ruminants has been carried out considering both the health status of the animals in terms of reduction in the incidence of and/or severity of diarrhea, carriage of pathogenic microorganisms and economic parameters. Also, studies can be grouped according to the age such as pre-ruminant life and life as adult ruminants. Most of the applications have been implemented on cows and calves, whereas little information is available for lambs, ewes and goats (Gaggia et al., 2010).

For ruminants, rumen has an important role in producing the source of proteins and energy with its complex microbial ecology. Therefore, manipulating the rumen ecosystem has a significant effect on improving animal productivity and reducing unwanted products like methane (FAO, 2016). Various strains of yeast, mainly *Saccharomyces cerevisiae*, are commonly preferred probiotics in ruminants, with their ability to affect the microbial population dynamics in the rumen and the breakdown of nutrients. After them, lactic acid bacteria are another important group of probiotics (Chaucheyras-Durand et al., 2008).

Probiotics ensure the production of organic acids, acetic acid and lactic acid in particular. So they have a strong inhibitory effect against gram-negative bacteria in ruminal and gastrointestinal microbiota and organic acids have been considered the main antimicrobial compounds responsible for the inhibitory activity of probiotics against pathogens such as *E. coli* (Bermudez-Brito et al., 2012).

One of the important causes of mortality and morbidity in young ruminants is neo-natal diarrhea generally caused by enterotoxigenic *E. coli*. Studies show that probiotics can reduce the incidence and number of days with diarrhea. Maldonado et al. (2017) reported lower morbidity and mortality in calves that were fed with fermented milk containing four different probiotic *Lactobacillus* strains. Morbidity was found to be 69.20% in animals without the probiotic, and 46.15% in probiotic-treated animals, mortality in the control group was 34.61% and 7.69% in animals fed with milk containing the probiotic. Similarly, Von Buenau et al. (2005) indicated that the administration of viable *E. coli* bacteria, strain Nissle 1917, has a clear beneficial effect on the prophylaxis and treatment of neonatal calf diarrhea. In their study from 335 newborn calves, the incidence of diarrhea was 65.2% under placebo and 26.5% under *E. coli* Nissle 1917 in the first step, and the rates were 63.0% and 12.2% in the second step, respectively. Galvao et al. (2005) added live yeasts, *S. cerevisiae*, and *S. cerevisiae*, spp. *Boulardii* into the feed of calves prior to and after weaning. They detected, calves receiving yeasts had fewer days with diarrhea prior to and after weaning than control calves. In contrary, according to He et al. (2017) no significant difference was observed in calf performance and health scores in calves fed with milk replacer supplemented with *S. cerevisiae* var *boulardii*. Masood and Mir (2018) reported that, the average fecal *E. coli* and total coliform count of experimental lambs fed with probiotic mix alone (*Saccharomyces cerevisiae* + *Lactobacillus acidophilus*) and in combination with enzyme mix were significantly lower as determined in 0, 30, 60 and 90 days of the experiment as compared to animals fed with enzyme mix alone. Also during the studies, the intermittent diarrheal condition was seen throughout the experimental period in control group animals but not in those given probiotics. It was thought that these results might be due to the probiotic mix having great potential to beneficially affect the gut microbiota such as *E. coli, Clostridium* spp. and coliforms. In a study, lactic acid bacteria inoculum was used to evaluate a level protection capacity in calves against a *Salmonella* Dublin infection. Results showed that the severity of *Salmonella* infection was reduced and milder microscopic lesions developed in the group treated with lactic acid bacteria (Frizzo et al., 2018).

Among foodborne pathogens *E. coli* O157: H7 is one of the most important causes of severe human diseases. The main reservoir of this pathogen is the gut of ruminants, mainly cattle. To reduce carriage and shedding of *E. coli* O157: H7 in cattle new approaches have been proposed, including feeding of antagonistic microorganisms, vaccination or bacteriophage treatment. In an *in vitro* experiment, it was indicated that in an *E. coli* O157: H7 inoculated rumen fluid, *Lactobacillus reuteri* has an antimicrobial activity and *E. coli* has a very limited growth in rectal contents (Bertin, 2017). According to the results of a meta-analysis, the prevalence of fecal *E. coli* O157 shedding in cattle is significantly reduced by direct fed microbial treatments (Wisener, 2013). Tabe et al. (2008) reported there was a 32% reduction in detectable levels of *E. coli* O157:H7 shed in feces of the naturally infected steers given *P. freudenreichii* and *L. acidophilus* supplemented feed, compared with controls. Similarly, in a two year trial, Peterson et al. (2007) showed a 35% decrease of *E. coli* O157: H7 shed in beef cattle, following the daily intake of *L. acidophilus* strain NP51.

The pH of the rumen is very crucial for the health of the ruminants. In some cases like consumption of a diet with a high proportion of non-structural carbohydrates (starch) and/or decreased proportion of fiber, the pH of the rumen may drop below the optimum range, in dairy and beef cattle. When the pH is below 6.0, the activity of cellulolytic bacteria is seriously decreased and the number of protozoa declines. As a result, rapid fermentation and rise of accumulation of lactate, short-chain fatty acids, and other volatile fatty acids occur in the rumen. This condition is referred to as subacute ruminal acidosis (SARA) when the pH drops below 5.6. As pH continues to drop, the lactobacilli population becomes dominant, and raises the lactate concentration and called lactic acidosis. This is a more severe form of ruminal acidosis where the pH drops below 5.2 and can lead to animal death if left untreated (Owens et al., 1998). Loss of appetite, diarrhea, dehydration, debilitation, impaired rumen motility, and impaired fiber digestibility are the results of ruminal acidosis and this condition is economically very important as productivity of the suffering animal is reduced due to these symptoms (Plaizier et al., 2012).

Studies have reported that probiotics are effective in preventing or treating ruminal acidosis. Reis et al. (2018) compared the efficacy of *S. cerevisiae* with monensin sodium in the prevention of SARA in sheep. They concluded in their study that the use of yeast culture can be beneficial in the prevention and treatment of SARA in sheep because it can effectively reduce the accumulation of lactic acid, and thereby increase ruminal pH and reduce ruminal osmolarity but monensin did not directly prevent these conditions. Similarly, Marden et al. (2008) tried to evaluate the capacity of *S. cerevisiae* (strain Sc 47), in optimizing ruminal pH in dairy cows. They reported that *S. cerevisiae* decreased the lactic acid concentration and allowed better fiber digestion which may result in the prevention of ruminal acidosis. In contrast, Hristov et al. (2010) did not find any effect of *S. cerevisiae* culture on ruminal fermentation. In order to maintain ruminal pH, lactic acid utilizing bacteria have also been used as probiotics. *Megasphaera elsdenii* utilize lactates and prevent rapid pH drops caused by their accumulation in the rumen and supplementation of *M. elsdenii* in the feed was recommended as a preventive for acute acidosis in transition animals (Teixeira de Magalhães et al., 2015). In a study by Goto et al. (2016) effects of a bacterial probiotic, consisting of *L. plantarum* strain 220, *E. faecium* strain 26 and *C. butyricum* strain Miyari, on ruminal fermentation and plasma metabolites was evaluated in Holstein cattle with induced SARA. The mean ruminal pH was found to be higher in the probiotic-treated groups compared with the control group during the SARA challenge. Researchers also, indicated that volatile ruminal fatty acid concentrations were not affected by probiotic treatment; however, the added probiotic groups had lower lactic acid levels compared with the control group.

The most consistent effects of probiotics in dairy animals are on the increase in dry matter (DM) intake and milk production. In addition, there are studies showing that probiotics have positive effects on the composition of milk. The possible mode of action for improved productivity could be a result of increased feed intake together with improved microbial digestion (Uyeno et al., 2015). It is emphasized that these effects depend on the modified ruminal fermentation and composition of ration used. Therefore, enhancement is provided in the rate of utilization of the animal from feed and it is observed that digestibility of some nutrients increases (Chaucheyras and Durand, 2010). In the results of a meta-analysis, commercial probiotics containing *S. cerevisiae* increased milk yield by 1.18 kg/day, milk fat yield by 0.06 kg/day and milk protein yield by 0.03 kg/day as reported. DM intake was increased by 0.62 kg/day during early lactation and 0.78 kg/day during late lactation (Poppy et al., 2012). Shakira et al. (2018) investigated the effect of indigenously isolated *S. cerevisiae* probiotics on milk production and some other parameters in lactating dairy cows. They observed a significant difference in milk production of lactating dairy cattle fed on *S. cerevisiae* probiotic feed than fed on control. Also, they reported that there was a significant value of milk fat has been recorded in the probiotic treated animal's milk. In Holstein cows a 7.6% increase was reported in average daily milk yield when a combination of *L. acidophilus* NP51 and *P. freudenreichii* NP24 were added to the feed (Boyd et al., 2011). Stella et al. (2007) stated that Saanen goats receiving *S. cerevisiae* at the rate of 4×10^9 cfu/day/animal have a 14% increase in average milk yield per day compared to the non-treated ones. In a study of Kritas et al. (2006), the effect of probiotic containing *Bacillus licheniformis* and *Bacillus subtilis* on sheep milk production was evaluated. Daily milk yield throughout the milking of ewes was significantly higher in the probiotic-treated group compared with the non-treated group. Also, fat and protein content of milk in ewes that received probiotics was significantly more compared with untreated ewes.

Effects of probiotics on rumen parameters were associated with the quality of roughage besides ration composition primarily (Opsi et al., 2012, Jurkovich et al., 2014). Apart from this, effects were expected on protecting the pH level of rumen by consuming oxygen produced by aerobic pathogens and transforming ammoniacal nitrogen into microbial protein by increasing the consistency of bacteria in rumen (Patra, 2012).

Feed is strongly related with the growth performance of young ruminants and weight gain in adults. By balancing the intestinal microbiota and ruminal ecosystem, probiotic supplementation in feed shows beneficial effects on feed intake and growth rate (Alugongo et al., 2017). Since growth performance parameters are more sensitive than health status parameters, it is difficult to assess the beneficial effect of probiotics applied to ruminants, therefore general results of the studies are contradictory (Khalid et al., 2011). Sharma et al. (2016) researched the effect of *S. cerevisiae* probiotic supplementation on growth performance in pre-ruminant buffalo calves. Results revealed that the effect of *S. cervisiae* was more effective during the first month of supplementation but could not sustain in the second month. In a study of Ghazanfar et al. (2015) the average daily weight gain (kg/d) of the heifers was found to be higher in the *S. cerevisiae* supplemented feed group than the control group. Also, the study revealed that the digestibility of dry matter, organic matter, crude protein, neutral and acidic detergent fibers, was higher in yeast- supplemented group compared with the control group. Lei et al. (2015) reported that *Bacillus amyloliquefaciens* when fed to dairy calves at the rate of 3.16×10^8 CFU per kg, improved growth rate by 39%, increased feed use efficiency by 14%. According to Titi et al.'s study (2008) calves which were fed a total mixed ration supplemented with yeast culture at a level of 20 kg/ yeast/ton of feed, showed no development in the parameters such as, final weight, average daily gain and feed conversion ratio. In another study, goats' kids feeding performance after addition of lactic probiotics to the rations was investigated. Results showed an increase in N intake (23%) and microbial protein synthesis, improvement in fiber degradation, NH_3 concentration, and body weight gain of growing kids (Galina, 2009). In contrast, Pienaar et al. (2012) reported that lambs fed with diets including rumen active live yeast products had no significant effect on daily live weight gain, feed conversion ratio and also feed intake. Likewise, Hernandez et al. (2009) and Haddad and Goussous (2005) reported similar results about the probiotic supplementation in the diets of lambs. In a meta-analysis of 21 publications, it was concluded that lactic acid probiotic bacteria (*L. acidophilus, L. plantarum, L. salivarius, E. faecium, L. casei/paracasei* or *Bifidobacterium* spp.) increase body weight gain and improved feed use efficiency in young calves compared with control groups when probiotics were added to replace milk, but were ineffective when added to whole milk (Frizzo et al., 2018).

5.2 Poultry

Microbial populations in the gastrointestinal tracts of poultry have an important impact on the normal digestive processes and in maintaining animal health. The cecum contains the highest microbial density and responsible for most of the fermentation in poultry (Brugman et al., 2018). The idea of using probiotics in poultry farms has come from the bird's behavior in nature. In the natural environment, the mother gives the hatching chicks a feed, which has been stored in her crop, fermented and mixed with beneficial microorganisms. The colonization is also facilitated by eating faeces. This vertical transmission allows protection of hatching chicks from pathogens and developing a microbiota in their intestines. However, in commercial production, this path is not possible since the young chickens lack contact with their mother and the natural environment. This results in prolonged colonization and a balance of intestinal microbiota taking around 21 days for broilers (Fuller, 2001).

The mechanism of actions of probiotic feed additives in poultry includes; maintaining normal intestinal microbiota by competitive exclusion and antagonism, altering metabolism by increasing digestive enzyme activity and decreasing harmful products like bacterial enzymes and ammonia, improving feed consumption and digestion and stimulating the immune system (Mirza, 2018).

In chickens, modern production systems may weaken immune functions and predisposing the animals to bacterial pathogens which pose a threat to animal health and food safety. The adaptation to the post-hatching period, feed changes or imbalances, transportation, processing at the hatchery and high stocking densities are some of the stress factors in modern production. During seasonal changes, fluctuation in temperature has also been documented as an additional stress factor for poultry (Traub-Dragatz et al., 2006, Gaggia et al., 2010).

Probiotics ranging from lactic acid bacteria to spore formers and yeast have been evaluated for their potential to improve growth rates in commercial poultry production. The improvement in growth rate is attributed to the increased feed intake and improved feed use efficiency in many cases (Afsharmanesh and Sadaghi, 2014). Studies revealed that the growth-promoting effects of probiotic bacteria were equal to or better than treatment with antibiotics in poultry. For instance, equal weight gain has been observed in animals treated with the antibiotic avilamycin, and better weight gain in animals treated with chloroxytetracycline or oxytetracycline following treatments with *Lactobacillus* sp. probiotics (Angelakis, 2017). Lei et al. (2015) discovered that, chickens fed diets supplemented with *Bacillus amyloliquefaciens* resulted with higher body weight gain and better food conversion ratio than the control, for the finisher phase and the overall period as well. Mountzouris et al. (2007) observed that a basal diet containing a probiotic mix *Lactobacillus, Bifidobacterium, Enterococcus,* and *Pediococcus* in a concentration of 10^8 CFU/g significantly increased body weight of broilers in comparison with the control group. Torres-Rodriguez et al. (2007) reported that administration of a *Lactobacillus* spp. based probiotic to turkeys' feed, increased the average daily gain and body weight and added that this gain is represents an economic alternative to improve turkey production. However, Olnood et al. (2015) reported that the addition of *L. johnsonii* did not improve body weight gain, feed intake and feed conversion ratio of broiler chickens raised on the litter during the 5-week experimental period. Similarly in another study, *S. cerevisiae* used as a dietary probiotic to assess performance and result, showed no overall weight gain difference (Karaoglu and Durdag, 2005). As seen from the results of different studies one of the interesting observations from probiotic feeding trials in poultry is that some probiotics are effective in growth promotion changes in the starter (early) phase while others affect the grower-finisher (later) phase and some studies found improved growth throughout the broiler production cycle. The main reason for this difference is not known, but it is thought to be related to the dynamics of the gut microbiota (FAO, 2016).

The oldest application of probiotics in poultry is generally associated with competitive exclusion. The point of CE is to control the colonization and shedding of pathogens by restoring a protective gut microbiota (Dhama et al., 2011). Presence of pathogens not only has an effect on animal health but they are also an important public health risk especially *Salmonella* and *Campylobacter*. In addition, enteric diseases like necrotic enteritis and coccidiosis can cause huge economic losses to the industry (Collins et al., 2009). Probiotic cultures have been used successfully to control and reduce *Salmonella* colonization. CE treatment also protects chicks against other microorganisms such as *C. jejuni, Listeria monocytogenes*, pathogenic *E. coli, Yersinia enterocolitica* and *C. perfrigens* (Gaggia et al., 2010). Lei et al. (2015) reported *E. coli* levels in the cecum were significantly reduced by dietary supplementation of *Bacillus amyloliquefaciens*. Also, they observed that compared with the control group the population of *Lactobacillus* was increased in treated groups, so it shows that probiotics not only control the pathogens but also help in the improving the presence of beneficial microorganisms. Increasing concerns about antibiotic-resistant strains of *Salmonella* led probiotics to be an emerging alternative control method for *Salmonella* (Tellez et al., 2012). Carter et al. (2017) observed a 2 log reduction of *Salmonella* Enteritidis colonization in the ileum, caecum and colon by using combined administration of the two probiotic bacteria, *L. salivarius* 59 and *E. faecium* PXN33. Probiotics also reduced the spread of

Salmonella from infected to healthy birds. Biloni et al. (2013) reported a slow horizontal transmission of *Salmonella* infection within the flock, by applying a probiotic containing *L. salivarius* and *Pediococcus parvulus*. In a study, it was reported that a commercial probiotic containing *Lactobacillus*, *Bifidobacterium*, and *Enterococcus* reduced the prevalence of *Campylobacter* infection in broiler chickens (Willis and Reid, 2008). Similarly, Guyard-Nicodeme et al. (2016) detected that *B. subtilis* reduces the shedding of *Campylobacter*. Jeong and Kim (2014) conducted a study to determine and confirm the effect of *B. subtilis* spore supplementation to feeds in broilers. In their results, they indicated that supplementation tended to reduce *C. perfringens* counts in the large intestine and excreta and the probiotic effect, was a significant increase in *Lactobacillus* counts in the cecum, ileal, and excreta compared with the control group.

Coccidiosis is a parasitic disease which is caused by different species of *Eimeria* protozoa that colonize different sections of the gastrointestinal tract of poultry. It is very important for poultry production, to have a high rate of resistance to anticoccidial drugs and severe economic consequences due to its ubiquitous nature (FAO, 2016). Although, inconclusive results were obtained from the literature research is reporting the effect of probiotics on coccidiosis. In the Dalloul et al. (2005) study data suggests a positive impact of the *Lactobacillus*-based probiotics on cellular immune responses of infected broilers as compared to control chickens resulting in enhanced resistance to *Eimeria acervulina* and reduced fecal oocyst shedding. Likewise, Lee et al. (2007) reported a reduced oocyst shedding in birds infected with 5,000 *E. acervulina* oocysts after they were fed with a *Pediococcus* based probiotic diet.

Poultry fed with probiotics had a tendency to display significant intestinal histological changes. These changes enhanced the absorption of nutrients, digestive enzyme secretion and keeping the microvilli intact (Kabir, 2009). According to Bai et al. (2013) when compared to chickens fed with or without antibiotics or probiotics, administration of *L. sakei* Probio-65 increased villi height and crypt depth in the jejunum of broilers. Hutsko et al. (2016) investigated the intestinal morphology of young turkeys which were fed with a commercial probiotic. Mean villus height, villus area, and crypt depth were significantly increased in the poults fed probiotic compared with control poults. It is known that the formation of villi in a wave like pattern in the jejunum enables better nutrient absorption than villi arranged in parallel or randomly positioned. Probiotics, promoted a wave-like arrangement of jejunum villi in broilers, while this wave-like pattern was absent in the jejunum of broilers fed with normal or antibiotic treated feed (Park et al., 2016).

There are numerous studies investigating the beneficial effects of probiotics on production performances in poultry such as meat quality, carcass yield, egg production and composition (Dhama et al., 2011). In the study of Hatab et al. (2016), the basal diet of layer chicks is supplemented with a probiotic mixture containing *B. subtilis* and *E. faecium*, until they are 10 weeks old. They demonstrated a significant increase in the relative weight of carcass, liver, heart, kidney, proventriculus, small intestine, thymus, spleen, bursa of Fabricius and small intestine length in the treated group as compared to the control group during the overall experimental period. Inatomi et al. (2015) assessed the carcass and meat quality of broilers in two groups, one fed with a conventional diet, and the second, fed with one supplemented with the probiotic mix (*B. mesentericus TO-A, C. butyricum TO-A, and S. faecalis T-110*) after being reared for 49 days. According to their results, the second group had significantly higher percentages of carcass yield than the first one and the first group had lower percentages of abdominal fat than the second. However, Afsharmanesh and Sadaghi (2014) did not find any difference in birds treated with commercial probiotics containing *B. subtilis* in carcass yield, growth rate, and feed use efficiency, on day 42. A pen trail conducted by Russell and Grimes (2009) researched whether a *Lactobacillus* mix was effective in improving turkey productive performance. They reported body weight was significantly greater for birds fed with probiotic feed compared with birds fed with the control feed for 12 weeks.

It is emphasized that probiotics can improve nutrient digestibility in poultry. According to a research, there is an increase in the apparent digestibility of dry matter, metabolizable energy, crude protein, amino acids, Ca and P in male broilers fed with maize soybean based diet supplemented with probiotic supplementation containing yeast (Li et al., 2008). Similarly, Apata (2008) also found that the probiotic *L. bulgaricus* could improve apparent ileal digestibility of dry matter and crude protein in broiler chicken fed with a maize-soybean-based diet. On the other hand the interaction with different feedstuffs used in poultry diets is little understood at present.

Broilers receiving *E. faecium* in the diet, showed significantly higher postmortem pH in the breast which was accompanied by lower drip loss and cooking loss (Zheng et al., 2014). Abdurrahman et al. (2016) and Haščík et al. (2015) tested the effect of *Lactobacillus* and reported an improvement in color characteristics in breast meat. Zhang et al. (2005) used *S. cerevisiae*, and found no improvement in tenderness in breast meat of commercial broilers. The research on the chemical composition of meat showed that the main components affected by probiotics in the poultry diet were protein and fat content (Popova, 2017). Zhao et al. (2013) observed a 3.6% increase in the intramuscular fat content in the breast of Ross broiler chicks treated with probiotic *C. butyricum*, while there was no effect detected with the probiotic *E. faecium*. Král et al. (2012) found improved protein content and reduced fat in birds fed with a probiotic containing *B. subtilis*. Yang et al. (2010) demonstrated that inclusion of *C. butyricum* in a diet modulated fatty acid composition in breast meat of broilers by increasing omega-3 fatty acids concentration.

The use of probiotics in the feed of laying hens has variable effects. Uphadaya et al. (2016) reported higher egg production when two different *Bacillus* strains were added to the diet experimentally. The Haugh unit and eggshell strength were found to be higher in treated groups. Similarly, Ribeiro et al. (2014) reported that supplementation of diet with *B. subtilis* upto a level of 10^5 CFU/g feed, improved egg production and egg mass in layers. Kurtoglu et al. (2004) reported that egg production rates of probiotic-supplemented groups were better than those of control groups. Also, they detected a significant decrease in the damaged egg ratio, egg yolk cholesterol, and serum cholesterol level parameters. Yousefi and Karkoodi (2007) stated that egg quality is improved from 63 to 75 weeks of age in layers when *S. cerevisiae* is added to the feed. Shell weight, shell thickness, yolk weight, and yolk cholesterol were found significantly different among treatment groups. The yolk cholesterol, especially, was found to be lower in the treated groups compared to the controlled ones.

5.3 Pigs

Raising healthy piglets has a great economic impact since pork is the world's most consumed meat (Brugman et al., 2018). To meet this demand intensive systems of pig production have been started. In intensive farming systems, in order to improve production, piglets are weaned at much earlier ages from 3 to 5 weeks, and this period is generally 17 weeks in the natural environment. However, it is accompanied by many adverse effects such as; reduced feed intake, body weight loss, impaired intestinal health, and diarrhea. The traditional approach to overcome this situation has been the use of in-feed antibiotics, but because of the previously mentioned concerns, new strategies have been researched in order to prevent the negative effects of weaning. Among them, a nutritional strategy that includes probiotics has received increasing attention (Patil et al., 2015). Weaning is a sudden, stressful, short, and complex event in a piglet's experience with many challenges such as separation from its mother, moving to another place, meeting new pen mates, establishing group hierarchy and the most important one is adapting to a new type of feed. All of this happens at a time when the animals still have an immature immune system and an unstable intestinal microbiota. Pathogens often profit from this reduced immune response of the animal and cause clinical or subclinical disease. In the post-weaning period diarrhea is the most common clinical manifestation (Gresse et al., 2017).

Pre-requisites for the healthy and efficient growth of young pigs are the rapid maturation of the gut mucosa and the mucosa-associated lymphoid tissue, and the formation of a local stable and complex bacterial community. In neonatal pigs, suckling and the maternal environment shape the gut microbiota (Teixeira de Magalhães, 2015). For many years, the weaning stage is the target age group for probiotic use research. Trevisi et al. (2015) reported that *S. cerevisiae* included feed, reduced the frequency of diarrhea and fecal ETEC shedding in pigs. In the Yin et al. (2014) study, two different *Lactobacillus* strains in the fermented feed, improved diarrhea scores, and fecal *Salmonella* shedding. Barba-Vidal et al. (2017) observed reduced colonization and fecal shedding of *Salmonella* in pigs fed with feed including *B. licheniformis*. Also, there are results contrary to these. Walsh et al. (2012) included *E. faecium, B. subtilis* and *B. licheniformis* into the drinking water of pigs but no effect was reported on *Salmonella shedding* scores, but increased coliform shedding was detected. Similarly, Trevisi et al. (2011) observed increased diarrhea scores during the weaning of piglets which were fed daily with *L. rhamnosus* GG strain added feed.

There are studies showing a benefit in body weight and carcass quality after probiotic administration in pigs. Kritas et al. (2015) applied sows and their litters *B. subtilis* C-3102 spore-based probiotic on a long-term basis. Significant benefits such as improved sow body condition during pregnancy, increased sow feed consumption, reduced sow weight loss during lactation and higher body weight of piglets at the weaning stage had been observed. Supplementation of weaned pigs with 2×10^9 CFU/kg feed with *S. cerevisiae* subsp. *boulardii* CNCM I-1079 for 6 weeks, followed by 1×109 CFU/kg feed of *P. acidilactici* CNCM MA 18/5 M for 3 weeks significantly improved the feed conservation rate without affecting intestinal structure (Le Bon et al., 2010). Böhmer et al. (2006) used a feed supplemented with *E. faecium* DSM 7134 in the feeding of sows between the 90th day of pregnancy and the 28th day of lactation. A significant improvement in feed consumption, offspring size and weight of studied animals was reported.

As adult pigs have a mature GI tract, with high digestive enzyme activity, immune capacity, and disease resistance, the influence of probiotics in these pigs is relatively limited. There is still a need to clarify the effectiveness of probiotics in pigs and the underlying mechanisms through which they function.

5.4 Aquaculture

Aquaculture is the fastest food producing sector in the world. In the last few years, probiotics became an integral part of the cultural practices for improving growth and disease resistance. One of the reasons is the wider range of probiotics evaluated for use when compared with those for terrestrial animals. Probiotics used in aquaculture are encompass both Gram-negative and Gram-positive bacteria, bacteriophages, yeasts and unicellular algae (Nayak, 2010). In aquaculture, there are different methods for probiotic application such as; adding directly into the culture water, bathing in bacterial suspension, and adding to the feed. However, studies showed that supplementation of probiotics as feed additives is the best method for aquatic animals (Hai, 2015).

Unlike terrestrial farm animals, the larval forms of almost all aquatic animals are released into the external environment. These larvae start feeding even though their digestive tract and the immune system is not fully developed. In addition, through the processes of osmoregulation and feeding, they constantly interact with the external microflora, which can contain pathogens or opportunist microorganisms. These conditions are considerable reasons for mortality and economic losses in aquaculture (Mingmongkolchai and Panbangred, 2018). Feeding of *L. rhamnosus* to larval zebrafish elevated larval growth performance by decreasing larval total body cholesterol and triglyceride content and elevated fatty acid levels (Rodiles et al., 2018). The significant increase in the mean weight and natural survival rate of larvae of *S. maximus* fed with enriched *Lactobacillus*, as well as high protection against a pathogenic *Vibrio* species were recorded (Nayak, 2010). The data revealed that the diets containing a yeast probiotic, *Phaffia rhodozyma*, led to a great improvement in larval survival (Irianto and Austin, 2002).

The water quality is a major determinant for enhancing the production rate and maintaining good health in fish. Alteration of quality parameters, encourages the growth of several pathogens. Among them *A. hydrophila* and *A. salmonicida* are considered to be the most common pathogens, while *V. anguillarum* and *V. parahaemolyticus* are the most familiar bacterial pathogens in the marine environment, which cause different types of fish disease (Newaj-Fyzul et al., 2014). The use of antibiotics is a common practice in aquaculture farming sectors. However, in the long term the antibiotics can create the emergence of drug resistant bacteria. Additionally, antibiotics inhibit or kill beneficial microbial flora and disturb the natural ecosystem that affects fish nutrition, physiology, and immunity. To avoid such adverse effects, probiotics have been introduced in fish farming industries for better health management practices (Banerjee and Ray, 2017). Safari et al. (2016) demonstrated that *E. casseliflavus* probiotic has the capability of improving growth performance and enhancing disease resistance in rainbow trout *Oncorhynchus mykiss* against *Streptococcus* by immunomodulation. Similarly, Cha et al. (2013) have reported a lower mortality rate in *Paralichthys Olivaceus* fed with a *B. subtilis* supplemented diet exposed to a pathogenic strain of *Streptoccocus iniae*. In a study of Liu et al. (2018) the cumulative mortality against *V. alginolyticus* was significantly reduced in the parrot fish, *Oplegnathus fasciatus*, fed with 10^{10} CFU/kg *B. subtilis* supplemented diet. Boonthai et al. (2011) examined the effect of *Bacillus* probiotic forms on the microbiota of black trigger shrimp. Shrimp supplemented with probiotics contained significantly lower concentrations of *Vibrio* in the hepatopancreas, intestine and culture water, compared to those in the control group. It is noteworthy that *S. cerevisiae*, and *P. rhodozyma* improved the resistance of juvenile penaeids to vibriosis (Irianto and Austin, 2002). Studies have shown that probiotics can not only be used for bacterial diseases, but can also be used against viral diseases in aquaculture. Liu et al. (2012) reported that *B. subtilis* E20 can effectively reduce the mortality rate in groups infected by Iridovirus as a feed supplement. Similarly, Harikrishnan et al. (2010) demonstrated that *Paralychthys olivaceus* fed with *Lactobacillus* significantly increase the survival rate in fish infected by lymphocystis disease virus (LCDV).

Probiotics are good producers of different types of extracellular enzymes like protease, amylase, and cellulose which enhance the host metabolism and increase the availability of nutrients. This is an important factor for maintaining fish health, and also for the fish reproductive period when female fish require more energy to enhance the fecundity rate. Recent investigations have revealed that there is a positive relationship between probiotics and fish reproduction (Banerjee and Ray, 2017). Dias et al. (2012) that *Brycon amazonicus* females fed with *B. subtilis* supplemented diet exhibited an increase in numbers of oocytes and had higher rates of fertilization and hatching of larvae. Barua et al. (2017) researched the response of prawn *M. rosenbergii* to a commercial probiotic. They concluded their study stating that, probiotic treatment, as compared to control, helped to sustain a higher number of eggs and hatchlings as well as improving relative fecundity and hatching rate per female. Vílchez et al. (2015) reported an enhancement in sperm count of *Anguilla anguilla* when treated with *L. rhamnosus*.

Besides nutritional and other health benefits, probiotics as water additives can also play a significant role in the composition of water by, decomposing organic matter, reducing the nitrogen and phosphorus levels as well as controlling ammonia, nitrite, and hydrogen sulfide (Pandiyan et al., 2013). Banerjee et al. (2016) reported that an addition of beneficial or probiotic bacteria reduces the pollutant load (heavy metal like Pb, Cd, Hg, Ni, etc.) and maintains a healthy condition for aquatic animals. Nimrat et al. (2012) have indicated that the addition of mixed *Bacillus* probiotic to white shrimp ponds enhances the water quality significantly by maintaining pH level, increasing ammonia and nitrite concentration. Likewise, Zokaeifar et al. (2014) detected a significant reduction of ammonia, nitrite and nitrate ions in tank water, added with probiotics.

6. Safety Aspects of Probiotics

The safety of probiotics is not specific to those used in animal feed and is discussed in general terms. Although micro-organisms used as probiotics in animal feed are relatively safe, precautions should be taken to protect animals, humans and the environment from potentially unsafe micro-organisms (FAO, 2016).

The main critical aspects of the relationship between probiotics and the host also can cause some limitations of their use. These are gastrointestinal or systemic infections of the animal fed with the probiotic, transfer of antibiotic resistance from probiotics to other pathogenic micro-organisms or vice versa, the release of infectious microorganisms or noxious compounds to the environment from the animal production system, and sensitization in the handlers (skin, eye or mucous membranes). Hyper-stimulation of the immune system in susceptible hosts are some of the safety concerns about probiotics (Doron and Syndman, 2015).

One of the important requirement for being a probiotic is safety. Probiotics must be free of any pathogenicity. Most known ones like *Bifidobacteria* and *Lactobacilli*, have a long history of safety with their use in fermented food and milk. Safety assessment of a particular microorganism in the USA is regulated by the FDA and microorganisms used for consumption purposes should have the Generally Regarded As Safe (GRAS) status. In Europe, EFSA introduced Qualified Presumption of Safety (QPS) term. The QPS concept is similar to GRAS status but it involves some additional criteria such as, including the history of safe usage and absence of risk of acquired resistance to antibiotics (Wright, 2005, Anadon et al., 2006). Nevertheless, probiotic strains have been implicated in infections, but only in rare cases. Generally, these infections are the result of translocation, which is the passage of the bacteria from the digestive tract to extra-intestinal sites. For example, bacteremia, endocarditis, or abscesses have been reported when using the *Lactobacillus* GG strain. Mostly these infections can occur in organisms with risk factors such as immunosuppression or recovery from oral or GI problems (Butel, 2014).

Another important risk is antibiotic resistance. A probiotic strain, should not have a resistance to antibiotics of clinical and veterinary interest. In reality, resistance to antibiotics is not a major safety issue in itself, but it becomes a problem when it is accompanied by a horizontal transfer of genetic determinants, especially in the host's digestive tract, between probiotic strains and the host's commensal bacteria. Resistance plasmids, such as genes *erm* and *tet* present in the genome of commensal bacteria, have also been found in a number of probiotic strains belonging to the *Lactobacillus* and *Bifidobacterium* genera, showing that probiotic strains may be vectors of resistance genes, acting as donors or recipients. Some probiotic strains such as *Enterococci* or *E. coli* have a higher risk because of belonging they belong to a genera or species naturally carrying factors of virulence and/or resistance to antibiotics. Despite the low risk, the possibility of transferring genetic materials has resulted in recommending the use of probiotic strains free of acquired and potentially transferable resistance genes (Aureli et al., 2011, Di Gioia and Biavati, 2018).

Other negative effects of probiotics could be related to the production of metabolites with a toxic potential. One of the possible risks concerns the production of D-lactate, responsible for lactic acidosis during bacterial fermentation (Snydman, 2008).

7. Conclusion and Future Perspectives

With progress over the past decade on the genetics of lactic acid bacteria and release of complete genome sequences for major probiotic species, the field is now armed with detailed information and sophisticated microbiological and bioinformatic tools. Similarly, advances in biotechnology could yield new probiotics designed for enhanced or expanded functionality.

Probiotics are the future of dietary supplements and medicine as they especially look like an important alternative for antibiotics. It can be argued that nutrition has done significantly more to improve the quality and duration of life during this past century than surgery and pharmaceuticals combined. As more controlled studies are carried out with probiotics, their physiological benefits and mechanisms are better understood, beyond these for direct intervention targets in recent years their expanded roles as live vehicles to deliver biological agents (vaccines, enzymes, and proteins) to targeted locations within the body has been considered and actively investigated.

References

Abdurrahman, Z.H., Y.B. Pramono and N. Suthama. 2016. Meat characteristic of crossbred local chicken fed inulin of Dahlia Tuber and *Lactobacillus* sp. Med. Pet. 39: 112-118.

Afsharmanesh, M. and B. Sadaghi. 2014. Effects of dietary alternatives (probiotic, green tea powder and Kombucha tea) as antimicrobial growth promoters on growth, ileal nutrient digestibility, blood parameters, and immune response of broiler chickens. Comp. Clin. Pathol. 23(3): 717-724.

Allen, H.K., U.Y. Levine, T. Looft, M. Bandrick and T.A. Casey. 2013. Treatment, promotion, commotion: Antibiotic alternatives in food-producing animals. Trends. Microbiol. 21(3): 114-119.

Alugongo, G.M., J. Xiao, Z. Wu, S. Li, Y. Wang, Z. Cao et al. 2017. Review: Utilization of yeast of *Saccharomyces cerevisiae* origin in artificially raised calves. J. Anim. Sci. Biotechno. 8: 34.

Anadón, A., M.R. Martínez-Larrañaga and M.A. Martínez. 2006. Probiotics for animal nutrition in the European Union. Regulation and safety assessment. Regul. Toxicol. Pharm 45: 91-95.

Angelakis, E. 2017. Weight gain by gut microbiota manipulation in productive animals. Microb. Pathogenesis 106: 162-170.

Apata, D. 2008. Growth performance, nutrient digestibility and immune response of broiler chicks fed diets supplemented with a culture of *Lactobacillus bulgaricus*. J. Sci. Food Agric. 88(7): 1253-1258.

Aureli, P., L. Capurso, A.M. Castellazzi, M. Clerici, M. Giovannini, L. Morelli et al. 2011. Probiotics and health: An evidence-based review. Pharm. Res. 63(5): 366-376.

Bai, S.P., A.M. Wu, X.M. Ding, Y. Lei, J. Bai, K.Y. Zhang et al. 2013. Effects of probiotic-supplemented diets on growth performance and intestinal immune characteristics of broiler chickens. Poult. Sci. 92: 663-670.

Banerjee, G., A. Nandi, S.K. Dan, P. Ghosh and A.K. Ray. 2016. Mode of association, enzyme producing ability and identification of autochthonous bacteria in the gastrointestinal tract of two Indian air-breathing fish, murrel (*Channa punctatus*) and stinging catfish (*Heteropneustes fossilis*). Proc. Zool. Soc. 70: 132-140.

Banerjee, G. and A.K. Ray. 2017. The advancement of probiotics research and its application in fish farming industries. Res. Vet. Sci. 115: 66-77.

Barba-Vidal, E., V.F.B. Roll, L. Castillejos, A.A. Guerra-Ordaz, X. Manteca, J.J. Mallo et al. 2017. Response to a *Salmonella typhimurium* challenge in piglets supplemented with protected sodium butyrate or *Bacillus licheniformis*: Effects on performance, intestinal health and behavior. Transl. Anim. Sci. 1: 186-200.

Barko, P.C., M.A. McMichael, K.S. Swanson and D.A. Williams. 2018. The gastrointestinal microbiome: A review. J. Vet. Intern. Med. 32: 9-25.

Barua, D., J. Das, I.A. Chowdhury, M.S. Hossain, H. Bhattacharjee, M.Z.R. Chowdhury et al. 2017. Response of berried prawn (*Macrobrachium rosenbergii*) to commercial probiotics. Aquac. Res. 48(12): 6016-6020.

Bermudez-Brito, M., Plaza-Díaz, J., Muñoz-Quezada, S., Gómez-Llorente, C. and Gil A. 2012. Probiotic mechanisms of action. Ann. Nutr. Metab. 61: 160-174.

Bertin, Y., C. Habouzit, L. Dunière, M. Laurier, A. Durand, D. Duchez et al. 2017. *Lactobacillus reuteri* suppresses *E. coli* O157: H7 in bovine ruminal fluid: Toward a pre-slaughter strategy to improve food safety? PLoS One 12(11): e0187229.

Biloni, A., C. Quintana, A. Menconi, G. Kallapura, J. Latorre, C. Pixley et al. 2013. Evaluation of effects of Early Bird associated with FloraMax-B11 on *Salmonella enteritidis*, intestinal morphology, and performance of broiler chickens. Poult. Sci. 92(9): 2337-2346.

Boonthai, T., V. Vuthiphandchai and S. Nimrat. 2011. Probiotic bacteria effects on growth and bacterial composition of black tiger shrimp (*Penaeus monodon*). Aquaclt. Nutr. 17: 634-644.

Boyd, J., J.W. West and J.K. Bernard. 2011. Effects of the addition of direct-fed microbials and glycerol to the diet of lactating dairy cows on milk yield and apparent efficiency of yield. J. Dairy Sci. 94: 4616-4622.

Böhmer, B.M., W. Kramer and D.A. Roth-Maier. 2006. Dietary probiotic supplementation and resulting effects on performance, health status, and microbial characteristics of primiparous sows. J. Anim. Physiol. Anim. Nutr. 90: 309-315.

Brugman, S., W. Ikeda-Ohtsubo, S. Braber, G. Folkerts, C.M.J. Pieterse and P.A.H.M. Bakker. 2018. A comparative review on microbiota manipulation: Lessons from fish, plants, livestock, and human research. Front Nutr. 5: 1-18.

Butel, M.J. 2014. Probiotics, gut microbiota and health. Médecine et maladies infectieuses 44(1): 1-8.

Carter, A., M. Adams, R.M. La Ragione and M.J. Woodward. 2017. Colonisation of poultry by *Salmonella Enteritidis* S1400 is reduced by combined administration of *Lactobacillus salivarius* 59 and *Enterococcus faecium* PXN-33. Vet. Microbiol. 199: 100-107.

Cha, J.H., S. Rahimnejad, S.Y. Yang, K.W. Kim and K.J. Lee. 2013. Evaluations of *Bacillus* spp. as dietary additives on growth performance, innate immunity and disease resistance of olive flounder (*Paralichthys olivaceus*) against *Streptococcus iniae* and as water additives. Aquacult. 402: 50-57.

Chan, Y.K., M. Estaki and D.L. Gibson. 2013. Clinical consequences of diet-induced dysbiosis. Ann. Nutr. Metabol. 63: 28-40.

Chaucheyras-Durand, F. and H. Durand. 2010. Probiotics in animal nutrition and health. Benef. Microbes. 1(1): 3-9.

Chaucheyras-Durand, F., N. Walker and A. Bach. 2008. Effects of active dry yeasts on the rumen microbial ecosystem: Past, present and future. Anim. Feed Sci. Tech. 145(1): 5-26.

Collins, J.W., R.M. La Ragione, M.J. Woodward and L.E.J. Searl. 2009. Application of prebiotics and probiotics in livestock. pp. 1123-1192. *In*: D. Charalampopoulos and R.A. Rastall (eds.). Prebiotics and Probiotics Science and Technology. Springer, Science+Business Media, LLC.

Dalloul, R.A., H.S. Lillehoj, N.M. Tamim, T.A. Shellem and J.A. Doerr. 2005. Induction of local protective immunity to *Eimeria acervulina* by a Lactobacillus-based probiotic. Compart. Immunol. Microbiol. Infect. Dis. 28(5-6): 351-361.

Dhama, K., V. Vinay, P.M. Sawant, T. Ruchi, R.K. Vaid, R.S. Chauhan et al. 2011. Applications of probiotics in poultry: Enhancing immunity and beneficial effects on production performances and health – A review. J. Immunol. Immunopathol. 13(1): 1-19.

Dias, D.C., A.F.G. Leonardo, L. Tachibana, C.F. Correa, I.C.A.C. Bordon et al. 2012. Effect of incorporating probiotics into the diet of matrinxã (*Brycon amazonicus*) breeders. J. Appl. Ichthyol. 28(1): 40-45.

Di Gioia, D. and B. Biavati. 2018. Probiotics and prebiotics in animal health and food safety: Conclusive remarks and future perspectives. pp. 269-273. *In*: D. Di Gioia and B. Biavati (eds.). Probiotics and Prebiotics in Animal Health and Food Safety. Springer Verlag, Switzerland.

Doron, S. and D.R. Snydman. 2015. Risk and safety of probiotics. Clin. Infect. Dis. 60: S129-S134.

Ducatelle, R., V. Eeckhaut, F. Haesebrouck and F. Van Immerseel. 2015. A review on prebiotics and probiotics for the control of dysbiosis: Present status and future perspectives. Animal 9(1): 43-48.

Eckles, C.H., V.M. Williams, J.W. Wilbur, L.S. Palmer and H.M. Harshaw. 1924. Yeast as a supplementary feed for calves. J. Dairy Sci. 7: 421-439.

Ezema, C. 2013. Probiotics in animal production: A review. J. Vet. Med. Anim. Health 5(11): 308-316.

Food and Agriculture Organization (FAO). 2009. Food and Agriculture Organization of the United Nations. Guidelines for the evaluation of probiotics in food. Available at: ftp://ftp.fao.org/es/esn/food/ wgreport2.pdf

Food and Agriculture Organization (FAO). 2016. Probiotics in animal nutrition – Production, impact and regulation by Y.S. Bajagai, A.V. Klieve, P.J. Dart and W.L. Bryden. Edit. Harinder P.S. Makkar. FAO Animal Production and Health Paper No. 179. Rome.

Frizzo, L.S., M.L. Signorini and M.R. Rosmin. 2018. Probiotics and prebiotics for the health of cattle. pp. 155-174. *In*: D. Di Gioia and B Biavati (eds). Probiotics and Prebiotics in Animal Health and Food Safety. Springer, Switzerland.

Fuller, R. 1989. Probiotics in man and animals. J. Appl. Bacteriol. 66: 365-378.

Fuller, R. 2001. The chicken gut microflora and probiotic supplements. J. Poult. Sci. 38(3): 189-196.

Gaggia, F., P. Mattarelli and B. Biavati. 2010. Probiotics and prebiotics in animal feding for sale food production. Int. J. Food Microbiol. 141: 15-28.

Galina, M., M. Ortiz-Rubio, M. Delgado-Pertiñez and L. Pineda. 2009. Goat kid's growth improvement with a lactic probiotic fed on a standard base diet. Options Méditerranéennes. Série A, Séminaires Méditerranéens. (85): 315-322.

Galvão, K.N., J.E.P. Santos, A. Coscioni, M. Villaseñor, W.M. Sischo, A.C.B. Berge et al. 2005. Effect of feeding live yeast products to calves with failure of passive transfer on performance and patterns of antibiotic resistance in fecal *Escherichia coli*. Reprod. Nutr. Dev. 45: 427-440.

Ghazanfar, S., M. Anjum, A. Azim and I. Ahmed. 2015. Effects of dietary supplementation of yeast (*Saccharomyces cerevisiae*) culture on growth performance, blood parameters, nutrient digestibility and fecal flora of dairy heifers. J. Anim. Plant Sci. 25(1): 53-59.

Goto, H., A.Q. Qadis, Y.H. Kim, K. Ikuta, T. Ichijo, S. Sato et al. 2016. Effects of a bacterial probiotic on ruminal pH and volatile fatty acids during subacute ruminal acidosis (SARA) in cattle. J. Vet. Med. Sci. 78(10): 1595-1600.

Gresse, R., F. Chaucheyras-Durand, M.A. Fleury, T. Van de Wiele, E. Forano, S. Blanquet-Diot et al. 2017. Gut microbiota dysbiosis in postweaning piglets: Understanding the keys to health. Trends Microbiol. 25(10): 851-873.

Guyard-Nicodème, M., A. Keita, S. Quesne, M. Amelot, T. Poezevara, B. Le Berre et al. 2016. Efficacy of feed additives against Campylobacter in live broilers during the entire rearing period. Poult. Sci. 95: 298-305.

Haddad, S.G. and S.N. Goussous. 2005. Effect of yeast culture supplementation on nutrient intake, digestibility and growth performance of Awassi lambs. J. Anim. Feed Sci. Tech. 118: 343-348.

Hai, N.V. 2015. The use of probiotics in aquaculture. J. Appl. Microbiol. 119(4): 917-935.

Hardy, B. 2002. The issue of antibiotic use in the livestock industry: What have we learned? Anim. Biotechnol. 13: 129-147.

Harikrishnan, R., C. Balasundaram and M.S. Heo. 2010. Effect of probiotics enriched diet on *Paralichthys olivaceus* infected with lymphocystis disease virus (LCDV). Fish Shellfish Immunol. 29: 868-874.

Haščík, P., L. Trembecká, M. Bobko, J. Čuboň, O. Bučko, J. Tkáčová et al. 2015. Evaluation of meat quality after application of different feed additives in diet of broiler chickens. Potravinarstvo. 9: 174-182.

Hatab, M.H., M.A. Elsayed and N.S. Ibrahim. 2016. Effect of some biological supplementation on productive performance, physiological and immunological response of layer chicks. J. Radiat. Res. Appl. Sci. 9: 185-192.

He, Z.X., B. Ferlisi, E. Eckert, H.E. Brown, A. Aguilar, M.A. Steele et al. 2017. Supplementing a yeast probiotic to pre-weaning Holstein calves: Feed intake, growth and fecal biomarkers of gut health. Anim. Feed Sci. Tech. 226: 81-87.

Hernandez, R., S.S. Gonzalez, J.M. Pinos-Rodrigues, M.A. Ortega, A. Hernandez, G. Bueno et al. 2009. Effect of yeast culture on nitrogen balance and digestion in lambs fed early, and mature orchard grass. J. Appl. Anim. Res. 32: 53-56.

Hristov, A., G. Varga, T. Cassidy, M. Long, K. Heyler, S. Karnati et al. 2010. Effect of *Saccharomyces cerevisiae* fermentation product on ruminal fermentation and nutrient utilization in dairy cows. J. Dairy Sci. 93(2): 682-692.

Hutsko, S.L., K. Meizlisch, M. Wick and M.S. Lilburn. 2016. Early intestinal development and mucin transcription in the young poult with probiotic and mannan oligosaccharide prebiotic supplementation. Poult. Sci. 95(5): 1173-1178.

Inatomi, T. 2015. Growth performance, gut mucosal immunity and carcass and intramuscular fat of broilers fed diet containing a combination of three probiotics. Sci. Postprint 1(2): e00052.

Iplikcioğlu Cil, G. 2018. Probiotics: Possibilities and Limitations of Their Application in Food and Feed. pp. 3-5. 3rd International Congress on Advances in Veterinary Science and Technics (ICAVST), September 5-9, 2018 Belgrade, Serbia.

Irianto, A. and B. Austin. 2002. Probiotics in aquaculture. J. Fish Dis. 25: 633-642.

Jeong, J.S. and I.H. Kim. 2014. Effect of *Bacillus subtilis* C-3102 spores as a probiotic feed supplement on growth performance, noxious gas emission, and intestinal microflora in broilers. Poult. Sci. 93: 3097-3103.

Jurkovich, V., E. Brydl, J. Kutasi, A. Harnos, P. Kovacs, L. Könyves et al. 2014. The Effects of *Saccharomyces Cerevisiae* strains on the rumen fermentation in sheep fed with diets of different forage to concentrate ratios. J. Appl. Anim. Res. 42(4): 481-486.

Kabir, S.M. 2009. The role of probiotics in the poultry industry. Int. J. Mol. Sci. 10(8): 3531-3546.

Karaoglu, M. and H. Durdag. 2005. The influence of dietary probiotic (*Saccharomyes cerevisiae*) supplementation and different slaughter age on the performance, slaughter and carcass properties of broilers. Int. J. Poult. Sci. 4: 309-316.

Khalid, M.F., M.A. Shahzad, M. Sarwar, A.U. Rehman, M. Sharif, N. Mukhtar et al. 2011. Probiotics and lamb performance: A review. Afr. J. Agric. Res. 6: 5198-5203.

Kogut, M.H. and R.J. Arsenault. 2016. Editorial: Gut health – The new pradigm in food animal production. Front. Vet. Sci. 3: 71.

Král, M., M. Angelovičová and Ľ. Mrázová. 2012. Application of probiotics in poultry production. Anim. Sci. Biotechnol. 45(1): 55-57.

Kritas, S.K., T. Marubashi, G. Filioussis, E. Petridou, G. Christodoulopoulos, A.R. Burriel et al. 2015. Reproductive performance of sows was improved by administration of a sporing bacillary probiotic (*Bacillus subtilis* C-3102). J. Anim. Sci. 93(1): 405-413.

Kritas, S.K., A. Govaris, G. Christodoulopoulos and A.R. Burriel. 2006. Effect of *Bacillus licheniformis* and *Bacillus subtilis* supplementation of ewe's feed on sheep milk production and young lamb mortality. J. Vet. Med. A. 53(4): 170-173.

Kurtoglu, V., F. Kurtoglu, E. Seker, B. Coskun, T. Balevi, E. Polat et al. 2004. Effect of probiotic supplementation on laying hen diets on yield performance and serum and egg yolk cholesterol. Food. Addit. Containm. 21: 817-823.

Le Bon, M., H.E. Davies, C. Glynn, C. Thompson, M. Madden, J. Wiseman et al. 2010. Influence of probiotics on gut health in the weaned pig. Livest. Sci. 133: 179-181.

Lee, S.H., H.S. Lillehoj, R.A. Dalloul, D.W. Park, Y.H. Hong, J.J. Lin et al. 2007. Influence of pediococcus-based probiotic on coccidiosis in broiler chickens. PMID: 17234844. Poult. Sci. 86: 63-66.

Lee, W.J. and K. Hase. 2014. Gut microbiota–generated metabolites in animal health and disease. Nat. Chem. Biol. 10: 416-424.

Lei, X., X. Piao, Y. Ru, H. Zhang, A. Péron, H. Zhang et al. 2015. Effect of *Bacillus amyloliquefaciens*-based direct-fed microbial on performance, nutrient utilization, intestinal morphology and cecal microflora in broiler chickens. Asian-Australas J. Anim. Sci. 28(2): 239-246.

Ley, R.E., C.A. Lozupone, M. Hamady, R. Knight and J.I. Gordon. 2009. Worlds within worlds: Evolution of the vertebrate gutmicrobiota. Nature Rev. Microbiol. 6: 776-788.

Li, L.L., Z.P. Hou, T.J. Li, G.Y. Wu, R.L. Huang, Z.R. Tang et al. 2008. Effects of dietary probiotic supplementation on ileal digestibility of nutrients and growth performance in 1- to 42-day-old broilers. J. Sci. Food Agric. 88(1): 35-42.

Liu, C.H., C.H. Chiu, S.W. Wang and W. Cheng. 2012. Dietary administration of the probiotic, *Bacillus subtilis* E20, enhances the growth, innate immune responses, and disease resistance of the grouper, Epinephelus coioides. Fish Shellfish Immunol. 33: 699-706.

Liu, C.H., K. Wu, T.W. Chu and T.M. Wu. 2018. Dietary supplementation of probiotic, *Bacillus subtilis* E20, enhances the growth performance and disease resistance against *Vibrio alginolyticus* in parrot fish (*Oplegnathus fasciatus*). Aquacult. Int. 26(1): 63-74.

Maldonado, N.C., J. Chiaraviglio, E. Bru, L. De Chazal, V. Santos and M.E.F. Nader-Macías. 2017. Effect of milk fermented with lactic acid bacteria on diarrheal incidence, growth performance and microbiological and blood profiles of newborn dairy calves. Probiotics & Antimicrob. Prot. 10: 668-676.

Marden, J., C. Julien, V. Monteils, E. Auclair, R. Moncoulon, C. Bayourthe et al. 2008. How does live yeast differ from sodium bicarbonate to stabilize ruminal pH in high-yielding dairy cows? J. Dairy Sci. 91(9): 3528-3535.

Markowiak, P. and K. Śliżewska. 2018. The role of probiotics, prebiotics and synbiotics in animal nutrition. Gut Pathog. 10: 21.

Masood, A.I. and S. Mir. 2018. Effect of feed additives supplementation alone and in effect of feeding probiotics mix alone and in combination with fibrolytic enzymes on gut health of Corriedale lambs. J. Pharmacogn. Phytochem. 7(1): 179-183.

Mingmongkolchai, S. and W. Panbangred. 2018. Bacillus probiotics: An alternative to antibiotics for livestock production. J. Appl. Microbiol. 124(6): 1334-1346.

Mirza, R.A. 2018. Probiotics and prebiotics for the health of poultry. pp. 127-155. *In*: D. Di Gioia and B. Biavati (eds.). Probiotics and Prebiotics in Animal Health and Food Safety. Springer, Switzerland.

Moharrery, A. and Asadi, E. 2009. Effects of supplementing malate and yeast culture (*Saccharomyces cerevisiae*) on the rumen enzyme profile and growth performance of lambs. J. Anim. Feed Sci. 18: 283-295.

Møllgaard, H. 1947. Resorption af calcium og fosforsyre. Bertning fra forsøgslab. 228: 1-55.

Mountzouris, K.C., P. Tsirtsikos, E. Kalamara, S. Nitsch, G. Schatzmayr, K. Fegeros et al. 2007. Evaluation of the efficacy of a probiotic containing *Lactobacillus, Bifidobacterium, Enterococcus,* and *Pediococcus* strains in promoting broiler performance and modulating cecal microflora composition and metabolic activities. Poult. Sci. 86: 309-317.

Musa, H.H., S.L. Wu, C.H. Zhu, H.I. Seri and G.Q. Zhu. 2009. The potential benefits of probiotics in animal production and health. J. Anim. Vet. Adv. 8: 313-321.

Nayak, S.K. 2010. Probiotics and immunity: A fish perspective. Fish Shellfish Immunol. 29: 2-14.

Newaj-Fyzul, A., A.H. Al-Harbi and B. Austin. 2014. Review: Developments in the use of probiotics for disease control in aquaculture. Aquacult. 431: 1-11.

Newbold, C.J. and K. Hillman. 2004. Feed Supplements: Enzymes, probiotics, yeasts. pp. 376-378. *In:* W.G. Pond (ed.). Encyclopedia of Animal Science. CRC Press.

Nimrat, S., S. Suksawat, T. Boonthai and V. Vuthiphandchai. 2012. Potential Bacillus probiotics enhance bacterial numbers, water quality and growth during early development of white shrimp (*Litopenaeus vannamei*). Vet. Microbiol. 159: 443-450.

Nurmi, E. and M. Rantala. 1973. New aspects of Salmonella infection in broiler production. Nature 241: 210-211.

Olnood, C.G., S.S.M. Beski, P.A. Lji and M. Choct. 2015. Delivery routes for probiotics: Effects on broiler performance, intestinal morphology and gut microflora. Anim. Nutr. 1: 192-202.

Opsi, F., R. Fortina, S. Tassone, R. Bodas and S. Lopez. 2012. Effects of inactivated and live cells of *Saccharomyces cerevisiae* on *in vitro* ruminal fermentation of diets with different Forage : Concentrate Ratio. J. Agr. Sci. 150: 271-283.

Owens, F.N., D.S. Secrist, W.J. Hill, and D.R. Gill. 1998. Acidosis in cattle: A review. J. Anim. Sci. 76: 275-286.

Pandiyan, P., D. Balaraman, R. Thirunavukkarasu, E.G.J. George, K. Subaramaniyan, S. Manikkam et al. 2013. Probiotics in aquaculture. Drug. Invent. Today 5: 55-59.

Park, Y.H., F. Hamidon, C. Rajangan, K.P. Soh, C.Y. Gan, T.S. Lim et al. 2016. Application of probiotics for the production of safe and high-quality poultry meat. Korean J. Food Sci. Anim. Resour. 36(5): 567-576.

Patil, A.K., S. Kumar, A.K. Verma and R.P.S. Baghel. 2015. Probiotics as feed additives in weaned pigs: A review. Livestock Res. Int. 3: 31-39.

Patra, A.K. 2012. The use of live yeast products as microbial feed additives in ruminant nutrition. Asian J. Anim. Vet. Adv. 7(5): 366-375.

Peterson, R.E., T.J. Klopfenstein, G.E. Erickson, J. Folmer, S. Hinkley, R.A. Moxley et al. 2007. Effect of *Lactobacillus acidophilus* strain NP51 on *Escherichia coli* O157:H7 fecal shedding and finishing performance in beef feedlot cattle. J. Food Prot. 70: 287-291.

Pienaar G.H., O.B. Einkamerer, H. Vander, J. Merwe, A. Hugo, G.D.J. Scholtz et al. 2012. The effects of an active live yeast product on the growth performance of finishing lambs. S. Afr. J. Anim. Sci. 42: 464-468.

Plaizier, J., E. Khafipour, S. Li, G. Gozho and D. Krause. 2012. Subacute ruminal acidosis (SARA), endotoxins and health consequences. Anim. Feed Sci. Technol. 172: 9-21.

Popova, T. 2017. Effect of probiotics in poultry for improving meat quality. Curr. Opin. Food Sci. 14: 72-77.

Poppy, G., A. Rabiee, I. Lean, W. Sanchez, K. Dorton and P. Morley. 2012. A meta-analysis of the effects of feeding yeast culture produced by anaerobic fermentation of *Saccharomyces cerevisiae* on milk production of lactating dairy cows. J. Dairy Sci. 95(10): 6027-6041.

Reis, L.F., R.S. Sousa, F.L.C. Oliveira, F.A.M.L. Rodrigues, C.A.S.C. Araújo et al. 2018. Comparative assessment of probiotics and monensin in the prophylaxis of acute ruminal lactic acidosis in sheep. BMC Vet. Res. 14: 9.

Ribeiro, V., L.F.T. Albino, H.S. Rostagno, S.L.T. Barreto, M.I. Hannas, D. Harrington et al. 2014. Effects of the dietary supplementation of *Bacillus subtilis* levels on performance, egg quality and excreta moisture of layers. Anim. Feed Sci. Technol. 195: 142-146.

Rodiles, A., M.D. Rawling, D.L. Peggs, G. do Vale Pereira, S. Voller, R. Yomla et al. 2018. Probiotic applications for finfish aquaculture. pp. 197-219 *In*: D. Di Gioia and B. Biavati (eds.). Probiotics and Prebiotics in Animal Health and Food Safety. Springer, Switzerland.

Russell, S.M. and J.L. Grimes. 2009. The effect of a direct-fed microbial (Primalac) on turkey live performance. J. Appl. Poult. Res. 18(2): 185-192.

Safari, R., M. Adel, C.C. Lazado, C.M. Caipang and M. Dadar. 2016. Host-derived probiotics *Enterococcus casseliflavus* improves resistance against *Streptococcus iniae* infection in rainbow trout (*Oncorhynchus mykiss*) via immunomodulation. Fish Shellfish Immunol. 52: 198-205.

Shakira, G., M. Qubtia, I. Ahmed, F. Hasan, M.I. Anjum, M. Imran et al. 2018. Effect of indigenously isolated *Saccharomyces cerevisiae* probiotics on milk production, nutrient digestibility, blood chemistry and fecal microbiota in lactating dairy cows. J. Anim. Plant Sci. 28(2): 407-420.

Sharma, P.K., K.A. Prajapati and M.K. Choudhary. 2016. Effect of probiotic supplementation on growth performance of pre-ruminant buffalo calves. J. Krishi Vigyan 4(2): 37-39.

Simon, O. 2005. Micro-organisms as feed additives – Probiotics. Adv Pork Product. 16: 161-167.

Simon, O., A. Jadamus and W. Vahje. 2001. Probiotic feed additives-effectiveness and expected modes of action. J. Anim. Feed Sci. 10: 51-67.

Snydman, D.R. 2008. The safety of probiotics. Clin. Infect. Dis. 46: 104-111.

Sommer, F., J.M. Anderson, R. Bharti, J. Raes and P. Rosenstiel. 2017. The resilience of the intestinal microbiota influences health and disease. Nat. Rev. Microbiol. 15: 630-638.

Stella, A., R. Paratte, L. Valnegri, G. Cigalino, G. Soncini, E. Chevaux et al. 2007. Effect of administration of live *Saccharomyces cerevisiae* on milk production, milk composition, blood metabolites, and faecal flora in early lactating dairy goats. Small Ruminant Res. 67(1): 7-13.

Tabe, E.S., J. Oloya, D.K. Doetkott, M.L. Bauer, P.S. Gibbs, M.L. Khaitsa et al. 2008. Comparative effect of direct-fed microbials on fecal shedding of *Escherichia coli* O157:H7 and salmonella in naturally infected feedlot cattle. J. Food Protect. 71: 539-544.

Teixeira de Magalhães, J., L.L. Bitencourt, M.C.T. Leite, A.P. do Carmo and C.A. de Moraes. 2015. The use of probiotics to enhance animal performance. pp. 459-468. *In*: K. Venema and A.P. do Carmo (eds.). Probiotics and Prebiotics. Caister Academic Press Norfolk, UK.

Tellez, G., C. Pixley, R.E. Wolfenden, S.L. Layton and B.M. Hargis. 2012. Probiotics/direct fed microbials for Salmonella control in poultry. Food Res. Int. 45(2): 628-633.

Titi, H., R.H. Dmour and A.Y. Abdullah. 2008. Growth performance and carcass characteristics of Awassi lambs and Shami goat kid culture in their finishing diet. J. Anim. Sci. 142: 375-383.

Torres-Rodriguez, A., A.M. Donoghue, D.J. Donoghue, J.T. Barton, G. Tellez, B.M. Hargis et al. 2007. Performance and condemnation rate analysis of commercial turkey flocks treated with a *Lactobacillus* spp.-based probiotic. Poult. Sci. 86: 444-446.

Traub-Dargatz, J.L., S.R. Ladely, D.A. Dargatz and P.J. Fedorka-Cray. 2006. Impact of heat stress on the fecal shedding patterns of *Salmonella enterica Typhimurium* DT104 and *Salmonella enterica Infantis* by 5-week-old male broilers. Foodborne Pathog. Dis. 3: 178-183.

Trevisi, P., L. Casini, F. Coloretti, M. Mazzoni, G. Merialdi, P. Bosi et al. 2011. Dietary addition of *Lactobacillus rhamnosus* GG impairs the health of *Escherichia coli* F4-challenged piglets. Animal 5: 1354-1360.

Trevisi, P., M. Colombo, D. Priori, L. Fontanesi, G. Galimberti, G. Calò et al. 2015. Comparison of three patterns of feed supplementation with live *Saccharomyces cerevisiae* yeast on postweaning diarrhea, health status, and blood metabolic profile of susceptible weaning pigs orally challenged with *Escherichia coli* F4ac. J. Anim. Sci. 93: 2225-2233.

Upadhaya, S.D., A. Hossiendoust and I.H. Kim. 2016. Probiotics in Salmonella-challenged Hy-Line brown layers. Poult. Sci. 95(8): 1894-1897.

Uyeno, Y., S. Shigemori and T. Shimosato. 2015. Effect of probiotics/prebiotics on cattle health and productivity. Microbes. Environ. 30: 126-132.

Vílchez, M.C., S. Santangeli, F. Maradonna, G. Gioacchini, C. Verdenelli, V. Gallego et al. 2015. Effect of the probiotic *Lactobacillus rhamnosus* on the expression of genes involved in European eel spermatogenesis. Theriogenol. 84: 1321-1331.

Von Buenau, R., L. Jaekel, E. Schubotz, S. Schwarz, T. Stroff, M. Krueger et al. 2005. *Escherichia coli* strain Nissle 1917: Significant reduction of neonatal calf diarrhea. J. Dairy Sci. 88(1): 317-323.

Walsh, M.C., M.H. Rostagno, G.E. Gardiner, A.L. Sutton, B.T. Richert, J.S. Radcliffe et al. 2012. Controlling Salmonella infection in weanling pigs through water delivery of direct-fed microbials or organic acids. Part I: Effects on growth performance, microbial populations, and immune status. J. Anim. Sci. 90: 261-271.

Willis, W. and L. Reid. 2008. Investigating the effects of dietary probiotic feeding regimens on broiler chicken production and *Campylobacter jejuni* presence. Poult. Sci. 87(4): 606-611.

Wisener, L.V., J.M. Sargeant, A.M. O'Connor, M.C. Faires and S.K. Glass-Kaastra. 2013. The use of direct-fed microbials to reduce shedding of *Escherichia coli* O157 in beef cattle: A systematic review and meta-analysis. Zoonoses Public Health 62: 75-89.

Wright, A.V. 2005. Regulating the Safety of Probiotics – The European Approach. Curr. Pharm. Design 11: 17-23.

Yang, X., B. Zhang, Y. Guo, P. Jiao and F. Long. 2010. Effects of dietary lipids and *Clostridium butyricum* on fat deposition and meat quality of broiler chickens. Poult. Sci. 89(2): 254-260.

Yin, F., A. Farzan, Q. (Chuck) Wang, H. Yu, Y. Yin, Y. Hou et al. 2014. Reduction of *Salmonella enterica Serovar Typhimurium* DT104 infection in experimentally challenged weaned pigs fed a Lactobacillus-fermented feed. Foodborne Pathog. Dis. 11: 628-634.

Yousefi, M. and K. Karkoodi. 2007. Effect of probiotic Thepax® and *Saccharomyces cerevisiae* supplementation on performance and egg quality of laying hens. Int. J. Poult. Sci. 6(1): 52-54.

Vyas, D., A. Uwizeye, R. Mohammed, W.Z. Yang, N.D. Walker et al. 2014. The effects of active dried and killed dried yeast on subacute ruminal acidosis, ruminal fermentation and nutrient digestibility in beef heifers. J. Anim. Sci. 92: 724-732.

Zhang, A., B. Lee, S. Lee, K. Lee, G. An, K. Song et al. 2005. Effects of yeast (*Saccharomyces cerevisiae*) cell components on growth performance, meat quality, and ileal mucosa development of broiler chicks. Poult. Sci. 84(7): 1015-1021.

Zhao, X., Y. Guo, S. Guo and J. Tan. 2013. Effects of *Clostridium butyricum* and *Enterococcus faecium* on growth performance, lipid metabolism, and cecal microbiota of broiler chickens. Appl. Microbiol. Biotechnol. 97(14): 6477-6488.

Zheng, A., J. Luo, K. Meng, J. Li, S. Zhang, G. Liu et al. 2014. Proteome changes underpin improved meat quality and yield of chickens (*Gallus gallus*) fed the probiotic *Enterococcus faecium*. BMC Genomics 15: 1167.

Zokaeifar, H., N. Babaei, C.R. Saad, M.S. Kamarudin, K. Sijam, J.L. Balcazar. 2014. Administration of *Bacillus subtilis* strains in the rearing water enhances the water quality, growth performance, immune response, and resistance against *Vibrio harveyi* infection in juvenile white shrimp, *Litopenaeus vannamei*. Fish Shellfish Immunol. 36: 68-74.

Insect Gut Bacteria and Iron Metabolism in Insects

Mahesh S. Sonawane[1,2], Rahul C. Salunkhe[3] and R.Z. Sayyed[1*]

[1] Department of Microbiology, PSGVPM's Arts, Science & Commerce College, Shahada - 425409, Maharashtra, India
[2] National Centre for Microbial Resource, National Centre for Cell Science, Pashan, Pune - 411021, Maharashtra, India
[3] RMR Biosciences LLP, 28, Shivanjali, New Ganesh Colony, Shirpur - 425405, Maharashtra, India

1. Introduction

Insects are the most flourishing life forms in the world having a variety of animal groups (Figure 1). Insects come under phylum Arthropods and class Insecta. Insects are a very diverse group on earth living in different environmental niches. Niches range from desert to marine ecosystems. Records of insects were noted from the Mahabharata. They had a record of insects like patanga, honey bee (Makshika) and bee (Bhramara) and many other similar references. In comparison with the total animals from the world, insects have the highest percentage being 75% of the total population (Ghosh, 1996). Although 1,020,169 species of insects have been identified in the world as per ZSI 2014, it contributes to 10-15% of the total estimated diversity of insects (Stork, 2007). In India, 63,706 species of insects were reported till Dec 2014 (Ghate et al., 2015). Furthermore, the key facts like

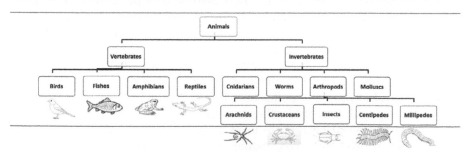

Fig. 1: Classification of the animal kingdom

*Corresponding author: sayyedrz@gmail.com

distribution, occurrence, and survival of insects in a varied environment are relatively unexplored. All insects have an external skeleton with the body generally divided into three parts i.e. head, thorax, and abdomen. The head has external mouthpart and a pair of antennae and a thorax which has wings and three pairs of legs.

2. Insects

2.1 Insects as Pests

Insects have positive as well as negative effects on our day to day life. We know insects as pests including grasshoppers which chew and feed on the external part of the plant. Aphids and plant bugs feed on plant sap and cause damage to the plant by sucking plant sap. Some insects found internally in plant tissues, for example, wood borers and seed weevils are very difficult to control. In some cases, insects act as a carrier for plant diseases. They carry pathogenic microorganisms or viruses with them and at the time of feeding on the plant they transmit these microorganisms from one to another. Due to insects feeding on plants or boring into them generating openings in different parts, through which disease-causing microbes can enter them and cause diseases. It ultimately damages the plant and reduces the crop yield in farms.

2.2 Insects as Beneficiary Arthropods

In nature, majority of the insects are beneficial but we are not able to differentiate such insects due to a lack of knowledge. They are involved in pollination, nutrient recycling and food webs of wildlife. They also act as biological indicators and biocontrol agents in controlling plant pathogens. Their most important function is in the pollination of plants.

Insects transfer pollen grains from one plant to another which helps in fertilization of floral plants. There are some reports which show that pollination through honey bees gives millions of tonnes of agricultural yield. Moths, butterflies wasps, among others also help in pollination. Larvae of fly maggots feed on animals and help the environment to decompose their material. The case is the same with of dung beetles and termites which help to decompose and recycle organic matter.

An insect is one of the main parts of the food web. Large insects, small mammals, fishes, and birds depend on insects for their diet. Insects also act as a biological indicator of pollution. For example, Mayflies are an aquatic insect. If the population of Mayflies is high in water then it is good and vice versa. Klamath weed beetles are used as biocontrol agents on large grass fields to control Klamath weed. Some insects act as predators or parasites. Predators are large in size as compared to pray insects and feed on them (Lady beetles feed on small aphids). Parasites (eg. wasps) are also insects which grow on or inside them feed on them host. Such predators and parasites are also used as a biocontrol for insect pests. It is important to study insects and their relationship with the environment.

2.3 Digestive System of Insects

Most of the insects have an entire digestive system consisting of a tube-like structure, called the alimentary canal, running along the length of the body throught it from the mouth to the anus (Figure 2). Ingested food is processed through this system and usually moves in only one direction. Evolutionary consequences are developing an adaptation of various parts according to functions like food digestion, nutrient absorption and waste excretion (Samaranayake and Wijekoon, 2011, Priya et al., 2012). In most insects, foregut (stomodeum), midgut (mesenteron) and hindgut (proctodeum) form the entire alimentary canal. The majority of food digested in the midgut once it comes from the foregut. In insects, gut structure and physiochemical conditions play a very vital role in digestion (Donovan et al., 2004).

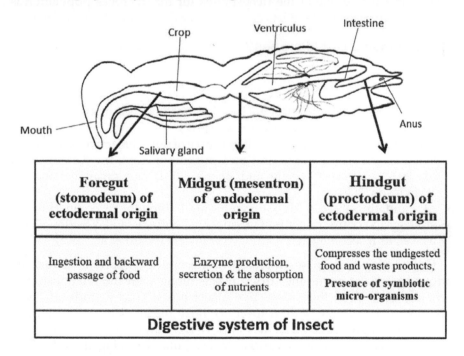

Foregut (stomodeum) of ectodermal origin	Midgut (mesentron) of endodermal origin	Hindgut (proctodeum) of ectodermal origin
Ingestion and backward passage of food	Enzyme production, secretion & the absorption of nutrients	Compresses the undigested food and waste products, **Presence of symbiotic micro-organisms**
Digestive system of Insect		

Fig. 2: The generalized digestive system of Insect and role of the alimentary canal

3. Insect Gut Microflora

Microbes are present everywhere. They are also present in and on insects. All the insect species are known to harbor a wealthy and complex community of microorganisms in their guts and other parts of the body, which comprises of interactions ranging from parasitism to obligate mutualism (Dillon and

Dillon, 2004). These microbes have a positive as well as a negative impact on insects. Insects live in diverse habitats and microbes help them to survive in such conditions. So it is important to study the insects' associated microbes (Provorov and Onishchuk, 2018).

Insects gut microbiota is majorly influenced by various environmental factors as well as their feeding habitats. Such variations also lead to the presence of different bacterial phylotypes present in the gut environment, making them an appealing system for study. An association between two or more organisms for a shorter or longer time is called symbiosis and the partners involved are called symbionts. In such associations, one, both, or neither partner is benefited. Symbioses are biologically important because they are pervasive and dominate the biota of many habitats. Depending on the interaction between these partners, symbiosis are of different types like mutualism, parasitism, commensalism among others (Daida et al., 1996, Dillon and Dillon, 2004). Symbionts are present in and on most of the body parts of insects and one of the richest sites for the microbes population is their digestive system.

3.1 Why Insect Gut Microbes?

In recent years there has been renewed interest in the understanding of insect gut microbes for the following mentioned reasons:

1. Gut microbes not only affect the health of the insects but also change the behavior of the insects (Ezenwa et al., 2012).
2. This diverse microbiota is a possible source of novel bioactive compounds such as antimalarial, antiviral and antitumor peptides (Chernysh et al., 2002), enzymes (Zhang and Brune, 2004) and novel metabolites (Wilkinson, 2001).
3. Manipulating these microbial symbionts is thought to be an effective strategy to control insect pests (Sanchez-Contreras and Vlisidou, 2008) as well as insect-mediated pathogen transfers (Beard et al., 2002).

Considering these observations, it is becoming important to identify various bacterial communities from insects along with the relationship of particular bacteria to their respective insect hosts. Thus a study of insect gut-associated bacteria with respect to host metabolism, development, defense, and reproduction becomes important. The occurrence of different microbial species present in various insect guts is cataloged in Table 1. Insect endosymbionts are also assisting in insect survival and reproduction. To move towards microbial association, host nutrition plays a crucial role. Hence this chapter also focuses on iron nutrition and its effect on insects after iron sequestration with the help of gut bacteria.

3.2 How to Study Insect Gut Microbes?

Bacterial diversity can be studied using two approaches namely culture-dependent and culture-independent, respectively.

3.2.1 Culture-dependent Approach

Regardless of the improvement in technology for studying biodiversity, cultivation methods still have an incredible significance. For the past 150 years, microbiologists have used culture-dependent methods for studying microbial diversity. Samples were spread on specific enrichment media for isolation of microbes. The growth of microbes was dependant on substances supplemented with agar. Total diversity was covered by using many general and specialized media. In most cases, many microbes are not cultured on the plate and hence there morphological and physiological study is not done. Many scientists are trying to design such media which are able to grow the non cultured microbes and hence the selection of media is very crucial for studying diversity. There is a necessity to maintain the microbial cultures in a pure and viable form for a long time period. So they are preserved in many ways using sub-culturing on respective media, storing in cold rooms at 0-4 °C, overlaying with oil, lyopreservation, and cryopreservation (Prakash et al., 2013). In cryopreservation pure cultures are preserved in 20% glycerol with phosphate buffer saline. Glycerol acts as cryoprotectant and phosphate buffer saline helps in osmoregulation. Vials are filled with culture suspension and stored at –80 °C in deep freezers for long term preservation.

3.2.2 Culture-independent Approach

It has been reported that only 1% of the total bacterial population from the earth has been studied and due to many reasons about 99% of the bacterial population has not been cultivated and studied yet. Vartoukian et al.,2010 reported cultivation of only 30 phyla out of various samples collected from 61 phyla. Researchers have tried to cultivate these uncultivated bacteria by modifying the growth conditions, however, a single approach for the cultivation of such bacteria is not sufficient for the growth of all types of microorganisms. Different physicochemical and nutritional factors like growth media, pH and temperature of incubation, growth rate, oxygen concentration, presence or absence of signaling molecules, antibiotic production, among others, vary from for each bacterial phylum. Hence, multiple approaches to the cultivation of such bacteria should be practiced to understand bacterial diversity. Thus a need to study uncultivable or culture-independent approach of the bacterial population from the insect samples can play a significant role in understanding the biodiversity (Vartoukian et al., 2010). To overcome these problems and know the biodiversity, uncultivable methods came into focus.

In this approach, the total DNA present in the sample is used for the identification of bacteria. This study of DNA from the environmental sample having an unculturable bacterial population is known as metagenomics. This involves preparation of clone library and sequencing for the identification of culture-independent bacteria. During the past few years, molecular techniques like 16S rRNA gene clone library, fluorescence in situ hybridization (FISH) and denaturing gradient gel electrophoresis (DGGE) were used to explore

Table 1: Insect gut-associated bacteria within different insects and insect order

Order	Insect	Flora	References
Blattodea	Cockroach	*Blattabacterium*	Douglas, 2007, Feldhaar and Gross, 2009
Coleoptera	Grain weevil	SOPE (*Sitophilus oryzae* primary endosymbiont), *Wolbachia*	Weiss and Aksoy, 2011
	Omnivorous beetle	*Enterococcus faecalis*	Lundgren and Lehman, 2010
	Sugarcane weevil	*Candidatus nardonella, Buchnera* sp*, Klebsiella* sp*, Citrobacter* sp*, Raoultella* sp*, Enterobacter* sp*, Acinetobacter* sp*, Burkholderia* sp*, Diaphorobacter* sp*, Methylophilus* sp*, Caulobacter* sp*, Streptococcus* sp*, Lactococcus* sp*, Leuconostoc* sp*, Sphingobacterium* sp.	Rinke et al., 2011
	Weevils	*Nardonella*	Hosokawa et al., 2006
	Grass grub	*Serratia entomophila & Serratia proteamaculans*	Sanchez-Contreras and Vlisidou, 2008
	Rice weevil	*P-endosymbiont SOPE*	Sanchez-Contreras and Vlisidou, 2008
	Longhorn beetle	*Aeromicrobium* sp*., Aurantimonas* sp*., Agrococcus* sp*., Brachybacterium* sp*., Brevibacterium* sp*., Chrysobacterium* sp*., Curtobacterium* sp*., Enterococcus* sp*., Flavobacterium* sp*., Leucobacter* sp*., Microbacterium* sp*., Olivibacter* sp*., Pseudomonas* sp*., Pseudonocardia* sp*., Sphingobacterium* sp*., Staphylococcus* sp*., Stenotrophomonas* sp*., Streptococcus* sp.	Scully et al., 2014
Diptera	Fruit fly (*Drosophila melanogaster*)	*Acetobacter pomorum, Bacillus firmus, Enterococcus faecalis, Lactobacillus plantarum, Wolbachia*	Sharon et al., 2010, Ringo, Sharon, and Segal, 2011

	Insect	Bacteria	References
	Mosquitoes	*Enterobacter* sp., *Microbacterium* sp., *Sphingomonas* sp., *Serratia* sp., *Chryseobacterium meningosepticum*, *Asaia bogorensis*, *Bacillus subtilis*, *Enterobacter aerogenes*, *Escherichia coli*, *Herbaspirillum* sp., *Pantoea agglomerans*, *Pseudomonas fluorescens*, *Pseudomonas stramirea*, *Pseudomonas aeruginosa*, *Phytobacter diazotrophicus*, *Serratia marcescens*, *Wolbachia*, *Klebsiella ozaenae*	Azambuja, Garcia, and Ratcliffe, 2005, Riehle and Jacobs-Lorena, 2005, Crotti et al., 2009, Dong, Manfredini, and Dimopoulos, 2009, Favia et al., 2007, Weiss and Aksoy, 2011
	Tsetse fly	*Wigglesworthia glossinidia, Sodalis glossinidius*	Akman et al., 2002, Riehle and Jacobs-Lorena, 2005, Hosokawa et al., 2006, Feldhaar and Gross, 2009, Weiss and Aksoy, 2011
Hemiptera	Mealybugs	*Tremblaya*	Hosokawa et al., 2006, Sanchez-Contreras and Vlisidou, 2008, Feldhaar and Gross, 2009
	Plataspid stinkbug	*Ishikawaella capsulate*	Nikoh et al., 2011
	Aphids	*Hamiltonella defensa, Dickeya dadantii, Buchnera* sp.	Akman et al., 2002, Hosokawa et al., 2006, Sanchez-Contreras and Vlisidou, 2008, Feldhaar and Gross, 2009
	Blood sucking bug	*Rhodococcus rhodnii*	Sanchez-Contreras and Vlisidou, 2008
	Kissing bug	*Rhodococcus* sp., *Serratia* sp., *Nocardia* sp., *Gordonia* sp.	Weiss and Aksoy, 2011
	Leafhopper	*Asaia* sp.	Crotti et al., 2009
	Pea aphid	*Buchnera* sp., *Serratia* sp., *Hamiltonella* sp.	Weiss and Aksoy, 2011

(Contd.)

Table 1: (*Contd.*)

Order	Insect	Flora	References
	Psyllids	*Carsonella* sp., *Hamiltonella defense*	Hosokawa et al., 2006, Douglas, 2007, Sanchez-Contreras and Vlisidou, 2008, Feldhaar and Gross, 2009, Weiss and Aksoy, 2011
	Sharpshooter	*Baumannia* sp., *Sulcia* sp.	Hosokawa et al., 2006, Sanchez-Contreras and Vlisidou, 2008, Feldhaar and Gross, 2009
	Stinkbugs	*Ishikawaella capsulate*	Sanchez-Contreras and Vlisidou, 2008
	Whiteflies	*Portiera aleyrodidarum*, *Hamiltonella defense*	Hosokawa et al., 2006, Douglas, 2007, Sanchez-Contreras and Vlisidou, 2008, Weiss and Aksoy, 2011
	Scale insects	*Tremblaya princeps*	Douglas, 2007
	Treehopper (*Auchenorrhyncha*)	*Baumannia cicadellinicola and Sulciamuelleri, Pyrenomycete fungi in some planthoppers*	Douglas, 2007
Hymenoptera	Carpenter ant	*Blochmannia* sp., *Wolbachia*	Hosokawa et al., 2006, Douglas, 2007, Sanchez-Contreras and Vlisidou, 2008, Feldhaar and Gross, 2009, Weiss and Aksoy, 2011

Isoptera	Termites	*Bacillus* sp., *Paenibacillus* sp., *Ruminococcus* sp., *Butyrivibrio* sp., *Bacteroides* sp., *Alcaligenes* sp., *Azospirillum* sp., *Brevibacillus* sp., *Cellulomonas*-related spp., *Clavibacter* sp., *Clostridium* sp., *Corynebacterium* sp., *Klebsiella* sp., *Kocuria* sp., *Microbacterium* sp., *Micromonospora* sp., *Nocardioforme* sp., *Rhizobia* sp., *Ochrobactrum* sp., *Sphingomonas* sp., *Spirosoma*-related spp., *Streptomyce* sp., *Mixotricha paradoxa*, *Pantoea agglomerans*, *Streptococcus* sp., *Desulfovibrio* sp., *Treponema* sp..	Ohkuma and Kudo, 1996, Dillon, Vennard and Charnley, 2002, Wenzel et al. 2003, König, 2006, Thong-On et al, 2012
Lepidoptera	Moth (*Agapeta zoegana*)	*Acinetobacter* sp., *Duganellazoo gloeodes*, *Escherichia coli*, *Pseudomonas* sp., and *Propionibacterium acnes*	Frederick and Caesar, 2000
	Moth (*Helicoverpa*)	*Bacillus firmus*, *Bacillus niabense*, *Paenibacillus jamilae*, *Cellulomonas variformis*, *Acinetobacter schindleri*, *Micrococcus yunnanesis*, *Enterobacter* sp., and *Enterococcus cassiliflavus*	Priya et al., 2012
	Moth Larvae	*Bacillus* sp.	Andrews and Spence, 1980
	Silkworm	*Bacillus circulans*, *Proteus vulgaris*, *Klebsiella pneumoniae*, *Escherichia coli*, *Citrobacter freundii*, *Serratia liquefaciens*, *Enterobacter* sp., *Pseudomonas fluorescens*, *P. aeruginosa*, *Aeromonas* sp. and *Erwinia* sp.	Anand et al., 2010
	Tobacco horn worm	*Photorhabdus luminescens*, *Photorhabdus asymbiotica*	Sanchez-Contreras and Vlisidou, 2008
	Moth (*Diatraea saccharalis*)	*Klebsiella* sp., *Stenotrophomonas* sp., *Microbacterium* sp., *Bacillus* sp. and *Enterococcus* sp.	Dantur et al., 2015
Neuroptera	Antlion	*Enterobacter aerogenes*, *Bacillus cereus*, *Bacillus sphaericus*, *Morganella morganii*, *Serratia marcescens*, *Klebsiella* spp	Sanchez-Contreras and Vlisidou, 2008

(Contd.)

Table 1: (*Contd.*)

Order	Insect	Flora	References
Orthoptera	Desert locust	*Escherichia coli, Enterobacter liquefaciens, Klebsiella pneumoniae, Enterobacter cloacae, Pseudomonas aeruginosa*	R. Dillon and Charnley, 2002
	Mole cricket	*Acidiphilium rubrum, Dienococcus proteolyticus, Sporosarcina ureae, Micrococcus varians, Micrococcus kristinae, Acetobactrium, Alcaligenes eutrophus, Micrococcus roseus, Micrococcus lylae, Sporosarcina, Citrobacter amalonaticus, Corynebacterium xerosis, Micrococcus kristinae, Acetobactrium, Alcaligenese utrophus, Micrococcus roseus, Micrococcus lylae*	Desai and Bhamre, 2012
	Grassland locusts	*Serratia marcescens* strain HR-3	Sanchez-Contreras and Vlisidou, 2008
Phthiraptera	Human body louse	*Rickettsia prowazekii*	Sanchez-Contreras and Vlisidou, 2008
	Anoplura (sucking lice)	*Riesiapediculicola*	Douglas, 2007
Psocoptera	Psocoptera (book lice)	*Rickettsia* sp.	Douglas, 2007
Siphonaptera	Human North America flea	*Yersinia pestis*	Sanchez-Contreras and Vlisidou, 2008

microbial biodiversity. Now, next-generation sequencing platforms like 454 pyrosequencing, ion torrent, and Illumina sequencing have come (Douterelo et al., 2014) into focus. Most of the uncultivable bacteria were identified by using these methods and accordingly strategies were designed to cultivate these microbes.

4. Association between Insect and the Gut Microbes

4.1 Host Metabolism

Insects are specific in their diet and therefore microbes related to their gut are also specific. Association is observed between insect and gut microbes. The wide range of mutualistic associations in insects and their gut microorganisms play a crucial role in food digestion (Daida et al., 1996, Scully et al., 2014). Genome analysis of various insects has revealed that the gut microflora of insects leads to genetic complementation required for digestion of diverse food, which contain sugars, fatty acids, amino acids, among others (Priya et al., 2012). Gut microbes initially convert these complex biomolecules into a simpler form which is subsequently used by the insect for nutrition and energy production (Ohkuma and Kudo, 1996). Furthermore, the insect gut provides a suitable environment for survival and establishment of microbes which in turn helps in the digestion of food (Rinke et al., 2011). Termites have extensively studied insects for mutualistic associations with their gut microbiota (Ohkuma and Kudo, 1996). It has been found that along with cellulolytic enzymes of termites, and their gut microbes play a crucial role in the digestion of woody material and subsequently are the main source of energy (Ohkuma and Kudo, 1996). Relatively recent studies showed that bacteria of genus *Bacillus* and *Paenibacillus* colonized in the termite gut, and played a key role in the digestion of polysaccharides and aromatic compounds (König, 2006). In other cases like *Carbidis* (ground beetles) which consume seeds, a bacterium *Enterococcus faecalis* associated with the gut contributes to the digestion of seeds (Rainio and Niemelä, 2003, König, 2006, Lundgren and Lehman, 2010). The role played by an earthworm in soil texture and fertility by recycling organic materials is well known (Samaranayake and Wijekoon, 2011). Earthworms harbor bacteria of the class Proteobacteria in their gut. These gut microflora may assist earthworms to emit nitrous oxide (N_2O) (Horn et al., 2005) which improves nitrogen content and hence soil fertility (König, 2006).

Some member of the order Lepidoptera, gut microflora is observed to help in the digestion, as well as the production of some biomolecules. Genus *Klebsiella, Bacillus, Enterococcus, Stenotrophomonas* and *Microbacterium* isolated from the gut of *Diatraea saccharalis* showed cellulase activity against sugarcane bagasse as a sole carbon source (Dantur et al., 2015). *Bombyx mori* fed on a mulberry leaf, which contains several polysaccharides like cellulose, xylan, pectin and starch (Dillon and Dillon, 2004). These polysaccharides are not easily digested by *B. mori*. However, their gut microflora helps them to digest it by providing digestive enzymes, enhance the digestion process and provide vitamins among other things.

There are reports which showed that production of the digestive enzyme of the gut microbiota varies from species to species and alongwith its location in the digestive tract (Anand et al., 2010). Mosquitoes are well known to interfere with the lifestyle of society. In 2008, Ponnusamy et al reported that the egg-laying behavior of *Aedes aegypti* was affected by bacterial cells kairomones. Their life cycle starts from aquatic followed by shifting to the terrestrial environment. This change in the environment effectively alters the microflora at different stages of the life cycles. *Cyanobacteria, Aeromonadaceae, Comamonadaceae, Erythrobacteraceae,* and *Rhodobacteraceae* in Proteobacteria are predominant in the initial stages (aquatic phase) like larva and pupa while the flora change drastically as *Enterobacteriaceae* and *Propionibacteriaceae* are in the adult stage (terrestrial phase) (Chakraborty et al., 2011). *Aedes aegypti* is the parasite vector of diseases like dengue fever, Chikungunya, and yellow fever. The gut microflora of these *A. aegypti* have an important role in sugar metabolism (Gaio et al., 2011).

In many cases, an association between microorganisms and insects is informal and temporary. Microbial association mostly depends on the diet of the insect and insect host (Frederick and Caesar, 2000). Perez-Cobas (2015) showed that a population of a few genera from the gut microflora of cockroaches varies with their composition of the diet. Insects depend on restricted diets, such as plant sap, vertebrate blood (Gaio et al., 2011) or woody materials (Ohkuma and Kudo, 1996). A number of fluid feeders (plant sap, blood) and predators of other arthropods are found in Hemiptera (Bug) (Fukatsu and Hosokawa, 2002). However, to fulfill other nutritional requirements insects depend on symbiotic microorganisms. Aphid, a known plant pest feeds on plant sap which is rich in carbohydrate but poor in amino acids. It is known that aphids do not synthesize all amino acids independently and endosymbiotic microflora play a crucial role to fulfill this requirement by amino acid synthesis (Clark et al., 1998). Some Locusts species consumed *Gramineae* grasses which have a cellulose content of nearly 30-50% for its digestion they may get help from gut microbes. These microbes assist in the digestion of cellulolytic material (Su et al., 2014). Gut microbiota in many insects also help in many ways, for example, convert H_2 and CO_2 in methane and acetate (Angerer et al., 1992), fix nitrogen (Odelson and Breznak, 1983) and recycle nitrogen present in uric acid (Thong-On et al., 2012).

4.2 Host Development and Defense

Microorganisms that reside in insect guts can play important roles in the host's development, of resistance to pathogens, reproduction, and survival (Brand et al., 1975, Zhang and Brune, 2004, Moran et al., 2005). Loss of gut microflora often results in an unusual development and reduced survival rate of the insect host (Eutick et al., 1978, Fukatsu and Hosokawa, 2002). These symbiotic microbes also defend against foreign pathogenic bacteria just like immune cells present in the body (Veivers et al., 1982, Ohkuma and Kudo, 1996). Recent reports show that the immune system of mosquitoes gets stronger due to gut bacteria and indirectly protect them from malarial parasites (Dong et al., 2009). But in the case of *Anopheline* mosquitoes, midgut microflora helps in arbovirus infections (Floch et al., 2007). There are reports to reduce parasite infections of sandflies against *Leishmania*. Gut microbes release many metabolites in the gut which can control the parasite infection. *S. marcescens, Serratia plymuthica, P. aeruginosa, Klebsiella* spp. and *Enterobacter* spp. synthesize prodigiosin, microbial pigment induced cellular DNA fragmentation and cell apoptosis. Cytolysin from and *Enterococcus faecalis*, siderophore from *Pseudomonas aeruginosa* and protease from *S. marcescens P. aeruginosa* have been reported to against insect parasites (Azambuja et al., 2005). Aphids have some symbionts like *Hamiltonella defensa, Serratia symbiotica*, and *Regiella insecticola* which act against bacteriophages, parasitoids, fungal pathogens and abiotic stresses (Guidolin and Consoli, 2017).

Some toxic chemicals like terpenes are released as a defense mechanism of the host in case of insect herbivores. Terpenes are aromatic compounds and toxic to living things like microbes, insects, fish, humans, etc. Terpenes in conifer plants help them to resist insect pests. They also attract insects e.g. bark and timber beetles of family Scolytidae choose their preferred host by releasing volatile terpenes (Rudinsky, 1966). In 1980, Andrews and Spence reported that insects with diet containing terpenes had changed in common gut microflora. The gut bacteria present might be assisting in degradation or conversion of terpenes into other components. Gut bacteria of *Schistocerca gregaria* locusts from order Orthoptera contribute to its immune system. Gut bacteria of *S. gregaria* plant phenolics are used for the production of guaiacol, a volatile aggregation pheromone. Aggregation pheromones are responsible for a swarm of locusts. Locusts have the presence of *Lactococcus* and *Raoultella* genera in a healthy gut but the application of *Paranosema locustae* reduces the gut flora of the *Locusta migratoria manilensis* which acts as a biocontrol to reduce locusts cloud (Tan et al., 2015). *S. gregaria* is susceptible to various infections if the gut bacterial diversity is less. Bacterial diversity also helps to resist pathogen colonization in the gut by producing phenolic compounds (Dillon et al., 2005). *Pantoea agglomerans* produced a phenolics compound from gut flora which acts as an antimicrobial agent for pathogens colonization (Tan et al., 2015).

4.3 Host Reproduction

Most insects are sexually reproduced and sexes are separated. They receive specific signals in visual, auditory or olfactory form from their partner which are species-specific. In insects, sex pheromones are responsible for mate selection and sexual performance. In some insects sexual performance, mating preferences (Sharon et al., 2010) and oviposition are influenced by gut microflora (Priya et al., 2012). Bacterial specific compounds may act as sex attractants in the grass grub beetle *Costelytra zealandica* (Brucker and Bordenstein, 2012) or induce insect genes code for sex pheromones (cuticular hydrocarbons) found in *Drosophila melanogaster* (Ringo et al., 2011). Terpenes from conifer plants help in the production of insect pheromones (Andrews and Spence, 1980). Gut microflora of insects synthesize pheromonal compounds from terpenes. Some symbiotic bacteria, such as genus *Asaia* were transmitted vertically in the mosquito (Favia et al., 2007, Crotti et al., 2009) and were reported to have helped in the fertility of the host mosquito (Gaio et al., 2011). In most of the cases endosymbionts like *Wolbachia* majorly interfere with insect reproduction. Such studies need to be taken on other insects as well especially crop pests which may help in the biological control of such insects.

4.4 Endosymbionts

Many obligate endosymbionts are responsible for the survival and reproduction of insects. These endosymbiotic bacteria cannot be grown outside the host environment, and,hence always found in strong association with their hosts. *Wolbachia* found in many insects (Werren et al., 1995), *Buchnera* of aphids (Shigenobu et al., 2000) and *Wiggles worthia* of tsetse flies (Akman et al., 2002) are some examples of such endosymbiotic bacteria. γ proteobacteria from the midgut of stinkbugs are essential for their growth and reproduction (Nikoh et al., 2011). These endosymbionts are vertically inherited from mothers to siblings, however, some studies show that some of them can be transmitted horizontally from one host to another (Nikoh et al., 2011). Reports suggest that γ proteobacteria vertically transmitted from generation to generation get their genome shrunk (Hosokawa et al., 2006). Some endosymbiotic microorganisms like *Wolbachia* and *Cardinium* have the ability to manipulate the biology of their host's reproduction. Endosymbiotic bacteria from insects are linked to shifting of the host insect population (Zchori-Fein et al., 2001, Zabalou et al., 2004, Favia et al., 2007). *Wolbachia* is one of the most intensively studied bacteria mainly found in reproductive tissues of insects. It is found in 40-52% species of all the terrestrial arthropods (Funkhouser-Jones et al., 2015). They are maternally inherited and hav the ability to alter insect sex by cytoplasmic incompatibility between strains (Zabalou et al., 2004, Werren, 1997), induction of parthenogenesis, feminization of males and male-killing (Werren, 1997). *Wolbachia* is responsible for the increase in fertility of insects and their effects differ with strain, host insect species and genetic systems (Vavre et al., 1999, Frederick and Caesar, 2000). These diverse effects are caused by endosymbionts making them an interesting candidate for pest insect control (Zchori-Fein et al., 2001, Zabalou et al., 2004, Favia et al., 2007). *Buchnera aphidicola* is an endosymbiont, living with almost all aphids. Its role is now known in aphids as a biosynthesiser of essential amino acids which are present in fewer amounts in the phloem sap (Guidolin and Consoli, 2017). Locusts also show the presence of mutualistic endosymbionts like *Arsenophonus* which are generally reported as male killers in insects. *Hamiltonella* protect their aphid hosts from parasitoids (Brady et al., 2014).

5. Iron (Fe) in Insects

Iron is the most important and essential metal ion on earth. It stands first in the rank of the most inexhaustible metals found on the Earth. Iron is place in the 26th position of the periodic table and in pure form it is soft, silvery and has high chemical reactivity. This is the reason why, iron is constantly found in its oxidized form. Presence of life on earth is because of iron and its magnetic property. Its presence in the core of the earth creates a magnetic field which helps in the formation of the environment and migration of living organisms. Iron plays a key role in all living organisms ranging from small microbes to giant animals. It is required for the activity of various enzymes and proteins and as essential cofactors in its ionic forms Fe^{2+} and Fe^{3+} (Miller et al., 2009). Fe^{3+} is insoluble, toxic and hence cells cannot use it readily (Raguzzi et al., 1988). In biological systems, it takes part in various key biochemical activities like catalysis, electron transport chain and oxygen transport (Papanikolaou and Pantopoulos, 2005). There is a very strong relationship between iron and iron-containing compounds with the gut microbiota (Selim et al., 2012). In mammals, enterocytes absorb iron in the form of heme or Fe^{2+} ion. Various pathways and enzymes are reported so far for absorption and metabolism of iron (Winzerling and Pham, 2006). In case of nutritional immunity (Liu et al., 2012) like hypoferraemia, iron concentration in eggs, milk, saliva, serum, and tears gets reduced, which ultimately limits the growth of bacteria (Messenger and Barclay, 1983). Hales (2010) showed that iron played a very important role in mitochondrial morphogenesis and fertility of vertebrates and invertebrates.

Even though the role of iron is well studied in humans and other mammals, very little is known about insects. Cytoplasmic ferritin is a polymer that acts as storage for iron in most organisms. Insect, hemolymph contains iron while ferritin acts as an iron transporter (Yoshiga et al., 1997, Dashti et al., 2016). Transferrins are biomolecules produced by insects' tissue, which bind with iron for its transport and subsequent metabolism (Yoshiga et al., 1997). There are some insects that feed on the blood of their host and thus get their iron nutrition. Blood-feeding insects utilize iron from blood and use this iron in their reproductive mechanism (Hales, 2010). Mosquito is an example of such insects in which a blood meal plays an important role in egg development as it contains haeme, the non-protein ferrous component of hemoglobin (Winzerling and Pham, 2006).

Excess of iron in the body is harmful to cells and tissues of all living organisms (Papanikolaou and Pantopoulos, 2005). Ionic iron is harmful to living beings since it forms free radicals (Yoshiga et al., 1999) which cause damage to tissues as well as biomolecules and subsequently lead to cell death. Such cases are mainly observed due to an iron overload in the body

(Papanikolaou and Pantopoulos, 2005). There are different mechanisms to overcome iron toxicity, in humans, for example, hepcidin protein is produced by the liver and then released into the blood for iron homeostasis (Collins 2008) while in insects transferrins help to maintain iron in water-soluble form so that its toxic property ceases (Yoshiga et al., 1999). Endosymbiotic *Wolbachia* helps *Asobara tabida* in the event of an iron overload. It acts as a buffer and keeps iron concentration in the required range and maintains iron homeostasis (Kremer et al., 2009).

Recently experiments suggested that transferrin levels increases in mosquito cell lines when they are allowed to be exposed to heat-killed bacteria (Yoshiga et al., 1997, Law, 2002). In *Drosophila melanogaster*, inoculums of *Escherichia coli*, significantly increase the transferrin mRNA synthesis as compared to in controls (Yoshiga et al., 1999). All these results strongly suggest that iron also takes part in the defense mechanism of insects (Yoshiga et al., 1997, Law, 2002). The following flow chart showed the possibility of iron metabolism in insect guts once it passes from a plant or animal to the insect gut (Figure 3) which decides the fate of the insect.

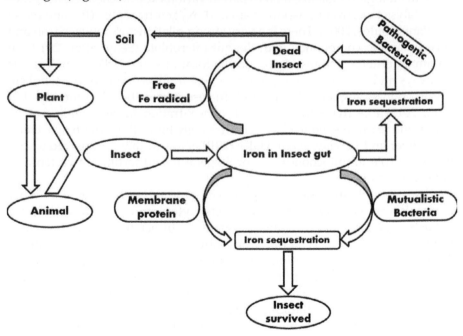

Fig. 3: Fate of Insect with respect to iron metabolism

5.1 Siderophores in Iron Sequestration

Along with mammals and insects, iron is also essential for microorganisms for their cell composition, intermediary metabolism, secondary metabolism, enzyme activity and host cell interactions (Messenger and Barclay, 1983). In the case of inflammations, iron transporters are blocked by the release of host hepcidin and prevents its absorption from the bloodstream (Deriu et al., 2013). Another mechanism is the targeting of siderophores by the secretion of lipocalin-2 involved in innate immunity which finally limits the iron in the gut (Flo et al., 2004, Deriu et al., 2013). These changes of iron concentration in the gut environment reshape microbial communities. *M. arborescens* is isolated from the lepidopteran gut produced Amino acid hydrolase (AAH). Iron concentration in AAH varies with the cultural environment in the gut (Ping et al., 2007). In microorganisms, iron helps against infection and colonization (Miller et al., 2009). So iron sequestration is requird. Sequestration is chelation, in which some complexes are formed between cell-secreted compounds and available metal ions so that these are easily available for cells or microorganisms. In iron sequestration, many proteins are involved, such as transferrin, lactoferrin, ferritin (Deriu et al., 2013) or siderophores which are present in eukaryotes while prokaryotes have bacterioferritin or siderophores (Kremer et al., 2009).

In pathogens, iron uptake is of two types i.e. direct or indirect. In a direct mechanism, pathogens take iron from different sources like lactoferrin, transferrin, ferritin, heme, etc (Miethke and Marahiel, 2007). But all pathogens do not have receptors for such direct absorption of iron and hence cannot survive. In such cases, the indirect iron uptake mechanism is followed. In the indirect uptake mechanism, this pathogen secretes some iron-chelating compounds for the acquisition of iron (Pal and Gokarn, 2010) and then absorb this complex with respective receptors. Some pathogens evade lipocalin-2 by producing large siderophores which cannot be sequestered by lipocalin-2 (Deriu et al., 2013). *Vibrio vulnificus* is a well known opportunistic pathogen in human beings. Its virulence is contributed by iron present in the host (Fouz et al., 1996, Miethke and Marahiel, 2007). *V. anguillarum* is responsible for hemolytic anemia in the presence of high iron availability, while this activity is suppressed in the absence of iron and making them avirulent (Miethke and Marahiel, 2007).

Microorganisms maintain their pathogenicity and virulence by mechanisms in which they produce low molecular weight substances for iron absorption which are known as siderophores (Sayyed and Chincholkar, 2010). Till date, more than 500 different siderophores have been identified. They are mainly classified into three groups i.e. catecholates, hydroxamates, and carboxylates. Siderophores are mainly classified on the basis of a chemical functional group to which iron attaches. Some microorganisms produce a mixed type of siderophore with the combination of catecholates, hydroxamates, and carboxylates groups. Siderophores are secreted by pathogens, and compete with host transferrin and absorb iron from them. This iron siderophore complex is then absorbed by bacterial cells and is used for its own cellular activity (Fouz et al., 1996). This complex formation mainly depends on the pH of the environment. The siderophore-iron complex is then transported into the cell via a cell membrane receptor protein. This protein is encoded by operon consisting of 5 genes. Once the amount of iron inside the cell is in between 10^{-7} to 10^{-5} M, operon gets turned off and uptake of Fe-siderophore complex stops (Ali and Vidhale, 2013).

Maximum pathogenic strains of bacteria produce siderophores, hence siderophore production is considered as a biochemical character for detection of virulent bacteria (Holden and Bachman, 2015). *Pseudomonas aeruginosa*, *Escherichia coli*, *Vibrio vulnificus*, and *Vibrio cholera* also produce siderophores for iron acquisition and virulence (Britigan et al., 2000). In some fungal infections, fungal virulence is dependent on iron availability hence they produce siderophores for iron acquisition. Ferricrocin is a well known intracellular siderophore isolated from fungus *Magnaporthe grisea* (Antelo et al., 2006).

6. Conclusion

The success of survival of the class Insecta is just an argument away and they are the most diverse organisms in the history of life. The human microbiome and its effects on human health are in boom nowadays. However, insect microbiomes is less studied as compared to the human microbiomes. The gut of many insects has been a significant reservoir for diverse microorganisms. However, lots of studies have been done in the world related to the association between insect species and their gut microbiota but many of them have microorganisms whose survival and functionality on or inside insects are still unknown and undefined. The contribution of India to insect-related studies is very less in the overall scenario of insect microbiomes. However, a majority of the study was done on researching specific groups or novel things. But only a few of them focused on total microbial diversity. Insect gut diversity studies were done by two approaches i.e. culture-dependent and culture-independent (metagenomics). The culture-dependent study has its own limitation as only one percent microbes can be cultured and about 99% cannot be cultured. This is because of lack of information in relation to growth parameters. Till the last few decades, most of the diversity study was done by using the culture-dependent approach. At that time the metagenomics approach was very expensive and bioinformatics tools were not well programmed. So, in the present study we focused on the metagenomics approach for supporting culture dependant study. Overall there isn't enough metagenomics data related insects in public datasets. So we tried to contribute metagenomics data of some insect orders to the dataset which later on helps in the study of that particular insect or insect order.

Iron is an important micro-nutrient essential for all biological systems. Iron concentration in the insect gut affects population dynamics, metabolism, pathogenicity of gut microflora as well as the production of secondary metabolites and enzymes by gut microflora. Iron metabolism in human beings is well studied but in the case of insects, it is poorly described. No direct study from my knowledge was done so far on the iron utilization in the insect gut and siderophore production of the gut bacteria. If we are trying to study iron concentration, insect and gut bacteria; it helps to understand the correlation among insect diets and gut microbes as well as the competition of iron uptake between bacterial siderophores and gut membrane iron transporters. A study of this relationship gives us insight into the relationship between insects and their microbiota. in the present study, we did the screening of all cultured bacterial isolates for siderophore production. This data may help to compute iron utilization in insects. One exciting area of study is targeting such microorganisms from insect pests, because an understanding of these interactions might pave the way for developing novel pest control strategies. Either target microbial symbionts directly to control the pest or try to alter gut microflora by other microbes which are harmful to insect pests.

References

Akman, L., A. Yamashita, H. Watanabe, K. Oshima, T. Shiba, M. Hattori et al. 2002. Genome sequence of the endocellular obligate symbiont of tsetse flies, *Wigglesworthia glossinidia*. Nature Genetics 32(3): 402-407.

Ali, S.S. and N.N. Vidhale 2013. Bacterial siderophore and their application: A review. Int. J. Curr. Microbiol. Appl. Sci. 2(12): 303-312.

Anand, P.A.A., S.J. Vennison, S.G. Sankar, D.I. Gilwax Prabhu, P.T. Vasan et al. 2010. Isolation and characterization of bacteria from the gut of *Bombyx mori* that degrade cellulose, xylan, pectin and starch and their impact on digestion. J. Insect Sci. 10(1): 107.

Andrews, R.E. and K.D. Spence. 1980. Action of douglas fir tussock moth larvae and their microflora on dietary terpenes. Appl. Environ. Microbiol. 40(5): 959-963.

Angerer, A., B. Klupp and V. Braun. 1992. Iron transport systems of *Serratia marcescens*. J. Bacteriol. 174(4): 1378-1387.

Antelo, L., C. Hof, K. Welzel, K. Eisfeld, O. Sterner and H. Anke. 2006. Siderophores produced by *Magnaporthe grisea* in the presence and absence of iron. Zeitschrift für Naturforschung - Section C: J. Biosci. 61(5-6): 461-464.

Azambuja, P., E.S. Garcia and N.A. Ratcliffe. 2005. Gut microbiota and parasite transmission by insect vectors. Trends Parasitol. 21(12): 568-572.

Beard, C.B., C. Cordon-Rosales and R.V. Durvasula. 2002. Bacterial symbionts of the *T riatominae* and their potential use in control of chagas disease transmission. Ann. Rev. Entomol. 47: 123-141.

Brady, C.M., M.K. Asplen, N. Desneux, G.E. Heimpel, K.R. Hopper, C.R. Linnen et al. 2014. Worldwide populations of the aphid *Aphis craccivora* are infected with diverse facultative bacterial symbionts. Microb. Ecol. 67(1): 195-204.

Brand, J.M., J.W. Bracke, A.J. Markovetz, D.L. Wood and L.E. Browne. 1975. Production of verbenol pheromone by a bacterium isolated from bark beetles. Nature 254(5496): 136-137.

Britigan, B.E., G.T. Rasmussen, O. Olakanmi and C.D. Cox. 2000. Iron acquisition from *Pseudomonas aeruginosa* siderophores by human phagocytes: An additional mechanism of host defense through iron sequestration. Infect. Immun. 68(3): 1271-1275.

Brucker, R.M. and S.R. Bordenstein. 2012. Speciation by symbiosis. Trends in Ecol. and Revol. 27(8): 443-451.

Chakraborty S., A. Khopade, R. Biao, W. Jian, X-Y. Liu, K. Mahadik et al. 2011. Characterization and stability studies on surfactant, detergent and oxidant stable α-amylase from marine haloalkaliphilic *Saccharopolyspora* sp. A9. J. Mol. Catal. B: Enzymatic 68(1): 52-58.

Chernysh, S., S.I. Kim, G. Bekker, V.A. Pleskach, N.A. Filatova, V.B. Anikin et al. 2002. Antiviral and antitumor peptides from insects. Proc. Natl. Acad. Sci. USA 99(20): 12628-12632.

Clark, M.A., L. Baumann and P. Baumann. 1998. *Buchnera aphidicola* (aphid endosymbiont) contains genes encoding enzymes of histidine biosynthesis. Curr. Microbiol. 37(5): 356-358.

Collins, J.F. 2008. Hepcidin regulation of iron transport. J. Nutr. (138): 2284-2288.

Crotti E., C. Damiani, M. Pajoro, E. Gonella, A. Rizzi, I. Ricci et al. 2009. *Asaia*, a versatile acetic acid bacterial symbiont, capable of cross-colonizing insects of phylogenetically distant genera and orders. Environ. Microbiol. 11(12): 3252-3264.

Daida, J.M., C.S. Grasso, S.A. Stanhope and S.J. Ross. 1996. Symbionticism and complex adaptive systems I: implications of having symbiosis occur in nature. Proc. Fifth. Ann. Conf. Evolut. Program., Cambridge. 1-10. https://pdfs.semanticscholar.org/055e/3f75d582382fc59f3057ce1b1d0bab841751. pdf

Dantur, K.I., R. Enrique, B. Welin and A.P. Castagnaro. 2015. Isolation of cellulolytic bacteria from the intestine of *Diatraea saccharalis* larvae and evaluation of their capacity to degrade sugarcane biomass. AMB Express 5(1): 15.

Dashti, Z.J.S., J. Gamieldien and A. Christoffels. 2016. Computational characterization of iron metabolism in the tsetse disease vector, *Glossina morsitans*: Ire stem-loops. BMC Genomics 17: 561.

Deriu, E., J.Z. Liu, M. Pezeshki, R.A. Edwards, R.J. Ochoa, H. Contreras et al. 2013. Probiotic bacteria reduce *Salmonella typhimurium* intestinal colonization by competing for iron. Cell Host Microbe 14(1): 26-37.

Desai, A.E. and P.R. Bhamre. 2012. Novel gut bacterial fauna of Gryllotalpa africana Beau. (Orthoptera: Gryllotalpidae). Int. J. Life Sci. 6(1): 50-55.

Dillon, R.J., C.T. Vennard and A.K. Charnley. 2002. A note: Gut bacteria produce components of a locust cohesion pheromone. J. Appl. Microbiol. 92(4): 759-763.

Dillon, R.J. and V.M. Dillon. 2004. The gut bacteria of insects: Nonpathogenic interactions. Ann. Rev. Entomol. 49(98): 71-92.

Dillon, R.J., C.T. Vennard, A. Buckling and A.K. Charnley. 2005. Diversity of locust gut bacteria protects against pathogen invasion. Ecol. Lett. 8(12): 1291-1298.

Dong, Y., F. Manfredini and G. Dimopoulos. 2009. Implication of the mosquito midgut microbiota in the defense against malaria parasites. PLoS Pathogens 5(5): e1000423.

Donovan, S.E., K.J. Purdy, M.D. Kane and P. Eggleton. 2004. Comparison of Euryarchaea strains in the guts and food-soil of the soil-feeding termite cubitermes fungifaber across different soil types. Appl. Envorn. Microbiol. 70(7): 3884-3892.

Douglas, A.E. 2007. Symbiotic microorganisms: Untapped resources for insect pest control. Trends Biotechnol. 25(8): 338-342.

Douterelo, I., J.B. Boxall, P. Deines, R. Sekar, K.E. Fish and C.A. Biggs. 2014. Methodological approaches for studying the microbial ecology of drinking water distribution systems. Water Res. 65: 134-156.

Eutick, M.L., R.W. O'Brien and M. Slaytor. 1978. Bacteria from the gut of Australian termites. Appl. Environ. Microbiol. 35(5): 823-828.

Ezenwa, V.O., N.M. Gerardo, D.W. Inouye, M. Medina and J.B. Xavier. 2012. Animal behavior and the microbiome. Science 338(6104): 198-199.

Feldhaar, H. and R. Gross. 2009. Insects as hosts for mutualistic bacteria. Int. J. Med. Microbiol. 299(1): 1-8.

Flo, T.H., K.D. Smith, S. Sato, D.J. Rodriguez, M.A. Holmes, R.K. Strong et al. 2004. Lipocalin 2 mediates an innate immune response to bacterial infection by sequestrating iron. Nature 432(7019): 917-921.

Floch, C., E. Alarcon-Gutiérrez and S. Criquet. 2007. ABTS Assay of phenol oxidase activity in soil. J. Microbiol. Methods 71(3): 319-324.

Fouz, B., E.G. Biosca, E. Alcaide and C. Amaro. 1996. Siderophore-mediated iron acquisition mechanisms in *Vibrio vulnificus* biotype 2. Appl. Environ. Microbiol. 62(3): 928-935.

Frederick, B.-A. and A.-J. Caesar. 2000. Analysis of bacterial communities associated with insect biological control agents using molecular techniques. Proc. of X Int. Symp. Biolog. Contr. Weeds 267: 261-267.

Fukatsu, T. and T. Hosokawa. 2002. Capsule-transmitted gut symbiotic bacterium of the japanese common plataspid stinkbug, *Megacopta punctatissima*. Appl. Environ. Microbiol. 68(1): 389-396.

Funkhouser-Jones, L.J., S.R. Sehnert, P. Martínez-Rodríguez, R. Toribio-Fernández, M. Pita, J.L. Bella et al. 2015. *Wolbachia* co-infection in a hybrid zone: Discovery of horizontal gene transfers from two *Wolbachia* supergroups into an animal genome. *Peer J.* 3: e1479.

Gaio, A. de O., D.S. Gusmão, A.V. Santos, M.A. Berbert-Molina, P.F.P. Pimenta et al. 2011. Contribution of midgut bacteria to blood digestion and egg production in *Aedes aegypti* (Diptera: Culicidae) (L.). Parasites & Vectors 4(1): 105.

Ghate, H.V., S.S. Jadhav, P.M. Sureshan and R.M. Sharma. 2015. Updated checklist of Indian Mantodea (Insecta). June: 1-31.

Ghosh, A.K. 1996. Insect biodiversity in India. Orient Insects 30(1): 1-10.

Guido, F., I. Ricci, C. Damiani, N. Raddadi, E. Crotti, M. Marzorati et al. 2007. Bacteria of the genus *Asaia* stably associate with *Anopheles stephensi*, an asian malarial mosquito vector. Proc. Natl. Acad. Sci. USA 104(21): 9047-9051.

Guidolin, A.S. and F.L. Cônsoli. 2017. Symbiont diversity of *Aphis* (Toxoptera) *citricidus* (Hemiptera: Aphididae) as influenced by host plants. Microb. Ecol. 73(1): 201-210.

Hales, K.G. 2010. Iron testes: Sperm mitochondria as a context for dissecting iron metabolism. BMC Biology 8: 8-10.

Holden, V.I. and M.A. Bachman. 2015. Diverging roles of bacterial siderophores during infection. Metallomics 7(6): 986-995.

Horn, M.A., J. Ihssen, C. Matthies, A. Schramm, G. Acker and H.L. Drake. 2005. *Dechloromonas denitrificans* sp. nov., *Flavobacterium denitrificans* sp. nov., *Paenibacillus anaericanus* sp. nov. and *Paenibacillus terrae* strain MH72, N_2O-producing bacteria isolated from the gut of the earthworm *Aporrectodea caliginosa*. Int. J. Syste. Evol. Microbiol. 55(3): 1255-1265.

Hosokawa, T., Y. Kicuchi, N. Nikoh, M. Shimada and T. Fukatsu. 2006. Strict host-symbiont cospeciation and reductive genome evolution in insect gut bacteria. PLoS Biology 4(10): 1841-1851.

König, H. 2006. *Bacillus* species in the intestine of termites and other soil invertebrates. J. Appl. Microbiol. 101(3): 620-627.

Kremer, N., D. Voronin, D. Charif, P. Mavingui, B. Mollereau and F. Vavre. 2009. *Wolbachia* interferes with ferritin expression and iron metabolism in insects. PLoS Pathogens 5(10): e1000630.

Law, J.H. 2002. Insects, oxygen, and iron. Biochem. Biophys. Res. Commun. 292(5): 1191-1195.

Liu, J.Z., S. Jellbauer, A.J. Poe, V. Ton, M. Pesciaroli, T.E. Kehl-Fie et al. 2012. Zinc sequestration by the neutrophil protein calprotectin enhances *Salmonella* growth in the inflamed gut. Cell Host Microb. 11(3): 227-239.

Lundgren, J.G. and R.M. Lehman. 2010. Bacterial gut symbionts contribute to seed digestion in an omnivorous beetle. PLoS One 5(5): e10831.

Messenger, A.J. and R. Barclay. 1983. Bacteria, iron and pathogenicity. Biochem. Edu. 11(2): 54-63.

Miethke, M. and M.A. Marahiel. 2007. Siderophore-based iron acquisition and pathogen control. Microbiol. Mol. Biol. Rev. 71(3): 413-451.

Miller, C.E., P.H. Williams and J.M. Ketley. 2009. Pumping iron: Mechanisms for iron uptake by *Campylobacter*. Microbiology 155(10): 3157-3165.

Moran, N.A., J.A. Russell, R. Koga and T. Fukatsu. 2005. Evolutionary relationships of three new species of *Enterobacteriaceae* living as symbionts of Aphids and other insects. Appl. Environ. Microbiol. 71(6): 3302-3310.

Nikoh, N., T. Hosokawa, K. Oshima, M. Hattori and T. Fukatsu. 2011. Reductive evolution of bacterial genome in insect gut environment. Genome Biol. Evol. 3(1): 702-714.

Odelson, D.A. and J.A. Breznak. 1983. Volatile fatty acid production by the hindgut microbiota of xylophagous termitest. Appl. Environ. Microbiol. 45(5): 1602-1613.

Ohkuma, M. and T. Kudo. 1996. Phylogenetic diversity of the intestinal bacterial community in the termite *Reticulitermes speratus*. Appl. Environ. Micrbiol. 62(2): 461-468.

Pal, R.B. and K. Gokarn. 2010. Siderophores and pathogenecity of microorganisms. J. Biol. Sci. Technol. 1(3): 127-134.

Papanikolaou G. and K. Pantopoulos. 2005. Iron metabolism and toxicity. Toxicol. Appl. Pharmacol. 202(2): 199-211.

Pérez-Cobas, A.E., E. Maiques, A. Angelova, P. Carrasco, A. Moya and A. Latorre. 2015. Diet shapes the gut microbiota of the omnivorous cockroach *Blattella germanica*. FEMS Microb. Ecol. 91(4): 1-14.

Ping, L., R. Büchler, A. Mithöfer, A. Svatos, D. Spiteller, K. Dettner et al. 2007. A Novel dps-type protein from insect gut bacteria catalyses hydrolysis and synthesis of N-acyl amino acids. Environ. Microbiol. 9(6): 1572-1583.

Ponnusamy, L., N. Xu, S. Nojima, D.M. Wesson, C. Schal and C.S. Apperson. 2008. Identification of bacteria and bacteria-associated chemical cues that mediate oviposition site preferences by *Aedes aegypti*. Proc. Natl. Acad. Sci. USA 105(27): 9262-9267.

Prakash, O., Y. Nimonkar and Y.S. Shouche. 2013. Practice and prospects of microbial preservation. FEMS Microbiol. Lett. 339(1): 1-9.

Priya, N.G., A. Ojha, M.K. Kajla, A. Raj and R. Rajagopal. 2012. Host plant induced variation in gut bacteria of *Helicoverpa armigera*. PLoS One 7(1): e30768.

Provorov, N.A. and O.P. Onishchuk. 2018. Microbial symbionts of insects: Genetic organization, adaptive role, and evolution. Microbiology 87(2): 151-163.

Raguzzi, F., E. Lesuisse and R.R. Crichton. 1988. Iron storage in *Saccharomyces cerevisiae*. FEBS Letters 231(1): 253-258.

Rainio, J. and J. Niemelä. 2003. Ground beetles (Coleoptera: Carabidae) as bioindicators. Biodiv. and Conserv. 12: 487-506.

Riehle, M.A. and M. Jacobs-Lorena. 2005. Using bacteria to express and display anti-parasite molecules in mosquitoes: Current and future strategies. Insect Biochem. Mol. Biol. 35(7): 699-707.

Ringo, J., G. Sharon and D. Segal. 2011. Bacteria-induced sexual isolation in drosophila. Fly (Austin). 5(4): 310-315.

Rinke, R., A.S. Costa, F.P.P. Fonseca, L.C. Almeida, I.D. Júnior and F. Henrique-Silva. 2011. Microbial diversity in the larval gut of field and laboratory populations of the sugarcane weevil *Sphenophorus levis* (Coleoptera, Curculionidae). Genet. Mol. Res. 10(4): 2679-2691.

Rudinsky, J.A. 1966. Scolytid beetles associated with *Douglas fir*: Response to terpenes. Science 152(3719): 218-219.

Samaranayake, J.W.K. and S.J. Wijekoon. 2011. Effect of selected earthworms on soil fertility, plant growth and vermicomposting. Trop. Agric. Res. Exten. 13(2): 34-40.

Sanchez-Contreras, M. and I. Vlisidou. 2008. The diversity of insect-bacteria interactions and its applications for disease control. Biotechnol. Genet. Eng. Rev. 25(1): 203-244.

Sayyed, R.Z. and S.B. Chincholkar. 2010. Growth and siderophores production in *Alcaligenes faecalis* is regulated by metal ions. Indian J. Microbiol. 50(2): 179-182.

Scully, E.D., S.M. Geib, J.E. Carlson, M. Tien, D. McKenna and K. Hoover. 2014. Functional genomics and microbiome profiling of the asian longhorned Beetle (*Anoplophora glabripennis*) reveal insights into the digestive physiology and nutritional ecology of wood feeding beetles. BMC Genomics 15(1): 1096.

Selim, S.A.H., S.M.E. Alfy, A.M. Diab, M.H. Abdel Aziz and M.F. Warrad. 2012. Intestinal bacterial flora that compete on the haem precursor iron fumarate in iron deficiency anemia cases. Malaysian J. Microbiol. 8(2): 92-96.

Sharon, G., D. Segal, J.M. Ringo, A. Hefetz, I. Zilber-Rosenberg and E. Rosenberg. 2010. Commensal bacteria play a role in mating preference of *Drosophila melanogaster*. Proc. Natl. Acad. Sci. USA 107(46): 20051-20056.

Shigenobu, S., H. Watanabe, M. Hattori, Y. Sakaki and H. Ishikawa. 2000. Genome sequence of the endocellular bacterial symbiont of Aphids *Buchnera* sp. APS. Nature 407(6800): 81-86.

Stork, N.E. 2007. Biodiversity: World of insects. Nature 448(7154): 657-658.

Su, L-J., H. Liu, Y. Li, H-F. Zhang, M. Chen, X-H. Gao et al. 2014. Cellulolytic activity and structure of symbiotic bacteria in locust guts. Genet Mol. Res. 13(3): 7926-7936.

Tan, S-q., K-q. Zhang, H-x. Chen, Y. Ge, R. Ji and Shi W-p. 2015. The mechanism for microsporidian parasite suppression of the hindgut bacteria of the migratory locust locusta *Migratoria manilensis*. Sci. Rep. 5: 17365.

Thong-On, A., K. Suzuki, S. Noda, J.-I. Inoue, S. Kajiwara and M. Ohkuma. 2012. Isolation and characterization of anaerobic bacteria for symbiotic recycling of uric acid nitrogen in the gut of various termites. Microbes Environ. 27(2): 186-192.

Vartoukian, S.R., R.M. Palmer and W.G. Wade. 2010. Strategies for culture of 'unculturable' bacteria. FEMS Microbiol. Lett. 309(1): 1-7.

Vavre, F., C. Girin and M. Bouletreau. 1999. Phylogenetic status of a fecundity-enhancing *Wolbachia* that does not induce the lytoky in trichogramma. Insect Mol. Biol. 8(1): 67-72.

Veivers, P.C., R.W. O'Brien and M. Slaytor. 1982. Role of bacteria in maintaining the redox potential in the hindgut of termites and preventing entry of foreign bacteria. J. Insect Physiol. 28(11): 947-951.

Weiss, B. and S. Aksoy. 2011. Microbiome influences on insect host vector competence. Trends Parasitol. 27(11): 514-522.

Wenzel, M., R. Radek, G. Brugerolle and H. Konig. 2003. Identification of the ectosymbiotic bacteria of *Mixotricha paradoxa* involved in movement symbiosis. Eur. J. Parasitol. 39(1): 11-23.

Werren, J.H., W. Zhang and L.R. Guo. 1995. Evolution and phylogeny of *Wolbachia*: Reproductive parasites of arthropods. Proc. Royal Soc. B: Biol. Sci. 261(1360): 55-63.

Werren, J.H. 1997. Biology of *Wolbachia*. Ann. Rev. Entomol. 24: 587-609.

Wilkinson, T. 2001. Disloyalty and treachery in bug-swapping shocker! Trends Ecol. Evol. 16(12): 659-661.

Winzerling, J.J. and D.Q.D. Pham. 2006. Iron metabolism in insect disease vectors: Mining the *Anopheles gambiae* translated protein database. Insect Biochem. Mol. Biol. 36(4): 310-321.

Yoshiga, T., V.P. Hernandez, A.M. Fallon and J.H. Law. 1997. Mosquito transferrin, an acute-phase protein that is up-regulated upon infection. Proc. Natl. Acad. Sci. USA 94(23): 12337-12342.

Yoshiga, T., T. Georgieva, B.C. Dunkov, N. Harizanova, K. Ralchev and J.H. Law 1999. *Drosophila melanogaster* transferrin: Cloning, deduced protein sequence, expression during the life cycle, gene localization and up-regulation on bacterial infection. Eur. J. Biochem. 260(2): 414-420.

Zabalou, S., M. Riegler, M. Theodorakopoulou, C. Stauffer, C. Savakis and K. Bourtzis. 2004. Wolbachia-induced cytoplasmic incompatibility as a means for insect pest population control. Proc. Natl. Acad. Sci. USA 101(42): 15042-15045.

Zchori-Fein, E., Y. Gottlieb, S.E. Kelly, J.K. Brown, J.M. Wilson, T.L. Karr et al. 2001. A newly discovered bacterium associated with parthenogenesis and a change in host selection behavior in parasitoid wasps. Proc. Natl. Acad. Sci. USA 98(22): 12555-12560.

Zhang, H. and A. Brune. 2004. Characterization and partial purification of proteinases from the highly alkaline midgut of the humivorous larvae of *Pachnoda ephippiata* (Coleoptera: Scarabaeidae). Soil. Biol. Biochem. 36(3): 435-442.

Index

Printed and bound by CPI Group (UK) Ltd, Croydon, CR0 4YY

24/10/2024

01778307-0013